T0302195

Mixture Models

Mixture models are a powerful tool for analyzing complex and heterogeneous datasets across many scientific fields, from finance to genomics. *Mixture Models: Parametric, Semiparametric, and New Directions* provides an up-to-date introduction to these models, their recent developments, and their implementation using R. It fills a gap in the literature by covering not only the basics of finite mixture models, but also recent developments such as semiparametric extensions, robust modeling, label switching, and high-dimensional modeling.

Features
- Comprehensive overview of the methods and applications of mixture models
- Key topics including hypothesis testing, model selection, estimation methods, and Bayesian approaches
- Recent developments, such as semiparametric extensions, robust modeling, label switching, and high-dimensional modeling
- Examples and case studies from such fields as astronomy, biology, genomics, economics, finance, medicine, engineering, and sociology
- Integrated R code for many of the models, with code and data available in the R Package MixSemiRob

Mixture Models: Parametric, Semiparametric, and New Directions is a valuable resource for researchers and postgraduate students from statistics, biostatistics, and other fields. It could be used as a textbook for a course on model-based clustering methods, and as a supplementary text for courses on data mining, semiparametric modeling, and high-dimensional data analysis.

MONOGRAPHS ON STATISTICS AND APPLIED PROBABILITY
Editors: F. Bunea, R. Henderson, L. Levina, N. Meinshausen, R. Smith

For more information about this series please visit: https://www.crcpress.com/Chapman--Hall
CRC-Monographs-on-Statistics--Applied-Probability/book-series/CHMONSTAAPP

Mixture Models
Parametric, Semiparametric, and New Directions

Weixin Yao and Sijia Xiang

CRC Press
Taylor & Francis Group
Boca Raton London New York

CRC Press is an imprint of the
Taylor & Francis Group, an **informa** business

A CHAPMAN & HALL BOOK

First edition published 2024
by CRC Press
2385 Executive Center Drive, Suite 320, Boca Raton, FL 33431, U.S.A.

and by CRC Press
4 Park Square, Milton Park, Abingdon, Oxon, OX14 4RN

CRC Press is an imprint of Taylor & Francis Group, LLC

© 2024 *Weixin Yao and Sijia Xiang*

Library of Congress Cataloging-in-Publication Data

Names: Yao, Weixin, author. | Xiang, Sijia, author.
Title: Mixture models : parametric, semiparametric, and new directions /
Weixin Yao and Sijia Xiang.
Description: First edition. | Boca Raton : CRC Press, 2024. | Series:
Champan & Hall/CRC monographs on statistics and applied probability |
Includes bibliographical references and index. | Summary: "Mixture
models are a powerful tool for analyzing complex and heterogeneous
datasets across many scientific fields, from finance to genomics.
Mixture Models: Parametric, Semiparametric, and New Directions provides
an up-to-date introduction to these models, their recent developments,
and their implementation using R. It fills a gap in the literature by
covering not only the basics of finite mixture models, but also recent
developments such as semiparametric extensions, robust modeling, label
switching, and high-dimensional modeling"-- Provided by publisher.
Identifiers: LCCN 2023047318 (print) | LCCN 2023047319 (ebook) | ISBN
9780367481827 (hardback) | ISBN 9781032728155 (paperback) | ISBN
9781003038511 (ebook)
Subjects: LCSH: Mixture distributions (Probability theory) | Parametric
modeling. | R (Computer program language)
Classification: LCC QA273.6 .Y36 2024 (print) | LCC QA273.6 (ebook) | DDC
519.2--dc23/eng/20240201
LC record available at https://lccn.loc.gov/2023047318
LC ebook record available at https://lccn.loc.gov/2023047319

ISBN: 978-0-367-48182-7 (hbk)
ISBN: 978-1-032-72815-5 (pbk)
ISBN: 978-1-003-03851-1 (ebk)

DOI: 10.1201/9781003038511

Typeset in Latin Modern Roman font
by KnowledgeWorks Global Ltd.

Publisher's note: This book has been prepared from camera-ready copy provided by the authors.

To Jian, Charlie, Kaylie, Jianhua, Zhangdi, and Weijin

To Zhaojun, Liyan, and Xi

Contents

Preface

Mixture models are very powerful and popular data analysis tools for heterogeneous data sets and have been widely used in many scientific fields, such as astronomy, biology, genomics, economics, finance, medicine, engineering, and sociology. There has been tremendous research development of mixture models to meet the demand of handling the new era of high-dimensional or complex data sets. Our main motivation for writing this book is to provide an up-to-date introduction of mixture models and their recent development.

Current mixture model books mainly cover parametric mixture models and do not cover in detail some recently developed mixture models such as high-dimensional mixture models, robust mixture models, semiparametric mixture models, and model selections for mixture models. This new book aims to fill this gap and would be a very timely reference for researchers and students who are interested in mixture models.

In this book, we mainly introduce finite mixture models and some of their recent developments, including semiparametric extensions, robust mixture modeling, and high-dimensional mixture modeling. More specifically, we introduce (1) the label-switching issue for mixture models and its possible solutions for both Bayesian mixture modeling and frequentist mixture modeling; (2) various hypothesis testing and model selection methods for mixture models; (3) robust estimation methods for mixture models and mixture regression models, (4) semiparametric extensions of parametric mixture models, and their estimation methods and asymptotic properties, and (5) mixture models for high-dimensional data.

The book will be targeted at MS- and PhD-level statisticians and biostatisticians, as well as sophisticated PhD-level researchers. It will be designed for both teaching and research. The text would be useful as a textbook for a course focused on model-based clustering methods. Additionally, it would be a suitable supplementary text for data mining, semiparametric modeling, and high-dimensional data analysis courses.

Symbols

Symbol Description

c	indicator of a component	$p(\cdot)$	general notation for pdf or pmf.
C	number of components		
i	indicator of an observation	$\phi(x; \mu, \sigma^2)$	normal density with mean μ and variance σ^2
n	number of observations		
d	dimension of the covariate		
Z_{ic}	component indicator	κ_l	$\kappa_l = \int t^l K(t) dt$
$K(\cdot)$	zero-symmetric kernel density function	ν_l	$\nu_l = \int t^l K^2(t) dt$
$K_h(\cdot)$	$K_h(\cdot) = h^{-1}K(\cdot/h)$ is the scaled kernel density function	π_c	proportion of the cth component
pdf	probability density function	μ_c	mean of the cth component
		σ_c^2	variance of the cth component
pmf	probability mass function	$\boldsymbol{\theta}$	collection of all the unknown parameters
cdf	cumulative distribution function	$\ell(\boldsymbol{\theta})$	log-likelihood function

In this book, we use a lowercase letter to denote a scalar, an uppercase letter to denote a random variable, a bold lowercase letter to denote a vector, which could be random or fixed depending on the context, and a bold uppercase letter to denote a matrix.

Authors

Dr. Weixin Yao is a professor and vice chair of the Department of Statistics at the University of California, Riverside. He earned his BS in statistics from the University of Science and Technology of China in 2002 and his PhD in statistics from Pennsylvania State University in 2007. His major research includes mixture models, nonparametric, and semiparametric modeling, robust data analysis, and high dimensional modeling. He has served as an associate editor for *Biometrics, Journal of Computational and Graphical Statistics, Journal of Multivariate Analysis*, and *The American Statistician*. In addition, Dr. Yao was also the guest editor of *Advances in Data Analysis and Classification*, for the special issue on "Models and Learning for Clustering and Classification," 2020-2021.

Dr. Sijia Xiang is a professor in statistics. She obtained her doctoral and master's degrees in statistics from Kansas State University in 2014 and 2012, respectively. Her research interests include mixture models, nonparametric/semiparametric estimation, robust estimation, and dimension reduction. Dr. Xiang has led several research projects, including "Statistical inference for clustering analysis based on high-dimensional mixture models" funded by the National Social Science Fund of China, "Semiparametric mixture model and variable selection research" funded by the National Natural Science Foundation of China, and "Research on the new estimation method and application of mixture model" funded by the Zhejiang Statistical Research Project. Dr. Xiang has also been selected as a "Young Discipline Leader" and a "Young Talented Person" in the Zhejiang Provincial University Leadership Program.

1

Introduction to mixture models

1.1 Introduction

Mixture models, also known as latent class models and unsupervised learning models, have experienced increasing interest over the last few decades. They are natural models for a heterogeneous population that consists of several unobserved homogeneous sub-populations. The homogeneous sub-populations are often called *components* of the population. Mixture models can be used for cluster analysis, latent class analysis, discriminant analysis, image analysis, survival analysis, disease mapping, meta-analysis, etc. Mixture models provide extremely flexible descriptive models for distributions in data analysis and inference, and are widely applied in many fields, including artificial intelligence, astronomy, biology, economics, engineering, finance, genetics, language processing, marketing, medicine, psychiatry, philosophy, physics, and social science; see, e.g., Frühwirth-Schnatter (2001), Grün and Hornik (2012), Liang (2008), and Kostantinos (2000).

For mixture models, the number of components is usually finite and the resulting mixture models are called *finite mixture models*. The finite mixture model is one of the most commonly used model-based clustering tools. Clustering is an exploratory data analysis tool and aims to divide data into groups/clusters such that observations in the same group, also called a cluster, are more similar to each other than to those in other groups/clusters. Organizing data into clusters can show some internal structures of the data. For example, we might want to cluster genes into clusters, based on their expression patterns, such that genes in the same cluster have similar functions; cluster customers into different groups with different advertisement strategies; segment patients enabling personalized treatments; cluster and summarize news; cluster/partition the market into different segmentations. If a document is composed of n different words from a total of N vocabularies and each word corresponds to one of C possible topics, then the distribution of n words could be modeled as a mixture of C components. In finance, financial returns and house prices usually behave differently in normal situations and during crisis times. In addition, even at a fixed period, the same type of houses might be priced differently in different neighborhoods. However, the prices of a particular type of house in the same neighborhood are expected to cluster around their mean. The naturally clustered data could also arise for the following data: wines from different types of grapes, healthy and sick individuals in medical services, infant birth weights – two types of pregnancies "normal" and "complicated," as well as low and high responses to stress.

Example 1.1.1. Below are the two examples of heterogeneous data sets, whose histograms are shown in Figure 1.1.

1. The Fishery Data (Titterington et al., 1985): This data set consists of the length of 256 snappers from a lake. The data is shown in Figure 1.1(a), where there are several subgroups due to different ages. However, it is very hard to measure the age of the fish, so the age groups remain unobserved. The mixture model can help determine the age group of each fish and their proportions, which would be useful for determining the relative mortality of the yearly cohorts.

2. Old Faithful dataset: This data set contains the time (in minutes) between eruptions of the Old Faithful geyser in Yellowstone National Park, USA. The data is shown in Figure 1.1(b), which clearly shows two modes. Hunter et al. (2007) analyzed this data set using a two-component mixture model.

□

Based on Figure 1.1, it can be seen that there are multiple modes in the histograms, which are usually (but not necessarily) the indications for heterogeneous populations. In Section 1.8.1, we will discuss the shapes of finite mixture models in more detail.

1.2   Formulations of mixture models

Let $(\mathbf{x}_1, \ldots, \mathbf{x}_n)$ be an independent identically distributed (iid) random sample of size n from a heterogeneous population with C homogeneous subpopulations/subgroups/clusters, usually called "components" in the literature of mixture models, where $\mathbf{x}_i$ is a p-dimensional random vector on $\mathbb{R}^p$, and the number of components, C, can be known or unknown. Denote the density function of the cth component by $g_c(\mathbf{x}; \lambda_c)$, $c = 1, \ldots, C$, where $g_c(\mathbf{x}; \lambda_c)$ is generally called the *component density*, and for the simplicity of the notation, the density function can be a probability density function (pdf) of a continuous variable or a probability mass function (pmf) of a discrete variable, and λ_c is a component-specific parameter, which can be a scalar or a vector. Let Z_i be the latent component label indicating which component $\mathbf{x}_i$ belongs to, and let $\pi_c = P(Z_i = c)$, where $0 \le \pi_c \le 1$, called the *component weight/mixing probability/proportion*, is the proportion of observations in the cth component

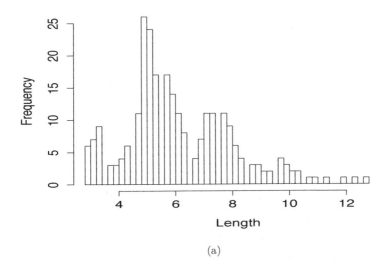

(a)

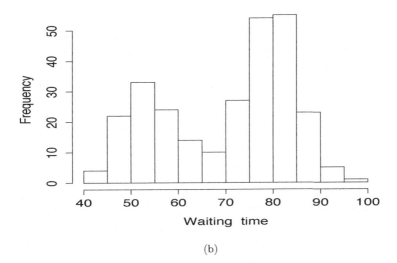

(b)

FIGURE 1.1
Histograms of heterogeneous data for (a) Fishery data; (b) Faithful data.

satisfying $\sum_{c=1}^{C} \pi_c = 1$. Therefore, we have

$$p(\mathbf{x}_i | Z_i = c) = g_c(\mathbf{x}_i; \lambda_c), \qquad i = 1, \dots, n; c = 1, \dots, C,$$

where $p(\cdot)$ is the general notation for the density function of both discrete and continuous random variables.

If we could observe the latent labels Z_i, $i = 1, \ldots, n$, then $\{(\mathbf{x}_i, Z_i),\ i = 1, \ldots, n\}$ is a random sample from the joint density

$$p(\mathbf{x}, c) = p(\mathbf{x} \mid Z = c)P(Z = c) = g_c(\mathbf{x}; \lambda_c)\pi_c .$$

If the component labels $(Z_1, \ldots, Z_n)$ are missing, then we could only observe $(\mathbf{x}_1, \ldots, \mathbf{x}_n)$ from the marginal density of X, which is the *mixture* density

$$f(\mathbf{x}; \boldsymbol{\theta}) = \sum_{c=1}^{C} p(\mathbf{x}, c) = \sum_{c=1}^{C} \pi_c g_c(\mathbf{x}; \lambda_c), \qquad (1.1)$$

where $\boldsymbol{\theta} = (\lambda_1, \ldots, \lambda_C, \pi_1, \ldots, \pi_C)$ collects all the unknown parameters. The family of distributions having density (1.1) is generally called a *C-component finite mixture model*. When $C = 1$, there is just one component, and $f(\cdot)$ will be called *unicomponent* mixture density. Usually, the component densities are from the same class of distributions but with different parameters, i.e., the mixture density (1.1) becomes

$$f(\mathbf{x}; \boldsymbol{\theta}) = \sum_{c=1}^{C} \pi_c g(\mathbf{x}; \lambda_c). \qquad (1.2)$$

The most commonly used mixture model is the finite normal mixture model, in which $g(\cdot)$ is a normal density. Other common choices for $g(\cdot)$ include the binomial distribution for the number of "successes" given a fixed number of total occurrences, multinomial distribution for counts of multi-way occurrences, Poisson distribution for counting data, negative binomial distribution for Poisson-type observations but with overdispersion, exponential distribution, log-normal distribution for skewed positive real numbers such as incomes or prices, multivariate normal distribution, and multivariate Student's t-distribution for vectors of heavy-tailed correlated outcomes.

The finite mixture model (FMM) is very flexible and can be used to approximate any smooth density functions arbitrarily well when the number of components is large. In fact, the kernel density estimator is a special case of finite mixture models. Marron and Wand (1992) and Janssen et al. (1995) used an example of 16 mixture densities, listed in Table 1.1, to demonstrate that the family of mixture densities can be very broad. The first 15 density examples were used in Marron and Wand (1992), and the last example was added by Janssen et al. (1995). The plots are shown in Figure 1.2.

In finite mixture model (1.2), the unknown parameter $\boldsymbol{\theta}$ can be also written as follows:

$$\begin{pmatrix} \pi_1 & \cdots & \pi_C \\ \lambda_1 & \cdots & \lambda_C \end{pmatrix}, \qquad (1.3)$$

where the cth column collects the unknown parameters of the cth component. Define a latent variable Λ with a probability distribution Q such that

TABLE 1.1

Sixteen normal mixture densities used by Marron and Wand (1992) and Janssen et al. (1995)

Density	$f(y)$
1. Gaussian	$N(0,1)$
2. Skewed unimodal	$\frac{1}{5}N(0,1) + \frac{1}{5}N(0.5, (\frac{2}{3})^2) + \frac{3}{5}N(\frac{13}{15}, (\frac{5}{9})^2)$
3. Strongly skewed	$\sum_{i=0}^{7} \frac{1}{8}N(3\{(\frac{2}{3})^i - 1\}, (\frac{2}{3})^{2i})$
4. Kurtotic unimodal	$\frac{2}{3}N(0,1) + \frac{1}{3}N(0, (\frac{1}{10})^2)$
5. Outlier	$\frac{1}{10}N(0,1) + \frac{9}{10}N(0, (\frac{1}{10})^2)$
6. Bimodal	$\frac{1}{2}N(-1, (\frac{2}{3})^2) + \frac{1}{2}N(1, (\frac{2}{3})^2)$
7. Separated bimodal	$\frac{1}{2}N(-\frac{3}{2}, (\frac{1}{2})^2) + \frac{1}{2}N(\frac{3}{2}, (\frac{1}{2})^2)$
8. Skewed bimodal	$\frac{3}{4}N(0,1) + \frac{1}{4}N(\frac{3}{2}, (\frac{1}{3})^2)$
9. Trimodal	$\frac{9}{20}N(-\frac{6}{5}, (\frac{3}{5})^2) + \frac{9}{20}N(\frac{6}{5}, (\frac{3}{5})^2) + \frac{1}{10}N(0, (\frac{1}{4})^2)$
10. Claw	$\frac{1}{2}N(0,1) + \sum_{i=0}^{4} \frac{1}{10}N(i/2 - 1, (\frac{1}{10})^2)$
11. Double claw	$\frac{49}{100}N(-1, (\frac{2}{3})^2) + \frac{49}{100}N(1, (\frac{2}{3})^2) + \sum_{i=0}^{6} \frac{1}{350}N((i-3)/2, (\frac{1}{100})^2)$
12. Asymmetric claw	$\frac{1}{2}N(0,1) + \sum_{i=-2}^{2} \frac{2^{1-i}}{31}N(i + \frac{1}{2}, (\frac{2^{-i}}{10})^2)$
13. Asymmetric double claw	$\sum_{i=0}^{1} \frac{46}{100}N(2i - 1, (\frac{2}{3})^2) + \sum_{i=1}^{3} \frac{1}{300}N(-i/2, (\frac{7}{100})^2)$ $+ \sum_{i=1}^{3} \frac{7}{300}N(i/2, (\frac{1}{100})^2)$
14. Smooth comb	$\sum_{i=0}^{5} \frac{2^{5-i}}{63}N(\frac{65-96/2^i}{21}, (\frac{32}{63})^2 \frac{1}{2^{2i}})$
15. Discrete comb	$\sum_{i=0}^{2} \frac{2}{7}N(\frac{12i-15}{7}, (\frac{2}{7})^2) + \sum_{i=8}^{10} \frac{1}{21}N(\frac{2i}{7}, (\frac{1}{21})^2)$
16. Distant Bimodal	$\frac{1}{2}N(-2.5, (\frac{1}{6})^2) + \frac{1}{2}N(2.5, (\frac{1}{6})^2)$

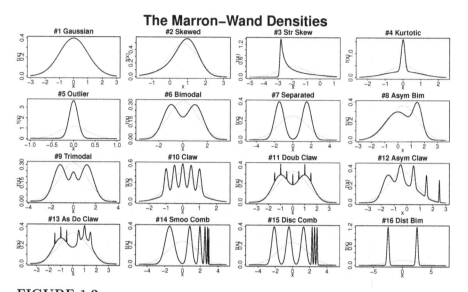

FIGURE 1.2
The plot of 16 normal mixture densities is listed in Table 1.1.

$Q(\lambda_c) = P(\Lambda = \lambda_c) = \pi_c$. Then, we can equate the set of unknown parameters in (1.3) uniquely with a discrete probability measure Q on the parameter space for λ, which puts a mass π_c at the support point λ_c. So estimating the unknown parameters in (1.3) for mixture models is the same as finding an unknown distribution Q. The probability distribution Q is also called the *mixing distribution* or the *latent distribution*, and the mixture density (1.2) can be formally rewritten as

$$f(\mathbf{x}; \boldsymbol{\theta}) = f(\mathbf{x}; Q) = \mathrm{E}[g(\mathbf{x}; \Lambda)] = \int g(\mathbf{x}; \lambda) dQ(\lambda). \qquad (1.4)$$

If Q is a discrete distribution with C points of support, then $f(\mathbf{x}; Q)$ is a C-component mixture model. A natural extension of the finite mixture model is to allow the mixing distribution Q to be continuous. In such a situation, the mixture model is called a *continuous mixture* or an *infinite mixture* with the density

$$f(\mathbf{x}; Q) = \int g(\mathbf{x}; \lambda) dQ(\lambda) = \int g(\mathbf{x}; \lambda) q(\lambda) d\lambda,$$

where $q(\cdot)$ is the probability density corresponding to Q. If Q is treated unspecified as to whether it is discrete, continuous or in any particular family of distributions, this will be called the *nonparametric mixture model*. We will discuss this and its estimation in more detail in Section 1.13.

In many situations, the latent variable Λ has a direct physical interpretation. For example, there might be a latent mathematical ability that determines the test score in mathematics on a certain exam, such as the SAT and

ACT. Given the mathematical ability, the test score has a normal distribution. Because of the existence of different mathematical abilities among the whole population, the test scores would have a mixture distribution. Referring back to the Fishery data in Example 1.1.1, the latent variable Λ represents the age of the fish. Within each age group, the length of the fish has a normal distribution. Due to the existence of different age groups, the overall length variable of the fish has a mixture distribution.

Many popular models can be considered as mixture models, such as random effects models, repeated measures models, latent class and latent trait models, missing covariates and data, etc. See Lindsay (1995) Section 1.13 for more details. Using the random effects model as an example, suppose we have data $\{x_{ij}, i = 1, \ldots, n, j = 1, \ldots, n_i\}$, where x_{ij} is the jth observation from ith subject/unit. Usually $\{x_{i1}, \ldots, x_{in_i}\}$ are assumed to be independent and identically distributed (iid) observations from $N(\mu_i, \sigma^2)$, where μ_i is the unknown mean for observations from the ith subject/unit. The random effects model assumes that μ_i is sampled from a large population with a latent distribution Q. Note that $\{x_{i1}, \ldots, x_{in_i}\}$ are marginally correlated because of sharing the common random effect μ_i but are conditionally independent after fixing μ_i. Therefore, the marginal density of observations from the ith subject is

$$f(\mathbf{x}_i; Q, \sigma^2) = \int \prod_{j=1}^{n_i} \phi(x_{ij}; \mu, \sigma^2) dQ(\mu), \qquad (1.5)$$

where $\mathbf{x}_i = (x_{i1}, \ldots, x_{in_i})^\top$, and $\phi(x; \mu, \sigma^2)$ is the probability density function for $N(\mu, \sigma^2)$, the normal distribution with mean μ and variance σ^2. Therefore, the marginal density of $\mathbf{x}_i$ in (1.5) has the mixture form (1.4). The usual random-effects model assumes that Q is a normal distribution. In order to relax the model assumption about the latent variable μ_i, we could estimate Q nonparametrically. This is also a natural approach when the data is discrete so that the normality of the observations is no longer a reasonable assumption. Please refer to Section 1.13 for more detail.

The most commonly used mixture models are finite normal mixture models with density

$$f(x) = \sum_{c=1}^{C} \pi_c \phi(x; \mu_c, \sigma_c^2). \qquad (1.6)$$

For a random variable X from the normal mixture model (1.6), it can be easily verified that

$$\mathrm{E}(X) = \mathrm{E}[\mathrm{E}(X \mid Z)] = \sum_{c=1}^{C} P(Z = c)\mathrm{E}(X \mid Z = c) = \sum_{c=1}^{C} \pi_c \mu_c \equiv \bar{\mu},$$

$$\mathrm{Var}(X) = \mathrm{E}[\mathrm{Var}(X \mid Z)] + \mathrm{Var}[\mathrm{E}(X \mid Z)] = \sum_{c=1}^{C} \pi_c \sigma_c^2 + \sum_{c=1}^{C} \pi_c (\mu_c - \bar{\mu})^2.$$

From the above results, we can see that the mean of X is a weighted average of the component means while the variance of X not only depends on the component variances but also depends on how separated the component means are. In Section 1.5, we will discuss how to estimate the unknown parameters in (1.6).

1.3   Identifiability

A very important concept associated with the mixture model is *identifiability*. All parameter estimations will become meaningless if the parameters in a parametric model are not identifiable.

Definition 1.3.1. We say a parametric density family of $f(x; \boldsymbol{\theta})$ is identifiable if $f(x; \boldsymbol{\theta}_1) = f(x; \boldsymbol{\theta}_2)$ for all x implies $\boldsymbol{\theta}_1 = \boldsymbol{\theta}_2$.

The parameter $\boldsymbol{\theta}$ is estimable only if the model is identifiable.

For the C-component finite mixture model (1.2), if $\pi_j = 0$ for some j or $\lambda_i = \lambda_j$ for some $i \neq j$, then the model is not identifiable. Let $\Theta \subset \mathbb{R}^d$ be the d-dimension subspace for λ. We define the parameter space for the finite mixture model (1.2) as

$$\Omega = \{\boldsymbol{\theta} : \sum_{c=1}^{C} \pi_c = 1, \pi_c > 0, \lambda_i \neq \lambda_j \in \Theta, 1 \leq i \neq j \leq C\}.$$

For any permutation $\boldsymbol{\omega} = (\boldsymbol{\omega}(1), \dots, \boldsymbol{\omega}(C))$ of the identity permutation $(1, \dots, C)$, define the corresponding permutation of the parameter vector $\boldsymbol{\theta}$ by

$$\boldsymbol{\theta}^{\boldsymbol{\omega}} = (\pi_{\boldsymbol{\omega}(1)}, \dots, \pi_{\boldsymbol{\omega}(C)}, \lambda_{\boldsymbol{\omega}(1)}, \dots, \lambda_{\boldsymbol{\omega}(C)})^{\top}.$$

Note that $f(x; \boldsymbol{\theta}) = f(x; \boldsymbol{\theta}^{\boldsymbol{\omega}})$, for any $\boldsymbol{\omega}$. Therefore, the mixture model can be defined to be identifiable only up to the permutation of the component parameters.

Definition 1.3.2. We say the finite mixture model (1.2) is identifiable if $\sum_{c=1}^{C} \pi_c g(x; \lambda_c) = \sum_{c=1}^{\tilde{c}} \tilde{\pi}_c f(x; \tilde{\lambda}_c)$ implies $C = \tilde{C}$, $(\pi_1, \dots, \pi_C) = (\tilde{\pi}_{\boldsymbol{\omega}(1)}, \dots, \tilde{\pi}_{\boldsymbol{\omega}(C)})$, and $(\lambda_1, \dots, \lambda_C) = (\tilde{\lambda}_{\boldsymbol{\omega}(1)}, \dots, \tilde{\lambda}_{\boldsymbol{\omega}(C)})$ for some permutation $\boldsymbol{\omega}$.

Example 1.3.1 (Examples of nonidentifiable mixture models). We give three examples of unidentifiable mixture distributions below. The first one is a mixture of Bernoulli distributions, followed by a mixture of uniform distributions, and the last one is a mixture of beta distributions.

1. Let $Ber(p)$ be the Bernoulli distribution with a success probability p. Note that the mixture of Bernoulli distribution $\pi Ber(p_1) + (1 - \pi)Ber(p_2)$ is still a Bernoulli distribution. Therefore, only $\pi p_1 + (1 - \pi)p_2$ is identifiable but not the whole parameter vector (π, p_1, p_2).

2. Let $U_{(a,b)}(x)$ be the uniform distribution on $[a, b]$. Note that

$$\frac{3}{4}U_{(-1,1)}(x) + \frac{1}{4}U_{(-3,3)}(x-4) = \frac{2}{3}U_{(-1,1)}(x) + \frac{1}{3}U_{(-4,4)}(x-3)$$

and $U_{(-\sigma,\sigma)}(x) = \sigma^{-1}U_{(-1,1)}(x/\sigma)$. Therefore,

$$f(\mu_1,\sigma_1,\mu_2,\sigma_2,\pi) = \pi\frac{1}{\sigma_1}U_{(-1,1)}\left(\frac{x-\mu_1}{\sigma_1}\right) + (1-\pi)\frac{1}{\sigma_2}U_{(-1,1)}\left(\frac{x-\mu_2}{\sigma_2}\right)$$

is not identifiable since $f(0,1,4,3,3/4) = f(0,1,3,4,2/3)$. Moreover, even if μ_1 and σ_1 are known, say $\mu_1 = 0$ and $\sigma_1 = 1$, the model is still unidentifiable.

3. Let's consider a two-component mixture of beta distributions in which each component has a beta distribution with the density

$$g(x;\boldsymbol{\lambda}) = \frac{x^{a-1}(1-x)^{b-1}}{B(a,b)}, 0 < x < 1,$$

where $\boldsymbol{\lambda} = (a,b), a > 0, b > 0$, and $B(a,b)$ is the beta function that normalizes the density function. Note that

$$\frac{2}{3}g(x;3,1) + \frac{1}{3}g(x;2,2) = \frac{2}{3}(3x^2) + \frac{1}{3}\{6x(1-x)\} = 2x$$
$$g(x;2,1) = 2x.$$

Therefore, mixtures of beta distributions are not identifiable either. □

Based on the above examples, we can see that not all mixture distributions are identifiable. Fortunately, for continuous variables, Teicher (1963) showed that except for mixtures of uniform densities and mixtures of beta distributions, most finite mixtures of continuous densities, such as mixtures of normals, mixtures of exponentials and Gamma distributions, are generally identifiable. Holzmann et al. (2006) established the identifiability results for finite mixtures of elliptical distributions such as multivariate t-distributions, Laplace distributions, symmetric stable laws, exponential power, and Kotz distributions. These results are extended to multivariate families such as multivariate mixtures of normals and the family of generalized Cauchy distributions, see Yakowitz and Spragins (1968). Titterington et al. (1985, Section 3.1) also provided some good discussions about the identifiability of mixture models.

Next, we provide some identifiability results about discrete mixture distributions.

Theorem 1.3.1. A mixture of discrete distributions over any finite number of categories is just another distribution over those categories.

Proof *Suppose $g_c(x)$ is a discrete distribution that takes values from 1 to K. The mixture distribution $f(x) = \sum_{c=1}^{C} \pi_c g_c(x)$ is still another discrete distribution over values from 1 to K with probability $f(k) = P(X = k) = \sum_{c=1}^{C} \pi_c g_c(k)$.*   □

Theorem 1.3.2. We have the following identifiability results for some discrete mixture models.

a. The mixture of Bernoulli distributions is not identifiable.

b. Mixtures of binomial distributions

$$\pi_1 Bin(N, \theta_1) + \cdots + \pi_C Bin(N, \theta_C)$$

are not identifiable if $N \leq 2C - 1$ (Teicher, 1960), where $Bin(n, p)$ is a binomial distribution with parameters (n, p). If the component parameters $\theta_1, \ldots, \theta_C$ are known, the mixing weights $(\pi_1, \ldots, \pi_C)$ are identifiable for $C \leq N + 1$ (Lindsay, 1995, Section 2.4.2).

c. Finite mixtures of Poisson distributions (Teicher, 1960) as well as finite mixtures of negative binomial distributions (Yakowitz and Spragins, 1968) are identifiable.

For the finite mixture model (1.2), when C is quite large, some of the mixing proportions become close to 0, and these mixture models are then close to nonidentifiable in the sense that the mixture density with C components might be empirically indistinguishable from one with fewer than C components. Because of this, in some applications, people use the mixture model with a continuous mixing distribution instead of a finite discrete mixing distribution. Please see Section 1.13 for more detail.

1.4   Maximum likelihood estimation

Four estimation methods are widely used in practice for parametric models: the method of moments, the minimum distance method, the maximum likelihood method, and the Bayesian method. Among these, maximum likelihood estimate (MLE) is one of the most popular ones partly because it provides an asymptotically efficient estimator.

Given observations $(x_1, \ldots, x_n)$ from the mixture density

$$f(x; \boldsymbol{\theta}) = \sum_{c=1}^{C} \pi_c g(x; \lambda_c),$$

where $\boldsymbol{\theta} = (\pi_1, \lambda_1, \ldots, \pi_C, \lambda_C)$, the MLE, denoted by $\hat{\boldsymbol{\theta}}_n$, is found by maximizing the following log-likelihood function over $\boldsymbol{\theta}$:

$$\log L(\boldsymbol{\theta}) = \sum_{i=1}^{n} \log \sum_{c=1}^{C} \pi_c g(x_i; \lambda_c), \qquad (1.7)$$

where $L(\boldsymbol{\theta}) \equiv L(\boldsymbol{\theta}; \mathbf{x})$ is the likelihood function

$$L(\boldsymbol{\theta}; \mathbf{x}) = \prod_{i=1}^{n} \sum_{c=1}^{C} \pi_c g(x_i; \lambda_c).$$

The score functions of (1.7) are

$$\partial \log L(\boldsymbol{\theta})/\partial \lambda_c = \sum_{i=1}^{n} \frac{\pi_c g'(x_i; \lambda_c)}{\sum_{c=1}^{C} \pi_c g(x_i; \lambda_c)}, c = 1, \ldots, C,$$

$$\partial \log L(\boldsymbol{\theta})/\partial \pi_c = \sum_{i=1}^{n} \frac{g(x_i; \lambda_c) - g(x_i; \lambda_C)}{\sum_{c=1}^{C} \pi_c g(x_i; \lambda_c)}, c = 1, \ldots, C-1.$$

There is no explicit formula for the MLE, but generally, we can use the EM algorithm (Dempster et al., 1977b) to find the MLE for finite mixture models. The EM algorithm is a general-purpose iterative algorithm for computing maximum likelihood estimates when the observations can be viewed as incomplete data. In Section 1.5, we will provide an introduction to the EM algorithm.

Feng and McCulloch (1996) and Cheng and Liu (2001) established the following asymptotic consistency results for the MLE $\hat{\boldsymbol{\theta}}_n$.

Theorem 1.4.1. Let $\mathbf{x} = (x_1, \ldots, x_n)$ be iid observations with density $f(x; \boldsymbol{\theta}_0)$. Assume that the parameter space Ω is a closed convex set. Under some regularity conditions, $dis\{\hat{\boldsymbol{\theta}}_n, \Omega_0\} \to 0$, with probability 1 (w.p.1.), where $\Omega_0 = \{\boldsymbol{\theta} : f(x; \boldsymbol{\theta}) = f(x; \boldsymbol{\theta}_0)\}$, and $dis\{x_n, \Omega\} \to 0$ if and only if there is a sequence of $\{y_n\}$ of points in Ω such that $\|x_n - y_n\| \to 0$ as $n \to \infty$, where $\|\cdot\|$ is the Euclidean distance.

If the true model is not a mixture model, then $f(x; \boldsymbol{\theta}_0)$ is the model that minimizes the *Kullback-Leibler distance* between the actual density, denoted by $f_0(x)$, and the postulated parametric mixture density family $f(x; \boldsymbol{\theta})$, i.e.,

$$\boldsymbol{\theta}_0 = \underset{\boldsymbol{\theta}}{\operatorname{argmin}} \int f_0(x) \log \left\{ \frac{f_0(x)}{f(x; \boldsymbol{\theta})} \right\} dx,$$

where $f_0(x)$ is the true density.

One can also use the moment's estimation by solving the equations that set the sample moments equal to the population moments. Minimum distance estimator is another commonly used parametric estimation method. For example, one can estimate the parameters by minimizing some distance between the empirical cumulative distribution function (cdf) and the estimated cdf or the distance between the nonparametric kernel density estimation and the mixture density. The maximum likelihood estimator can also be viewed as a special case of minimum-distance estimators because it minimizes the Kullback-Leibler distance between the empirical distribution and the mixture distribution. In Section 4.1, we will introduce how to use Bayesian methods to estimate mixture models.

1.5   EM algorithm

1.5.1   Introduction of EM algorithm

In Section 1.4, we introduced the maximum likelihood estimator for finite mixture models. However, there is no explicit solution to maximize (1.7). In this section, we introduce the Expectation-Maximization (EM) algorithm which is commonly used to maximize (1.7).

The EM algorithm (Dempster et al., 1977b) is an iterative computation method to find the maximum likelihood or maximum a posteriori (MAP) estimator of parameters in statistical models where

- Some parts of the data are missing/hidden, and the analysis of the incomplete data is somewhat complicated or nonlinear.

- It is possible to "fill in" the missing data, and the analysis of the complete data is relatively simple.

The notation of "missing data" does not have to involve real missing data, but any incomplete information that we can assume to simplify a model and its likelihood function.

In general, suppose the observed data $\mathbf{x} = (x_1, \ldots, x_n)$ are from $f(x; \boldsymbol{\theta})$. The MLE of $\boldsymbol{\theta}$ is $\hat{\boldsymbol{\theta}} = \arg\max_{\boldsymbol{\theta}} \ell(\boldsymbol{\theta})$, where

$$\ell(\boldsymbol{\theta}) = \log L(\boldsymbol{\theta}; \mathbf{x}) = \sum_{i=1}^{n} \log f(x_i; \boldsymbol{\theta}). \tag{1.8}$$

Sometimes, it is hard to find $\hat{\boldsymbol{\theta}}$ when there is no closed form. Suppose for the complete data $\mathbf{y} = (\mathbf{x}, \mathbf{z})$ with the log-likelihood $\ell_c(\boldsymbol{\theta})$, there is a simple or closed form of the MLE, where $\mathbf{z} = (z_1, \ldots, z_n)$ are the unobserved (or missing) data. We can then use the EM algorithm to find $\hat{\boldsymbol{\theta}}$. The EM algorithm iterates between performing an expectation (E) step, which computes the expectation of the complete log-likelihood evaluated at the current estimators, and a maximization (M) step, which updates parameters by maximizing the expected complete log-likelihood found on the E step. More specifically, the EM algorithm is performed as follows.

Algorithm 1.5.1 EM Algorithm

---

1. Start with initial parameter values $\boldsymbol{\theta}^{(0)}$.

2. Expectation Step (E step): At the $(k+1)$th step, compute the conditional expectation of the complete-data log-likelihood $\ell_c(\boldsymbol{\theta})$ given the observed data $\mathbf{x}$, using current estimate $\boldsymbol{\theta}^{(k)}$:

$$Q(\boldsymbol{\theta} \mid \boldsymbol{\theta}^{(k)}) = E(\ell_c(\boldsymbol{\theta}) \mid \mathbf{x}, \boldsymbol{\theta}^{(k)}).$$

3. Maximization Step (M step): Update the new estimate $\boldsymbol{\theta}^{(k+1)}$ as the maximizer of $Q(\boldsymbol{\theta} \mid \boldsymbol{\theta}^{(k)})$ over $\boldsymbol{\theta}$:

$$\boldsymbol{\theta}^{(k+1)} = \arg \max_{\boldsymbol{\theta}} Q(\boldsymbol{\theta} \mid \boldsymbol{\theta}^{(k)}).$$

4. Iterate Steps 2 and 3 until some convergence criterion is attained.

We will discuss the choice of initial values and convergence criteria in Sections 1.5.7 and 1.5.8, respectively. One main appealing feature of using the EM algorithm to maximize the likelihood is that derivatives of log-likelihood are not needed. Instead, the complete data log-likelihood is predicted and then maximized.

Dempster et al. (1977b) showed that the log-likelihood function $\ell(\boldsymbol{\theta}; \mathbf{x})$ is nondecreasing after each EM iteration; that is, $\ell(\boldsymbol{\theta}^{(k+1)}; \mathbf{x}) \geq \ell(\boldsymbol{\theta}^{(k)}; \mathbf{x})$ for $k = 0, 1, 2, \ldots$.

Theorem 1.5.1. One of the most important properties of the EM algorithm is that after each iteration the likelihood is nondecreasing:

$$\ell(\boldsymbol{\theta}^{(k+1)}) \geq \ell(\boldsymbol{\theta}^{(k)}),$$

where $\ell(\boldsymbol{\theta}) = \log L(\boldsymbol{\theta}; \mathbf{x})$ is the mixture log-likelihood.

Proof *Note that*

$$\ell_c(\boldsymbol{\theta}) = \log L(\boldsymbol{\theta}; \boldsymbol{y}) = \log L(\boldsymbol{\theta}; \boldsymbol{x}) + \log L(\boldsymbol{\theta}; \boldsymbol{z} \mid \boldsymbol{x}).$$

Taking the conditional expectation of both sides with respect to $f(\boldsymbol{z} \mid \boldsymbol{x}, \boldsymbol{\theta}^{(k)})$, we have

$$Q(\boldsymbol{\theta} \mid \boldsymbol{\theta}^{(k)}) = \log L(\boldsymbol{\theta}; \boldsymbol{x}) + h(\boldsymbol{\theta} \mid \boldsymbol{\theta}^{(k)}),$$

where $h(\boldsymbol{\theta} \mid \boldsymbol{\theta}^{(k)}) = E_{\boldsymbol{z}}\{\log L(\boldsymbol{\theta}; \boldsymbol{z} \mid \boldsymbol{x}) \mid \boldsymbol{x}, \boldsymbol{\theta}^{(k)}\}$. Hence,

$$\log L(\boldsymbol{\theta}; \boldsymbol{x}) = Q(\boldsymbol{\theta} \mid \boldsymbol{\theta}^{(k)}) - h(\boldsymbol{\theta} \mid \boldsymbol{\theta}^{(k)}).$$

From the information inequality, for any two densities $f(x) \neq g(x)$ we have

$$E_g \log f(X) \leq E_g \log g(X).$$

Apply this to the conditional density of $x \mid y$, we have

$$h(\boldsymbol{\theta}^{(k+1)} \mid \boldsymbol{\theta}^{(k)}) \leq h(\boldsymbol{\theta}^{(k)} \mid \boldsymbol{\theta}^{(k)}).$$

In addition, the next iteration, $\boldsymbol{\theta}^{(k+1)}$, satisfies the inequality

$$Q(\boldsymbol{\theta}^{(k+1)} \mid \boldsymbol{\theta}^{(k)}) \geq Q(\boldsymbol{\theta}^{(k)} \mid \boldsymbol{\theta}^{(k)}).$$

Hence,

$$\log L(\boldsymbol{\theta}^{(k+1)}; \boldsymbol{x}) \geq \log L(\boldsymbol{\theta}^{(k)}; \boldsymbol{x}).$$

$\square$

Converging to a critical point of the likelihood occurs under mild regularity conditions (Wu, 1983). Note that the above monotone property applies to the original likelihood function $L(\boldsymbol{\theta}; \mathbf{x})$ in (1.8), not the complete data log-likelihood $\ell_c(\boldsymbol{\theta})$. This monotone property makes EM a numerically stable procedure as it climbs up the likelihood surface after each iteration; in contrast, no such guarantee exists for the Newton-Raphson algorithm. Moreover, the Newton-Raphson algorithm might even converge to a local minimum or not converge at all. Another practical advantage of the EM algorithm is that it usually handles parameter constraints automatically. This is because each M step produces an MLE-type estimate. For example, estimates of probabilities are naturally constrained to be between zero and one. In general, there can only be a guarantee that the EM algorithm converges to a stationary value (i.e. a local/global maximum or a saddle point), not necessarily the global optimum. In complex cases, it is important to try several initial values or to start with a sensible estimate.

There is a similarity between the Newton-Raphson and EM algorithms. With the Newton-Raphson algorithm, we obtain a quadratic approximation of the objection function $f(\boldsymbol{\theta})$ around an initial estimate $\boldsymbol{\theta}^{(0)}$:

$$f(\boldsymbol{\theta}) \approx f(\boldsymbol{\theta}^{(0)}) + \frac{\partial f(\boldsymbol{\theta}^{(0)})}{\partial \boldsymbol{\theta}^{\top}}(\boldsymbol{\theta} - \boldsymbol{\theta}^{(0)}) + \frac{1}{2}(\boldsymbol{\theta} - \boldsymbol{\theta}^{(0)})^{\top}\frac{\partial^2 f(\boldsymbol{\theta}^{(0)})}{\partial \boldsymbol{\theta} \partial \boldsymbol{\theta}^{\top}}(\boldsymbol{\theta} - \boldsymbol{\theta}^{(0)}) := q(\boldsymbol{\theta}),$$

and find the update $\boldsymbol{\theta}^{(1)}$ as the maximizer of $q(\boldsymbol{\theta})$. The algorithm converges quickly if $f(\boldsymbol{\theta})$ is well approximated by $q(\boldsymbol{\theta})$. With the EM algorithm, the objective function $\log L(\boldsymbol{\theta}; \mathbf{x})$ is approximated by $Q(\boldsymbol{\theta} \mid \boldsymbol{\theta}^{(k)})$.

From the proof of M step in Theorem 1.5.1, we can see that as long as $\boldsymbol{\theta}^{(k+1)}$ satisfies $Q(\boldsymbol{\theta}^{(k+1)} \mid \boldsymbol{\theta}^{(k)}) \geq Q(\boldsymbol{\theta}^{(k)} \mid \boldsymbol{\theta}^{(k)})$, not necessarily the one that maximizes the $Q(\boldsymbol{\theta} \mid \boldsymbol{\theta}^{(k)})$, we can have the monotone result stated in Theorem 1.5.1. Such a modified algorithm is called the Generalized EM (GEM) algorithm (Dempster et al., 1977b). In some applications, it might not be easy to maximize Q, and it is much less computationally demanding to increase Q. Clearly, the EM algorithm is a special case of the GEM algorithm.

Example 1.5.1. The famous genetic example from Rao (1973, page 369) assumes that the phenotype data

$$\mathbf{x} = (x_1, x_2, x_3, x_4) = (125, 18, 20, 34)$$

are distributed according to the multinomial distribution with probabilities

$$\left\{ \frac{1}{2} + \frac{\theta}{4}, \frac{1-\theta}{4}, \frac{1-\theta}{4}, \frac{\theta}{4} \right\}.$$

The log-likelihood based on $\mathbf{x}$ is

$$\log L(\theta; \mathbf{x}) = x_1 \log(2 + \theta) + (x_2 + x_3) \log(1 - \theta) + x_4 \log \theta,$$

which does not yield a closed-form estimate of θ.

Now we treat $\mathbf{x}$ as the incomplete data from $\mathbf{y} = (y_1, \ldots, y_5)$ with multinomial probabilities

$$\left\{ \frac{1}{2}, \frac{\theta}{4}, \frac{1-\theta}{4}, \frac{1-\theta}{4}, \frac{\theta}{4} \right\}.$$

Here $x_1 = y_1 + y_2, x_2 = y_3, x_3 = y_4$ and $x_4 = y_5$. The log-likelihood based on $\mathbf{y}$ is

$$\log L(\theta; \mathbf{y}) = (y_2 + y_5) \log \theta + (y_3 + y_4) \log(1 - \theta),$$

which readily yields

$$\hat{\theta} = \frac{y_2 + y_5}{y_2 + y_3 + y_4 + y_5},$$

so the "complete data" $\mathbf{y}$ is simpler than $\mathbf{x}$.

In this example, the E step is to find

$$Q(\theta) = \mathrm{E}(y_2 + y_5 \mid \mathbf{x}, \theta^0) \log \theta + \mathrm{E}(y_3 + y_4 \mid \mathbf{x}, \theta^0) \log(1 - \theta)$$
$$= \{\mathrm{E}(y_2 \mid \mathbf{x}, \theta^0) + y_5\} \log \theta + (y_3 + y_4) \log(1 - \theta),$$

so we only need to compute

$$\hat{y}_2 = E(y_2 \mid \mathbf{x}, \theta^0).$$

Since $y_1 + y_2 = x_1$, the conditional distribution of $y_2 \mid x_1$ is binomial with parameter $x_1 = 125$ and probability

$$p^0 = \frac{\theta^0/4}{1/2 + \theta^0/4},$$

and so

$$\hat{y}_2 = x_1 \frac{\theta^0/4}{1/2 + \theta^0/4}.$$

The M step yields an update

$$\theta^1 = \frac{\hat{y}_2 + y_5}{\hat{y}_2 + y_3 + y_4 + y_5}.$$

The algorithm iterates between the above two steps. From the last category of $\mathbf{x}$, we may obtain an initial value: $\theta^0/4 = 34/197$ or $\theta^0 = 0.69$. The MLE $\hat{\theta} = 0.627$. □

1.5.2   EM algorithm for mixture models

Recall that the mixture log-likelihood based on the observed data $\mathbf{x} = (x_1, \ldots, x_n)$ is

$$\log L(\boldsymbol{\theta}; \mathbf{x}) = \sum_{i=1}^{n} \log \left\{ \sum_{c=1}^{C} \pi_c g_c(x_i; \lambda_c) \right\}.$$

We can define the missing data $(z_{ic}, i = 1, \ldots, n, c = 1, \ldots, C)$, where

$$z_{ic} = \begin{cases} 1, & \text{if } i\text{th observation is from the } c\text{th component;} \\ 0, & \text{otherwise} \end{cases}$$

and $\mathbf{z}_i = (z_{i1}, \ldots, z_{iC})^\top$. We define the "complete data" $\mathbf{y} = (\mathbf{y}_1, \ldots, \mathbf{y}_n)$, where $\mathbf{y}_i = (x_i, \mathbf{z}_i^\top)$. Now the complete log-likelihood for $\mathbf{y}$ is

$$\log L_c(\boldsymbol{\theta}; \mathbf{y}) = \sum_{i=1}^{n} \sum_{c=1}^{C} z_{ic} \log\{\pi_c g_c(x_i; \lambda_c)\}. \tag{1.9}$$

In the E step, we need to calculate

$$Q(\boldsymbol{\theta} \mid \boldsymbol{\theta}^{(k)}) = E\{\log L_c(\boldsymbol{\theta}; \mathbf{y}) \mid \mathbf{x}, \boldsymbol{\theta}^{(k)}\}.$$

Notice that $\log L_c(\boldsymbol{\theta}; \mathbf{y})$ is a linear function of the missing data $\mathbf{z}$. Therefore, in the E step, we can simply calculate the conditional expectation of the missing data $\mathbf{z}$ and then input the expected value to $\log L_c(\boldsymbol{\theta}; \mathbf{Y})$, i.e., the E step is simplified to calculating

$$p_{ic}^{(k+1)} = E\{z_{ic} \mid x_i, \boldsymbol{\theta}^{(k)}\} = P(z_{ic} = 1 \mid x_i, \boldsymbol{\theta}^{(k)}) = \frac{\pi_c^{(k)} g_c(x_i; \lambda_c^{(k)})}{\sum_{l=1}^{C} \pi_l^{(k)} g_l(x_i; \lambda_l^{(k)})},$$

which can be also considered as the posterior probability that x_i is from the cth component. In clustering problems, it is also the quantity of interest.

In the M step, we need to maximize

$$Q(\boldsymbol{\theta} \mid \boldsymbol{\theta}^{(k)}) = E\{\log L_c(\boldsymbol{\theta}; \mathbf{y}) \mid \mathbf{x}, \boldsymbol{\theta}^{(k)}\} = \sum_{c=1}^{C} p_{ic}^{(k+1)} \{\log g_c(x_i; \lambda_c) + \log \pi_c\}.$$

The EM algorithm for the above general mixture model is described in the following Algorithm 1.5.2.

Algorithm 1.5.2 EM algorithm for general mixture models

Starting with an initial value $\boldsymbol{\theta}^{(0)}$, iterate the following two steps until convergence.

E-step: Find the conditional probabilities

$$
p_{ic}^{(k+1)} = \frac{\pi_c^{(k)} g_c(x_i; \lambda_c^{(k)})}{\sum_{l=1}^{C} \pi_l^{(k)} g_l(x_i; \lambda_l^{(k)})},
$$

where $i = 1, \ldots, n, c = 1, \cdots, C$.

M-step: Update each λ_c by

$$
\lambda_c^{(k+1)} = \arg\max_{\lambda_c} \sum_{i=1}^{n} p_{ic}^{(k+1)} \log g_c(x_i \mid \lambda_c)
$$

and π_c by

$$
\pi_c^{(k+1)} = \frac{\sum_{i=1}^{n} p_{ic}^{(k+1)}}{n}, \quad c = 1, \ldots, C.
$$

Note that if $\log L_c(\boldsymbol{\theta}; \mathbf{y})$ is not a linear function of the missing data $\mathbf{z}$, we can't simply compute the expectation of $\mathbf{z}$ in the E step, and instead, we still need to compute $E\{\log L_c(\boldsymbol{\theta}; \mathbf{Y}) \mid \mathbf{X}, \boldsymbol{\theta}^{(k)}\}$ in order to get $Q(\boldsymbol{\theta} \mid \boldsymbol{\theta}^{(k)})$.

Let $\hat{\boldsymbol{\theta}}$ be the converged value. A probabilistic clustering of the observations is provided in terms of their posterior probabilities of the component membership:

$$
\hat{p}_{ic} = \frac{\hat{\pi}_c g(x_i; \hat{\lambda}_c)}{\sum_{l=1}^{C} \hat{\pi}_l g(x_i; \hat{\lambda}_l)},
$$

which is the estimated probability that the data point x_i belongs to the cth component. Instead of fuzzy classification, each data point can also be assigned to a particular population with the "maximum a posteriori rule (MAP)":

$$
\hat{z}_{ic} = \begin{cases} 1, & \hat{p}_{ic} \geq \hat{p}_{ik} \text{ for all } k \neq c, \\ 0, & \text{otherwise.} \end{cases} \tag{1.10}
$$

More specifically, for the normal mixture model

$$
f(x; \boldsymbol{\theta}) = \sum_{c=1}^{C} \pi_c \phi(x; \mu_c, \sigma_c^2),
$$

where $\boldsymbol{\theta} = (\pi_1, \mu_1, \sigma_1, \cdots, \pi_C, \mu_C, \sigma_C)$, the EM algorithm for the above normal mixture model is described in the following Algorithm 1.5.3.

Algorithm 1.5.3 EM algorithm for normal mixture models

Given initial values $\hat{\mu}_c^{(0)}, \hat{\sigma}_c^{2(0)}, \hat{\pi}_c^{(0)}, c = 1, \ldots, C$, at the $(k+1)^{th}$ step, the EM algorithm computes the following two steps.

E Step: Compute the classification probability that the observation x_i comes from the cth component based on the current estimates:

$$p_{ic}^{(k+1)} = \frac{\pi_c^{(k)} \phi(x_i; \mu_c^{(k)}, \sigma_c^{2(k)})}{\sum_{c=1}^{C} \pi_c^{(k)} \phi(x_i; \mu_c^{(k)}, \sigma_c^{2(k)})},$$

where $i = 1, \ldots, n, c = 1, \cdots, C$.

M Step: Update the parameter estimates:

$$\mu_c^{(k+1)} = \sum_{i=1}^{n} p_{ic}^{(k+1)} x_i / n_c^{(k+1)},$$

$$\sigma_c^{2(k+1)} = \sum_{i=1}^{n} p_{ic}^{(k+1)} (x_i - \mu_c^{(k+1)})^2 / n_c^{(k+1)},$$

$$\pi_c^{(k+1)} = n_c^{(k+1)} / n,$$

where $c = 1, \ldots, C$ and $n_c^{(k+1)} = \sum_{i=1}^{n} p_{ic}^{(k+1)}$.

If we assume the variances are equal, i.e., $\sigma_1 = \sigma_2 = \cdots = \sigma_C = \sigma$, then in the M step, σ is updated by

$$\sigma^{2(k+1)} = \frac{1}{n} \sum_{i=1}^{n} \sum_{c=1}^{C} p_{ic}^{(k+1)} (x_i - \mu_c^{(k+1)})^2.$$

1.5.3   Rate of convergence of the EM algorithm

In practice, the convergence of the EM algorithm has been observed to be very slow in some applications. As noted in Dempster et al. (1977b), the convergence rate of EM is linear and governed by the fraction of missing information in the data. If the mixture components are similar, then the convergence rate will be slow. For a more detailed description of the EM algorithm, please refer to McLachlan and Krishnan (2007).

The EM algorithm defines a mapping $\boldsymbol{\theta} \to M(\boldsymbol{\theta})$ such that

$$\boldsymbol{\theta}^{(k+1)} = M(\boldsymbol{\theta}^{(k)}), k = 0, 1, 2, \ldots.$$

If $\boldsymbol{\theta}^{(k)}$ converges to $\boldsymbol{\theta}^*$ and M is continuous, then $\boldsymbol{\theta}^*$ must satisfy $\boldsymbol{\theta}^* = M(\boldsymbol{\theta}^*)$. By a Taylor expansion of $\boldsymbol{\theta}^{(k+1)} = M(\boldsymbol{\theta}^{(k)})$ about $\boldsymbol{\theta}^*$, we have that

$$\boldsymbol{\theta}^{(k+1)} - \boldsymbol{\theta}^* \approx J(\boldsymbol{\theta}^*)(\boldsymbol{\theta}^{(k)} - \boldsymbol{\theta}^*),$$

where $J(\boldsymbol{\theta}) = \partial M(\boldsymbol{\theta})/\partial \boldsymbol{\theta}$. Since $J(\boldsymbol{\theta}^*)$ is usually nonzero, the EM algorithm is essentially a linear iteration with the linear rate matrix $J(\boldsymbol{\theta}^*)$. The matrix $J(\boldsymbol{\theta}^*)$ is also called the rate of convergence.

A measure of the actual observed convergence rate is defined as

$$r = \lim_{k \to \infty} \frac{||\boldsymbol{\theta}^{(k+1)} - \boldsymbol{\theta}^*||}{||\boldsymbol{\theta}^{(k)} - \boldsymbol{\theta}^*||} = \lim_{k \to \infty} \frac{||\boldsymbol{\theta}^{(k+1)} - \boldsymbol{\theta}^{(k)}||}{||\boldsymbol{\theta}^{(k)} - \boldsymbol{\theta}^{(k-1)}||},$$

where $|| \cdot ||$ is any norm. It is well known that $r = \lambda_{max}(J(\boldsymbol{\theta}^*))$, the largest eigenvalue of $J(\boldsymbol{\theta}^*)$. Note that the smaller the r value is, the quicker the algorithm converges.

Dempster et al. (1977b) established how the rate of convergence $J(\boldsymbol{\theta}^*)$ depends on the information matrices

$$J(\boldsymbol{\theta}^*) = \mathcal{I}_c^{-1}(\boldsymbol{\theta}^*)\mathcal{I}_m(\boldsymbol{\theta}^*),$$

where $\mathcal{I}_c(\boldsymbol{\theta})$ is the information matrix for the complete data with log-likelihood $\ell_c(\boldsymbol{\theta})$ and $\mathcal{I}_m(\boldsymbol{\theta})$ is the information matrix for the unobservable data $\mathbf{z}$ conditional on $\mathbf{x}$, i.e.,

$$\mathcal{I}_m(\boldsymbol{\theta}) = -E_{\boldsymbol{\theta}}\{\partial^2 \log L(\boldsymbol{\theta}; \mathbf{z}|\mathbf{x})/\partial\boldsymbol{\theta}\partial\boldsymbol{\theta}^\top |\mathbf{x}\}.$$

Therefore, the rate of convergence of EM algorithm is determined by the largest eigenvalue of the information ratio matrix $\mathcal{I}_c^{-1}(\boldsymbol{\theta}^*)\mathcal{I}_m(\boldsymbol{\theta}^*)$, which measures the proportion of information about $\boldsymbol{\theta}$ which is missing due to the "missing" data $\mathbf{z}$.

Note that

$$J(\boldsymbol{\theta}^*) = I_p - \mathcal{I}_c^{-1}(\boldsymbol{\theta}^*)\mathcal{I}(\boldsymbol{\theta}^*),$$

where I_p denotes a $p \times p$ identity matrix, and

$$\mathcal{I}(\boldsymbol{\theta}^*) = -\frac{\partial^2 \log L(\boldsymbol{\theta}; \mathbf{x})}{\partial\boldsymbol{\theta}\partial\boldsymbol{\theta}^\top}$$

is the negative of the Hessian of the log-likelihood function. Therefore, the rate of convergence of the EM algorithm is also determined by the smallest eigenvalue of $\mathcal{I}_c^{-1}(\boldsymbol{\theta}^*)\mathcal{I}(\boldsymbol{\theta}^*)$.

1.5.4 Classification EM algorithms

In many applications, the Q function might be difficult to compute in E-step, for example, when the missing data is high-dimensional or there are incomplete observations such as censored data, the conditional expectation contains a high-dimensional integral or an integral over an irregular region. For those situations, there are also two variants of the EM algorithm: Classification EM algorithms (CEM) and Stochastic EM algorithms (SEM) that can help simplify the computation.

The CEM algorithm treats the latent component indicators $\mathbf{z}$ as unknown parameters and aims to maximize the complete data log-likelihood (1.9), instead of the log-likelihood. The CEM consists of the E step, C step, and M step, and converges in a finite number of iterations. The E step is the same as (1.5.5) in the standard EM algorithm. The C step then creates a partition of the data by assigning each point to the component maximizing the classification probabilities found in the E step as in (1.10). The M step is similar to the one used in the standard EM algorithm but replaces the classification probabilities with the clustering partition of (1.10) found in the C step. We describe the above procedures in more detail in Algorithm 1.5.4.

Algorithm 1.5.4 Classification EM algorithm (CEM)

Given initial parameters $\hat{\mu}_c^{(0)}, \hat{\sigma}_c^{2(0)}, \hat{\pi}_c^{(0)}, c = 1, \ldots, C$, at $(k+1)^{th}$ step, the CEM algorithm computes the following three steps:

E Step: Compute classification probabilities

$$p_{ic}^{(k+1)} = E(Z_{ic} \mid \mathbf{x}, \boldsymbol{\theta}^{(k)}) = \frac{\pi_c^{(k)} \phi(x_i; \mu_c^{(k)}, \sigma_c^{2(k)})}{\sum_{c=1}^{C} \pi_c^{(k)} \phi(x_i; \mu_c^{(k)}, \sigma_c^{2(k)})},$$

where $i = 1, \ldots, n, c = 1, \cdots, C$.

C Step: Create a partition of the data based on the following hard clustering labels for $i = 1, \ldots, n, c = 1, \cdots, C$:

$$z_{ic}^{(k+1)} = \begin{cases} 1, & p_{ic}^{(k+1)} \geq p_{ik}^{(k+1)} \text{ for all } k \neq c; \\ 0, & \text{otherwise.} \end{cases}$$

M Step: Update the parameter estimates based on the partition in C Step.

$$\mu_c^{(k+1)} = \sum_{i=1}^{n} z_{ic}^{(k+1)} x_i / n_c^{(k+1)},$$

$$\sigma_c^{2(k+1)} = \sum_{i=1}^{n} z_{ic}^{(k+1)} (x_i - \mu_c^{(k+1)})^2 / n_c^{(k+1)},$$

$$\pi_c^{(k+1)} = n_c^{(k+1)}/n,$$

where $c = 1, \ldots, C$ and $n_c^{(k+1)} = \sum_{i=1}^{n} z_{ic}^{(k+1)}$.

If we assume that the variance is equal, i.e. $\sigma_1 = \sigma_2 = \cdots = \sigma_C = \sigma$, then in the M step σ is updated by

$$\sigma^{2(k+1)} = \sum_{i=1}^{n} \sum_{c=1}^{C} z_{ic}^{(k+1)} (x_i - \mu_c^{(k+1)})^2 / n.$$

The CEM usually converges much faster than the regular EM algorithm and can provide an easily interpretable clustering of the data. However, it yields inconsistent estimates of the model parameters (Bryant and Williamson, 1978; Bryant, 1991) and has poor performance when the mixture components are overlapping or in disparate proportions (McLachlan and Peel, 2000, Section 2.21). In addition, the CEM tends to underestimate the variances due to its ignorance of classification uncertainty. However, for the purpose of clustering, it is generally believed that it performs similarly to the regular EM algorithm.

One interesting fact is that the above CEM algorithm with equal component variance is the same as the well-known K-means clustering (MacQueen et al., 1967). Similar equivalence also holds for multivariate data. Let $x_{c1}, \ldots, x_{cn_c}$ be the points of $x_1, \ldots, x_n$ in cth cluster, where $c = 1, \ldots, C$, $\sum_{c=1}^{C} n_c = n$. Then, K-means clustering optimizes the following average distance to members of the same cluster

$$\sum_{c=1}^{C} \frac{1}{n_c} \sum_{i=1}^{n_c} \sum_{j=1}^{n_c} (x_{ci} - x_{cj})^2,$$

which is twice the total distance to centers, also called squared error

$$\sum_{c=1}^{C} \sum_{i=1}^{n_c} (x_{ci} - \mu_c)^2,$$

where μ_c is the center of the cth cluster.

Compared to the traditional clustering algorithm, model-based clustering using Gaussian mixture models has the following advantages:

- Flexibility in Cluster Shape: Model-based clustering can capture clusters of arbitrary shapes and sizes. Traditional clustering methods like k-means assume that clusters are spherical and equally sized, which might not always align with the underlying data distribution. Model-based approaches, like Gaussian Mixture Models (GMMs), can model clusters with different shapes, variances, and orientations.

- Probabilistic Framework: Model-based clustering provides a probabilistic framework for clustering. It assigns a probability distribution to each data point, indicating the likelihood of it belonging to each cluster. This can be more informative than a hard assignment to a single cluster, as it captures uncertainty and partial membership.

- Automatic Model Selection: Model-based clustering can automatically determine the number of clusters based on model selection criteria like the Bayesian Information Criterion (BIC) or the Akaike Information Criterion (AIC). This alleviates the need for users to specify the number of clusters a priori, which is a common challenge in traditional clustering.

- Handling Mixed Data Types: Model-based clustering can handle datasets with mixed data types (e.g., continuous, categorical) and can incorporate dependencies between features within clusters. This makes it suitable for more complex and diverse datasets.

1.5.5   Stochastic EM algorithms

If between the E and M steps, a random generation of the unknown component labels Z_{ic} is added by drawing them at random from the multinomial distribution with C categories specified by the $\{p_{ic}^{(k+1)}, c = 1, \ldots, C\}$, then such algorithm is called *stochastic EM algorithm*. It essentially assigns each observation to one of the C components. The stochastic EM algorithm gives the algorithm a chance of escaping from the current path of convergence to a local maximizer to other paths, which is desirable if a poor initial value is used. This algorithm will generate a Markov Chain, $\theta^{(k)}$, which converges to a stationary distribution. In practice, 100–200 iterations are often used for burn-in to allow the sequence to approach its stationary regime. The detailed description of the SEM is given in Algorithm 1.5.5.

Algorithm 1.5.5 Stochastic EM algorithms (SEM)

Given initial parameters $\hat{\mu}_c^{(0)}, \hat{\sigma}_c^{2(0)}, \hat{\pi}_c^{(0)}, c = 1, \ldots, C$, at the $(k+1)^{th}$ step, the SEM algorithm computes the following three steps:

E Step: Compute classification probabilities

$$p_{ic}^{(k+1)} = E(Z_{ic} \mid \mathbf{x}, \boldsymbol{\theta}^{(k)}) = \frac{\pi_c^{(k)} \phi(x_i; \mu_c^{(k)}, \sigma_c^{2(k)})}{\sum_{c=1}^{C} \pi_c^{(k)} \phi(x_i; \mu_c^{(k)}, \sigma_c^{2(k)})},$$

where $i = 1, \ldots, n, c = 1, \cdots, C$.

S Step: Generate hard clustering labels $\{z_{i1}^{(k+1)}, \ldots, z_{iC}^{(k+1)}\}$ from the multinomial distribution with probabilities $\{p_{i1}^{(k+1)}, \ldots, p_{iC}^{(k+1)}\}, i = 1, \ldots, n$.

M Step: Update the parameter estimates based on the partition in S Step.

$$\mu_c^{(k+1)} = \sum_{i=1}^{n} z_{ic}^{(k+1)} x_i / n_c^{(k+1)},$$

$$\sigma_c^{2(k+1)} = \sum_{i=1}^{n} z_{ic}^{(k+1)} (x_i - \mu_c^{(k+1)})^2 / n_c^{(k+1)},$$

$$\pi_c^{(k+1)} = n_c^{(k+1)} / n,$$

where $c = 1, \ldots, C$ and $n_c^{(k+1)} = \sum_{i=1}^{n} z_{ic}^{(k+1)}$.

From the above description, we can see that the SEM is similar to CEM except that the hard labels Z_{ic} are randomly generated from multinomial distributions instead of being computed by maximizing the classification probabilities. The SEM is a data augmentation algorithm (Wei and Tanner, 1990) and does not converge pointwise. Instead, it generates a Markov chain whose stationary distribution is more or less concentrated around the MLE. To get the parameter estimate, one can either use the average of the iterated values after the initial burn-in period or use the parameter value that has the largest log-likelihood.

In the S step, if multiple replicates of the hard labels are generated and the M step estimates are averaged over all replicates, then the algorithm becomes the so-called *Monte Carlo EM algorithm*. Therefore, the SEM algorithm is the same as the Monte Carlo EM algorithm with a single replication (McLachlan and Krishnan, 2007, Chapter 6).

1.5.6 ECM algorithm and some other extensions

Meng and Rubin (1993) developed an *expectation-conditional maximization (ECM)* algorithm, which replaced the M-step of the EM algorithm with a number of computationally simpler conditional maximization steps. The monotone property of the ECM algorithm can also be established. Meng and Dyk (1997); McLachlan and Peel (2000); McLachlan et al. (2003) further extended the ECM algorithm and proposed an alternating expectation-conditional maximization (AECM) algorithm, where the specification of the complete data is allowed to be different in each CM-step.

1.5.7 Initial values

It is well known that there exist multiple local maximums for the mixture log-likelihood. Jin et al. (2016) proved that the multi-modality of mixture log-likelihood exists even on well-separated mixtures of Gaussians with large sample sizes, and these local maximums can be arbitrarily worse than those of the global maximums. Améndola et al. (2016) proved that, for suitably small sample sizes, the samples could be constructed such that the likelihood function has an unbounded number of local maximums. The solution found by the EM algorithm depends on the initial point, and there is no guarantee that the EM algorithm will converge to the global maximum. Therefore, it is prudent to run the algorithm from several initial points and choose the best local optima solution.

There are a variety of methods available for choosing the initial/starting values. One natural way is to initialize the EM algorithm with a partition of the data by some clustering methods, such as the K-means (MacQueen et al., 1967) or hierarchical clustering, where initial parameter values are calculated based on the partitioned data. One can also use other estimation methods, such as the method of moments (Furman and Lindsay, 1994; Lindsay and Basak, 1993), as an initial value. Böhning et al. (1994) proposed using well-separated initial values and demonstrated numerically that the algorithm can then converge faster. Böhning (1999) proposed an initial partition of the data by maximizing the within-sum of squares criterion. The initial parameter values can then be easily calculated based on the initial partition.

Another popular method to broaden the search of the parameter space is to use random initial values. Random initial values can be derived by first randomly partitioning the data into C groups and then calculating the component parameters, such as component means and variances, based on each partitioned data. The mixing proportions can be set to equal or can be uniformly generated from $U(0,1)$. However, if the sample size is large, such random partition tends to produce similar component parameter estimates based on the central limit theorem. Let p be the number of unknown parameters for each component. To mitigate this problem, instead of partitioning the whole data, we can randomly assign kp points to each component and then estimate the component parameters based on the assigned points, where k is a very small value say 1 to 5. We can also just start from the best of several random points (for each initial value, the likelihood is calculated and the set with the largest likelihood is considered the best and used as the initial point). Repeating the above steps, we can then get a set of random initial values. For other methods to choose initial values, see McLachlan and Basford (1988); Finch et al. (1989); Seidel et al. (2000). Karlis and Xekalaki (2003) provided a good review and comparison of different methods for choosing initial values.

Biernacki et al. (2003) proposed a search/run/select (S/R/S) strategy, also known as emEM, to improve the chance of finding the highest likelihood from the limited computation. Specifically, the S/R/S strategy consists of the following three steps:

1. Search m initial values based on the methods mentioned in the previous paragraph.

2. Run the EM algorithm from each initial value with a fixed number of iterations, such as 5 or 10.

3. Select the solution providing the largest likelihood among the m initial values.

If we repeat the three-step S/R/S strategy, we can then get multiple initial points. Michael and Melnykov (2016) proposed using the model averaging of all short EM runs rather than only using the best initial value in Step 3.

1.5.8   Stopping rules

Several convergence criteria/stopping rules have been proposed. Lindstrom and Bates (1988) proposed stopping the algorithm based on the relative change of the parameters and/or of the log-likelihood, i.e., we can stop the algorithm if

$$\frac{||\boldsymbol{\theta}^{(k+1)} - \boldsymbol{\theta}^{(k)}||}{||\boldsymbol{\theta}^{(k)}||} < \epsilon \text{ or } \frac{\log L(\boldsymbol{\theta}^{(k+1)}) - \log L(\boldsymbol{\theta}^{(k)})}{\log L(\boldsymbol{\theta}^{(k)})} < \epsilon.$$

The above criteria only indicates a lack of progress rather than the actual convergence, and hence have the risk of stopping the algorithm too early when the algorithm is slow. The gradient function (Lindsay, 1983; Lindsay et al., 1983; Lindsay, 1995; Pilla and Lindsay, 2001) can be used to check if we found the MLE.

Böhning et al. (1994) proposed using Aitken's acceleration scheme as a stopping rule. Let $\ell^{(k)} = \log L(\boldsymbol{\theta}^{(k)})$. Due to the linear convergence of the EM algorithm, there exists a constant $0 < a < 1$, such that $\ell^{(k+1)} - \ell^{\infty} \approx a(\ell^{(k)} - \ell^{\infty})$ when k is large enough, where ℓ^{∞} is the converged log-likelihood. It can be derived that

$$\ell^{(k+1)} - \ell^{(k)} \approx (1 - a)(\ell^{\infty} - \ell^{(k)}). \tag{1.11}$$

From (1.11), if the algorithm is slow, then the constant a is close to 1, and a very small increase of the log-likelihood function, $\ell^{(k+1)} - \ell^{(k)}$, does not necessarily imply that $\ell^{(k)}$ is very close to ℓ^{∞}. Therefore, when the convergence rate is close to 1, using the difference of two consecutive log-likelihoods as the stopping rule, we might stop the algorithm early before it actually converges.

Based on (1.11), the predicted final value at the kth iteration is

$$\ell^{\infty} = \ell^{k-1} + \frac{1}{1-a}(\ell^{(k)} - \ell^{(k-1)}).$$

Then, we can stop the algorithm if $\ell^{\infty} - \ell^{(k)} < c$. The rate a can be estimated by

$$a \approx \frac{\ell^{(k+1)} - \ell^{(k)}}{\ell^{(k)} - \ell^{(k-1)}},$$

and (1.11) can be also derived based on this linear rate:

$$\begin{aligned} \ell^{\infty} - \ell^{(k-1)} &= \ell^{(k)} - \ell^{(k-1)} + \ell^{(k+1)} - \ell^{(k)} + \ell^{(k+2)} - \ell^{(k+1)} \\ &= \ell^{(k)} - \ell^{(k-1)} + a(\ell^{(k)} - \ell^{(k-1)}) + a^2(\ell^{(k)} - \ell^{(k-1)}) + \cdots \\ &= (\ell^{(k)} - l^{(k-1)})(1 + a + a^2 + \cdots) \\ &= (\ell^{(k)} - \ell^{(k-1)})\frac{1}{1-a}. \end{aligned}$$

Let $\hat{\boldsymbol{\theta}}$ be the estimate corresponding to ℓ^{∞}. Based on Lindsay (1995, Sec. 6.4.1),

$$2\{\ell^{\infty}(\hat{\boldsymbol{\theta}}) - \ell^{(k)}(\boldsymbol{\theta}^{(k)})\} \approx (\hat{\boldsymbol{\theta}} - \boldsymbol{\theta}^{(k)})^{\top}\{\text{Cov}(\hat{\boldsymbol{\theta}})\}^{-1}(\hat{\boldsymbol{\theta}} - \boldsymbol{\theta}^{(k)}),$$

for $\boldsymbol{\theta}^{(k)}$ in the neighborhood of the $\hat{\boldsymbol{\theta}}$. Therefore, if the stopping threshold $c = 0.005$ is chosen, i.e., $\ell^{\infty} - \ell^{(k)} \leq 0.005$, approximately,

$$(\hat{\boldsymbol{\theta}} - \boldsymbol{\theta}^{(k)})^{\top}\{\text{Cov}(\hat{\boldsymbol{\theta}})\}^{-1}(\hat{\boldsymbol{\theta}} - \boldsymbol{\theta}^{(k)}) \leq 0.01,$$

which implies that $\boldsymbol{\theta}^{(k)}$ deviates from $\hat{\boldsymbol{\theta}}$ by at most 0.1 standard units. Based on the above arguments, practically, $c = 0.005$ is recommended for Aitken's acceleration stopping rule.

1.6   Some applications of EM algorithm

1.6.1   Mode estimation

Li et al. (2007) proposed the Modal EM (MEM) algorithm, a type of EM algorithm, to find the mode of the mixture density with known parameters and successfully apply it to nonparametric clustering by grouping data points into one cluster if they are associated with the same hilltop. Li et al. (2007) have successfully applied the MEM algorithm to do nonparametric clustering by locating the local modes of kernel density (1.13) when starting from each observation, assuming that the observations converged to the same mode are in the same cluster.

Suppose we are interested in finding the mode of a mixture density

$$f(x) = \sum_{c=1}^{C} \pi_c g_c(x), \tag{1.12}$$

where π_cs are known component proportions and $g_c(x)$s are completely specified component densities without any unknown parameters. Note that the objective of finding the mode of a known mixture density is different from finding the MLE for a mixture model given some observations. Li et al. (2007) proposed the following EM algorithm to find the mode of (1.12).

Algorithm 1.6.1 MEM algorithm (Li et al., 2007) to find the mode of (1.12): given any initial value $x^{(0)}$, iterate the following two steps until convergence.

E Step: Calculate

$$p_c^{(k+1)} = \frac{\pi_c g_c(x^{(k)})}{\sum_{c=1}^{C} \pi_c g_c(x^{(k)})}.$$

M Step: Update the estimation of the mode by

$$x^{(k+1)} = \arg\max_{x} \sum_{c=1}^{C} p_c^{(k+1)} \log g_c(x),$$

which simplifies to $x^{(k+1)} = \sum_{c=1}^{C} p_c^{(k+1)} \mu_c$ if g_cs are normal densities with mean μ_c and equal covariance matrix Σ.

Algorithm 1.6.1 can be also used to estimate the modes for any unknown density function. Given the observation $(x_1, \ldots, x_n)$ from the population X

with density $f(x)$, suppose we want to estimate the mode of $f(x)$. Parzen (1962) and Eddy (1980) proposed to estimate the mode of $f(x)$ by maximizing the kernel density estimator of $f(x)$:

$$\hat{f}(x) = \frac{1}{n} \sum_{i=1}^{n} K_h(x_i - x), \tag{1.13}$$

where $K_h(t) = h^{-1}K(t/h)$ and $K(t)$ is a kernel density with mean 0 such as the standard Gaussian distribution. Here x can be a scalar or a vector. If x is a vector, then we can use a multivariate normal kernel. Note that the above kernel density estimator (1.13) can be considered as a special case of (1.12) with n components and equal component proportions $\pi_c = 1/n$. Therefore, we can apply Algorithm 1.6.1 to find the mode of $\hat{f}(x)$ in (1.13). In the $(k+1)^{\text{th}}$ step, the mode can be updated by

$$x^{(k+1)} = \sum_{j=1}^{n} p_j^{(k+1)} x_j,$$

where

$$p_j^{(k+1)} = \frac{K_h(x_j - x^{(k)})}{\sum\limits_{i=1}^{n} K_h(x_i - x^{(k)})}, \ j = 1, \dots, n \, .$$

Therefore, the estimated mode can be considered as a weighted average of the observations and the weights that depend on the distance between each observation and the mode.

1.6.2 Maximize a mixture type objective function

Yao (2013a) proposed a generalized MEM algorithm (GMEM) to maximize the following general mixture type objective function

$$f(x) = \sum_{k=1}^{K} w_k \left[\log \left\{ \sum_{l=1}^{L} a_{kl} f_{kl}(x) \right\} \right], \tag{1.14}$$

where K, L, w_ks, and a_{kl}s are known positive constants, $f_{kl}(x)$s are positive known functions (not necessary density functions), and x can be a scalar or a vector. Unlike the MEM algorithm (Li et al., 2007), in (1.14), neither $\sum_{l=1}^{L} a_{kl}$ is required to be 1 nor $f_{kl}(x)$s are required to be density functions.

In addition, when $K = 1$, the objective function (1.14) becomes

$$f(x) = w_1 \log \left\{ \sum_{l=1}^{L} a_{1l} f_{1l}(x) \right\} \propto \sum_{l=1}^{L} a_{1l} f_{1l}(x),$$

which simplifies to (1.12) if a_{1l}s are component proportions and $f_{1l}(x)$s are component densities. Therefore, the MEM algorithm (Li et al., 2007) can be considered as a special case of the GMEM (Yao, 2013a). A detailed description of GMEM to maximize (1.14) is given below. If $f_{kl}(x)$ is a normal density with

---

Algorithm 1.6.2 GMEM

Given the initial value $x^{(0)}$, in the $(t+1)^{\text{th}}$ step, compute the following two steps.

E Step: Calculate

$$\pi_{kl}^{(t+1)} = \frac{a_{kl} f_{kl}(x^{(t)})}{\sum_{l=1}^{L} a_{kl} f_{kl}(x^{(t)})}, \quad k = 1, \ldots, K; \, l = 1, \ldots, L.$$

M Step: Update

$$x^{(t+1)} = \arg\max_x \sum_{k=1}^{K} \sum_{l=1}^{L} \left\{ w_k \pi_{kl}^{(t+1)} \log f_{kl}(x) \right\}. \tag{1.15}$$

---

mean μ_{kl} and variance σ_{kl}^2, then the above M step has an explicit formula, i.e.,

$$x^{(t+1)} = \frac{\sum_{k=1}^{K} \sum_{l=1}^{L} w_k \pi_{kl}^{(t+1)} \mu_{kl} \sigma_{kl}^{-2}}{\sum_{k=1}^{K} \sum_{l=1}^{L} w_k \pi_{kl}^{(t+1)} \sigma_{kl}^{-2}}.$$

Yao (2013a) called the above algorithm the generalized modal EM algorithm (GMEM) and proved that the GMEM algorithm monotonically increases the objective function (1.14) after each iteration.

Theorem 1.6.1. (Yao, 2013a) The objective function (1.14) is nondecreasing after each iteration of GMEM Algorithm 1.6.2, i.e., $f(x^{(t+1)}) \geq f(x^{(t)})$, until a fixed point is reached. The GMEM is strictly monotonically increasing at the $(t+1)$th step, i.e., $f(x^{(t+1)}) > f(x^{(t)})$ if either of the following two conditions is satisfied:

1. There exists $1 \leq k \leq K$ and $1 \leq l_1 < l_2 \leq L$ such that

$$\frac{f_{kl_1}(x^{(t+1)})}{f_{kl_1}(x^{(t)})} \neq \frac{f_{kl_2}(x^{(t+1)})}{f_{kl_2}(x^{(t)})}.$$

2. In the M step of (1.15), $Q(x^{(t+1)} \mid x^{(t)}) > Q(x^{(t)} \mid x^{(t)})$.

   Proof *Let $Y_k^{(t+1)}$ be a discrete random variable such that*

$$P\left(Y_k^{(t+1)} = \frac{f_{kl}(x^{(t+1)})}{f_{kl}(x^{(t)})}\right) = \frac{a_{kl} f_{kl}(x^{(t)})}{\sum_{l=1}^{L} a_{kl} f_{kl}(x^{(t)})} \triangleq \pi_{kl}^{(t+1)}, l = 1, \ldots, L.$$

Then,

$$
\begin{aligned}
f(x^{(t+1)}) - f(x^{(t)}) &= \sum_{k=1}^{K} w_k \log \left\{ \frac{\sum_{l=1}^{L} a_{kl} f_{kl}(x^{(t+1)})}{\sum_{l=1}^{L} a_{kl} f_{kl}(x^{(t)})} \right\} \\
&= \sum_{k=1}^{K} w_k \log \left\{ \sum_{l=1}^{L} \frac{a_{kl} f_{kl}(x^{(t)})}{\sum_{l=1}^{L} a_{kl} f_{kl}(x^{(t)})} \frac{a_{kl} f_{kl}(x^{(t+1)})}{a_{kl} f_{kl}(x^{(t)})} \right\} \\
&= \sum_{k=1}^{K} w_k \log \left\{ \sum_{l=1}^{L} \pi_{kl}^{(t+1)} \frac{f_{kl}(x^{(t+1)})}{f_{kl}(x^{(t)})} \right\} \\
&= \sum_{k=1}^{K} w_k \log \left\{ E\left(Y_k^{(t+1)}\right) \right\}.
\end{aligned}
$$

Based on Jensen's inequality, we have

$$
\begin{aligned}
f(x^{(t+1)}) - f(x^{(t)}) &\geq \sum_{k=1}^{K} w_k E \left\{ \log(Y^{(k+1)}) \right\} \\
&= \sum_{k=1}^{K} w_k \sum_{l=1}^{L} \pi_{kl}^{(t+1)} \log \frac{f_{kl}(x^{(t+1)})}{f_{kl}(x^{(t)})} \\
&= \sum_{k=1}^{K} \sum_{l=1}^{L} w_k \pi_{kl}^{(t+1)} \log \frac{f_{kl}(x^{(t+1)})}{f_{kl}(x^{(t)})}.
\end{aligned}
$$

The equality occurs if and only if $f_{kl}(x^{(t+1)})/f_{kl}(x^{(t)})$ is the same for all ls given any $k = 1, \ldots, K$. Based on the property of the M-step of (1.15), we have

$$
\sum_{k=1}^{K} \sum_{l=1}^{L} w_k \pi_{kl}^{(t+1)} \log\{f_{kl}(x^{(t+1)})\} \geq \sum_{k=1}^{K} \sum_{l=1}^{L} w_k \pi_{kl}^{(t+1)} \log\{f_{kl}(x^{(t)})\}
$$

and

$$
f(x^{(t+1)}) - f(x^{(t)}) \geq 0.
$$

The strict inequality holds if either of the two conditions in Theorem 1.6.1 holds. □

1.6.2.1 Adaptive estimation for regression models

Linear regression models are widely used to investigate the relationship between several variables. Suppose $(\mathbf{x}_1, y_1), \ldots, (\mathbf{x}_n, y_n)$ are sampled from the regression model

$$
y = \mathbf{x}^\top \boldsymbol{\beta} + \epsilon, \tag{1.16}
$$

where $\mathbf{x}$ is a p-dimensional vector of covariates and is independent of the error ϵ with $\mathrm{E}(\epsilon) = 0$. The well-known least squares estimate (LSE) of β is

$$\hat{\beta}_{LSE} = \underset{\beta}{\mathrm{argmin}} \sum_{i=1}^{n} (y_i - \mathbf{x}_i^\top \beta)^2.$$

For normally distributed errors, $\hat{\beta}_{LSE}$ is exactly the maximum likelihood estimate (MLE). However, $\hat{\beta}_{LSE}$ will lose some efficiency when the error is not normally distributed. Therefore, it is desirable to have an estimate that can be adaptive to the unknown error distribution.

Suppose that $f(t)$ is the marginal density of ϵ in (1.16). If we know $f(t)$, then we can obtain a more accurate estimate of β in (1.16) by maximizing the log-likelihood, rather than using the least squares estimator (LSE):

$$\sum_{i=1}^{n} \log f(y_i - \mathbf{x}_i^\top \beta). \tag{1.17}$$

In practice, $f(t)$ is often unknown and thus (1.17) is not directly applicable. To attenuate this, Yao and Zhao (2013) proposed an adaptive linear regression to efficiently estimate β in (1.16) with unknown error density. Denote by $\tilde{\beta}$, an initial estimate of β, such as the LSE in (7.2). Based on the residuals $\tilde{\epsilon}_i = y_i - \mathbf{x}_i^\top \tilde{\beta}$, we can estimate $f(t)$ by the nonparametric kernel density estimate, denoted by $\tilde{f}(t)$, as

$$\tilde{f}(t) = \frac{1}{n} \sum_{j=1}^{n} K_h(t - \tilde{\epsilon}_j),$$

where $K_h(t) = h^{-1}K(t/h)$, $K(\cdot)$ is a kernel density, and h is the tuning parameter.

Replacing $f(\cdot)$ in (1.17) with $\tilde{f}(\cdot)$, Yao and Zhao (2013) proposed the following kernel density-based regression parameter estimate (KDRE) as

$$\hat{\beta} = \underset{\beta}{\mathrm{argmax}} \sum_{i=1}^{n} \log \left\{ \frac{1}{n} \sum_{j=1}^{n} K_h \left(y_i - \mathbf{x}_i^\top \beta - \tilde{\epsilon}_j \right) \right\}. \tag{1.18}$$

Yao and Zhao (2013) proved that $\hat{\beta}$ in (1.18) is asymptotically the most efficient estimator and has the same asymptotic variance as that of the infeasible "oracle" MLE, which assumes $f(\cdot)$ were known. Therefore, $\hat{\beta}$ can be automatically adaptive to different unknown error distributions; it is as efficient as $\hat{\beta}_{LSE}$ when the error exactly has Gaussian distribution and is much more efficient than $\hat{\beta}_{LSE}$ when the error is not normally distributed.

Note that the objective function (1.18) has the mixture form of (1.14). Therefore, we can use the GMEM Algorithm 1.6.2 to maximize it. At the

$(k+1)$th step, we can update β by

$$
\begin{aligned}
\beta^{(k+1)} &= \underset{\beta}{\operatorname{argmax}} \sum_{i=1}^{n} \sum_{j=1}^{n} \left\{ p_{ij}^{(k+1)} \log K_h(y_i - \mathbf{x}_i^\top \beta - \tilde{\epsilon}_j) \right\} \\
&= \underset{\beta}{\operatorname{argmin}} \sum_{i=1}^{n} \sum_{j=1}^{n} \left\{ p_{ij}^{(k+1)} (y_i - \mathbf{x}_i^\top \beta - \tilde{\epsilon}_j)^2 \right\},
\end{aligned} \qquad (1.19)
$$

which has an explicit solution if $K_h(\cdot)$ is a Gaussian kernel density, where

$$
p_{ij}^{(k+1)} = \frac{K_h(y_i - \mathbf{x}_i^\top \beta^{(k)} - \tilde{\epsilon}_j)}{\sum_{l=1}^{n} K_h(y_i - \mathbf{x}_i^\top \beta^{(k)} - \tilde{\epsilon}_l)} \propto K_h(y_i - \mathbf{x}_i^\top \beta^{(k)} - \tilde{\epsilon}_j). \qquad (1.20)
$$

Based on (1.19), the KDRE (Yao and Zhao, 2013) can be viewed as a weighted least squares estimate. Specifically, the estimate involves the weighted squared difference between the new residual $y_i - \mathbf{x}_i^\top \beta$ and the initial residual $\tilde{\epsilon}_j$ for all $1 \le i, j \le n$. The weights in (1.20) indicate that isolated outliers (i.e., cases where $\tilde{\epsilon}_j$ is large) will have little effect on updating $\beta^{(k+1)}$, as the corresponding weights $p_{ij}^{(k+1)}$ will be small.

The idea of the above GMEM algorithm can also be applied to find an adaptive nonlinear regression estimator if the linear regression function in (1.16) is replaced by a parametric nonlinear regression function. Similarly, we can use the GMEM to compute the adaptive nonparametric regression (Linton and Xiao, 2007) or some other nonparametric or semiparametric regression models.

1.6.2.2 Edge-preserving smoothers for image processing

Given iid observations $(x_1, y_1), \ldots, (x_n, y_n)$, suppose we want to estimate $m(x) = \mathrm{E}(Y \mid x)$ at data points, i.e., $m(x_1), \ldots, m(x_n)$. Traditional nonparametric regression estimators are often unsuitable for image processing due to their tendency to blur sharp "edges." The edge-preserving smoother of $m(x_i)$ proposed by Chu et al. (1998) is the local maximizer of

$$
\sum_{j=1}^{n} \phi_g(y_j - \theta) \phi_h(x_i - x_j), \qquad (1.21)
$$

when starting from y_i, where ϕ_h is the Gaussian kernel with a bandwidth h.

Since (1.21) has a mixture-type objective function (1.14) with $K = 1$, we can apply the following GMEM algorithm to maximize it.

Algorithm 1.6.3 GMEM for (1.21).

Given the initial value $\theta^{(0)} = y_i$, in the $(k+1)^{\text{th}}$ step,
E Step: Calculate

$$\pi(j \mid \theta^{(k)}) = \frac{\phi_g(y_j - \theta^{(k)})\phi_h(x_i - x_j)}{\sum_{j=1}^{n} \phi_g(y_j - \theta^{(k)})\phi_h(x_i - x_j)}, \ j = 1, \ldots, n.$$

M Step: Update

$$\theta^{(k+1)} = \arg\max_{\theta} \sum_{j=1}^{n} \pi(j \mid \theta^{(k)}) \log\{\phi_g(y_j - \theta)\} = \sum_{j=1}^{n} \pi(j \mid \theta^{(k)})y_j.$$

1.6.2.3 Robust generalized M estimator for linear regression

For the linear regression model (1.16), it is well known that the LSE $\hat{\boldsymbol{\beta}}_{LSE}$ in (7.2) is very sensitive to outliers. Many robust regression estimators have been proposed, among which the M-estimator (Huber, 1981; Andrews and Mallows, 1974) is one of the most commonly used ones. The M-estimator replaces the square loss in (7.2) by a robust loss function $\rho(\cdot)$, i.e., estimates $\boldsymbol{\beta}$ by minimizing

$$\sum_{i=1}^{n} \rho(y_i - \mathbf{x}_i^\top \boldsymbol{\beta}).$$

The M estimator works well if there are only outliers in the y direction. However, it is not robust against high-leverage outliers (the outliers in both x and y directions) (Maronna et al., 2019).

Generalized M estimators (GM estimators for short) (Hampel et al., 1986) can instead handle the high-leverage outliers. The GM estimators find $\boldsymbol{\beta}$ by minimizing

$$\ell(\boldsymbol{\beta}) = \sum_{i=1}^{n} w(\mathbf{x}_i)\rho(y_i - \mathbf{x}_i^\top \boldsymbol{\beta}), \tag{1.22}$$

where $w(\cdot)$ is a weight function used to down-weight the high leverage points, such as $w(\mathbf{x}_i) = \sqrt{1 - h_{ii}}$ with h_{ii} being the leverage of the ith observation. Note that $h_{ii} = \partial \hat{y}_i / \partial y_i$ and $\text{Var}(e_i) = (1 - h_{ii})\sigma^2$, $0 \leq h_{ii} \leq 1$. We focus on redescending functions $\rho'(\cdot)$, which reject gross outliers entirely, in contrast to the Huber estimator which treats them similarly to moderate outliers. Moreover, for the Cauchy distribution, redescending M-estimators are approximately 20% more efficient than Huber's estimator. According to Chu et al. (1998), minimizing (1.22) is equivalent to maximizing

$$Q(\boldsymbol{\beta}) = \sum_{i=1}^{n} w(\mathbf{x}_i)K_h(y_i - \mathbf{x}_i^\top \boldsymbol{\beta}), \tag{1.23}$$

where K_h is a kernel density. Note that (1.23) has the mixture form (1.14) with $K = 1$. Therefore, we can use the GMEM Algorithm 1.6.2 to maximize (1.23).

Note that the GMEM still applies if the weight function $w(\cdot)$ in (1.23) also depends on the response variable y, such as the initially estimated residuals.

1.7 Multivariate normal mixtures

1.7.1 Introduction

A C-component p-dimensional multivariate normal mixtures has the probability density function

$$f(\mathbf{x}; \boldsymbol{\theta}) = \sum_{c=1}^{C} \pi_c \phi(\mathbf{x}; \boldsymbol{\mu}_c, \boldsymbol{\Sigma}_c), \mathbf{x} \in \mathbb{R}^p, \tag{1.24}$$

where π_c is the mixing proportion of the cth component, $\pi_c \in [0, 1]$, $\sum_{c=1}^{C} \pi_c = 1$, and $\phi(\mathbf{x}; \boldsymbol{\mu}, \boldsymbol{\Sigma})$ is the density of a multivariate normal distribution with mean $\boldsymbol{\mu}$ and variance $\boldsymbol{\Sigma}$.

Given observations $\{\mathbf{x}_1, \dots, \mathbf{x}_n\} \in \mathbb{R}^p$ from the multivariate normal mixtures (1.24), the log-likelihood is

$$\ell(\boldsymbol{\theta}) = \sum_{i=1}^{n} \log \left(\sum_{c=1}^{C} \pi_c \phi(\mathbf{x}_i; \boldsymbol{\mu}_c, \boldsymbol{\Sigma}_c) \right). \tag{1.25}$$

The EM algorithm to maximize (1.25) is summarized in Algorithm 1.7.1. If we assume that $\boldsymbol{\Sigma}_1 = \cdots = \boldsymbol{\Sigma}_C = \boldsymbol{\Sigma}$, then the update of $\boldsymbol{\Sigma}$ in M-step is

$$\boldsymbol{\Sigma}^{(k+1)} = \frac{\sum_{i=1}^{n} \sum_{c=1}^{C} p_{ic}^{(k+1)} (\mathbf{x}_i - \boldsymbol{\mu}_c^{(k+1)})(\mathbf{x}_i - \boldsymbol{\mu}_c^{(k+1)})^{\top}}{n}.$$

It is easy to see that (1.24) leads to over-parameterized solutions and computationally intensive estimation procedures when p is large. Indeed, a C-component GMM with unrestricted component-covariance matrices is highly parameterized, with

$$(C - 1) + Cp + Cp(p+1)/2$$

number of parameters, which is a quadratic function of p. In Section 1.7.2, we will introduce some parsimonious multivariate normal mixture models to reduce the number of parameters for the component covariance matrices. In Chapter 8, we will further introduce other dimension reduction methods for high-dimensional multivariate normal mixture models.

1.7.2 Parsimonious multivariate normal mixture modeling

To reduce the dimension of parameters in the covariance matrix, we can consider adding parsimonious assumptions on component-covariance matrices.

Algorithm 1.7.1 EM algorithm to maximize (1.25)

Starting with an initial value $\boldsymbol{\theta}^{(0)}$, in $(k+1)$th step, compute the following two steps.

E-step: Find the conditional probabilities:

$$p_{ic}^{(k+1)} = \frac{\pi_c^{(k)} \phi(\mathbf{x}_i; \boldsymbol{\mu}_c^{(k)}, \boldsymbol{\Sigma}_c^{(k)})}{\sum_{l=1}^{C} \pi_l^{(k)} \phi(\mathbf{x}_i; \boldsymbol{\mu}_l^{(k)}, \boldsymbol{\Sigma}_l^{(k)})}, \quad i = 1, \ldots, n, \quad c = 1, \ldots, C.$$

M-step: Update parameters by

$$\pi_c^{(k+1)} = \frac{\sum_{i=1}^{n} p_{ic}^{(k+1)}}{n},$$

$$\boldsymbol{\mu}_c^{(k+1)} = \frac{\sum_{i=1}^{n} p_{ic}^{(k+1)} \mathbf{x}_i}{\sum_{i=1}^{n} p_{ic}^{(k+1)}},$$

$$\boldsymbol{\Sigma}_c^{(k+1)} = \frac{\sum_{i=1}^{n} p_{ic}^{(k+1)} (\mathbf{x}_i - \boldsymbol{\mu}_c^{(k+1)})(\mathbf{x}_i - \boldsymbol{\mu}_c^{(k+1)})^\top}{\sum_{i=1}^{n} p_{ic}^{(k+1)}},$$

where $c = 1, \ldots, C$.

Banfield and Raftery (1993) and Celeux and Govaert (1995) proposed to parameterize covariance matrices through the eigenvalue/spectral decomposition in the form

$$\Sigma_c = \lambda_c D_c A_c D_c^\top, \quad c = 1, \ldots, C, \tag{1.26}$$

where $\lambda_c = |\Sigma_c|^{1/p}$, D_c is the orthogonal matrix of eigenvectors, A_c is a diagonal matrix such that $|A_c| = 1$ with normalized eigenvalues on the diagonal. The parameter λ_c determines the volume of the cth cluster, D_c the orientation, and A_c the shape. Let $A_c = \text{diag}\{a_{c1}, \ldots, a_{cp}\}$, where $a_{c1} \geq a_{c2} \geq \cdots \geq a_{cp}$. If $a_{c1} \approx a_{c2} \cdots \approx a_{cp}$, then the cth cluster will tend to be hyperspherical. If the largest diagonal value of A_c dominates all others, i.e., a_{c1}/a_{cp} is very large, then the cth cluster will be concentrated on a line. If the first k largest values of a_{cl}s are much larger than all remaining ones, i.e., $a_{c1} \approx a_{c2} \cdots \approx a_{ck} >> a_{c,k+1}$, where $k \leq p$, then the cth cluster will be concentrated on a k-dimensional plane in a p-dimensional space.

Many special cases of the above models have been considered in the literature. For example, if $\lambda_c = \sigma^2$ and $A_c = I$, then $\Sigma_c = \sigma^2 I$. If $\lambda_c = \lambda, D_c = D$, and $A_c = A$, then $\Sigma_c = \Sigma$, the equal variance case. More generally, by allowing some of the volumes, shapes, and orientations of clusters but not all of these quantities to vary between clusters, we obtain parsimonious and interpretable models that are appropriate to describe various clustering situations. For instance, we can assume equal shapes and orientations but with different volumes, i.e., $A_c = A$ and $D_c = D, c = 1, \ldots C$. We will denote

TABLE 1.2
Fourteen special models based on the matrix decomposition (1.26)

Model	Number of parameters	Closed M-step form
$[\lambda DAD']$	$Cp + C - 1 + p(p+1)/2$	Yes
$[\lambda_c DAD']$	$C(p+2) - 2 + p(p+1)/2$	No
$[\lambda DA_cD']$	$p(4C + p - 1)/2$	No
$[\lambda_c DA_cD']$	$p(4C + p - 1)/2 + C - 1$	No
$[\lambda D_c AD'_c]$	$Cp(p+1)/2 + p + C - 1$	Yes
$[\lambda_c D_c AD'_c]$	$Cp(p+1)/2 + p + 2(C - 1)$	No
$[\lambda D_c A_c D'_c]$	$Cp(p+3)/2$	Yes
$[\lambda_c D_c A_c D'_c]$	$Cp(p+3)/2 + C - 1$	Yes
$[\lambda A]$	$p(C+1) + C - 1$	Yes
$[\lambda_c A]$	$p(C+1) + 2(C - 1)$	No
$[\lambda A_c]$	$2Cp$	Yes
$[\lambda_c A_c]$	$2Cp + C - 1$	Yes
$[\lambda I_p]$	$C(p+1)$	Yes
$[\lambda_c I_p]$	$p(C+1) + C - 1$	Yes

this model by $[\lambda_c DAD']$. With this notation, we can have the following fourteen different models: $[\lambda DAD'], [\lambda_c DAD'], [\lambda DA_cD'], [\lambda D_c AD'_c], [\lambda_c DA_cD'],$ $[\lambda D_c A_c D'_c], [\lambda_c D_c AD'_c], \quad [\lambda_c D_c A_c D'_c], [\lambda A], [\lambda_c A], [\lambda A_c], [\lambda_c A_c], [\lambda I_p], [\lambda_c I_p].$ Adapted from Celeux and Govaert (1995), in Table 1.2, we summarize the above-mentioned 14 special models of the matrix decomposition (1.26) along with their number of parameters and whether there exist closed-form solutions in the M step. If there are no closed-form solutions in the M step, then an iterative procedure is necessary. See Celeux and Govaert (1995) for more detail about how to estimate the above-constrained mixture models. The computations can be also implemented by the R package *"mclust."* In practice, the information criteria such as AIC and BIC can be used to choose an appropriate parsimonious model.

The above parsimonious matrix decomposition techniques have also been used by many other multivariate mixture models, such as the parsimonious multivariate mixture regression models (Murphy and Murphy, 2017).

1.8 The topography of finite normal mixture models

1.8.1 Shapes of some univariate normal mixtures

One interesting and somewhat surprising feature of finite normal mixture models is that the number of modes is not the same as the number of components. It is even possible that the mixture density has only one mode when the

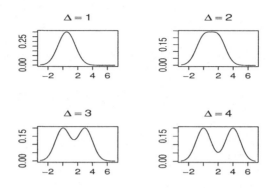

FIGURE 1.3
The plots of the mixture density $f(x) = 0.5\phi(x; 0, 1) + 0.5\phi(x; \Delta, 1)$ when $\Delta = 1, 2, 3$, and 4.

components are close enough. For multivariate mixture models, it is also possible that the number of modes is more than the number of components. In this section, we will introduce the shapes of finite univariate normal mixture models. In the next section, we will then investigate the topography of multivariate normal mixture models.

Let's first consider a simple two-component normal mixture with equal component variance

$$f(x) = \pi_1 \phi(x; \mu_1, \sigma^2) + \pi_2 \phi(x; \mu_2, \sigma^2). \tag{1.27}$$

Let $\Delta = |\mu_2 - \mu_1|/\sigma$ be the Mahalanobis distance between the homoscedastic components of the normal mixture density. Then, we have the following results.

Theorem 1.8.1. (Titterington et al., 1985, Section 5.5)
When $\pi_1 = \pi_2 = 0.5$, then $f(x)$ in (1.27) has two modes if $\Delta > 2$ but only one unique mode if $\Delta \leq 2$.

Therefore, the two-component normal mixture does not necessarily have two modes. In fact, it only has one mode when the distance between two component means is very small, say less than two standard deviations when the mixing proportions and component variances are equal.

Example 1.8.1. Let $\mu_1 = 0, \sigma = 1, \pi_1 = 0.5$. Figure 1.3 shows the plots of $f(x)$ versus x when $\Delta = 1, 2, 3, 4$. It can be seen that the two-component mixture density with equal proportions only has one mode when $\Delta = 1$ and 2 and has two modes when $\Delta = 3$ and 4. $\qquad\square$

Note that the results in Theorem 1.8.1 are only valid when $\pi_1 = \pi_2 = 0.5$. When the component proportions are not equal, the situation is much

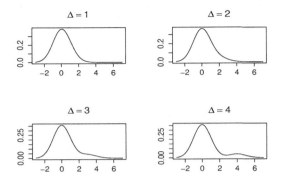

FIGURE 1.4
The plots of the mixture density $f(x) = 0.75\phi(x; 0, 1) + 0.25\phi(x; \Delta, 1)$ when $\Delta = 1, 2, 3$, and 4.

more complicated and the shape of the density depends on both Δ and the proportions.

Let $r = \max(\pi_1, \pi_2) / \min(\pi_1, \pi_2)$, the ratio of the larger mixing proportion to the smaller. Preston (1953) derived how the skewness γ_1 and kurtosis γ_2 depend on the separation Δ between two components and the relative proportion ratio r

$$\gamma_1 = \frac{r(r-1)\Delta^3}{\{r\Delta^2 + (r+1)^2\}^{3/2}}$$

and

$$\gamma_2 = \frac{r(r^2 - 4r + 1)\Delta^4}{\{r\Delta^2 + (r+1)^2\}^2}.$$

Therefore, when $r = 1$, equal proportions for two components, $\gamma_1 = 0$, and the mixture density is symmetric about its mean/median.

Holzmann and Vollmer (2008) extended the result in Theorem 1.8.1 to the situation when $\pi_1 \neq \pi_2$.

Theorem 1.8.2. (Holzmann and Vollmer, 2008)
The mixture density $f(x)$ in (1.27) is unimodal if and only if $\Delta \leq 2$ or if $\Delta > 2$ and

$$\left| \log \left(\frac{1 - \pi_1}{\pi_1} \right) \right| \geq 2\log \left(\Delta - \sqrt{\Delta^2 - 4} \right) + \frac{\Delta}{2}\sqrt{\Delta^2 - 4} - 2\log 2, \quad (1.28)$$

otherwise, it is bimodal.

Theorem 1.8.1 is a special case of Theorem 1.8.2 when taking $\pi_1 = \pi_2 = 0.5$. Note that the left side of (1.28) goes to infinity if π_1 goes to 0 or 1. Therefore, even if $\Delta > 2$ the mixture density (1.27) is still unimodal if the component proportion is close to 0 or 1.

Example 1.8.2. Let $\mu_1 = 0, \sigma = 1, \pi_1 = 0.75, \pi_2 = 0.25$. Figure 1.4 shows the plots of $f(x)$ versus x when $\Delta = 1, 2, 3, 4$. It can be seen that the shape of the density in Figure 1.4 is much different from those in Figure 1.3; for example, the $\Delta = 3$ still gives an unimodal mixture density. □

Theorem 1.8.3. (Eisenberger, 1964) Consider a general two-component normal mixture with *unequal* component variance

$$f(x) = \pi_1\phi(x; \mu_1, \sigma_1^2) + \pi_2\phi(x; \mu_2, \sigma_2^2). \tag{1.29}$$

We have the following results.

(a). If

$$(\mu_2 - \mu_1)^2 < \frac{27\sigma_1^2\sigma_2^2}{4(\sigma_1^2 + \sigma_2^2)},$$

the mixture density $f(x)$ is unimodal for all $0 < \pi_1 < 1$.

(b). If

$$(\mu_2 - \mu_1)^2 > \frac{8\sigma_1^2\sigma_2^2}{\sigma_1^2 + \sigma_2^2},$$

there exist $0 < \pi_1 < 1$ such that $f(x)$ is bimodal.

(c). For any parameter set $(\mu_1, \mu_2, \sigma_1, \sigma_2)$, there exists $0 < \pi_1 < 1$ such that $f(x)$ is unimodal.

Based on Theorem 1.8.3 (a), the two-component normal mixture (1.29) is unimodal when the two components means are close enough no matter what the value of π_1 is. Note, however, that even when two-component means are very separate, the mixture density (1.29) could still be unimodal for some proportion π_1 (usually when it is close to 0 or 1). This can be easily understood by noting that the mixture density (1.29) reduces to a homogeneous one-component normal distribution when π_1 goes to 0 or 1.

The next two theorems provide more detailed conditions when the mixture density (1.29) is unimodal.

Theorem 1.8.4. (Behboodian, 1970)
A sufficient condition for the unimodality of $f(x)$ in (1.29) is $|\mu_1 - \mu_2| \leq 2\min(\sigma_1, \sigma_2)$. In addition, if $\sigma_1 = \sigma_2 = \sigma$, a sufficient condition for the unimodality is

$$|\mu_1 - \mu_2| \leq 2\sigma\sqrt{1 + \frac{|\log \pi_1 - \log(1 - \pi_1)|}{2}}.$$

Theorem 1.8.5. (Schilling et al., 2002)
Let $r = \sigma_1^2/\sigma_2^2$ and define

$$\delta = \frac{\sqrt{-2 + 3r + 3r^2 - 2r^3 + 2(1 - r + r^2)^{1.5}}}{\sqrt{r}(1 + \sqrt{r})},$$

which is equal to 1 if the variances are equal. When $\pi_1 = 1/2$, the mixture density $f(x)$ in (1.29) is unimodal if and only if $|\mu_1 - \mu_2| < \delta|\sigma_1 + \sigma_2|$.

1.8.2 The topography of multivariate normal mixtures

In Section 1.8.1, we discussed the shape of the two-component univariate normal mixture density. The topography of a multivariate mixture (1.24) is much more complicated; unlike the univariate case, the number of modes of a multivariate mixture can be significantly more than the number of components. The seminal paper by Ray and Lindsay (2005) provided a very good insight into the topography of multivariate normal mixtures. We introduce and discuss some of their results in this section.

Define a $(C-1)$ – dimensional unit simplex set:

$$S_C = \left\{ \boldsymbol{\alpha} = (\alpha_1,\ldots,\alpha_C)^\top \in \mathbb{R}^C : \alpha_c \in [0,1], \sum_{c=1}^{C} \alpha_c = 1 \right\}.$$

Following Ray and Lindsay (2005), we define the ridgeline function to be

$$r(\boldsymbol{\alpha}) = \left(\alpha_1 \boldsymbol{\Sigma}_1^{-1} + \alpha_2 \boldsymbol{\Sigma}_2^{-1} + \cdots + \alpha_C \boldsymbol{\Sigma}_C^{-1} \right)^{-1}$$
$$\times \left(\alpha_1 \boldsymbol{\Sigma}_1^{-1} \boldsymbol{\mu}_1 + \alpha_2 \boldsymbol{\Sigma}_2^{-1} \boldsymbol{\mu}_2 + \cdots + \alpha_C \boldsymbol{\Sigma}_C^{-1} \boldsymbol{\mu}_C \right),$$

which will play a very crucial role when characterizing the topography of multivariate normal mixtures (1.24). Let

$$\mathcal{M} = \{ r(\boldsymbol{\alpha}) : \boldsymbol{\alpha} \in S_C \}, \tag{1.30}$$

which is called the ridgeline surface or manifold. Note that the ridgeline function $r(\boldsymbol{\alpha})$ and the manifold $\mathcal{M}$ depend only on the means and variances of the component densities, but not the component proportions π_js.

If $C = 2$,

$$r(\alpha) = \left[\alpha \boldsymbol{\Sigma}_1^{-1} + (1-\alpha)\boldsymbol{\Sigma}_2^{-1} \right]^{-1} \left[\alpha \boldsymbol{\Sigma}_1^{-1} \boldsymbol{\mu}_1 + (1-\alpha)\boldsymbol{\Sigma}_2^{-1} \boldsymbol{\mu}_2 \right],$$

where $\alpha \in [0,1]$. The resulting image $\mathcal{M}$ is a one-dimensional curve and will be called the ridgeline, which defines a curve from $\boldsymbol{\mu}_1$ to $\boldsymbol{\mu}_2$. It can be proved that all points on the ridgeline curves are the "kissing points" of the two ellipses formed by the contour $\{\mathbf{x} : \phi_1(\mathbf{x}; \boldsymbol{\mu}_1, \boldsymbol{\Sigma}_1) = c\}$ and $\{\mathbf{x} : \phi_2(\mathbf{x}; \boldsymbol{\mu}_2, \boldsymbol{\Sigma}_2) = d\}$ for some constants c and d. Now we are ready to present the main result by Ray and Lindsay (2005).

Theorem 1.8.6. Let $f(\mathbf{x})$ be the density of a C-component multivariate normal density as given by (1.24). Then all of $f(\mathbf{x})$'s critical values, including modes, antimodes (the lowest point between the modes), and saddle points (zero slopes in orthogonal directions but not a local extremum), are points in the ridgeline surface $\mathcal{M}$.

Proof :

$$0 = \nabla f(\boldsymbol{x}) = \pi_1 \phi_1(\boldsymbol{x}) \frac{\nabla \phi_1(\boldsymbol{x})}{\phi_1(\boldsymbol{x})} + \pi_2 \phi_2(\boldsymbol{x}) \frac{\nabla \phi_2(\boldsymbol{x})}{\phi_2(\boldsymbol{x})} + \cdots + \pi_C \phi_C(\boldsymbol{x}) \frac{\nabla \phi_C(\boldsymbol{x})}{\phi_C(\boldsymbol{x})},$$

where we also use the simple notation $\phi_c(\boldsymbol{x})$ to denote $\phi(\boldsymbol{x}; \boldsymbol{\mu}_c, \boldsymbol{\Sigma}_c)$. Let

$$\alpha_i = \frac{\pi_i \phi_i(\boldsymbol{x})}{\sum_{c=1}^{C} \pi_c \phi_c(\boldsymbol{x})}.$$

Note that $0 \le \alpha_i \le 1$ and $\sum_{c=1}^{C} \alpha_c = 1$. In addition, we have

$$\frac{\nabla \phi_c(\boldsymbol{x})}{\phi_c(\boldsymbol{x})} = -\boldsymbol{\Sigma}_c^{-1}(\boldsymbol{x} - \boldsymbol{\mu}_c).$$

Therefore, for every critical value $\boldsymbol{x}$ of $f(\boldsymbol{x})$, there exists an $\boldsymbol{\alpha}$ such that

$$\alpha_1 \boldsymbol{\Sigma}_1^{-1}(\boldsymbol{x} - \boldsymbol{\mu}_1) + \alpha_2 \boldsymbol{\Sigma}_2^{-1}(\boldsymbol{x} - \boldsymbol{\mu}_2) + \cdots + \alpha_C \boldsymbol{\Sigma}_C^{-1}(C) = 0.$$

Solving the above equation for $\boldsymbol{x}$, we have

$$\boldsymbol{x} = \left(\alpha_1 \boldsymbol{\Sigma}_1^{-1} + \alpha_2 \boldsymbol{\Sigma}_2^{-1} + \cdots + \alpha_C \boldsymbol{\Sigma}_C^{-1}\right)^{-1}$$
$$\times \left(\alpha_1 \boldsymbol{\Sigma}_1^{-1} \boldsymbol{\mu}_1 + \alpha_2 \boldsymbol{\Sigma}_2^{-1} \boldsymbol{\mu}_2 + \cdots + \alpha_C \boldsymbol{\Sigma}_C^{-1} \boldsymbol{\mu}_C\right).$$

□

Next, we extend the above results for multivariate normal mixtures to multivariate mixtures of elliptical distributions based on the work of Alexandrovich et al. (2013).

Definition 1.8.1. A random vector **x** is said to have an elliptical distribution if its density function f has the form

$$f(\mathbf{x}) = k \cdot g((\mathbf{x} - \boldsymbol{\mu})^\top \boldsymbol{\Sigma}^{-1}(\mathbf{x} - \boldsymbol{\mu})),$$

where k is the normalizing constant, **x** is a p-dimensional random vector with the location parameter $\boldsymbol{\mu}$ (i.e., median also the mean if the latter exists), and $\boldsymbol{\Sigma}$ is a positive definite scatter matrix, which is proportional to the covariance matrix if the latter exists.

Remark 1.8.1. (Elliptical distributions)

1. Examples of elliptical distribution are multivariate normal distribution, multivariate t-distribution, symmetric multivariate stable distribution, multivariate logistic distribution, and multivariate symmetric general hyperbolic distribution.

2. If $\boldsymbol{\mu} = 0$ and $\boldsymbol{\Sigma} = cI_p$, where c is a positive scalar and I_p is the identity matrix, the distribution is called a *spherical* distribution.

Theorem 1.8.7. Let $\boldsymbol{\alpha}$ be a k-dimensional vector and B be a $k \times p$ matrix. If X is an elliptical distribution with a location parameter $\boldsymbol{\mu}$ and a scatter matrix $\boldsymbol{\Sigma}$, then $\boldsymbol{\alpha} + B\mathbf{X}$ is also an elliptical distribution with a location parameter $\boldsymbol{\alpha} + B\boldsymbol{\mu}$ and a scatter matrix $B\boldsymbol{\Sigma}B^\top$.

Theorem 1.8.8. Let $f(\mathbf{x})$ be the density of a C-component elliptical multivariate distribution. Then all critical values of $f(\mathbf{x})$, including modes, antimodes, and saddle points, are points in the ridge line surface $\mathcal{M}$ as defined in (1.30).

Theorem 1.8.8 is an extension of Theorem 1.8.6. Note that when component variances $\boldsymbol{\Sigma}_c$s are equal,

$$r(\boldsymbol{\alpha}) = \alpha_1 \boldsymbol{\mu}_1 + \alpha_2 \boldsymbol{\mu}_2 + \cdots + \alpha_C \boldsymbol{\mu}_C.$$

Therefore, the manifold $\mathcal{M}$ is a convex hull of the means $\boldsymbol{\mu}_j$. Then we have the following result.

Proposition 1.8.1. If $\boldsymbol{\Sigma}_c = \boldsymbol{\Sigma}$ for all $c = 1, \ldots, C$, then the convex hull of the means of $\boldsymbol{\mu}_j$ of the component densities contains all critical points of the density $f(\mathbf{x})$.

Example 1.8.3. (Ray and Lindsay, 2005) (Two components, three modes, unequal variances). We consider the following two-component bivariate normal mixture density $f(\mathbf{x})$ with the parameters $\pi_1 = \pi_2 = 0.5$,

$$\boldsymbol{\mu}_1 = \begin{pmatrix} 0 \\ 0 \end{pmatrix}, \qquad \boldsymbol{\Sigma}_1 = \begin{pmatrix} 1 & 0 \\ 0 & 0.05 \end{pmatrix},$$

$$\boldsymbol{\mu}_2 = \begin{pmatrix} 1 \\ 1 \end{pmatrix}, \qquad \boldsymbol{\Sigma}_2 = \begin{pmatrix} 0.05 & 0 \\ 0 & 1 \end{pmatrix}.$$

Figure 1.5 shows the contours of the density given above, along with the ridgeline curve. It can be seen that the ridgeline curve passes through the three modes and saddle points of the mixture density $f(\mathbf{x})$. □

Let $h(\boldsymbol{\alpha}) = f(r(\boldsymbol{\alpha}))$ be the so-called ridgeline elevation function. Note that when $C = 2$, $\boldsymbol{\alpha}$ is one-dimensional. If $p > C - 1$, the $h(\boldsymbol{\alpha})$ plot is a dimension reduction plot.

Theorem 1.8.9. Every critical point $\boldsymbol{\alpha}$ of $h(\boldsymbol{\alpha})$ corresponds to a critical point $r(\boldsymbol{\alpha})$ of $f(\mathbf{x})$. A critical point $\boldsymbol{\alpha}$ of $h(\boldsymbol{\alpha})$ gives a local maximum of $f(\mathbf{x})$ if and only if it is a local maximum of $h(\boldsymbol{\alpha})$. If $p > C - 1$, then $f(\mathbf{x})$ has no local minima but only saddle points and local maxima.

Based on Theorem 1.8.9, finding modes of the p-dimensional mixture density $f(\mathbf{x})$ can be simplified to finding modes of a $(C-1)$-dimensional ridgeline elevation function $h(\boldsymbol{\alpha})$, which could be much easier when C is small compared to p. For example, if $C = 2$, it does not matter how large p is, finding the modes of p-dimensional multivariate mixtures density $f(\mathbf{x})$ can be simplified to finding modes of a one-dimensional curve $h(\boldsymbol{\alpha})$. The modes of $h(\boldsymbol{\alpha})$ can be done numerically by using any nonlinear optimization software, such as the "*nlm*" function in R.

We conclude this section by presenting a result about two-component multivariate normal mixtures (Ray and Lindsay, 2005).

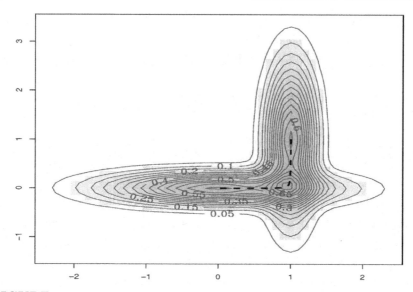

FIGURE 1.5

Contour plot and ridgeline curve $(--)$ for the mixture density given in Example 1.8.3.

Theorem 1.8.10. Let $f(\mathbf{x})$ be the mixture of two multivariate normal densities with means $\boldsymbol{\mu}_1$ and $\boldsymbol{\mu}_2$ and variances $\boldsymbol{\Sigma}_1$ and $\boldsymbol{\Sigma}_2 = k^2 \boldsymbol{\Sigma}_1$.

(a) The density of $f(\mathbf{x})$ is unimodal for any mixing proportion π if

$$(\boldsymbol{\mu}_2 - \boldsymbol{\mu}_1)^\top \boldsymbol{\Sigma}_1^{-1} (\boldsymbol{\mu}_2 - \boldsymbol{\mu}_1) \leq \frac{2(1 - k^2 + k^4)^{1.5} - (2k^6 - 3k^4 - 3k^2 + 2)}{k^2}.$$

(b) If the parameters do not satisfy the above condition, $f(\mathbf{x})$ is bimodal if and only if $\pi_1 < \pi < \pi_2$, where

$$\frac{1}{\pi_i} = 1 + \frac{\alpha_i}{1 - \alpha_i} \frac{\phi_1(r(\alpha_i))}{\phi_2(r(\alpha_i))}, \quad i = 1, 2,$$

and the $\alpha_i \in [0, 1]$ are the two solutions of

$$\{k^2(1 - \alpha) + \alpha\}^3 - \alpha(1 - \alpha)k^2(\boldsymbol{\mu}_2 - \boldsymbol{\mu}_1)^\top \boldsymbol{\Sigma}_1^{-1}(\boldsymbol{\mu}_2 - \boldsymbol{\mu}_1) = 0.$$

Based on Theorem 1.8.10 (a), when $k = 1$, i.e., $\boldsymbol{\Sigma}_1 = \boldsymbol{\Sigma}_2 = \boldsymbol{\Sigma}$, the multivariate normal mixture density of $f(x)$ is unimodal for any mixing proportion π if $(\boldsymbol{\mu}_2 - \boldsymbol{\mu}_1)^\top \boldsymbol{\Sigma}^{-1}(\boldsymbol{\mu}_2 - \boldsymbol{\mu}_1) \leq 4$, i.e., the Mahalanobis distance between $\boldsymbol{\mu}_1$ and $\boldsymbol{\mu}_2$ is no larger than 4. This result can be considered a natural extension of the univariate result discussed in Section 1.8.1 (the two-component normal mixture density is unimodal if $|\mu_2 - \mu_1|/\sigma \leq 2$).

1.9 Unboundedness of normal mixture likelihood

One special characteristic and challenging issue of the normal mixture model is its unbounded log-likelihood. The unboundedness of the likelihood function also occurs for finite gamma mixture models, finite location-scale mixture models, and the corresponding multivariate versions and mixtures of regression models introduced in Section 3. For simplicity of explanation, we mainly focus on normal mixture models, but the introduced solutions also apply to other mixture models.

Recall that the normal mixture log-likelihood function for iid observations $\{x_1, \ldots, x_n\}$ is

$$\log L(\boldsymbol{\theta}) = \sum_{i=1}^{n} \log\{\sum_{c=1}^{C} \pi_c \phi(x_i; \mu_c, \sigma_c^2)\}. \tag{1.31}$$

Note that

$$\log L(\boldsymbol{\theta}) \geq \log\left\{\pi_1 \frac{1}{\sqrt{2\pi}\sigma_1} \exp\{-\frac{(x_1-\mu_1)^2}{2\sigma_1^2}\}\right\} + \sum_{i=2}^{n} \log\left\{\sum_{c=2}^{C} \pi_c \phi(x_i; \mu_c, \sigma_c^2)\right\}. \tag{1.32}$$

When $\mu_1 = x_1$, the left side of (1.32) tends to ∞ as σ_1 goes to zero when other parameters are fixed. Therefore, $\log L(\boldsymbol{\theta})$ in (1.31) is unbounded; the maximum likelihood estimator (MLE) does not exist as a global maximizer of the mixture likelihood function without any restriction on the component variances, but still exists as a local maximizer, and its consistency is proved by Kiefer (1978); Peters and Walker (1978); Redner and Walker (1984). In this case, our interest will be the consistent local maximizer instead of the global maximizer. Note that if there are multiple local maxima in the interior of the parameter space, there is also a problem of identifying the consistent root, which is a very difficult problem itself even for nonmixture models. However, under some regularity conditions, Hathaway (1985) proved that if we define the MLE as the maximum interior/local mode, then it is still consistent.

To solve the issue of the unbounded likelihood, the simplest solution is to assume that the component variances are equal, i.e., $\sigma_1 = \sigma_2 = \cdots = \sigma_C = \sigma$ so that the likelihood function is bounded. Then, the update formula for σ in the M step is

$$\hat{\sigma}^{2(k+1)} = \frac{\sum_{i=1}^{n} \sum_{c=1}^{C} \hat{p}_{ij}^{(k+1)} (x_i - \hat{\mu}_c^{(k+1)})^2}{n}.$$

However, such constraints might not be reasonable for main applications. In order to relax such assumptions, many methods have been proposed. For example, Hathaway (1985) proposed the restricted MLE by maximizing the likelihood function over a restricted/constrained parameter space. Chen et al. (2008) proposed a penalized maximum likelihood estimator (PMLE) by adding a penalty function for the component variances to prevent them from going to 0. Yao (2010) proposed a profile log-likelihood method to avoid the difficulty of choosing a cut point for the restricted MLE (Hathaway, 1985). Seo and Kim (2012) proposed a likelihood-based k-deleted maximum likelihood estimator and a score-based k-deleted maximum likelihood estimator, while the score-based k-deleted MLE is slightly better than the likelihood-based k-deleted MLE. By extending the methods of Seo and Kim (2012), Kim and Seo (2014) proposed a new type of likelihood-based estimator and a gradient-based k-deleted maximum likelihood estimator whose performance is as good as the score-based k-deleted likelihood method but with more efficient computation. Liu et al. (2015) proposed a correct likelihood function based on interval censoring to fix the unbounded likelihood problem. Generalizing the constraint method proposed by Ingrassia (2004), Rocci et al. (2017) proposed an affine equivalent constrained method by shrinking the class conditional covariance matrices towards a pre-specified matrix. Note that all of the above-mentioned methods require some selection of tuning parameters. We introduce some of them in more detail in the next few sections.

1.9.1 Restricted MLE

One of the commonly used methods to avoid the unboundedness of the normal mixture log-likelihood is to use the restricted MLE, which is the maximizer of the likelihood function over a restricted/constrained parameter space (Hathaway, 1985):

$$\Omega_\epsilon = \{\boldsymbol{\theta} \in \Omega : \sigma_h/\sigma_j \geq \epsilon > 0, 1 \leq h \neq j \leq C\}, \tag{1.33}$$

where $\epsilon \in (0,1]$, Ω denotes the unconstrained parameter space. Note that the large spike of the normal mixture log-likelihood mainly happens when one of the components tries to fit only one or very few points with its variance going to 0. The above-constrained parameter space Ω_ϵ aims to avoid such singular components by restricting the component variance from being too small. Hathaway (1985) showed that the restricted maximizer $\hat{\boldsymbol{\theta}}_\epsilon$ of $L(\boldsymbol{\theta})$ over Ω_ϵ exists and that $\hat{\boldsymbol{\theta}}_\epsilon$ is strongly consistent for the true value $\boldsymbol{\theta}$ provided that the true value of $\boldsymbol{\theta}$ lies in Ω_ϵ. See Hathaway (1985, 1986) and Bezdek et al. (1985) for more details.

For the univariate normal distribution, in order to incorporate the constraint (1.33), Hathaway (1983, 1986) and Bezdek et al. (1985) proposed a constrained version of the EM algorithm. For a multivariate normal mixture with unequal covariance matrix, $\Sigma_c(c = 1 \ldots, C)$, the likelihood function is also unbounded. Similar to the univariate case, we can put some constraints on the covariance matrix to get the constrained global maximizer. For example, we can constrain all the eigenvalues of $\Sigma_h \Sigma_j^{-1}$ $(1 \leq h \neq j \leq C)$ to be greater than or equal to some minimum value $\epsilon > 0$ or $|\Sigma_h|/|\Sigma_j| \geq \epsilon > 0$ $(1 \leq h \neq j \leq C)$, where $|\Sigma|$ is the determinant of the matrix Σ.

However, a big challenge for the restricted MLE is to choose an appropriate cut point ϵ in (1.33). If ϵ is too large, the consistent local maxima might not belong to the constrained parameter space Ω_ϵ, and thus the restricted MLE $\hat{\theta}_\epsilon$ will not be consistent. In addition, Ω_ϵ might also miss some interior modes worthy of consideration. On the other hand, if ϵ is too small, it is possible that some boundary points, satisfying $\sigma_h/\sigma_j = \epsilon$ for some h and j, have larger log-likelihood than the interior modes. In this situation, the resulting restricted MLE $\hat{\theta}_\epsilon$ is on the boundary of Ω_ϵ, and hence different choices of ϵ provide us with different estimates.

1.9.2   Penalized likelihood estimator

Chen et al. (2008) proposed a penalized maximum likelihood estimator (PMLE), denoted by $\tilde{\theta}$, by maximizing the following penalized log-likelihood function for normal mixtures:

$$pl_n(\boldsymbol{\theta}) = \sum_{i=1}^{n} \log \left\{ \sum_{c=1}^{C} \pi_c \phi(x_i; \mu_c, \sigma_c) \right\} + p_n(\boldsymbol{\theta}),$$

where the penalty function $p_n(\boldsymbol{\theta})$ is used to prevent σ from going to 0 and satisfies the following conditions:

(C1): $p_n(\boldsymbol{\theta}) = \sum_{c=1}^{C} \tilde{p}_n(\sigma_c)$.

(C2): $\sup_{\sigma>0} \max\{0, \tilde{p}_n(\sigma)\} = o(n)$ and $\tilde{p}_n(\sigma) = o(n)$ at any fixed $\sigma > 0$.

(C3): When $0 < \sigma \leq 8/(nM)$, where $M = \max\{\sup_x f(x; \boldsymbol{\theta}_0), 8\}$ and $\boldsymbol{\theta}_0$ is the true value, we have $\tilde{p}_n(\sigma) \leq 4(\log n)^2 \log \sigma$ for large enough n.

The condition (C2) prevents the penalty function from dominating the likelihood function at any parameter value so that the estimates are still mainly driven by the likelihood function. Based on the condition (C3), $p_n(\boldsymbol{\theta})$ satisfies the condition that $p_n(\boldsymbol{\theta}) \to -\infty$ as $\min\{\sigma_c, c = 1, \ldots, C\} \to 0$ to prevent the singular component variances. Chen et al. (2008) proposed the following penalty function

$$p_n(\boldsymbol{\theta}) = -a_n \sum_{c=1}^{C} \{S_x/\sigma_c^2 + \log(\sigma_c^2)\},$$

where $a_n = n^{-1/2}$ or n^{-1} is recommended, and S_x is the sample variance. From a Bayesian point of view, the penalty function puts an Inverse Gamma distribution prior to σ^2, where S_x is the mode of the prior distribution or a prior estimate of σ_c^2, and a large value of a_n implies a strong conviction in the prior estimate. Chen et al. (2008) and Chen and Tan (2009) established the consistency of the PMLE $\tilde{\boldsymbol{\theta}}$ for both univariate and multivariate cases. We provide the computation in Algorithm 1.9.1.

Algorithm 1.9.1 PMLE

Given an initial value $\boldsymbol{\theta}^{(0)}$, iterate the following E-step and M-step until convergence.

E step: Find the classification probabilities:

$$p_{ic}^{(k+1)} = \frac{\pi_c^{(k)}\phi(x_i; \mu_c^{(k)}, \sigma_c^{2(k)})}{\sum_{l=1}^{C}\pi_l^{(k)}\phi(x_i; \mu_l^{(k)}, \sigma_l^{2(k)})}, \quad i = 1, \ldots, n, \quad c = 1, \ldots, C.$$

M step: Update parameter estimates:

$$\pi_c^{(k+1)} = \frac{1}{n}\sum_{i=1}^{n}\pi_{ic}^{(k+1)},$$

$$\mu_c^{(k+1)} = \frac{\sum_{i=1}^{n}p_{ic}^{(k+1)}x_i}{n\pi_c^{(k+1)}},$$

$$\sigma_c^{2(k+1)} = \frac{2a_n S_x + S_c^{(k+1)}}{2a_n + n\pi_c^{(k+1)}},$$

where $S_c^{(k+1)} = \sum_{i=1}^{n}p_{ic}^{(k+1)}(x_i - \mu_c^{(k+1)})^2$.

Chen and Tan (2009) further extended the method of Chen et al. (2008) to multivariate normal mixture models with the penalty

$$p_n(\boldsymbol{\theta}) = -n^{-1}\sum_{c=1}^{C}\{tr(S_x\Sigma_c) + \log|\Sigma_c|\}, \tag{1.34}$$

where Σ_c is the cth component variance and $|\Sigma|$ is the determinant of Σ. The computations are similar to the above algorithm.

From a Bayesian point of view, the penalty function (1.34) puts a Wishart distribution prior on Σ_c, and S_x is the mode of the prior distribution. Increasing the value of a_n implies a stronger conviction on S_x as the possible value of Σ_c. Chen et al. (2008) and Chen and Tan (2009) also proved the consistency of the penalized estimator of the mixing distribution even when C is larger than the true number of components. Tanaka (2009) and Chen et al. (2016b) extended the idea of the above-Penalized MLE to finite location-scale mixture models and finite gamma mixture models, respectively.

1.9.3 Profile likelihood method

To mitigate the difficulty and the potential issues of choosing a cut point for the restricted MLE (Hathaway, 1985), Yao (2010) proposed a profile log-likelihood method to find the maximum interior mode for the normal mixture with unequal variances. Unlike the constrained EM algorithm (Hathaway, 1985, 1986), the profile likelihood method does not need to specify a cut-point ϵ. In addition, this method can help find out whether there are some other minor interior modes and investigate how the choice of ϵ in (1.33) affects the restricted MLE.

Ideally, if we could plot the log-likelihood versus the parameters, we could better understand which modes are on the boundary and which modes are in the interior parameter space. However, such plots are usually not available since the dimension of unknown parameters for mixture models is (much) larger than two. Yao (2010) proposed using a profile likelihood to reduce the dimension of unknown parameters to one so that we can visualize the likelihood surface.

Let $\sigma_{(1)} \leq \sigma_{(2)} \leq \cdots \leq \sigma_{(C)}$ be ordered sequence of $(\sigma_1, \ldots, \sigma_C)$. Define $k = \sigma_{(1)}/\sigma_{(C)}$, the ratio of the smallest component standard deviation to the largest one. Let

$$\Theta_k = \{\boldsymbol{\theta} \in \Omega : \sigma_{(1)} = k\sigma_{(C)}\}.$$

Note that the mixture log-likelihood is bounded over Θ_k for each fixed k and thus the MLE is well defined in Θ_k.

We define the profile log-likelihood estimator as

$$p(k) = \max_{\boldsymbol{\theta} \in \Theta_k} \log L(\boldsymbol{\theta}), \quad k \in (0, 1], \tag{1.35}$$

where $L(\boldsymbol{\theta})$ is defined in (1.31). We present some properties of the profile likelihood $p(k)$ in the following theorem.

Theorem 1.9.1. Below are the properties of $p(k)$, defined in (1.35) (Yao, 2010).

(a) The profile likelihood $p(k)$ is unbounded and goes to infinity when k goes to zero.

(b) Suppose the restricted MLE over the constrained parameter space Ω_ϵ in (1.33) is $\hat{\boldsymbol{\theta}}_\epsilon$ with $\hat{\sigma}_1 \leq \hat{\sigma}_2 \leq \cdots \leq \hat{\sigma}_C$. Let $\hat{k} = \hat{\sigma}_1/\hat{\sigma}_C$. Then $p(\hat{k})$ maximizes the profile log-likelihood $p(k)$ in K_ϵ, where $K_\epsilon = \{k \in (0, 1] : k \geq \epsilon\}$. The reverse is also true.

(c) Suppose $\tilde{k}$ is a local mode for the profile log-likelihood $p(k)$ with the corresponding $\tilde{\boldsymbol{\theta}}$. Then $\tilde{\boldsymbol{\theta}}$ is a local mode for the original mixture log-likelihood.

Based on Theorem 1.9.1 (a), $p(k)$ is also unbounded. Therefore, we cannot estimate k by maximizing $p(k)$ directly. However, based on Theorem 1.9.1 (b) and (c), finding the maximum interior mode of the mixture likelihood $\log L(\boldsymbol{\theta})$ can be simplified to finding the maximum interior mode of $p(k)$. Since k is a one-dimensional parameter, the profile likelihood transfers the problem of finding the maximum interior mode for a high-dimensional likelihood function $\log L(\boldsymbol{\theta})$ into locating the maximum interior model for a one-dimensional profile likelihood function $p(k)$. In addition, we can easily use the plot of $p(k)$ versus k to locate the maximum interior mode of $p(k)$ without requiring to choose a cut point ϵ. This is one of the main benefits of the profile likelihood method. Based on the plot of $p(k)$ versus k, we can also better understand how the cut point ϵ in (1.33) affects the restricted MLE.

Let $\hat{k}$ be the maximum interior mode of (1.35). Then, fixing k at $\hat{k}$, we can find the MLE of (1.31) over the parameter space $\Theta_{\hat{k}}$, denoted by $\hat{\boldsymbol{\theta}}(\hat{k})$. Then, the $\hat{\boldsymbol{\theta}}(\hat{k})$ is the proposed maximum interior mode of (1.31) by Yao (2010).

Similar to regular mixture models, the profile log-likelihood $p(k)$ does not have an explicit form and the EM algorithm needs to be applied to find $p(k)$ for any fixed k. Algorithm 1.9.2 provides the computation of $p(k)$ for two components $(C = 2)$. Please refer to Yao (2010) for the computation of $p(k)$ when $C > 2$.

Algorithm 1.9.2 Profile likelihood estimation (Yao, 2010)

Starting with the initial parameter values

$$\{\hat{\pi}_1^{(0)}, \hat{\mu}_1^{(0)}, \hat{\mu}_2^{(0)}, \hat{\sigma}_1^{(0)}, \hat{\sigma}_2^{(0)}\},$$

where $\hat{\sigma}_1^{(0)} = k\hat{\sigma}_2^{(0)}$, iterate the following two steps until convergence.

E Step: Compute the classification probabilities:

$$\hat{p}_{ij}^{(t+1)} = \frac{\hat{\pi}_j^{(t)} \phi(x_i; \hat{\mu}_j^{(t)}, \hat{\sigma}_j^{2(t)})}{\sum\limits_{l=1}^{2} \hat{\pi}_l^{(t)} \phi(x_i; \hat{\mu}_l^{(t)}, \hat{\sigma}_l^{2(t)})}, i = 1, \dots, n, j = 1, 2.$$

M step: Update the component parameters:

$$\hat{\mu}_j^{(t+1)} = \frac{\sum_{i=1}^n \hat{p}_{ij}^{(t+1)} x_i}{n_j^{(t+1)}}, \ \hat{\pi}_j^{(t+1)} = \frac{n_j^{(t+1)}}{n}, \ j = 1, 2,$$

$$\hat{\sigma}_1^{2(t+1)} = \frac{\sum\limits_{i=1}^{n} \left[\hat{p}_{i1}^{(t+1)} (x_i - \hat{\mu}_1^{(t+1)})^2 + k^2 \hat{p}_{i2}^{(t+1)} (x_i - \hat{\mu}_2^{(t+1)})^2 \right]}{n},$$

$$\hat{\sigma}_2^{(t+1)} = \hat{\sigma}_1^{(t+1)}/k,$$

where $n_j^{(t+1)} = \sum_{i=1}^n \hat{p}_{ij}^{(t+1)}$.

TABLE 1.3

Local maximizers for Example 1.9.1

Local maximizer	$\log L$	π_1	μ_1	μ_2	σ_1	σ_2
$k = 0.1891$	-153.2144	0.0934	-0.1700	0.8280	0.2175	1.1503
$k = 0.4378$	-152.9230	0.2199	-0.0567	0.9578	0.5092	1.1629
$k = 0.8209$	-153.0170	0.2796	2.0455	0.2260	0.6791	0.8273

Example 1.9.1. (Yao, 2010). 100 observations are generated from

$$0.3N(0, 0.5^2) + 0.7N(1, 1).$$

Figure 1.6 is the profile log-likelihood plot of $p(k)$ vs. k. From the plot, we can see that $p(k)$ goes to infinity when k goes to zero. Figure 1.6 (b) provides the plot excluding the area where k is very close to zero with a very large log-likelihood. Based on Figure 1.6 (b), there are three interior modes, which are reported in Table 1.3. By comparing the values of $\log L$, we know that the maximum interior mode is at $k = 0.4378$.

Note that when $k < 0.07$, $p(k) > -152.9230$ (the profile log-likelihood value of the maximum interior mode). Therefore, if $\varepsilon < 0.07$ in Ω_ϵ of (1.33), the restricted MLE $\hat{\boldsymbol{\theta}}_\epsilon$ (Hathaway, 1985, 1986) is on the boundary of the parameter space Ω_ε, which also implies that $\hat{\boldsymbol{\theta}}_\epsilon$ depends on the cut point ϵ. If $0.07 < \epsilon < 0.4378$, the restricted MLE can successfully find the maximum interior mode and provide the same estimate as our profile likelihood method. However, if ϵ is too large, the restricted MLE might miss some interior modes. For example, if $0.1891 < \epsilon < 0.4378$, the restricted MLE will miss the first interior mode ($k = 0.1891$). As argued by McLachlan and Peel (2000, Section 8.3.2), other minor modes might also provide useful information, especially for clustering applications.

Example 1.9.2. *The Crab Data:* Let's also consider the famous crab data set (Pearson, 1894) with the histogram shown in Figure 1.7. The data set consists of the measurements on the ratio of the forehead to body length of 1000 crabs sampled from the Bay of Naples. We fit a two-component normal mixture model to this data set based on Pearson (1894).

Figure 1.8 is the profile log-likelihood plot. As mentioned by Yao (2010), when k is from 10^{-4} to 10^{-2}, the corresponding log-likelihood is too large. Therefore, only the profile log-likelihood plot for k values from 10^{-2} to 1 is shown. Based on the plot, there is only one interior mode with $k = 0.6418$, and the corresponding MLE of $(\pi_1, \mu_1, \mu_2, \sigma_1, \sigma_2)$ is $(0.5360, 0.6563, 0.6355, 0.0126, 0.0196)$.

When the cut point $\epsilon < 0.05$ in Ω_ϵ of (1.33), the restricted MLE $\hat{\boldsymbol{\theta}}_\epsilon$ occurs on the boundary of Ω_ϵ. When $\epsilon > 0.05$, the restricted MLE is the same as the maximum interior mode found by the profile log-likelihood method.

$\square$

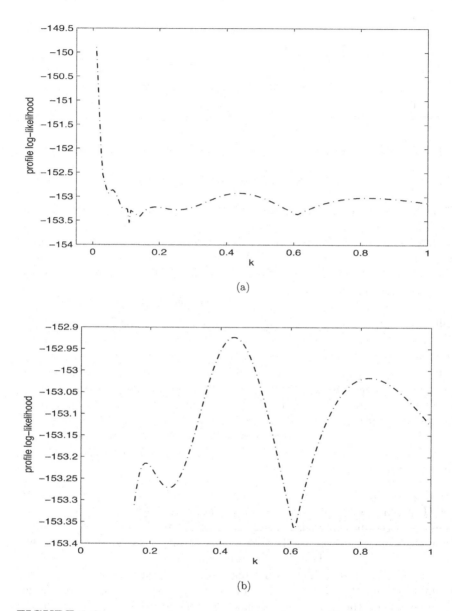

FIGURE 1.6
Profile log-likelihood plot for Example 1.9.1: (a) for all k values from 10^{-4} to 1; (b) for k values from 0.15 to 1.

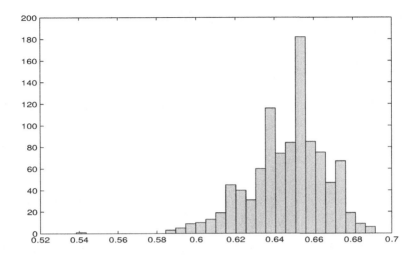

FIGURE 1.7
Histogram of crab data. The number of bins used is 30.

1.10 Consistent root selections for mixture models

It is well known that the mixture likelihood equation has multiple roots and it has long been challenging to select a consistent root for finite mixture models. Simply using the root with the largest likelihood often does not work due to the unbounded likelihood and spurious roots (McLachlan and Peel, 2000). Spurious roots often contain components with scales (or spreads) too different to be credible and usually happen when a few data points are relatively close together. The methods introduced in Section 1.9 aim to solve the unbounded mixture likelihood issues. However, those methods usually still require the choice of some tuning parameters. In addition, they might choose spurious root as the final solution and can't provide clear guidelines on how to choose the "right" root among a few candidates.

Wichitchan et al. (2019b) proposed a simple method to choose the consistent root by minimizing the distance between the corresponding estimated cdf and the empirical distribution function (edf). If the true model were known, one would naturally choose the root such that the corresponding estimated model is closest to the true model based on some distance measure. When the true model is unknown in practice, we could replace it with some consistent nonparametric model estimator. Note that the empirical distribution function

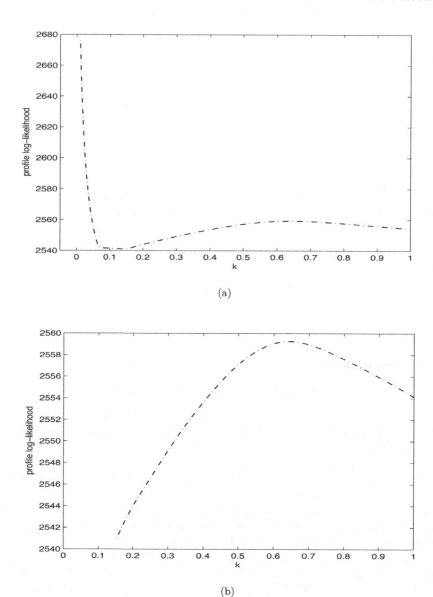

FIGURE 1.8
Profile log-likelihood plot for crab data: (a) for all k values from 10^{-2} to 1;
(b) for k values from 0.15 to 1.

(edf) based on $\{x_1, \ldots, x_n\}$, defined as

$$F_n(x) = \frac{1}{n} \sum_{i=1}^{n} I(x_i \le x), \qquad (1.36)$$

is a consistent nonparametric estimate of the true cumulative distribution function (cdf), where $I(A)$ is an indicator function and equal to 1 if A is correct and 0 otherwise. Given any root $\hat{\boldsymbol{\theta}}$, the corresponding estimated cumulative distribution function is

$$F(x; \hat{\boldsymbol{\theta}}) = \sum_{c=1}^{C} \hat{\pi}_c \Phi(x; \hat{\mu}_c, \hat{\sigma}_c),$$

where $\Phi(x; \mu, \sigma)$ is the cdf of $N(\mu, \sigma^2)$. Given a sequence of found roots $\hat{\boldsymbol{\theta}}_j, j = 1, \ldots, k$, Wichitchan et al. (2019b) proposed choosing the "best" root by minimizing $L[F(\cdot; \hat{\boldsymbol{\theta}}_j), F_n(\cdot)]$ over j, where L can be any reasonable loss/distance function to compare two cdfs, such as the goodness of fit (GOF) test statistics: Kolmogorov-Smirnov statistic (Chakravarty et al., 1967), Cramér-Von Mises statistic (Cramér, 1928), Watson statistic (Watson, 1961), and Anderson Darling statistic (Anderson and Darling, 1952). We provide their definitions below for convenience. Please also refer to Section 6.7 for more explanations about those test statistics.

Kolmogorov-Smirnov statistic: $D(\hat{\boldsymbol{\theta}}) = \sup_x |F_n(x) - F(x; \hat{\boldsymbol{\theta}})|.$

Cramér-Von Mises statistic: $W^2(\hat{\boldsymbol{\theta}}) = \sum_{i=1}^{n} \left[F_n(x_i) - F(x_i; \hat{\boldsymbol{\theta}}) \right]^2.$

Watson statistic: $U^2(\hat{\boldsymbol{\theta}}) = W^2(\hat{\boldsymbol{\theta}}) - n(\bar{z}(\hat{\boldsymbol{\theta}}) - 1/2)^2$, where W^2 is Cramér Von Mises statistic, and $\bar{z}(\boldsymbol{\theta}) = n^{-1} \sum_{i=1}^{n} F(x_i; \boldsymbol{\theta})$.

Anderson Darling statistic:

$$A^2(\boldsymbol{\theta}) = -n - \frac{1}{n} \sum_{i=1}^{n} (2i - 1) \left\{ \ln z_i(\boldsymbol{\theta}) + \ln(1 - z_{n+1-i}(\boldsymbol{\theta})) \right\},$$

where $z_i(\boldsymbol{\theta}) = F(x_{(i)}; \boldsymbol{\theta})$ and $x_{(1)} \leq x_{(2)} \leq \cdots \leq x_{(n)}$.

Wichitchan et al. (2019b) proved that the above-selected roots are consistent.

Note that the above-selected roots are not distance estimators. Instead, Wichitchan et al. (2019b) used the distance method to choose the root of the maximum likelihood estimate so that the efficiency property of MLE is kept. The idea of the proposed method can be extended to many other mixture models when the component density is not normal.

Based on numerical studies of Wichitchan et al. (2019b), the proposed root selection method has better performance than the choices of roots based on global maximum likelihood, likelihood-based k-deleted (MLE^L_{-k}), score-based k-deleted (MLE^S_{-k}), gradient-based k-deleted (MLE^G_{-k}) (Seo and Kim, 2012; Kim and Seo, 2014), and the correct likelihood method (CL) with $\Delta = \hat{\sigma}/100$ or $\Delta = \hat{\sigma}/10$ (Liu et al., 2015). In addition, the new method based on the Cramér Von Mises (CVM) statistic overall has the best performance and is recommended in practice.

1.11 Mixture of skewed distributions

Finite mixtures of symmetric distributions, particularly normal mixtures, are important tools in statistical modeling and analysis. However, if the data contains asymmetric outcomes, then the normality assumptions for component densities could be violated. In addition, if we use Gaussian mixture models (GMM) to fit skewed data sets, then the model tends to overfit the data since they need to include additional components to capture the skewness; increasing the number of pseudo-components may lead to difficulties and inefficiencies in computations. As a result, several mixture models with skewed component distributions have been proposed to model and cluster data that consists of asymmetric clusters or with outliers, which are commonly seen in a variety of fields, including biostatistics, bioinformatics, finance, image analysis, and medical science, etc. Mixtures of skew-normal distributions play without a doubt a central role in mixture models of skew distributions. In addition, the mixture of skew t-distributions, as a natural extension of the mixture of t-distributions, has received much attention, since it retains reasonable tractability and is more robust against outliers than the mixture of skew-normal distributions. In addition, there are mixtures of normal-inverse-Gaussian distributions, mixtures of shifted asymmetric Laplace distributions, and mixtures of generalized hyperbolic distributions, among others. Most of these parametric families of skew distributions are closely related and can be classified into four forms under a recently proposed scheme (Lee and McLachlan, 2013a,b), namely, the restricted, unrestricted, extended, and generalized forms.

1.11.1 Multivariate skew-symmetric distributions

1.11.1.1 Multivariate skew normal and skew t-distributions

Let us begin with a brief introduction of multivariate skew-symmetric distributions, including the skew normal and skew t-distributions.

First, let us look at a skew-normal distribution studied by Allman et al. (2009), which is so fundamental that many of the other skew-normal distributions can be considered as special cases of it. Let us call it FSN for short. Let $\mathbf{X}_1$ be a p-dimensional normally distributed random variable, that is, $\mathbf{X}_1 \sim N_p(\mathbf{0}, \mathbf{\Sigma})$, and $\mathbf{X}_0$ be a q-dimensional random vector, where $q \geq 1$. Then, $\mathbf{X} = \boldsymbol{\mu} + (\mathbf{X}_1|\mathbf{X}_0 + \boldsymbol{\tau} > \mathbf{0})$ follows an FSN distribution, where $\mathbf{X}_1|\mathbf{X}_0 + \boldsymbol{\tau} > \mathbf{0}$ is equal to $\mathbf{X}_1$ if all the elements of $\mathbf{X}_0 + \boldsymbol{\tau}$ are positive, and $-\mathbf{X}_1$ otherwise, for some q-dimensional vector $\boldsymbol{\tau}$, known as the location parameter of $\mathbf{X}_0$. Note that although $\mathbf{X}_0$ is assumed to be normally distributed in most cases, it is not a necessary condition for this definition. If the joint distribution of $\mathbf{X}_1$

and $\mathbf{X}_0$ is a multivariate normal:

$$\begin{pmatrix} \mathbf{X}_0 \\ \mathbf{X}_1 \end{pmatrix} \sim N_{p+q} \left(\begin{pmatrix} \tau \\ \mu \end{pmatrix}, \begin{pmatrix} \Gamma & \Delta^\top \\ \Delta & \Sigma \end{pmatrix} \right), \tag{1.37}$$

then $\mathbf{X} = \mathbf{X}_1 | \mathbf{X}_0 > \mathbf{0}$ becomes a location-scale variant of the canonical FSN distribution.

Restricted multivariate skew distributions: The univariate skew normal distribution was first introduced by Azzalini (1985) and has been extended to the multivariate case by Azzalini and Dalla Valle (1996). This type of **restricted skew normal distribution (rMSN)** can be categorized into the restricted multivariate skew distributions and is a special case of (1.37) by letting $q = 1, \tau = 0$, and $\Gamma = 1$. There have been several versions of multivariate skew-normal distributions studied, among those, the one proposed by Pyne et al. (2009) would be the most popularly used one, which has the following density:

$$f_{rMSN}(\mathbf{x}; \mathbf{u}, \Sigma, \delta) = 2\phi_p(\mathbf{x}; \mathbf{u}, \Sigma)\Phi_1\left(\delta^\top \Sigma^{-1}(\mathbf{x} - \mu); 0, 1 - \delta^\top \Sigma^{-1}\delta\right), \tag{1.38}$$

where $\phi_p(\cdot; \mu, \Sigma)$ is a p-variate normal distribution with mean μ and covariance Σ, and $\Phi_p(\cdot; \mu, \Sigma)$ denotes the corresponding distribution function.

The conditional-type stochastic representation of (1.38) is:

$$\mathbf{X} = \mu + (\mathbf{X}_1 | X_0 > 0), \tag{1.39}$$

where

$$\begin{pmatrix} X_0 \\ \mathbf{X}_1 \end{pmatrix} \sim N_{p+1} \left(\begin{pmatrix} 0 \\ \mu \end{pmatrix}, \begin{pmatrix} 1 & \delta^\top \\ \delta & \Sigma \end{pmatrix} \right),$$

where $\delta \in \mathbb{R}^p$ controls the correlation between $\mathbf{X}_1$ and X_0. There is a convenient stochastic representation corresponding to rMSN, given by

$$\mathbf{X} = \mu + \delta|\tilde{X}_0| + \tilde{\mathbf{X}}_1,$$

where $\tilde{X}_0$ and $\tilde{\mathbf{X}}_1$ are independent, $\tilde{X}_0 \sim N(0, 1)$ and $\tilde{\mathbf{X}}_1 \sim N_p(\mathbf{0}, \Sigma - \delta\delta^\top)$, and $|X_0|$ denotes the vector whose ith element is the absolute value of the ith element of X_0.

The **restricted multivariate skew t-distribution (rMST)** (Branco and Key, 2001; Azzalini and Capitanio, 2003; Pyne et al., 2009) is a natural extension of the rMSN. The conditional-type stochastic representation of a rMST is given by

$$\mathbf{X} = \mu + (\mathbf{X}_1 | X_0 > 0),$$

where

$$\begin{pmatrix} X_0 \\ \mathbf{X}_1 \end{pmatrix} \sim t_{p+1} \left(\begin{pmatrix} 0 \\ \mathbf{0} \end{pmatrix}, \begin{pmatrix} 1 & \delta^\top \\ \delta & \Sigma \end{pmatrix}, \nu \right),$$

and the equivalent convolution-type representation is

$$\mathbf{X} = \mu + \tilde{\delta}|\tilde{X}_0| + \tilde{\mathbf{X}}_1,$$

where

$$\begin{pmatrix} \tilde{X}_0 \\ \tilde{\mathbf{X}}_1 \end{pmatrix} \sim t_{p+1}\left(\begin{pmatrix} 0 \\ \mathbf{0} \end{pmatrix}, \begin{pmatrix} 1 & \mathbf{0}^\top \\ \mathbf{0} & \boldsymbol{\Sigma} - \boldsymbol{\delta}\boldsymbol{\delta}^\top \end{pmatrix}, \nu \right).$$

The density of the aforementioned rMST is given by

$$f_{rMST}(\mathbf{x}; \mathbf{u}, \boldsymbol{\Sigma}, \boldsymbol{\delta}, \nu)$$
$$= 2t_p(\mathbf{x}; \mathbf{u}, \boldsymbol{\Sigma}, \nu) T_1\left(\boldsymbol{\delta}^\top \boldsymbol{\Sigma}^{-1}(\mathbf{x} - \boldsymbol{\mu})\sqrt{\frac{\nu + p}{\nu + d(\mathbf{x})}}; 0, \lambda, \nu + p \right), \qquad (1.40)$$

where $\lambda = 1 - \boldsymbol{\delta}^\top \boldsymbol{\Sigma}^{-1}\boldsymbol{\delta}$, and $d(\mathbf{x}) = (\mathbf{x} - \boldsymbol{\mu})^\top \boldsymbol{\Sigma}^{-1}(\mathbf{x} - \boldsymbol{\mu})$ is the squared Mahalanobis distance between $\mathbf{x}$ and $\boldsymbol{\mu}$ with respect to $\boldsymbol{\Sigma}$. Also, in this case, $t_p(\cdot; \boldsymbol{\mu}, \boldsymbol{\Sigma}, \nu)$ and $T_p(\cdot; \boldsymbol{\mu}, \boldsymbol{\Sigma}, \nu)$ represent the density and distribution function of a p-dimensional t-distribution. There are various extensions of the multivariate skew t-distributions, and the readers are referred to, see, for example, Lee and McLachlan (2013a); Branco and Key (2001); Azzalini and Capitanio (2003); Lachos et al. (2010).

Similar to the MSN and MST, the skewing mechanism can be readily applied to other parametric distributions. For example, by applying the mechanism to the elliptically-contoured (EC) distributions, the restricted multivariate skew-elliptical (rMSE) has a density

$$f_{rMSE}(\mathbf{x}; \boldsymbol{\mu}, \boldsymbol{\Sigma}, \boldsymbol{\delta}) = 2f_p(\mathbf{x}; \boldsymbol{\mu}, \boldsymbol{\Sigma})F_{d(\mathbf{x})}(\omega; 0, \lambda), \qquad (1.41)$$

where $f_p(\cdot)$ denotes the density of an EC random variable defined by

$$f_p(\mathbf{x}; \boldsymbol{\mu}, \boldsymbol{\Sigma}, \tilde{f}) = a|\boldsymbol{\Sigma}|^{-1/2}\tilde{f}\{(\mathbf{x} - \boldsymbol{\mu})^\top \boldsymbol{\Sigma}^{-1}(\mathbf{x} - \boldsymbol{\mu})\}, \qquad (1.42)$$

with $\tilde{f}$ being a suitable parametric function from $\mathbb{R}^+$ to $\mathbb{R}^+$, known as the density generator of f_p, and the constant a being a normalizing constant. In addition, $\omega = \boldsymbol{\delta}^\top \boldsymbol{\Sigma}^{-1}(\mathbf{x} - \boldsymbol{\mu})$, $F_{d(\mathbf{x})}(\cdot; \boldsymbol{\mu}, \boldsymbol{\Sigma})$ is the distribution function corresponding to $f_p(\mathbf{x}; \boldsymbol{\mu}, \boldsymbol{\Sigma}, \tilde{f}_{d(\mathbf{x})})$, and

$$\tilde{f}_{d(\mathbf{x})}(\omega) = \frac{\tilde{f}(\omega + d(\mathbf{x}))}{\tilde{f}(d(\mathbf{x}))}.$$

Unrestricted multivariate skew normal and skew t-distribution: The unrestricted skew distribution is quite similar to the restricted form, except that in (1.39) the univariate latent variable X_0 is replaced by a p-dimensional latent vector $\mathbf{X}_0$. In this case, $\mathbf{X}_0 > 0$ means a set of componentwise inequalities, i.e., $X_{0i} > 0$ for $i = 1, \ldots, p$. As a result, a family of unrestricted multivariate skew distributions was proposed.

Sahu et al. (2003) studied a family of unrestricted multivariate skew-elliptical (uMSE) class, defined by

$$f_{uMSE}(\mathbf{x}; \boldsymbol{\mu}, \boldsymbol{\Sigma}, \boldsymbol{\delta}) = 2^p f_p(\mathbf{x}; \boldsymbol{\mu}, \boldsymbol{\Sigma})F_{d(\mathbf{x})}(\boldsymbol{\Delta}\boldsymbol{\Sigma}^{-1}(\mathbf{x} - \boldsymbol{\mu}); 0, \boldsymbol{\Lambda}), \qquad (1.43)$$

where $\boldsymbol{\Delta}$ is a diagonal matrix whose elements are given by $\boldsymbol{\delta}$, and $\boldsymbol{\Lambda} = \mathbf{I}_p - \boldsymbol{\Delta}\boldsymbol{\Omega}^{-1}\boldsymbol{\Delta}$. Note that (1.41) and (1.43) only match when $p = 1$. Both the unrestricted multivariate skew normal and the unrestricted multivariate skew $t-$distributions were proposed in a similar manner, and their density functions are defined by

$$f_{uMSN}(\mathbf{x}; \boldsymbol{\mu}, \boldsymbol{\Sigma}, \boldsymbol{\delta}) = 2^p \phi_p(\mathbf{x}; \boldsymbol{\mu}, \boldsymbol{\Sigma}) \Phi_p(\boldsymbol{\Delta}\boldsymbol{\Sigma}^{-1}(\mathbf{x} - \boldsymbol{\mu}), \mathbf{0}, \boldsymbol{\Lambda}),$$

and

$$f_{uMST}(\mathbf{x}; \mathbf{u}, \boldsymbol{\Sigma}, \boldsymbol{\delta}, \nu)$$
$$= 2^p t_p(\mathbf{x}; \mathbf{u}, \boldsymbol{\Sigma}, \nu) T_p \left(\boldsymbol{\Delta}\boldsymbol{\Sigma}^{-1}(\mathbf{x} - \boldsymbol{\mu}) \sqrt{\frac{\nu + p}{\nu + d(\mathbf{x})}}; \mathbf{0}, \lambda, \nu + p \right),$$

respectively. In this case, the stochastic representations $\mathbf{X} = \boldsymbol{\mu} + \boldsymbol{\delta}|\tilde{\mathbf{X}}_0| + \tilde{\mathbf{X}}_1$, and $\mathbf{X} = \boldsymbol{\mu} + (\mathbf{X}_1|\mathbf{X}_0 > 0)$ still hold, but in the uMSN case, $\tilde{\mathbf{X}}_0$ and $\tilde{\boldsymbol{\Sigma}}^{1/2}\tilde{\mathbf{X}}_1$ (with $\tilde{\boldsymbol{\Sigma}} = \boldsymbol{\Sigma} - \boldsymbol{\Delta}^2$) are independent and p-dimensional normally distributed, $\mathbf{X}_0 \sim N_p(\mathbf{0}, \mathbf{I}_p)$, $\mathbf{X}_1 \sim N_p(\mathbf{0}, \boldsymbol{\Sigma})$ and $\text{cov}(\mathbf{X}_1, \mathbf{X}_0) = \boldsymbol{\Delta}$.

Extended multivariate skew normal and skew t-distributions: By simultaneously imposing a set of q constraints on the latent variable as $\mathbf{X}_0 + \boldsymbol{\tau} > 0$, meaning that $X_{0i} + \tau_i > 0$ for $i = 1, \ldots, q$, a more general extension of the skewed distributions, named the unified skew-normal (uMSN), were proposed by Arellano-Valle and Azzalini (2006). It is constructed by allowing $\tilde{\mathbf{X}}_0 \sim TN_q(\boldsymbol{\tau}, \boldsymbol{\Lambda})$ and $\tilde{\mathbf{X}}_1 \sim N_p(\mathbf{0}, \boldsymbol{\Sigma})$, where $TN_q(\boldsymbol{\tau}, \boldsymbol{\Lambda})$ denotes a q-dimensional normal variable with mean $\boldsymbol{\tau}$ and covariance $\boldsymbol{\Lambda}$ truncated to the positive hyperplane. The corresponding density is

$$f_{uMSN}(\mathbf{x}; \boldsymbol{\mu}, \boldsymbol{\Sigma}, \boldsymbol{\Lambda}, \boldsymbol{\Delta}, \boldsymbol{\tau})$$
$$= \phi_p(\mathbf{x}; \boldsymbol{\mu}, \boldsymbol{\Sigma}) \frac{\Phi_q(\boldsymbol{\tau} + \boldsymbol{\Delta}^\top \boldsymbol{\Sigma}^{-1}(\mathbf{x} - \boldsymbol{\mu}); \mathbf{0}, \boldsymbol{\Gamma} - \boldsymbol{\Delta}^\top \boldsymbol{\Sigma}^{-1}\boldsymbol{\Delta})}{\Phi_q(\boldsymbol{\tau}; \mathbf{0}, \boldsymbol{\Gamma})}. \tag{1.44}$$

Note that the uMSE can be considered as a form that incorporates the restricted and unrestricted classes with appropriate restrictions on the parameters.

Generalized multivariate skew normal and skew t-distributions: If the latent variable $\mathbf{X}_0$ is allowed to have a distribution different from $\mathbf{X}_1$, then a generalized form of the multivariate skew distribution (Arellano-Valle and Genton, 2005) is obtained. In general, the density of this class takes the form

$$f_g(\mathbf{x}; \boldsymbol{\mu}, \boldsymbol{\Sigma}) = f_p(\mathbf{x}; \boldsymbol{\mu}, \boldsymbol{\Sigma}) Q(\omega(\mathbf{x})),$$

where $Q(\cdot)$ is a distribution function and $\omega(\mathbf{x})$ is an odd function of $\mathbf{x}$. The family of skew-normal symmetric and the family of skew t-symmetric distributions are good examples of this class. The fundamental skew normal distribution (FSN) mentioned earlier is another special case of it, with density

$$f_{FSN}(\mathbf{x}; \boldsymbol{\mu}, \boldsymbol{\Sigma}, Q_q) = K_q^{-1} \phi_p(\mathbf{x}; \boldsymbol{\mu}, \boldsymbol{\Sigma}) Q_q(\mathbf{x}),$$

where $Q_q(\mathbf{x})$ is a skewing function, and $K_q = E\{Q_q(\mathbf{X})\}$ is a normalizing constant. When $p = 1$, $f_p(\cdot)$ is the standard normal density, and the skewing function $Q(\cdot)$ is the distribution function of some elliptical distributions, such as the t, Cauchy, logistic, Laplace or uniform, then we obtain skew normal-t, skew normal-Cauchy, skew normal-Laplace, etc. By applying a similar approach to the density of a t-distribution, a family of skew t-symmetric distributions was proposed, which include the skew t-normal, skew t-Cauchy, skew t-Laplace, skew t-logistic, and skew t-uniform distributions.

1.11.1.2   Multivariate generalized hyperbolic distribution

In the past few years, a family of the normal variance-mean mixture models has also been developed. Widely used in financial data analysis, the generalized hyperbolic (GH) distribution family has become increasingly popular.

This family of distributions can be obtained as a mean-variance mixture of normal distributions. Assume

$$\mathbf{X} = \boldsymbol{\mu} + W\boldsymbol{\alpha} + \sqrt{W}\mathbf{U},$$

where $\mathbf{U} \sim N(\mathbf{0}, \boldsymbol{\Sigma})$ and

$$\mathbf{X}|W = w \sim N_p(\boldsymbol{\mu} + w\boldsymbol{\alpha}, w\boldsymbol{\Sigma}).$$

Now, if the latent variable W is assumed to follow an inverse Gaussian distribution $W \sim IG(\xi, \sigma)$, then the resulting distribution of X is the normal inverse Gaussian (NIG) distribution (Karlis and Santourian, 2009), with density:

$$f_{NIG}(\mathbf{x}; \boldsymbol{\mu}, \boldsymbol{\Sigma}, \boldsymbol{\alpha}, \xi, \sigma) = 2^{-(\frac{p-1}{2})}\sigma \left(\frac{\sqrt{\xi^2 + r(\boldsymbol{\alpha}, \boldsymbol{\Sigma})}}{\pi\sqrt{\sigma^2 + d(\mathbf{x})}} \right)^{\frac{p+1}{2}} e^{\xi\sigma + \boldsymbol{\alpha}^\top(\mathbf{x}-\boldsymbol{\mu})}$$

$$\times K_{\frac{p+1}{2}}(\sqrt{(\xi^2 + r(\boldsymbol{\alpha}, \boldsymbol{\Sigma}))(\sigma^2 + d(\mathbf{x}))}),$$

where $r(\boldsymbol{\alpha}, \boldsymbol{\Sigma}) = \boldsymbol{\alpha}^\top\boldsymbol{\Sigma}^{-1}\boldsymbol{\alpha}$, $d(\mathbf{x}) = (\mathbf{x} - \boldsymbol{\mu})^\top\boldsymbol{\Sigma}^{-1}(\mathbf{x} - \boldsymbol{\mu})$ is the squared Mahalanobis distance and K_λ is the modified Bessel function of the third kind.

If $W \sim \text{Exp}(1)$, then $\mathbf{X}$ follows a multivariate shifted asymmetric Laplace (SAL) distribution (Morris and McNicholas, 2003), and its density is

$$f_{SAL}(\mathbf{x}; \boldsymbol{\mu}, \boldsymbol{\Sigma}, \boldsymbol{\alpha}) = \frac{2\exp\{(\mathbf{x} - \boldsymbol{\mu})^\top\boldsymbol{\Sigma}^{-1}\boldsymbol{\alpha}\}}{(2\pi)^{p/2}|\boldsymbol{\Sigma}|^{1/2}} \left(\frac{d(\mathbf{x})}{2 + r(\boldsymbol{\alpha}, \boldsymbol{\Sigma})} \right)^{\frac{2-p}{4}}$$

$$\times K_{\frac{2-p}{2}}(\sqrt{(2 + r(\boldsymbol{\alpha}, \boldsymbol{\Sigma}))d(\mathbf{x})}).$$

If W is assumed to follow $W \sim Gamma(\gamma, \gamma)$, then it results in a variance-gamma (VG) distribution (McNicholas et al., 2017) and its density is

$$f_{VG}(\mathbf{x}; \boldsymbol{\mu}, \boldsymbol{\Sigma}, \boldsymbol{\alpha}, \gamma) = \frac{2\gamma^\gamma \exp\{(\mathbf{x}-\boldsymbol{\mu})\boldsymbol{\Sigma}^{-1}\boldsymbol{\alpha}\}}{(2\pi)^{p/2}|\boldsymbol{\Sigma}|^{1/2}\Gamma(\gamma)} \left(\frac{d(\mathbf{x})}{r(\boldsymbol{\alpha}, \boldsymbol{\Sigma}) + 2\gamma} \right)^{\frac{\gamma}{2}-\frac{p}{4}}$$
$$\times K_{\gamma-\frac{p}{2}}(\sqrt{(r(\boldsymbol{\alpha}, \boldsymbol{\Sigma}) + 2\gamma)d(\mathbf{x})}).$$

In addition, if W follows a generalized inverse Gaussian distribution (as defined in Browne and McNicholas (2015)) as $W \sim GIG(\omega, 1, \lambda)$, then $\mathbf{X}$ follows a generalized hyperbolic distribution with density:

$$f_{GH}(\mathbf{x}; \boldsymbol{\mu}, \boldsymbol{\Sigma}, \boldsymbol{\alpha}, \lambda, \omega) = \frac{\exp\{(\mathbf{x}-\boldsymbol{\mu})\boldsymbol{\Sigma}^{-1}\boldsymbol{\alpha}^\top\}}{(2\pi)^{np/2}|\boldsymbol{\Sigma}|^{1/2}K_\lambda(\omega)} \left(\frac{\omega + d(\mathbf{x})}{\omega + r(\boldsymbol{\alpha}, \boldsymbol{\Sigma})} \right)^{\frac{\lambda}{2}-\frac{p}{4}}$$
$$\times K_{\lambda-p/2}\left(\sqrt{(\omega + r(\boldsymbol{\alpha}, \boldsymbol{\Sigma}))(\omega + d(\mathbf{x}))} \right).$$

Note that by the presence of the index parameter λ, the family of generalized hyperbolic distributions is quite flexible, and with appropriate parameterizations, the GH distribution can replicate the density of Gaussian, variance-gamma, as well as t-distributions.

1.11.2 Finite mixtures of skew distributions

First proposed by Pyne et al. (2009), the **finite mixture of restricted skew-normal distributions** has been studied in several articles, with different variants. The density of a finite mixture of restricted skew-normal distributions is

$$f(\mathbf{x}; \boldsymbol{\Psi}) = \sum_{c=1}^{C} \pi_c f_{rMSN}(\mathbf{x}; \boldsymbol{\mu}_c, \boldsymbol{\Sigma}_c, \boldsymbol{\delta}_c),$$

where $f_{rMSN}(\cdot)$ denotes the p-dimensional restricted skew normal distribution defined in (1.38).

In a similar manner, a **finite mixture of unrestricted skew-normal distributions** can also be defined as

$$f(\mathbf{x}; \boldsymbol{\Psi}) = \sum_{c=1}^{C} \pi_c f_{uMSN}(\mathbf{x}; \boldsymbol{\mu}_c, \boldsymbol{\Sigma}_c, \boldsymbol{\delta}_c),$$

where $f_{uMSN}(\cdot)$ is the density of the unrestricted skew normal distribution defined in (1.44). The EM algorithm corresponding to this is discussed in detail in Lin (2009).

As a natural combination of the mixture of skew-normal distributions and the mixture of t-distributions, several versions of **finite mixtures of skew t-distributions** have also been studied by several articles. Generally speaking, the mixture of skew t-distributions has a density defined as

$$f(\mathbf{x}; \boldsymbol{\Psi}) = \sum_{c=1}^{C} \pi_c f_{rMST}(\mathbf{x}; \boldsymbol{\mu}_c, \boldsymbol{\Sigma}_c, \boldsymbol{\delta}_c, \nu_c),$$

where $f_{rMST}(\cdot)$ denotes the density of a restricted multivariate skew t-distribution, defined in (1.40). Please see Pyne et al. (2009) for more details about the estimation procedure.

The mixture of the unrestricted skew t-distribution is defined in a similar manner by

$$f(\mathbf{x}; \mathbf{\Psi}) = \sum_{c=1}^{C} \pi_c f_{uMST}(\mathbf{x}; \boldsymbol{\mu}_c, \mathbf{\Sigma}_c, \boldsymbol{\delta}_c, \nu_c).$$

However, due to the p-dimensional skewing function, the mixture of unrestricted skew t-distributions has been used less by researchers. The EM algorithms can be found in Lee and McLachlan (2011).

The finite mixture of multivariate normal inverse Gaussian distributions (O'Hagan et al., 2016), generalized hyperbolic distributions (Browne and McNicholas, 2015), variance-gamma distributions (McNicholas et al., 2017), and multivariate shifted asymmetric Laplace (Morris and McNicholas, 2003) are defined, similarly.

1.12   Semi-supervised mixture models

For many applications, it is possible that we have component labels for some, but not all of the observations. For such situations, we can easily adapt an existing EM algorithm to incorporate those known labels. Suppose $\{x_1, \ldots, x_{n_1}\}$ have known labels z_{ij}, $i = 1, \ldots, n_1, c = 1, \ldots, C$, while $x_{n_1+1}, \ldots, x_n$ have unknown labels from the mixture density

$$f(x; \boldsymbol{\theta}) = \sum_{c=1}^{C} \pi_c f(x; \lambda_c).$$

Then we can use a semi-supervised EM algorithm, as described in Algorithm 1.12.1, to estimate the unknown parameters.

Algorithm 1.12.1 Semi-supervised EM algorithm

Starting with an initial value $\theta^{(0)}$, in the $(k+1)$th step, letting $p_{ij}^{(k+1)} = z_{ij}$, $i = 1, \ldots, n_1, c = 1, \ldots, C$, we compute the following two steps.

E-step: Find the conditional probabilities:

$$
\begin{aligned}
p_{ic}^{(k+1)} &= E\{z_{ic} \mid x_i, \theta^{(k)}\} \\
&= P(z_{ic} = 1 \mid x_i, \theta^{(k)}) \\
&= \frac{\pi_c^{(k)} f(x_i; \lambda_c^{(k)})}{\sum_{l=1}^{C} \pi_l^{(k)} f(x_i; \lambda_l^{(k)})}, i = n_1 + 1, \ldots, n, c = 1, \ldots, C.
\end{aligned}
$$

M-step: Update each λ_c by

$$
\lambda_c^{(k+1)} = \arg \max_{\lambda_c} \sum_{i=1}^{n} \{p_{ic}^{(k+1)} \log f(x_i; \lambda_c)\}.
$$

Update π_c by

$$
\pi_c^{(k+1)} = \frac{\sum_{i=1}^{n} p_{ic}^{(k+1)}}{n},
$$

if the labeled data has been obtained by sampling from the mixture, and by

$$
\pi_c^{(k+1)} = \frac{\sum_{i=n_1+1}^{n} p_{ic}^{(k+1)}}{n - n_1},
$$

if the labeled data does not provide any information on the π_i (for example, where the labeled data have been obtained by sampling separately from each mixture component).

1.13 Nonparametric maximum likelihood estimate

1.13.1 Introduction

When the order of the mixture model is unknown or the mixing distribution is not discrete, we may employ a nonparametric assumption on the mixing distribution. That is the mixture model has the following form:

$$
f(x; Q) = \int g(x; \lambda) dQ(\lambda), \tag{1.45}
$$

where Q is an unspecified distribution function. Please also refer to Section 1.2 for more explanation about why mixture models can be expressed in the format of (1.45). A *nonparametric maximum likelihood estimate* (NPMLE) of Q is a distribution function which maximizes the log-likelihood function:

$$\ell(Q) = \sum_{i=1}^{n} \log f(x_i; Q), \tag{1.46}$$

over all possible mixing distributions (Laird, 1978). In Section 9.7, we will also introduce a Bayesian method to estimate Q based on Bayesian nonparametric mixture models.

Conditions for the existence of NPMLE were established by Kiefer and Wolfowitz (1956). The identifiability problems for the NPMLE of a mixing distribution have been studied by Teicher (1963); Barndorff (1965); Chandra (1977); Jewell (1982); Lindsay and Roeder (1993). The consistency of the NPMLE was established under very general conditions (Leroux and Puterman, 1992; Kiefer and Wolfowitz, 1956). Its computation and properties were defined by Laird (1978) and Lindsay (1983); Lindsay et al. (1983). Full discussions were given in Lindsay (1995) and Böhning (1999).

In order to determine whether a given distribution function Q_0 is the NPMLE or not, a useful tool is called the directional derivative. Given two mixing distributions Q_0 and Q_1, the directional derivative of $\ell(Q)$ at Q_0 towards Q_1 is defined to be

$$\begin{aligned}
D_{Q_0}(Q_1) &= \lim_{\epsilon \to 0^+} \frac{\ell\{(1-\epsilon)Q_0 + \epsilon Q_1\} - \ell(Q_0)}{\epsilon} \\
&= \sum_{i=1}^{n} \frac{f(x_i; Q_1) - f(x_i; Q_0)}{f(x_i; Q_0)} = \sum_{i=1}^{n} \left(\frac{f(x_i; Q_1)}{f(x_i; Q_0)} - 1 \right).
\end{aligned}$$

If Q_1 is a point mass function at λ, denoted as δ_λ, we get the *gradient function*

$$D_{Q_0}(\lambda) = \sum_{i=1}^{n} \frac{f(x_i; \lambda) - f(x_i; Q_0)}{f(x_i; Q_0)} = \sum_{i=1}^{n} \left(\frac{f(x_i; \lambda)}{f(x_i; Q_0)} - 1 \right).$$

The gradient function is an important function for determining the properties of NPMLE. If $\hat{Q}$ is the NPMLE, then $\ell(Q)$ can not increase in any direction starting from $\hat{Q}$, hence the gradient function $D_{\hat{Q}}(\lambda)$ should be nonpositive.

The fundamental theorem of nonparametric maximum likelihood estimation is listed in the following (Lindsay, 1983; Lindsay et al., 1983).

Theorem 1.13.1. *Existence, discreteness, and uniqueness.*
Let k be the number of different values among $f(x_1; Q), \ldots, f(x_n; Q)$. Under the condition[1] that the component density $g(x_i, \lambda)$ is both nonnegative and bounded as a function of λ. There exists a NPMLE Q which is discrete with

---

[1]Technically, there are two other assumptions, which generally do not present a genuine difficulty. Please see Lindsay (1995) for more detail.

no more than k distinct support points. In addition, the fitted log-likelihood values, namely,

$$\{\log f(x_1; \hat{Q}), \dots, \log f(x_n; \hat{Q})\}$$

are unique, where $\hat{Q}$ is the NPMLE of Q.

Comment: The uniqueness property implies that if two distributions maximize the log-likelihood (1.46), then the corresponding log-likelihood vectors must be equal.

Theorem 1.13.2. *Gradient characterization and support point properties.* The distribution function $\hat{Q}$ is the NPMLE of Q if and only if

$$D_{\hat{Q}}(\lambda) \le 0, \forall \lambda.$$

Further, the supports of $\hat{Q}$ are contained in the set of λ such that $D_{\hat{Q}}(\lambda) = 0$.

1.13.2 Computations of NPMLE

Vertex-Direction-Method (VDM) is one of the earliest methods to compute the NPMLE of Q by utilizing Theorem 1.13.2. VDM first searches the maximizer λ^* of $D_{\hat{Q}_n}(\lambda)$ with the current estimator $\hat{Q}_n$. If $D_{\hat{Q}_n}(\lambda^*) \le 0$, the algorithm stops and returns $\hat{Q}_n$ as the NPMLE of Q. Otherwise, VDM updates $\hat{Q}_n$ to $(1 - \alpha)\hat{Q}_n + \alpha\delta_{\lambda^*}$ for some $0 < \alpha < 1$ (such as finding α by maximizing the log-likelihood (1.46)). This type of algorithm was first proposed in the literature of optimal design theory (Wynn, 1970, 1972; Atwood et al., 1976; Wu, 1978). Bohning (1982), Lindsay (1983), and Lindsay et al. (1983) showed their connection to the NPMLE problem in continuous mixtures.

Although VDM simplifies the estimation procedure and guarantees the convergence to the NPMLE of Q, it is generally too slow and requires too many iterations until the gradient condition is satisfied. This slow convergence is mainly caused by too many support points because the number of support points in VDM always increases in each iteration. Several algorithms have been suggested to speed up the convergence such as Vertex-Exchange-Method (VEM) (Böhning, 1986), Intra-Simplex-Direction-Method (ISDM) (Lesperance and Kalbfleisch, 1992), and Constrained-Newton method for Multiple supports (CNM) (Wang, 2007). Wang (2007) demonstrated that the CNM algorithm is much faster than most other existing algorithms.

1.13.3 Normal scale of mixture models

If the component density $f(y; \lambda)$ in (1.45) is a normal density with mean 0 and scale parameter σ, and Q is an unknown distribution of the scale parameter σ, then the mixture model becomes

$$f(x; Q) = \int \frac{1}{\sigma} \phi(x/\sigma) dQ(\sigma), \tag{1.47}$$

where $\phi(x)$ is a standard normal density. The above model (1.47) is called *normal scale mixture model*. Let $\mathcal{F}$ be the class of all densities which can be written as in (1.47). $\mathcal{F}$ contains almost all symmetric unimodal continuous probability densities such as normal, Laplace, t, double exponential, exponential power family, stable distributions, and so on (Andrews and Mallows, 1974). Efron and Olshen (1978) and Basu (1996) discussed how many distributions are contained in $\mathcal{F}$. Kelker (1971) and Andrews and Mallows (1974) also studied necessary and sufficient conditions for a probability density to be a member of $\mathcal{F}$.

The well-known *contaminated normal mixture model*

$$f(x) = (1 - \varepsilon)\phi(x; \mu, \sigma^2) + \varepsilon\phi(x; \mu, k\sigma^2), \qquad (1.48)$$

where $k \gg 1$ and ε is small, is a special case of (1.47) if Q is a discrete probability distribution that only has two support points σ^2 and $k\sigma^2$ with probability $(1 - \varepsilon)$ and ε, respectively. The above contaminated normal mixture model (1.48) can be used to model the normal population with few atypical observations/outliers with the second component representing the outlier component. Because of this, it is popularly used to model the regression error density to provide a robust regression estimator. Next, we further use two examples to illustrate the applications of the normal scale of mixture models.

Example 1.13.1. Semiparametric mixture models: Bordes et al. (2006b) and Hunter et al. (2007) considered the following location-shifted semiparametric model with symmetric nonparametric component densities:

$$f(x; \boldsymbol{\theta}, g) = \sum_{c=1}^{C} \pi_c g(x - \mu_c), \qquad (1.49)$$

where $\boldsymbol{\theta} = (\pi_1, \mu_1, \ldots, \pi_C, \mu_C)$, and g is an unknown but symmetric density about zero. Bordes et al. (2006b) proved the identifiability of the model (1.49) for $C = 2$. Hunter et al. (2007) further established the identifiability of model (1.49) for both $C = 2$ and $C = 3$.

Xiang et al. (2016) proposed modeling the unspecified g as a continuous normal-scale mixture by assuming that g has the form of (1.47). Under this model class, (1.49) can be expressed as

$$
\begin{aligned}
f(x; \boldsymbol{\theta}, Q) &= \sum_{c=1}^{C} \pi_c \left\{ \int \frac{1}{\sigma} \phi\left(\frac{x - \mu_c}{\sigma} \right) dQ(\sigma) \right\} \\
&= \int \sum_{c=1}^{C} \frac{\pi_c}{\sigma} \phi\left(\frac{x - \mu_c}{\sigma} \right) dQ(\sigma).
\end{aligned}
\qquad (1.50)
$$

The identifiability of (1.50) can be seen by combining the identifiability results of (1.49) and $\mathcal{F}$.

In the first displayed expression of (1.50), if Q is known, $\int \frac{1}{\sigma} \phi \left(\frac{x - \mu_j}{\sigma} \right) dQ(\sigma)$ can be considered as a parametric component density in a finite location mixture model. On the other hand, in the last displayed expression of (1.50), if (π_j, μ_j)s are known, $\sum_{j=1}^{C} \frac{\pi_j}{\sigma} \phi \left(\frac{x - \mu_j}{\sigma} \right)$ plays a role of a component density in the nonparametric mixture with unknown mixing distribution Q. Therefore, the estimation of (1.50) can be done by alternatively estimating Q given $\boldsymbol{\theta}$ based on the NPMLE introduced in Section 1.13.2 and estimating $\boldsymbol{\theta}$ given Q based on the regular EM algorithm for finite mixture models. □

Example 1.13.2. Dealing with outlier modeling: For the linear regression model

$$y = \mathbf{x}^\top \boldsymbol{\beta} + \epsilon,$$

where $\mathrm{E}(\epsilon) = 0$, the traditional least squares estimate of the regression coefficient $\boldsymbol{\beta}$ is equivalent to the maximum likelihood estimate if the error is assumed to be normal. However, it may lose some efficiency when the error is not normal. In addition, it is sensitive to outliers or heavy-tailed data.

Seo et al. (2017) proposed modeling the error density by a normal scale mixture density as in (1.47). Given $\boldsymbol{\beta}$, Q can be estimated by NPMLE. Given Q, $\boldsymbol{\beta}$ can be estimated by MLE. Such estimated $\boldsymbol{\beta}$ can be adaptive to different error distributions due to the semiparametric estimation of the error density by the continuous normal scale mixture density. □

Example 1.13.3. Measurement error models: $Y = x\beta + \epsilon$ and x can't be observed. But we have a surrogate measure z such that $z = x + \varepsilon$. Suppose (Y, X) has a multivariate normal distribution with zero means and covariance matrix

$$\begin{pmatrix} \sigma_Y^2 & \rho \sigma_Y \sigma_X \\ \rho \sigma_Y \sigma_X & \sigma_X^2 \end{pmatrix},$$

then $\beta = \rho \sigma_Y / \sigma_X$. Since (Y, Z) has zero means and covariance matrix

$$\begin{pmatrix} \sigma_Y^2 & \rho \sigma_Y \sigma_X \\ \rho \sigma_Y \sigma_X & \sigma_X^2 + \sigma_\varepsilon^2 \end{pmatrix},$$

the regression coefficient for (Y, Z) is

$$b = \rho \sigma_Y \sigma_X / (\sigma_X^2 + \sigma_\varepsilon^2) = \beta \frac{\sigma_X^2}{\sigma_X^2 + \sigma_\varepsilon^2} \leq \beta.$$

Therefore, if we ignore the measurement error of the predictor, we will get a biased and inconsistent estimate of the true regression coefficient β.

Next, we also discuss how to use NPMLE to estimate the regression model with measurement error. Suppose X has the distribution Q, then the density of (z, y) is

$$f(y, z) = \int f(y, z \mid x) dQ(x)$$

$$= \int f(y \mid z,x) f(z \mid x) dQ(x)$$

$$= \int f(y \mid x) f(z \mid x) dQ(x),$$

which is a mixture distribution. Note that the unknown parameters in $f(y \mid x)$ are our interest. With a parametric assumption for $f(z|x)$, we can estimate all unknown parameters by maximizing the log-likelihood for the observed data (z, y) based on NPMLE. □

Besides the above applications, the normal scale mixture density (1.47) can be used to relax the traditional normality assumption for many other models. For example, Seo and Lee (2015) also utilized the model (1.47) to efficiently estimate the distribution of innovations as well as parameters in semiparametric generalized autoregressive conditional heteroskedasticity models. For random/mixed effects models, we can assume that the random effects have the normal scale mixture density (1.47) to relax the traditional normality assumption.

1.14 Mixture models for matrix data

Realizations from random matrices are generated by the simultaneous observation of variables in different situations or locations and are commonly arranged in three-way data structures. These types of data are commonly seen in longitudinal data on multiple response variables, multivariate spatial data, or spatial-temporal data. Other examples arise when some objects are rated on multiple attributes by multiple experts or from experiments in which individuals provide multiple ratings for multiple objects. The data that is characterized by three classes of entities or modes can all be summarized in a three-way structure.

Specifically speaking, a p-variate response is observed repeatedly in r locations, or a univariate response is observed in p locations r times. Let $\mathbf{Y}_i$ be a $r \times p$ matrix for each statistical unit, with $i = 1, \ldots, n$. We further assume that a random sample of n individual matrices $\mathbf{Y}_1, \ldots, \mathbf{Y}_n$ is independent and identically distributed. Suppose we are interested in clustering these n observed matrices into C groups, with $C < n$. Since correlations between variables could change across locations, and correlations across different occasions or situations can be quite different for each response, clustering three-way data can be quite difficult. A simple way to deal with this clustering problem is through some dimension reduction techniques, such as principal component analysis. By converting a three-way data set into two-way data, we can then apply classical techniques for clustering. However, as common knowledge of PCA, the first principal component, which is the direction that explains the major part of the total variance, does not necessarily preserve all the information needed for clustering. Another way to deal with it is through the least-square approach (Gordon, 1998; Vichi, 1999; Vichi et al., 2007), which is completely free of the distributional assumption of the clusters.

When the mixture likelihood approach is adapted to deal with the three-way data, it is assumed that each unit i belongs to one of the C possible groups in proportions $\pi_1, \ldots, \pi_C$, respectively. Given occasions $l = 1, \ldots, r$, it is assumed that

$$\mathbf{Y}_{il} \sim \phi^{(p)}(\boldsymbol{\mu}_{cl}, \boldsymbol{\Sigma}_c),$$

with a probability $\pi_c, c = 1, \ldots, C$, where $\phi^{(p)}$ represents a p-dimensional normal distribution, $\boldsymbol{\mu}_{cl}$ is a group-specific and location-specific mean, and $\boldsymbol{\Sigma}_c$ is the component covariance matrix that does not depend on the occasion. Then, the mixture model assumed in this case is

$$f(\mathbf{Y}_i) = \sum_{c=1}^{C} \pi_c \prod_{l=1}^{r} \phi^{(p)}(\boldsymbol{\mu}_{cl}, \boldsymbol{\Sigma}_c). \tag{1.51}$$

In Model (1.51), the correlations between occasions are assumed to be zero, and correlations between variables are assumed to be constant across the third mode. Hunt and Jorgensen (1999) further extended Model (1.51) to account for mixed observed variables, and Vermunt (2007) extended Model (1.51) to a hierarchical approach, which allows units to belong to different classes in various situations.

There are other ways to analyze matrix variate data, that is to vectorize the data first and then apply multivariate methodologies. However, performing the analysis using a matrix variate model has the benefit of simultaneously considering the temporal covariances ($\boldsymbol{\Psi}$) as well as the covariances for the variables ($\boldsymbol{\Sigma}$). In addition, by vectorizing the data, a $p \times d$-dimensional data results in a pd-dimensional vector, and therefore, the number of free parameters increases from $(p^2 + d^2 + p + d)/2$ to $(p^2 d^2 + pd)/2$.

1.14.1 Finite mixtures of matrix normal distributions

Viroli (2011) presented a new approach, which models the distribution of observed matrices, instead of units, so that the full information of variables and situations can be taken into account simultaneously.

In addition to the multivariate normal distribution, matrix normal distribution, which is a multivariate analysis tool for modeling random matrices, also plays an important role. Again, suppose $\mathbf{Y}_1, \ldots, \mathbf{Y}_n$ represents n independently and identically distributed random matrices of the dimension $p \times d$, where d represents the number of occasions and p is the number of attributes. Assume $\mathbf{M}$ is a $p \times d$ matrix of means, $\boldsymbol{\Psi}$ is a $d \times d$ covariance matrix containing the variances and covariances between the d occasions or times, and $\boldsymbol{\Sigma}$ is a $p \times p$ covariance matrix containing the variances and covariances of the

p variables or locations. Then, the $p \times d$ matrix normal distribution is defined as

$$f(\mathbf{Y}; \mathbf{M}, \mathbf{\Phi}, \mathbf{\Sigma})$$

$$= (2\pi)^{-\frac{pd}{2}} |\mathbf{\Sigma}|^{-\frac{d}{2}} |\mathbf{\Psi}|^{-\frac{p}{2}} \exp\left[-\frac{1}{2} \mathrm{tr}\left\{ \mathbf{\Sigma}^{-1}(\mathbf{Y} - \mathbf{M})\mathbf{\Psi}^{-1}(\mathbf{Y} - \mathbf{M})^{\top} \right\} \right],$$

denoted as

$$\mathbf{Y} \sim \phi^{(p \times d)}(\mathbf{M}, \mathbf{\Sigma}, \mathbf{\Psi}). \tag{1.52}$$

An equivalent definition of (1.52) is $\mathrm{vec}(\mathbf{Y}) \sim \phi^{(pd)}(\mathrm{vec}(\mathbf{M}), \mathbf{\Psi} \otimes \mathbf{\Sigma})$, where $\otimes$ is the Kronecker product.

Suppose that heterogeneity is observed in the data, so we assume that observed matrices belong to C sub-populations with probabilities $\pi_1, \ldots, \pi_C$. Then, the observed matrices have finite mixtures of matrix normal distributions with the following density:

$$f(\mathbf{Y}_i; \pi_1, \ldots, \pi_C, \mathbf{\Theta}_1, \ldots, \mathbf{\Theta}_C) = \sum_{c=1}^{C} \pi_c \phi^{(p \times d)}(\mathbf{Y}_i; \mathbf{M}_c, \mathbf{\Sigma}_c, \mathbf{\Psi}_c), \tag{1.53}$$

where $\mathbf{\Theta}_c = \{\mathbf{M}_c, \mathbf{\Psi}_c, \mathbf{\Sigma}_c\}$. The EM algorithm for the maximum likelihood estimation is summarized in Algorithm 1.14.1.

Algorithm 1.14.1 EM algorithm for the model (1.5.3).

Starting with some initial parameter values, in the $(k+1)$th step, we perform the following two steps.

E-step: Find the conditional probabilities:

$$p_{ic}^{(k+1)} = \frac{\pi_c^{(k)} \phi^{(r \times p)}(\mathbf{Y}_i; \mathbf{M}_c^{(k)}, \mathbf{\Sigma}_c^{(k)}, \mathbf{\Psi}_c^{(k)})}{\sum_{h=1}^{C} \pi_h^{(k)} \phi^{(r \times p)}(\mathbf{Y}_i; \mathbf{M}_h^{(k)}, \mathbf{\Sigma}_h^{(k)}, \mathbf{\Psi}_h^{(k)})}, \quad c = 1, \ldots, C.$$

M-step: Update parameters by

$$\pi_c^{(k+1)} = \frac{\sum_{i=1}^{n} p_{ic}^{(k+1)}}{n},$$

$$\mathbf{M}_c^{(k+1)} = \frac{\sum_{i=1}^{n} p_{ic}^{(k+1)} \mathbf{Y}_i}{\sum_{i=1}^{n} p_{ic}^{(k+1)}},$$

$$\mathbf{\Sigma}_c^{(k+1)} = \frac{\sum_{i=1}^{n} p_{ic}^{(k+1)} (\mathbf{Y}_i - \mathbf{M}_c^{(k+1)}) \mathbf{\Psi}_c^{(k)-1} (\mathbf{Y}_i - \mathbf{M}_c^{(k+1)})^{\top}}{p \sum_{i=1}^{n} p_{ic}^{(k+1)}},$$

$$\mathbf{\Psi}_c^{(k+1)} = \frac{\sum_{i=1}^{n} p_{ic}^{(k+1)} (\mathbf{Y}_i - \mathbf{M}_c^{(k+1)})^{\top} \mathbf{\Sigma}_c^{(k+1)-1} (\mathbf{Y}_i - \mathbf{M}_c^{(k+1)})^{\top}}{r \sum_{i=1}^{n} p_{ic}^{(k+1)}},$$

where $c = 1, \ldots, C$.

In the vector-valued model world, contributions have been made to build alternative mixture models capable of accommodating skewness and thus robust deviations from normality. Similarly, Melnykov and Zhu (2018) suggested a matrix transformation mixture model that is related to Model (1.53), but by adding additional skewness parameters, the new model is robust to deviations from normality. To be more specific, let $\mathcal{T}(y; \lambda)$, for some λ, be a transformation leading to near-normality in the univariate case. Then, the distribution of the original data can be approximated by $\phi(\mathcal{T}(y; \lambda); \mu, \sigma^2)|d\mathcal{T}(y; \lambda)/dy|$. Based on the assumption that the coordinate-wise transformation leads to the joint normality, for the matrix-valued data $\mathbf{Y}$, define a transformation of $\mathbf{Y}$ as

$$\mathcal{T}(\mathbf{Y}, \mathbf{\Lambda}) = \begin{pmatrix} \mathcal{T}(Y_{11}, \Lambda_{11}) & \cdots & \mathcal{T}(Y_{1d}, \Lambda_{1d}) \\ \vdots & \ddots & \vdots \\ \mathcal{T}(Y_{p1}, \Lambda_{p1}) & \cdots & \mathcal{T}(Y_{pd}, \Lambda_{pd}) \end{pmatrix},$$

where $\mathbf{\Lambda}$ is a $p \times d$-dimensional matrix of transformation parameters. Even for small p and d, the value pd can be high enough to prohibit the new model from being useful. Fortunately, through an additive effect assumption for transformation parameters associated with rows and columns, which is convenient to use and easy to interpret, a new transformation matrix is given by

$$\mathcal{T}(\mathbf{Y}, \boldsymbol{\lambda}, \boldsymbol{\nu}) = \begin{pmatrix} \mathcal{T}(Y_{11}, \lambda_1 + \nu_1) & \cdots & \mathcal{T}(Y_{1d}, \lambda_1 + \nu_d) \\ \vdots & \ddots & \vdots \\ \mathcal{T}(Y_{p1}, \lambda_p + \nu_1) & \cdots & \mathcal{T}(Y_{pd}, \lambda_p + \nu_d) \end{pmatrix},$$

where $\boldsymbol{\lambda} = (\lambda_1, \ldots, \lambda_p)^\top$ and $\boldsymbol{\nu} = (\nu_1, \ldots, \nu_d)^\top$ denote the transformation parameters associated with rows and columns, respectively. Then, the authors proposed to use the following to model the original skewed matrix data:

$$f(\mathbf{Y}; \boldsymbol{\theta}) = \sum_{c=1}^{C} \pi_c \phi^{(p \times d)}(\mathcal{T}(\mathbf{Y}; \boldsymbol{\lambda}_c, \boldsymbol{\nu}_c); \mathbf{M}_c, \mathbf{\Sigma}_c, \mathbf{\Psi}_c) J(\mathbf{Y}; \boldsymbol{\lambda}_c, \boldsymbol{\nu}_c), \qquad (1.54)$$

where $\boldsymbol{\theta}$ collects all unknown parameters, and

$$J(\mathbf{Y}; \boldsymbol{\lambda}, \boldsymbol{\nu}) = |\partial \text{vec}(\mathcal{T}(\mathbf{Y}, \boldsymbol{\lambda}, \boldsymbol{\nu})) / \partial (\text{vec} \mathbf{Y})^\top|$$

denotes the Jacobian associated with the transformation $\mathcal{T}$. There are plenty of possible transformations, among which, the power transformation proposed by Yeo and Johnson (2000) is an attractive option, defined as:

$$\mathcal{T}(y; \lambda) = I(y \geq 0, \lambda \neq 0)\{(1+y)^\lambda\}/\lambda + I(y \geq 0, \lambda = 0)\log(1+y)$$
$$+ I(y < 0, \lambda \neq 2)\{(1-y)^{2-\lambda} - 1\}/(2-\lambda) + I(y < 0, \lambda = 2)\log(1-y).$$

The EM algorithm for the maximum likelihood estimator of the model (1.54) is summarized in Algorithm 1.14.2.

Algorithm 1.14.2 EM algorithm for the model (1.54)

Starting with some initial parameter values, in the $(k+1)$th step, we perform the following two steps.

E-step: Find the conditional probabilities:

$$p_{ic}^{(k+1)} = \frac{\pi_c^{(k)} \phi^{(p\times d)}(\mathcal{T}(\mathbf{Y}_i; \boldsymbol{\lambda}_c^{(k)}, \boldsymbol{\nu}_c^{(k)}); \mathbf{M}_c^{(k)}, \boldsymbol{\Sigma}_c^{(k)}, \boldsymbol{\Psi}_c^{(k)}) J(\mathbf{Y}_i; \boldsymbol{\lambda}_c^{(k)}, \boldsymbol{\nu}_c^{(k)})}{\sum_{h=1}^{C} \pi_h^{(k)} \phi^{(p\times d)}(\mathcal{T}(\mathbf{Y}_i; \boldsymbol{\lambda}_h^{(k)}, \boldsymbol{\nu}_h^{(k)}); \mathbf{M}_h^{(k)}, \boldsymbol{\Sigma}_h^{(k)}, \boldsymbol{\Psi}_h^{(k)}) J(\mathbf{Y}_i; \boldsymbol{\lambda}_h^{(k)}, \boldsymbol{\nu}_h^{(k)})},$$

where $c = 1, \dots, C$.

M-step: Update parameters by

$$\pi_c^{(k+1)} = \frac{\sum_{i=1}^{n} p_{ic}^{(k+1)}}{n},$$

$$\mathbf{M}_c^{(k+1)} = \frac{\sum_{i=1}^{n} p_{ic}^{(k+1)} \mathcal{T}(\mathbf{Y}_i; \boldsymbol{\lambda}_c^{(k+1)}, \boldsymbol{\nu}_c^{(k+1)})}{\sum_{i=1}^{n} p_{ic}^{(k+1)}},$$

$$\boldsymbol{\Sigma}_c^{(k+1)} = \frac{\sum_{i=1}^{n} p_{ic}^{(k+1)} r_{ic}^{(k+1)} \boldsymbol{\Psi}_c^{(k)-1} (r_{ic}^{(k+1)})^\top}{d \sum_{i=1}^{n} p_{ic}^{(k+1)}},$$

$$\boldsymbol{\Psi}_c^{(k+1)} = \frac{\sum_{i=1}^{n} p_{ic}^{(k+1)} r_{ic}^{(k+1)} \boldsymbol{\Sigma}_c^{(k+1)-1} (r_{ic}^{(k+1)})^\top}{p \sum_{i=1}^{n} p_{ic}^{(k+1)}},$$

where $r_{ic}^{(k+1)} = \mathcal{T}(\mathbf{Y}_i; \boldsymbol{\lambda}_c^{(k+1)}, \boldsymbol{\nu}_c^{(k+1)}) - \mathbf{M}_c^{(k+1)}$ and $c = 1, \dots, C$.

In addition, Gallaugher and McNicholas (2018) proposed four skewed matrix variate distributions, including the matrix variate skew-t (MVST), the matrix variate variance-gamma (MVVG), the matrix variate normal inverse Gaussian (MVNIG), and the matrix variate generalized hyperbolic distribution (MVGH). The densities are

$$f_{MVST}(\mathbf{Y}|\boldsymbol{\theta}) = \frac{2(\frac{\nu}{2})^{\nu/2} \exp\{\text{tr}(\boldsymbol{\Sigma}^{-1}(\mathbf{Y}-\mathbf{M})\boldsymbol{\Psi}^{-1}\mathbf{A}^\top)\}}{(2\pi)^{pd/2}|\boldsymbol{\Sigma}|^{p/2}|\boldsymbol{\Psi}|^{d/2}\Gamma(\frac{\nu}{2})} \left\{ \frac{\delta(\mathbf{X}; \mathbf{M}, \boldsymbol{\Sigma}, \boldsymbol{\Psi}) + \nu}{\rho(\mathbf{A}, \boldsymbol{\Sigma}, \boldsymbol{\Psi})} \right\}^{-\frac{\nu+pd}{4}}$$

$$\times K_{-\frac{\nu+pd}{4}}\left(\sqrt{\rho(\mathbf{A}, \boldsymbol{\Sigma}, \boldsymbol{\Psi})\{\delta(\mathbf{Y}; \mathbf{M}, \boldsymbol{\Sigma}, \boldsymbol{\Psi}) + \nu\}}\right),$$

$$f_{MVVG}(\mathbf{Y}|\boldsymbol{\theta}) = \frac{2\gamma^\gamma \exp\{\text{tr}(\boldsymbol{\Sigma}^{-1}(\mathbf{Y}-\mathbf{M})\boldsymbol{\Psi}^{-1}\mathbf{A}^\top)\}}{(2\pi)^{pd/2}|\boldsymbol{\Sigma}|^{p/2}|\boldsymbol{\Psi}|^{d/2}\Gamma(\gamma)} \left\{ \frac{\delta(\mathbf{X}; \mathbf{M}, \boldsymbol{\Sigma}, \boldsymbol{\Psi})}{\rho(\mathbf{A}, \boldsymbol{\Sigma}, \boldsymbol{\Psi}) + 2\gamma} \right\}^{\frac{\gamma-pd/2}{2}}$$

$$\times K_{\gamma-pd/2}\left(\sqrt{\{\rho(\mathbf{A}, \boldsymbol{\Sigma}, \boldsymbol{\Psi}) + 2\gamma\}\delta(\mathbf{Y}; \mathbf{M}, \boldsymbol{\Sigma}, \boldsymbol{\Psi})}\right),$$

$$f_{MVNIG}(\mathbf{Y}|\boldsymbol{\theta}) = \frac{2\exp\{\text{tr}(\boldsymbol{\Sigma}^{-1}(\mathbf{Y}-\mathbf{M})\boldsymbol{\Psi}^{-1}\mathbf{A}^\top)\}}{(2\pi)^{(pd+1)/2}|\boldsymbol{\Sigma}|^{p/2}|\boldsymbol{\Psi}|^{d/2}} \left\{ \frac{\delta(\mathbf{X}; \mathbf{M}, \boldsymbol{\Sigma}, \boldsymbol{\Psi}) + 1}{\rho(\mathbf{A}, \boldsymbol{\Sigma}, \boldsymbol{\Psi}) + \tilde{\gamma}^2} \right\}^{-\frac{1+pd}{4}}$$

$$\times K_{-(1+pd)/2}(\sqrt{\{\rho(\mathbf{A}, \mathbf{\Sigma}, \mathbf{\Psi}) + \tilde{\gamma}^2\}\{\delta(\mathbf{Y}; \mathbf{M}, \mathbf{\Sigma}, \mathbf{\Psi}) + 1\}}),$$

$$f_{MVGH}(\mathbf{Y}|\theta) = \frac{\exp\{\mathrm{tr}(\mathbf{\Sigma}^{-1}(\mathbf{Y} - \mathbf{M})\mathbf{\Psi}^{-1}\mathbf{A}^\top)\}}{(2\pi)^{(pd)/2}|\mathbf{\Sigma}|^{p/2}|\mathbf{\Psi}|^{d/2}K_\lambda(\omega)} \left\{ \frac{\delta(\mathbf{X}; \mathbf{M}, \mathbf{\Sigma}, \mathbf{\Psi}) + \omega}{\rho(\mathbf{A}, \mathbf{\Sigma}, \mathbf{\Psi}) + \omega} \right\}^{\frac{\lambda - pd/2}{2}}$$

$$\times K_{(\lambda - pd)/2}(\sqrt{\{\rho(\mathbf{A}, \mathbf{\Sigma}, \mathbf{\Psi}) + \omega\}\{\delta(\mathbf{Y}; \mathbf{M}, \mathbf{\Sigma}, \mathbf{\Psi}) + \omega\}}),$$

where

$$\delta(\mathbf{Y}; \mathbf{M}, \mathbf{\Sigma}, \mathbf{\Psi}) = \mathrm{tr}\{\mathbf{\Sigma}^{-1}(\mathbf{Y} - \mathbf{M})\mathbf{\Psi}^{-1}(\mathbf{Y} - \mathbf{M})^\top\},$$

$$\rho(\mathbf{A}, \mathbf{\Sigma}, \mathbf{\Psi}) = \mathrm{tr}\{\mathbf{\Sigma}^{-1}\mathbf{A}\mathbf{\Psi}^{-1}\mathbf{A}^\top\},$$

and

$$K_\lambda(u) = \frac{1}{2} \int_0^\infty y^{\lambda - 1} \exp\left\{ -\frac{u}{2}\left(y + \frac{1}{y} \right) \right\}$$

is the modified Bessel function of the third kind with index λ. $\mathbf{A}$ is a $p \times d$-dimensional matrix representing skewness, $\nu > 0, \gamma > 0, \tilde{\gamma} > 0, \lambda \in \mathbb{R}$ and $\omega > 0$. Given the four skewed matrix variate distributions, then, the authors proposed mixture models whose components follow one of the four skewed matrix variate distributions. The parameter estimation is performed by an expectation conditional maximization (ECM) algorithm. The readers are referred to Gallaugher and McNicholas (2018) for detailed steps of each of the four mixtures of skewed matrix variate distributions.

1.14.2 Parsimonious models for modeling matrix data

The finite mixture of matrix normal distributions (1.53) is easy to understand, but tends to suffer from overparameterization due to a high number of parameters involved in the model, and thus leads to issues like overfitting and underestimation of the mixture order. The overparameterization issue is well-documented for multivariate mixture models, that is, for vector-valued data. Readers are referred to Chapter 8 for a more detailed discussion on this topic. One popularly used way is to consider different kinds of parsimonious models, which reduces the number of free parameters by putting constraints on the mean and covariance matrices.

When it comes to the mixture of matrix Gaussian distributions, the issue of overparameterization could be even more severe, due to the coexistence of two covariance matrices, related to rows and columns. To address this issue, Sarkar et al. (2020) derived a series of parsimonious mixture models for matrix Gaussian distributions and transformation mixtures. Note that in vector-valued models, the covariance matrix is decomposed as $\mathbf{\Sigma}_c = \psi_c \mathbf{\Gamma}_c \mathbf{\Delta}_c \mathbf{\Gamma}_c^\top$, where $\psi_c = |\boldsymbol{\Sigma}_c|^{1/p}$, $\boldsymbol{\Lambda}_c$ is the matrix whose elements are eigenvectors of $\boldsymbol{\Sigma}_c$, and $\boldsymbol{\Delta}_c$ is a diagonal matrix whose elements are eigenvalues of $\boldsymbol{\Sigma}_c$. Geometrically, ψ_c reflects the volume, $\boldsymbol{\Gamma}_c$ determines the orientation, and $\boldsymbol{\Delta}_c$ indicates the shape. Different combinations of these three characteristics lead to a total of fourteen parsimonious models. Note that due to the identifiability issue, the column covariance matrix $\boldsymbol{\Psi}_c$ is restricted by $|\boldsymbol{\Psi}_c| = 1$, and hence reduces the number of models related to $\boldsymbol{\Psi}_c$ from 14 to 7. Then, the column matrix $\boldsymbol{\Psi}_c$ and row matrix $\boldsymbol{\Sigma}_c$ produce a total of $14 \times 7 = 98$ different parsimonious models.

In addition to the covariance matrices $\boldsymbol{\Psi}_c$ and $\boldsymbol{\Sigma}_c$, the mean matrix $\mathbf{M}_c$ is also parameterized by general form (with no constraints) and additive form, where an additive model of $\mathbf{M}_c$ is defined as $\mathbf{M}_c = \boldsymbol{\alpha}_c \mathbf{1}_d^\top + \mathbf{1}_p \boldsymbol{\beta}_c^\top$, where $\boldsymbol{\alpha}_c$ is a mean vector of length p associated with the rows, and $\boldsymbol{\beta}_c$ is a mean vector of length d associated with the columns. Note that due to the identifiability issue, it is required that $\beta_{cd} = 0$ for $c = 1, \ldots, C$.

To sum up, the 98 possible parameterizations of the covariance matrices and the 2 possible parameterizations of the mean matrices give rise to a total of 196 mixture models of matrix data. In addition, note that the rows can be imposed with skewness parameters or not, and so can the columns. These 4 options, together with the 98 possible parameterizations mentioned above, give rise to a total of $196 \times 4 = 784$ matrix mixture models in total. Please see Sarkar et al. (2020) for the list of matrix parsimonious models and a detailed description of the estimation procedure of the various parameterizations for the row-driven and column-driven covariance matrices $\boldsymbol{\Sigma}_c$ and $\boldsymbol{\Psi}_c$.

1.15   Fitting mixture models using R

There are several packages available to fit mixture models, such as `mixtools`, `mclust`, `EMCluster`, etc. The most recent R package for mixture models is `MixSemiRob` which can estimate most of the mixture models introduced in this book. We apply a finite mixture model to the famous Old Faithful Geyser Data, which contains 272 observations on 2 variables, the waiting time between eruptions and the duration of the eruption for the Old Faithful Geyser in Yellowstone National Park, Wyoming, USA. Here we only consider the eruption time in minutes as our variable of interest. If you run the following code, the result in Figure 1.9 will be shown, which clearly indicates a two-component mixture model.

```
data(faithful)
library(MixSemiRob)
res = mixnorm(faithful$waiting)
```

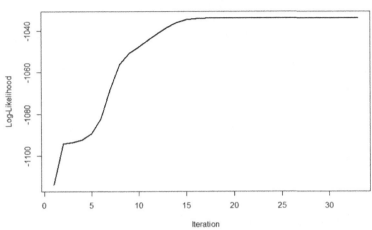

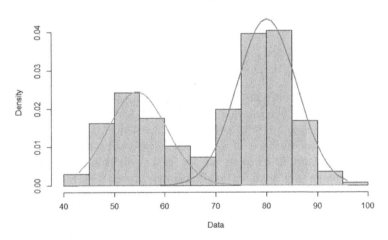

FIGURE 1.9
Histogram and estimated densities of the faithful data.

In Section 1.9.2, Yao (2010) proposed a profile likelihood method to find the maximum interior mode for the normal mixture with unequal variances. We apply this method to the faithful data, and the results are shown in Figure 1.10.

```
x = faithful$eruptions
hist(x, xlab = "eruptions", main = "")
grid = seq(from = 0.01, to = 1, length = 200)
```

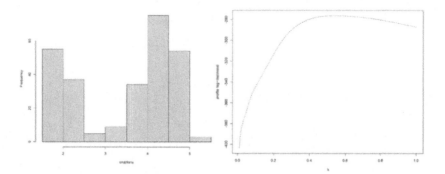

FIGURE 1.10
Profile log-likelihood of faithful data.

```
likelihood = numeric()
for(i in 1:200){
  k = grid[i]
  est = mixpf(x, k)
  likelihood[i] = est$lik
}
plot(grid, likelihood, type = "l", lty = 2, xlab = "k",
     ylab = "profile log-likelihood")
```

The figure shows the profile log-likelihood of eruption data at different k values, and there is only one interior mode at $k = 0.5373$, and the corresponding MLE is $(\hat{\pi}_1, \hat{\mu}_1, \hat{\mu}_2, \hat{\sigma}_1, \hat{\sigma}_2) = (0.65, 4.27, 2.02, 0.438, 0.24)$.

In Section 1.11, a mixture of generalized hyperbolic distribution is introduced, and the MixGHD package can be used to fit this distribution. The bankruptcy dataset contains the ratio of retained earnings (RE) to total assets, and the ratio of earnings before interests and taxes (EBIT) to total assets of 66 American firms recorded in the form of ratios. Half of the selected firms had filed for bankruptcy. We fit the bankruptcy data with a mixture of generalized hyperbolic distribution.

```
library(MixGHD)
data(bankruptcy, package = "MixGHD")
bankruptcyY = bankruptcy[, "Y"]
bankruptcyX = bankruptcy[, c("RE", "EBIT")]
label = bankruptcyY + 1
a = round(runif(20) * 65 + 1)
label[a] = 0
model = MGHD(data = bankruptcyX, G = 2, label = label)
table(model@map, bankruptcyY + 1)
plot(model)
```

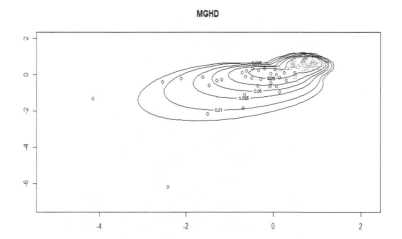

FIGURE 1.11
Contour of the bankruptcy data.

The table shows the classification results to be 100% correct, and the contour of the corresponding results is shown in Figure 1.11.

In Section 1.13, nonparametric maximum likelihood estimators are introduced to fit the unknown mixing distributions. Xiang et al. (2016) applied this NPMLE to the semiparametric location-shifted mixture models. We apply this method to the elbow diameter data as follows based on the R package MixSemiRob. The dataset contains the elbow diameters of 507 physically active people, and due to the gender difference, there are likely two clusters of observations, as shown in Figure 1.12.

```
library(MixSemiRob)
data(elbow, package = "MixSemiRob")
ini = mixnorm(elbow)
res = mixScale(elbow, ini)
```

In Section 1.14, Sarkar et al. (2020) derived a series of parsimonious mixture models for matrix Gaussian distributions and transformation mixtures. The MatTransMix package is available to apply these methods, and we now apply them to the crime data. The Y in crime data is a $10 \times 13 \times 236$ data frame, which contains 236 crime rates on the following 10 variables from 2000 to 2012: population of each city, violent crime rate (total number of violent crimes), murder and nonnegligent manslaughter rate (number of murders), forcible rape rate (number of rape crimes), robbery rate (number of robberies), aggravated assault rate (number of assaults), property crime rate total (number of property crimes), burglary rate (number of burglary crimes), larceny-theft

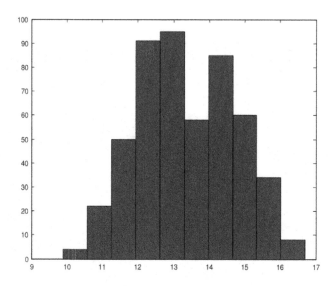

FIGURE 1.12
Histogram of the elbow diameter data.

rate (number of theft crimes), motor vehicle theft rate (number of vehicle theft crimes).

```
data(IMDb)
Y=IMDb$Y/100
p=dim(Y)[1]
T=dim(Y)[2]
n=dim(Y)[3]
K=3
init=MatTrans.init(Y, K = K, n.start = 2)
M=MatTrans.EM(Y, initial = init, model = "X-VVV-VV",
                long.iter = 1000, silent = FALSE)
print.EM(M)
```

Note that `model="X-XXX-XX"` specifies the model being fitted. To be more specific, they denote the **mean structure** (2 options: "G" for general; and "A" for additive), the **variance-covariance** Σ (14 options: "EII" for spherical equal volume; "VII" for spherical, unequal volume; "EEI" for diagonal, equal volume and shape; "VEI" for diagonal, varying volume, equal shape; "EVI" for diagonal, equal volume, varying shape; "VVI" for diagonal, varying volume and shape; "EEE" for ellipsoidal, equal volume, shape, and orientation; "EVE" for ellipsoidal, equal volume and orientation; "VEE" for ellipsoidal, equal shape and orientation; "VVE" for ellipsoidal, equal orientation; "EEV" for ellipsoidal, equal volume and equal shape; "VEV" for ellipsoidal, equal shape; "EVV" ellipsoidal, equal volume; and "VVV" for ellipsoidal, varying volume, shape, and orientation), and **variance-covariance** Ψ (8 options: "II," "EI," "VI," "EE," "VE," "EV," "VV," and "AR").

2

Mixture models for discrete data

In this chapter, we introduce finite mixture models for discrete data, including categorical data (nominal or ordinal), counts (univariate or multivariate), rankings, and so on. Both univariate and multivariate cases are considered.

In fact, the finite mixture models for discrete data are similar to those designed for continuous data, since many types of discrete data can be considered as discretized versions of some continuous latent data. Therefore, the idea of the EM algorithm can be readily applied to this chapter for model estimation. However, as the dimension of the data increases, it becomes much more difficult to model discrete multivariate data than continuous data. For example, consider the Poisson distribution, which is the most widely used distribution to model counting data, it is far more difficult to generalize from the univariate to even the bivariate case than to generalize the univariate Gaussian to the multivariate Gaussian. As a result, finite mixture models for discrete data are not as plentiful as for continuous data.

2.1   Mixture models for categorical data

Categorical data has a distinct structure from continuous data. For this reason, the distance functions used in the continuous case might not be applicable to the categorical one. Also, algorithms that are designed to cluster continuous data can not be applied directly to categorical data.

2.1.1   Mixture of ranking data

Ranking data arises in several contexts, especially when objective and precise measurements of the phenomena of interest are impossible or deemed unreliable, and the observer gathers ordinal information in terms of orderings, preferences, judgments, and relative or absolute rankings among competitors. Ranking data is quite commonly seen in social and behavioral sciences, political sciences, psychology, etc. For example, in social and behavioral sciences, observations are often required to rank a finite set of K items according to certain criteria, such as their personal preferences or attitudes. In psychological experiments and political surveys, people are asked to rank consumer goods, political candidates, or goals. In sports, teams, horses, and cars compete; the final outcome is a ranking among competitors.

In many examples, since the population of the ranking data is heterogeneous, it is natural to model the data with finite mixture models, rather than a simple model for ranking data. Applications of such mixture models in election voting and political studies can be found in Murphy and Martin (2003); Gormley and Murphy (2008); Busse et al. (2007); Lee and Philip (2012).

Let R be a central ranking, and λ be a precision parameter. Then, the probability density of a ranking r is

$$g(r; \lambda, R) = c(\lambda)e^{-\lambda d(r,R)},$$

where $c(\lambda)$ is a normalizing constant and $d(r, R)$ measures the distance between rankings R and r, such as the Kendall (d_K) and Spearman (d_S) distances listed below:

$$d_K(r, s) = \#\{(i, j), i < j, (r_i - r_j)(s_i - s_j) < 0\},$$

$$d_S(r, s) = \sqrt{\sum_{i=1}^{M}(r_i - s_i)^2},$$

where $r = (r_1, \ldots, r_M)$ and $s = (s_1, \ldots, s_M)$ are two rankings of M objects, and $\#A$ is the cardinality of the set A.

Then, a finite mixture model for ranking data is

$$f(r; \boldsymbol{\theta}) = \sum_{c=1}^{C} \pi_c g(r; \lambda_c, R_c) = \sum_c \pi_c c(\lambda_c)e^{-\lambda_c d(r,R_c)}, \qquad (2.1)$$

where $\boldsymbol{\theta} = (\pi_1, \ldots, \pi_{C-1}, R_1, \ldots, R_C, \lambda_1, \ldots, \lambda_C)^\top$. Similar to Gaussian mixture models, different constraints can be put on the precision parameter λ_c to achieve parsimonious models.

Similar to regular mixture models, an EM-type algorithm for model (2.1) is proposed by Murphy and Martin (2003) described in Algorithm 2.1.1.

Algorithm 2.1.1 EM algorithm for mixtures of rankings model

Starting with an initial value $\boldsymbol{\theta}^{(0)}$, in $(k+1)$th step, we compute the following two steps.

E-step: Find the conditional probabilities:

$$p_{ic}^{(k+1)} = \frac{\pi_c^{(k)} g(r_i; \lambda_c^{(k)}, R_c^{(k)})}{\sum_{c'=1}^{C} \pi_{c'}^{(k)} g(r_i; \lambda_{c'}^{(k)}, R_{c'}^{(k)})},$$

where $i = 1, \ldots, n, c = 1, \cdots, C$.

M-step: Update the parameter estimates as follows.

- Compute the values of R_c as

$$R_c^{(k+1)} = \arg\min_R \sum_{i=1}^{n} p_{ic}^{(k+1)} d(r_i, R).$$

- Update π_c as

$$\pi_c^{(k+1)} = \sum_{i=1}^{n} p_{ic}^{(k+1)} / n.$$

- Update λ_c as the solution of

$$\sum_r d(r, R_c^{(k+1)}) g(r; \lambda_c, R_c^{(k+1)}) = \frac{\sum_{i=1}^{n} p_{ic}^{(k+1)} d(r_i, R_c^{(k+1)})}{\sum_{i=1}^{n} p_{ic}^{(k+1)}},$$

where the summation in the left-hand side is taken over all possible rankings r. If $\lambda_c = \lambda$ for all c, then λ can be updated by the solution of

$$\sum_r d(r, R_c^{(k+1)}) g(r; \lambda, R_c^{(k+1)}) = \frac{\sum_{c=1}^{C} \sum_{i=1}^{n} p_{ic}^{(k+1)} d(r_i, R_c^{(k+1)})}{\sum_{c=1}^{C} \sum_{i=1}^{n} p_{ic}^{(k+1)}}.$$

In order to account for the order of the ranking elicitation process, Mollica and Tardella (2014, 2017) generalized the Plackett-Luce (PL) model to mixture modeling and illustrated its maximum likelihood estimation. PL's probabilistic expression moves from the decomposition of the ranking process in independent stages, one for each rank that has to be assigned, combined with the underlying assumption of a standard forward procedure on the ranking elicitation. In fact, a ranking can be elicited through a series of sequential comparisons in which a single item is preferred and, after being selected, is removed from the next comparisons.

Other mixture models for ranking data can be found in D'Elia and Piccolo (2005), where the ranking r is considered as the realization of a shifted binomial random variable, and in Jacques and Biernacki (2014), where an insertion sorting rank model suited for multivariate and partial rankings is proposed.

2.1.2   Mixture of multinomial distributions

The multinomial distribution describes the probability of obtaining a specific number of counts for d different outcomes when each outcome has a fixed probability of occurring. Applications of such a model can be found in Kamakura and Russell (1989), where a multinomial logit mixture regression model is applied to marketing research, in Jorgensen (2004) in an internet traffic clustering application, in Jorgensen (2013) in demography, etc.

The probability mass function (pmf) of a multinomial distribution $M(N, \mathbf{p})$ is

$$g(\mathbf{x}; N, \boldsymbol{\rho}) = \frac{N!}{x_1! \ldots, x_d!} \prod_{j=1}^{d} \rho_j^{x_j},$$

where $\sum_{j=1}^{d} \rho_j = 1$, $\boldsymbol{\rho} = (\rho_1, \ldots, \rho_d)^\top$, $\mathbf{x} = (x_1, \ldots, x_d)^\top$ and $N = \sum_{j=1}^{d} x_j$. Then, a finite mixture of multinomial distributions takes the form:

$$f(\mathbf{x}; \boldsymbol{\theta}) = \sum_{c=1}^{C} \pi_c g(\mathbf{x}; N, \boldsymbol{\rho}_c),$$

where $\boldsymbol{\theta} = (\pi_1, \ldots, \pi_{C-1}, \boldsymbol{\rho}_1, \ldots, \boldsymbol{\rho}_C)^\top$, and an EM-type algorithm as described in Algorithm 2.1.2 is easily constructed to fit the model.

Algorithm 2.1.2 EM algorithm for mixture of multinomial model

---

Starting with an initial value $\boldsymbol{\theta}^{(0)}$, in the $(k+1)$th step, we compute the following two steps.

E-step: Find the conditional probabilities:

$$p_{ic}^{(k+1)} = \frac{\pi_c^{(k)} \prod_{j=1}^{d} (\rho_{cj}^{(k)})^{x_{ij}}}{\sum_{c'=1}^{C} \pi_{c'}^{(k)} \prod_{j=1}^{d} (\rho_{c'j}^{(k)})^{x_{ij}}}, i = 1, \ldots, n, c = 1, \cdots, C.$$

M-step: Update parameters by

$$\pi_c^{(k+1)} = \sum_{i=1}^{n} p_{ic}^{(k+1)} / n,$$

$$\rho_{cj}^{(k+1)} = \frac{\sum_{i=1}^{n} p_{ic}^{(k+1)} \mathbf{x}_{ij}}{\sum_{i=1}^{n} \sum_{j'=1}^{d} p_{ic}^{(k+1)} \mathbf{x}_{ij'}}.$$

Meilă and Heckerman (2001) systematically studied several model-based clustering methods with the mixture of multinomial distributions and experimentally compared them using different criteria such as clustering accuracy, computation time, and numbers of selected clusters. Hasnat et al. (2015) proposed a model-based clustering method using a mixture of multinomial distributions, which performs clustering and model selection simultaneously.

2.2 Mixture models for counting data

Counting data, for example, the number of accidents reported or the number of clients waiting in front of an ATM, which describes the number of occurrences of some events in a unit of time, are commonly seen in marketing, sports, geophysics, epidemiology, and so on. Poisson distribution is surely the most commonly used distribution to model such counting data. However, since the mean and variance of the Poisson distribution are equal, it is not suited to model counting data that is over-dispersed. For example, in actuarial applications, observed data on the number of claims often exhibits variance that noticeably exceeds their mean. As a result, finite mixtures of Poisson distributions are usually considered.

2.2.1 Mixture models for univariate counting data

Following the previous notations, consider the following probability mass function:

$$f(x; \boldsymbol{\theta}) = \sum_{c=1}^{C} \pi_c g(x; \lambda_c),$$

where $x \in \mathbb{N}_0 = \{0, 1, 2, \ldots\}$. Now, if $g(x; \lambda_c)$ is a Poisson distribution, then the mixture distribution becomes

$$f(x; \boldsymbol{\theta}) = \sum_{c=1}^{C} \pi_c \frac{\exp(-\lambda_c)\lambda_c^x}{x!}, \tag{2.2}$$

where λ_c is the mean of the c-th component, such that $0 \leq \lambda_1 < \lambda_2 < \ldots < \lambda_c$ to make the mixture of Poisson distributions identifiable. Figure 2.1 shows the pmf of several mixtures of Poisson distributions. From the plot, it is clear that the mixture model is more suitable to model over-dispersed counting data. In order to make a bimodal or trimodal distribution, the mean values λ_cs need to be distant from each other. Through the EM algorithm, a closed-form M-step update can be obtained, which makes the model more user-friendly. Karlis and Xekalaki (2003) did a systematic review and simulation comparison of several methods for choosing initial values for the EM algorithm of the mixture of Poisson distributions.

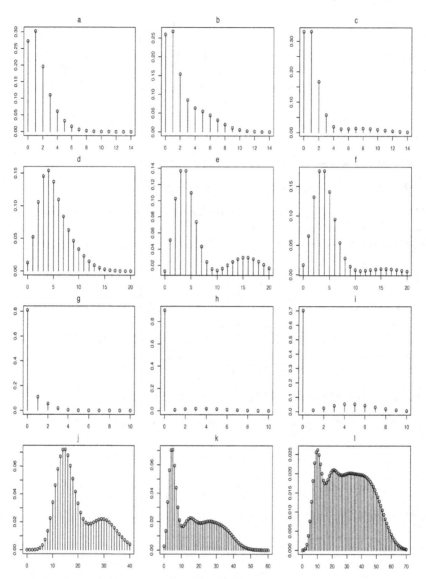

FIGURE 2.1
Finite mixture of Poisson distributions with different settings of the mixing proportions and mean parameters: (a) $\boldsymbol{\pi} = (0.7, 0.3), \boldsymbol{\lambda} = (1, 3)$, (b) $\boldsymbol{\pi} = (0.7, 0.3), \boldsymbol{\lambda} = (1, 5)$, (c) $\boldsymbol{\pi} = (0.7, 0.3), \boldsymbol{\lambda} = (1, 8)$, (d) $\boldsymbol{\pi} = (0.7, 0.3), \boldsymbol{\lambda} = (4, 8)$, (e) $\boldsymbol{\pi} = (0.7, 0.3), \boldsymbol{\lambda} = (4, 16)$, (f) $\boldsymbol{\pi} = (0.9, 0.1), \boldsymbol{\lambda} = (4, 16)$, (g) $\boldsymbol{\pi} = (0.7, 0.3), \boldsymbol{\lambda} = (0, 1)$, (h) $\boldsymbol{\pi} = (0.9, 0.1)^{\top}, \boldsymbol{\lambda} = (0, 4)$, (i) $\boldsymbol{\pi} = (0.7, 0.3), \boldsymbol{\lambda} = (0, 5)$, (j) $\boldsymbol{\pi} = (0.7, 0.3), \boldsymbol{\lambda} = (15, 30)$, (k) $\boldsymbol{\pi} = (0.4, 0.2, 0.2, 0.2), \boldsymbol{\lambda} = (5, 15, 25, 35)$, (l) $\boldsymbol{\pi} = (0.2, 0.2, 0.2, 0.2, 0.2), \boldsymbol{\lambda} = (10, 20, 30, 40, 50)$.

If a binomial distribution is used to replace the Poisson distribution in (2.2), then a finite mixture of binomial distributions is defined as

$$f(x; \boldsymbol{\theta}) = \sum_{c=1}^{C} \pi_c \binom{N}{x} \eta_c^x (1 - \eta_c)^{N-x}, \qquad (2.3)$$

where $x = 0, 1, \ldots, N$, η_c is the probability of success of the cth component which follows a binomial distribution $B(N, \eta_c)$. The finite mixture of binomial distributions (2.3) can be traced back to Pearson (1915), where the yeast cell count is modeled by a mixture of two binomial distributions. Compared to the mixture of Poisson distributions, model (2.3) has an obvious drawback, that is, x can only take on a finite number of values. The conditions required for model (2.3) to be identifiable have been studied by Teicher (1963).

2.2.2 Count data with excess zeros

In application, many datasets contain a large number of zeros, and cannot be modeled directly using a single distribution. Then, a natural way is to consider a two-component mixture model, where one component has a degenerate distribution at zero. This type of model is also known as a zero-inflated model.

Let $C = 2$ and $\lambda_1 = 0$, then model (2.2) becomes a zero-inflated Poisson distribution, which is commonly used to model counting data with excess zeros. The plots in the third row of Figure 2.1 give illustrations of what zero-inflated Poisson models might look like.

Since introduced by Lambert (1992), quite a few zero-inflated regression models on several other distributions have been proposed, such as zero-inflated negative binomial regression models and zero-inflated binomial regression models. When the data is collected over time, Cao and Yao (2012) proposed a semiparametric mixture of binomial regression models to incorporate inflated zeros in the data while also allowing the success probability and component proportions to vary over time t. More specifically, the model is stated as

$$f(x(t); \boldsymbol{\theta}) = \pi(t)\mathrm{Bin}(x(t); N, 0) + (1 - \pi(t))\mathrm{Bin}(x(t); N, p(t)).$$

There are other cases, in practice, where there might be excess ones in the data. For example, in population size estimation, many capture–recapture-type data exhibit a preponderance of "1"-counts. Then, a one-inflated positive Poisson model is proposed by Godwin and Böhning (2017), where one component of a two-component mixture model is a degenerate distribution that puts all its mass at exactly one.

2.2.3 Mixture models for multivariate counting data

To handle heterogeneous multivariate counting data, mixture models for multivariate counting data are proposed. In continuous cases, we might just need to generalize a univariate Gaussian distribution to a multivariate one. However, this generalization is quite difficult in the discrete world. In fact, even for the Poisson distribution, which is the most commonly used and well-studied discrete distribution, the multivariate generalization can not be easily done.

Consider the following bivariate Poisson distribution. For $\mathbf{x} = (x_1, x_2)^\top \in \mathbb{N} \times \mathbb{N}$, consider the following joint pmf

$$g(\mathbf{x}; \boldsymbol{\lambda}) = e^{-(\lambda_1 + \lambda_2 + \lambda_0)} \frac{\lambda_1^{x_1}}{x_1!} \frac{\lambda_2^{x_2}}{x_2!} \sum_{t=0}^{\min(x_1, x_2)} \binom{x_1}{t} \binom{x_2}{t} t! \left(\frac{\lambda_0}{\lambda_1 \lambda_2} \right)^t, \qquad (2.4)$$

where $\boldsymbol{\lambda} = (\lambda_0, \lambda_1, \lambda_2)^\top$ are nonnegative parameters. It is easy to see that λ_0 controls the covariance between X_1 and X_2, and $\lambda_0 + \lambda_1$ and $\lambda_0 + \lambda_2$ are the means (and variances) of the marginal distributions of X_1 and X_2, respectively. Based on this bivariate Poisson distribution, Karlis and Meligkotsidou (2007) and Chib and Winkelmann (2001) have extended it to finite mixtures and infinite mixtures. However, due to the summation sign in (2.4), this pmf is not suitable for large counts, hence the application of the aforementioned mixture of bivariate Poisson distributions is quite limited.

Karlis and Meligkotsidou (2007) studied a finite mixture of multivariate Poisson distributions for a general dimension. To start with, let $\mathbf{X} = (X_1, \ldots, X_m)^\top$ be a vector of discrete random variables, and define a mapping $u : \mathbb{N}^q \to \mathbb{N}^m, q \geq m$, such that $\mathbf{X} = u(\mathbf{Y}) = \mathbf{AY}$, where $\mathbf{Y} = (Y_1, \ldots, Y_q)^\top$, $Y_r \sim Poi(\theta_r)$ and Y_rs are independent, and $\mathbf{A}$ is an $m \times q$ binary matrix with no duplicate columns. Then, $\mathbf{X}$ is said to follow a multivariate Poisson distribution with parameter $\boldsymbol{\theta} = (\theta_1, \ldots, \theta_q)$, with mean and covariance:

$$\mathrm{E}(\mathbf{X}|\boldsymbol{\theta}) = \mathbf{A}\boldsymbol{\theta} \text{ and } \mathrm{Var}(\mathbf{X}|\boldsymbol{\theta}) = \mathbf{A}\Sigma\mathbf{A}^\top,$$

where $\Sigma = \mathrm{diag}(\theta_1, \ldots, \theta_q)$. Obviously, X_i follows a univariate Poisson distribution. Recalling the mapping u, if $\mathbf{x} \in \mathbb{N}^m$, let the set $u^{-1}(\mathbf{x})$ denote the inverse image of $\mathbf{x}$ under u. Then, the pmf of $\mathbf{X}$ is defined as

$$MP_m(\mathbf{x}|\boldsymbol{\theta}) = \sum_{\mathbf{y} \in u^{-1}(\mathbf{x})} P(\mathbf{Y} = \mathbf{y}|\boldsymbol{\theta}) = \sum_{\mathbf{y} \in u^{-1}(\mathbf{x})} \prod_{r=1}^{q} PO(y_r|\theta_r), \qquad (2.5)$$

where $PO(y|\theta) = e^{-\theta}\theta^y/y!$ is the pmf of the univariate Poisson distribution with parameter θ.

Based on this multivariate Poisson distribution defined in (2.5), it is natural for Karlis and Meligkotsidou (2007) to further propose a finite mixture of multivariate Poisson distributions, whose density is assumed by

$$f(\mathbf{x}; \boldsymbol{\theta}) = \sum_{c=1}^{C} \pi_c MP_m(\mathbf{x}; \boldsymbol{\theta}_c), \qquad (2.6)$$

where $MP_m(\mathbf{x}; \boldsymbol{\theta}_c)$ is defined in (2.5). Then, the mean and variance-covariance matrix of (2.6) is

$$E(\mathbf{X}) = \sum_{c=1}^{C} \pi_c \mathbf{A} \boldsymbol{\theta}_c,$$

and

$$\text{Cov}(\mathbf{X}) = \mathbf{A} \left[\sum_{c=1}^{C} \pi_c (\boldsymbol{\Sigma}_c + \boldsymbol{\theta}_c \boldsymbol{\theta}_c^{\top}) - \left(\sum_{c=1}^{C} \pi_c \boldsymbol{\theta}_c \right) \left(\sum_{c=1}^{C} \pi_c \boldsymbol{\theta}_c \right)^{\top} \right] \mathbf{A}^{\top},$$

where $\boldsymbol{\Sigma}_c = \text{diag}(\theta_{1c}, \dots, \theta_{qc})$. As expected, parameter estimators can be deducted in closed forms through an EM algorithm (Karlis and Meligkotsidou, 2007), but the computation is very expensive as the number of dimensions increases, due to the increased number of parameters and the summation in the joint pmf.

When it comes to mixtures of multivariate Poisson distributions, another class of methods is based on the conditional independence assumption. Given that the component label is $Z_i = c$, the elements in $\mathbf{x}_i$ are assumed to be conditionally independent and the j-th variable of $\mathbf{x}_i$ is distributed as $X_{ij}|Z_i = c \sim \text{Poi}(\lambda_{jc})$. Then, the mixture distribution of the i-th observation is

$$f(\mathbf{x}_i; \boldsymbol{\theta}) = \sum_{c=1}^{C} \pi_c \prod_{j=1}^{p} \frac{e^{-\lambda_{jc}} \lambda_{jc}^{x_{ij}}}{x_{ij}!}.$$

Due to the simple and clear distribution of Poisson, a closed-form expression of the parameter estimators based on an EM algorithm can be readily derived. In a similar manner, in order to model the success of students in two separate tests, Hoben et al. (1993) proposed a mixture of bivariate binomial distributions assuming conditional independence. However, the assumption of conditional independence is quite hard to satisfy in many real applications. Hence, although simple, the above model is not popularly used in real life.

As a compromise, in order to model the number of purchases of some products, Brijs et al. (2004) considered a mixture model, where conditional on component labels, correlations are considered within blocks while independence is assumed between blocks. It is a compromise between the simple assumption of independence and the computation complexity caused by having too many parameters due to assuming a full structure. However, the difficult task at hand is determining the blocks of the variables because it is generally difficult for researchers to determine whether or not a correlation exists between variables.

2.3   Hidden Markov models

The hidden Markov model (HMM) has a close connection with mixture models and provides a convenient way of formulating an extension of a mixture model to allow for dependent data arising from heterogeneous populations, and thus has become increasingly popular in applications for sequences or spatial data. One very successful practical use of the HMM has been in speech processing applications, including speech recognition and its noise robustness, speaker recognition, speech synthesis, and speech enhancement. In these applications, the HMM is used as a powerful model to characterize the temporally non-stationary, spatially variable, but regular, learnable patterns of the speech signal.

To start with, in a classic mixture model, the component-indicator vectors $(\mathbf{Z}_i, i = 1, \ldots, n)$ are vectors of either zero or one, based on whether $\mathbf{x}_i$ belongs to the c-th component or not. In other words, $\mathbf{Z}_1, \ldots, \mathbf{Z}_n$ are i.i.d. variables from a single multinomial trial with C categories and probabilities $(\pi_1, \ldots, \pi_C)$. If the component labels $(\mathbf{Z}_1, \ldots, \mathbf{Z}_n)$ are missing, we only observe $(\mathbf{x}_1, \ldots, \mathbf{x}_n)$ from the marginal mixture density

$$f(\mathbf{x}_i; \boldsymbol{\theta}) = \sum_{c=1}^{C} p(\mathbf{x}_i \mid Z_{ic} = 1)P(Z_{ic} = 1) \tag{2.7}$$

$$= \sum_{c=1}^{C} \pi_c g_c(\mathbf{x}_i; \lambda_c),$$

where $\mathbf{Z}_i = (Z_{i1}, \ldots, Z_{iC})^{\top}$, and $g_c(\mathbf{x}; \lambda_c)$ is the cth component density with unknown parameter λ_c. Note that, in this case, the observations $\mathbf{x}_i$s are assumed to be independent of each other.

However, the independence assumption is difficult to satisfy in some cases, such as the time-series data. As a result, the Hidden Markov model relaxes this independence assumption on $\mathbf{x}_i$ by taking successive observations to be correlated through their components of origin. Instead of a multinomial distribution, the component-indicator vectors $(\mathbf{Z}_i, i = 1, \ldots, n)$ are modeled by a stationary Markovian model with a transition probability matrix $\mathbf{A} = (\pi_{jc})_{j,c=1,\ldots,C}$. That is,

$$P(Z_{i+1,c} = 1 | Z_{ij} = 1, \mathbf{Z}_{i-1}, \ldots, \mathbf{Z}_0) = P(Z_{i+1,c} = 1 | Z_{ij} = 1) = \pi_{jc},$$

for $i = 1, \ldots, n-1, c, j = 1, \ldots, C$. The distribution of the observation $\mathbf{x}$ still depends on the value of $\mathbf{Z}$ through (2.7), but the variables $(\mathbf{x}_1, \ldots, \mathbf{x}_n)$ are correlated through their component indicators $(\mathbf{Z}_i, i = 1, \ldots, n)$.

2.3.1 The EM algorithm

The parameter estimation in hidden Markov models is mostly done by maximum likelihood through an EM algorithm, known as the Baum-Welch algorithm.

Let $P(z_{1c} = 1) = \pi_{0c}$, $c = 1, \ldots, C$. Then, the distribution of $\mathbf{Z}$ can be written as

$$p(\mathbf{z}; \boldsymbol{\theta}) = p(\mathbf{z}_1; \boldsymbol{\theta}) \prod_{i=1}^{n-1} p(\mathbf{z}_{i+1}|\mathbf{z}_i; \boldsymbol{\theta}),$$

where $p(\mathbf{z}_1; \boldsymbol{\theta}) = \prod_{c=1}^{C} \pi_{0c}^{z_{1c}}$, and

$$p(\mathbf{z}_{i+1}|\mathbf{z}_i; \boldsymbol{\theta}) = \prod_{j=1}^{C} \prod_{c=1}^{C} \pi_{jc}^{z_{i,j} z_{i+1,c}}.$$

Then the joint distribution of $\mathbf{x}$, i.e., the likelihood function, is

$$f(\mathbf{x}; \boldsymbol{\theta}) = \sum_{\mathbf{z}} p(\mathbf{z}; \boldsymbol{\theta}) p(\mathbf{x}|\mathbf{z}, \boldsymbol{\theta}) = \sum_{\mathbf{z}} p(\mathbf{z}; \boldsymbol{\theta}) \prod_{i=1}^{n} \prod_{c=1}^{C} g_c(\mathbf{x}_i; \lambda_c)^{z_{ic}}.$$

Note that the above likelihood contains C^n summands due to different possibilities of $\mathbf{z}$ and is hence analytically very difficult to handle. Next, we introduce an EM algorithm to maximize the above likelihood function.

Let $\mathbf{y}_i = (\mathbf{x}_i^\top, \mathbf{z}_i^\top)^\top$. The complete log-likelihood of the complete data $(\mathbf{y}_i, i = 1, \ldots, n)$ is

$$\log L_c(\boldsymbol{\theta}; \mathbf{Y}) = \log p(\mathbf{z}; \boldsymbol{\theta}) + \sum_{c=1}^{C} \sum_{i=1}^{n} z_{ic} \log g_c(\mathbf{x}_i; \lambda_c)$$

$$= \sum_{c=1}^{C} z_{1c} \log \pi_{0c} + \sum_{c=1}^{C} \sum_{j=1}^{C} \sum_{i=1}^{n-1} z_{i,j} z_{i+1,c} \log \pi_{jc} + \sum_{c=1}^{C} \sum_{i=1}^{n} z_{ic} \log g_c(\mathbf{x}_i; \lambda_c).$$

The EM algorithm is described in Algorithm 2.3.1.

Algorithm 2.3.1 EM algorithm for HMM

Start with initial parameter values $\boldsymbol{\theta}^{(0)}$, at the $(k+1)$-th step,

E-step: Compute the conditional expectation:

$$Q(\boldsymbol{\theta}|\boldsymbol{\theta}^{(k)}) = \sum_{c=1}^{C} \tau_{1c}^{(k)} \log \pi_{0c} + \sum_{j=1}^{C} \sum_{c=1}^{C} \sum_{i=1}^{n-1} \tau_{ijc}^{(k)} \log \pi_{jc}$$

$$+ \sum_{c=1}^{C} \sum_{i=1}^{n} \tau_{ic}^{(k)} \log g_c(\mathbf{x}_i; \lambda_c),$$

where $\tau_{ijc}^{(k)}$ and $\tau_{ic}^{(k)}$ are defined as

$$\tau_{ijc} = P(Z_{ij} = 1, Z_{i+1,c} = 1|\mathbf{x}), \quad i = 1, \ldots, n-1, \quad (2.8)$$

and

$$\tau_{ic} = P(Z_{ic} = 1|\mathbf{x}), \quad i = 1, \ldots, n.$$

From (2.8), it can be shown that

$$\tau_{ic} = \sum_{j=1}^{C} \tau_{i-1,jc}, \quad i = 2, \ldots, n,$$

and

$$\tau_{1c} = \pi_{0c} g_c(\mathbf{x}_1) / \sum_{j=1}^{C} \pi_{0j} g_j(\mathbf{x}_1).$$

M-step: Determine the new estimate $\boldsymbol{\theta}^{(k+1)}$ by

$$\pi_{0c}^{(k+1)} = \tau_{1c}^{(k)},$$

$$\pi_{jc}^{(k+1)} = \sum_{i=1}^{n-1} \tau_{ijc}^{(k)} / \sum_{i=1}^{n-1} \tau_{ij}^{(k)},$$

and

$$g_c^{(k+1)}(\mathbf{x}_i) = \sum_{t=1}^{n-1} \tau_{tc}^{(k)} \delta(\mathbf{x}_t - \mathbf{x}_i) / \sum_{t=1}^{n-1} \tau_{tc}^{(k)}, \quad (2.9)$$

where $\delta(\mathbf{x} - \mathbf{y})$ is one if $\mathbf{x} = \mathbf{y}$, and zero otherwise.

Note that if the MLE of the component densities $g_c(\mathbf{x}; \lambda_c)$ is available in closed form, then the M-step (2.9) can be implemented in closed form.

2.3.2   The forward-backward algorithm

The forward-backward algorithm is designed to calculate the conditional probabilities τ_{ijc} and τ_{ic}. To be more specific, τ_{ijc} can be expressed as

$$\tau_{ijc} = \frac{P(Z_{ij} = 1, Z_{i+1,c} = 1, \mathbf{x})}{P(\mathbf{x})} = \frac{a_{ij}\pi_{jc}g_c(\mathbf{x}_{i+1})b_{i+1,c}}{\sum_{j=1}^{C}\sum_{c=1}^{C} a_{ij}\pi_{jc}g_c(\mathbf{x}_{i+1})b_{i+1,c}},$$

where

$$a_{ic} = P(\mathbf{X}_1 = \mathbf{x}_1, \ldots, \mathbf{X}_i = \mathbf{x}_i, Z_{ic} = 1), \quad i = 1, \ldots, n,$$

and

$$b_{ic} = P(\mathbf{X}_{i+1} = \mathbf{x}_{i+1}, \ldots, \mathbf{X}_n = \mathbf{x}_n | Z_{ic} = 1),$$
$$i = n - 1, n - 2, \ldots, 1, c = 1, \ldots, C.$$

The forward probability a_{ic} is calculated by a forward recursion. At the $(k + 1)$-th iteration:

Initialization:

$$a_{1c}^{(k)} = \pi_{0c}^{(k)} g_c^{(k)}(\mathbf{x}_1), \qquad c = 1, \ldots, C.$$

Induction:

$$a_{i+1,c}^{(k)} = \left(\sum_{j=1}^{C} a_{ij}^{(k)}\pi_{jc}^{(k)}\right) g_c^{(k)}(\mathbf{x}_{i+1}), \qquad i = 1, \ldots, n - 1.$$

Termination:

$$P_{\boldsymbol{\theta}^{(k)}}(\mathbf{X}_1 = \mathbf{x}_1, \ldots, \mathbf{X}_n = \mathbf{x}_n) = \sum_{c=1}^{C} a_{nc}^{(k)}.$$

On the other hand, the value of b_{ic} is updated through backward recursion. At the $(k + 1)$-th iteration:

Initialization:
$$b_{nj}^{(k)} = 1, \qquad j = 1, \ldots, C.$$

Induction:

$$b_{ij}^{(k)} = \sum_{c=1}^{C} \pi_{jc}^{(k)} g_c^{(k)}(\mathbf{x}_{i+1})b_{i+1,c}^{(k)}, \qquad i = n - 1, \ldots, 1; j = 1, \ldots, C.$$

Then, the final calculation in the E step is

$$\tau_{ijc}^{(k)} = \frac{a_{ij}^{(k)}\pi_{jc}^{(k)} g_c^{(k)}(\mathbf{x}_{i+1})b_{i+1,c}^{(k)}}{\sum_{j=1}^{C}\sum_{c=1}^{C} a_{ij}^{(k)}\pi_{jc}^{(k)} g_c^{(k)}(\mathbf{x}_{i+1})b_{i+1,c}^{(k)}}.$$

In fact, the HMM is an important extension of the finite mixture model to the time series data. Assuming the component indicator variable constitutes a Markov chain, one can easily construct HMMs for discrete data by selecting the state-specific distributions to be any discrete distribution. For example, Leroux and Puterman (1992) proposed an HMM based on the Poisson distribution to model the counts of fetal movement. Spezia et al. (2014) modeled mussel counts in a river in Scotland using an HMM based on the negative binomial distribution.

2.4 Latent class models

In this section, we introduce latent class models (LCMs) (Lazarsfeld, 1950), a statistical method used for identifying unobserved or latent subgroups or classes within a population based on their patterns of responses to a set of observed variables or indicators. In these models, each individual is assumed to belong to one and only one latent class and the goal is to estimate the parameters of the model, such as the proportion of individuals in each class and the probabilities of each class responding in a certain way to each of the observed variables. Latent class models can be utilized in various fields such as psychology, sociology, marketing, epidemiology, and public health to identify unobserved subgroups or classes within a population. Their primary purpose is to uncover patterns of behavior, preferences, or risks that may not be immediately apparent through observable variables (also called manifest variables). For example, in the behavioral and social sciences, LCM can be used to analyze data related to disorders like antisocial personality disorder and eating disorders, to determine whether there are subtypes with distinct sets of symptoms.

We introduce various types of Latent Class Models (LCMs) and their applications in statistical modeling. The traditional LCM assumes that each observation belongs to a single latent class, while the LCM regression model includes covariates for predicting class membership. The random effect LCM is designed to manage locally dependent data by introducing random effects for each cluster. The multilevel LCM is a variation of LCM that examines multilevel data, such as nested data, whereas the Latent transition analysis (LTA) models transitions between latent classes over time.

2.4.1 Introduction

Latent class models (LCM), proposed by Lazarsfeld (Lazarsfeld, 1950), are widely used in biostatistics, mixture samples, and other fields to extract meaningful latent groups from data. Latent class variables cannot be directly observed but affect the behavior of other observed variables. Failure to identify

TABLE 2.1

Example: 2 × 2 contingency table

		Variable A	
		0	1
Variable B	0	1750	3500
	1	2250	2000

TABLE 2.2

Example: three-way table

		Variable C						
		0			1			
		Variable A					Variable A	
		0	1				0	1
Variable B	0	1000	500	Variable B		0	750	3000
	1	2000	1000			1	250	1000

TABLE 2.3

Latent variable models

		Lat. Var.	
		Continuous	Categorical
Obs. Var.	Continuous	Factor Analysis	Latent Profile Analysis
	Categorical	Latent Trait Analysis	Latent Class Analysis

latent class variables can lead to a misunderstanding of the relationship among manifest variables. For example, consider the contingency table in Table 2.1, where we want to examine the independence between variables A and B. If we perform a chi-square test and obtain a chi-square value of 36.24, we will reject the null hypothesis that A and B are independent at a 95% level. However, if there exists a latent class variable C that can split the sample into two subgroups, as shown in Table 2.2, we can see that in every subgroup, variables A and B are independent. Thus, the extraction of latent variables can reduce the dependency between manifest variables.

Latent variable models can be categorized based on the types of observed variables and latent variables, as shown in Table 2.3. The table presents four categories of models based on whether the observed variables are continuous or categorical and whether the latent variables are continuous or categorical. The four categories are factor analysis, latent trait analysis, latent profile analysis, and latent class analysis (LCA). This section focuses on the LCA, which is a type of latent variable model where both observed and latent variables are categorical.

The concept of latent class was first introduced by Lazarsfeld in his work on latent structure analysis (Lazarsfeld, 1950). Goodman provided a formal mathematical model for both unrestricted and restricted latent class models in his pioneering work (Goodman, 1974). He also proposed an iterative method to estimate the model parameters using maximum likelihood methods. Goodman's work on solving the LCM problem coincided with the Expectation-Maximization (EM) algorithm proposed by Dempster et al. (1977a), which has since become a mainstay for estimating parameters in LCMs. Bayesian estimation is another popular method for estimating parameters in LCMs, and has been discussed in articles such as Elliott et al. (2005) and Asparouhov and Muthén (2010). A number of articles on the review and applications of LCMs have also been published. For example, Hagenaars (1990) provided a general framework for categorical longitudinal data analysis, while Vermunt and Magidson (2004) gave a detailed review of traditional LCMs. Please also refer to Magidson et al. (2020) and Nylund-Gibson and Choi (2018) for further review papers on LCM.

The latent class model was built on the locally independent assumption, which assumes that the conditional probabilities of manifest variables given the latent class are mutually independent. However, this assumption may not always hold true in practice. To address this issue, Qu et al. (1996) proposed a more advanced version of the LCM by incorporating random effects. This model assumes that the manifest variables are determined by both the latent class and a random variable, relaxing the locally independent assumption and allowing for more flexible modeling of the relationship between manifest and latent variables.

In the presence of covariates, the locally independent assumption may be violated, as covariates are also manifest variables and assumed to be ancestors of other manifest variables. For example, if we want to examine whether there is an independent relationship between a group of people's working hours and their salary, the latent variable might be their working efficiency. However, other covariates, such as education level and age, might also affect the independence or behavior of the working efficiency latent variable. To address this, an advanced version of the traditional LCM is to incorporate covariates into the model. Furthermore, we can also perform LC regression, which involves regressing the proportion of latent class on covariates (Agresti, 2003).

Another type of LCM we will discuss is multilevel LCA (Vermunt, 2003; Asparouhov and Muthén, 2008; Henry and Muthén, 2010). This type of model is designed to handle cases where samples have manifest subgroups. For example, in the working time and salary case, if we group individuals by their companies or year of entry, the proportion of different working efficiencies may vary across these groups. In this scenario, we can label the group variable with another latent class, and use a 2-level latent class model. If the group variable represents time, and we incorporate a transition probability matrix, the problem becomes Latent Transition Analysis (Collins and Wugalter, 1992; Humphreys and Janson, 2000; Lanza et al., 2005). Finally, we also introduce the LC Tree model (van den Bergh et al., 2017), which is useful for analyzing sparse samples.

TABLE 2.4

Topics in LCM

Purpose	Model	Package
Sample with covariates	LC regression	poLCA
Not satisfied with local independence	random effect LCA	randomLCA
Sample comes from multiple groups	multilevel LCA	glca
For longitudinal data	LTA, LMM	LMMest

Table 2.4 provides a summary of different models in LCA, their purposes, and the corresponding R packages. In the following, we will provide more detailed introductions of these models.

2.4.2 Latent class models

2.4.2.1 Model

We begin with the traditional latent class model (LCM), introduced by Goodman (1974). Let $\boldsymbol{Y} = (Y_1, \ldots, Y_r)^\top$ denote the outcome vector, containing r categorical variables with c categories. We use capital letters for variables and small letters for individual realizations. The latent variables are denoted by $U_1, \ldots, U_l$, each of which is categorical with k categories. In most LC models, we focus on the case when $l = 1$ for the simplicity of explanation, but the ideas can be easily generalized to the cases when $l > 1$. Then, the probability density function of the manifest variable $\boldsymbol{Y}$ is given by

$$f_{\boldsymbol{Y}}(\boldsymbol{y}) = \sum_u f_{\boldsymbol{Y}|U}(\boldsymbol{y} \mid u) f_U(u),$$

where $f_{\boldsymbol{Y}|U}(\boldsymbol{y} \mid u)$ is the conditional probability distribution of $\boldsymbol{Y}$ given $U = u$, and $f_U(u)$ is the prior distribution of the latent variable U.

As mentioned in the introduction, the LCM typically assumes local independence, which means that given the latent class, the category variables Y_js are independent of each other. Therefore, we have

$$f_{\boldsymbol{Y}|U}(\boldsymbol{y} \mid u) = \prod_{j=1}^r f_{Y_j|U}(y_j \mid u), \tag{2.10}$$

where $\boldsymbol{y} = (y_1, \ldots, y_r)^\top$, and $f_{Y_j|U}(y_j \mid u)$ is the conditional density of Y_j given U. Note that the conditional independence assumption used by (2.10) greatly reduces the dimension of density function from r to 1 and hence avoids the "curse of dimensionality." The traditional LC model is shown in Figure 2.2.

In the traditional latent class model, the parameters to be estimated are $f_U(u)$ and $f_{\boldsymbol{Y}|U}(\boldsymbol{y} \mid u)$. For convenience, we define $P_{jy|u} = f_{Y_j|U}(y \mid u)$, where $j = 1, \ldots, r$, and $y = 0, \ldots, c-1$, and $\pi_u = f_U(u)$, where $u = 1, \ldots, k$. These

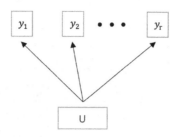

FIGURE 2.2
Traditional LCM diagram.

parameters represent probabilities, therefore satisfying certain constraints: $\sum_{u=1}^{k} \pi_u = 1$, and $\sum_{y=0}^{c-1} P_{jy|u} = 1$ for all $j = 1, \ldots, r$ and $u = 1, \ldots, k$. The degree of freedom in this model is $k - 1 + kr(c - 1) = k - 1 + kr$.

Given observations $(\mathbf{y}_1, \ldots, \mathbf{y}_n)$, the model log-likelihood is equal to

$$\ell(\boldsymbol{\theta}) = \sum_{i=1}^{n} \log f_{\mathbf{Y}}(\mathbf{y}_i) = \sum_{\mathbf{y}} n_{\mathbf{y}} \log f_{\mathbf{Y}}(\mathbf{y}), \tag{2.11}$$

where $n_{\mathbf{y}}$ is the number of occurrences of the configuration $\mathbf{y}$ in the sample, that is,

$$n_{\mathbf{y}} = \sum_{i=1}^{n} I(\mathbf{y}_i = \mathbf{y}).$$

2.4.2.2   EM algorithm

Instead of directly maximizing the $\ell(\boldsymbol{\theta})$ in Equation (2.11), which is challenging due to the existence of latent variables, the EM Algorithm focuses on the complete log-likelihood function:

$$\ell_c(\boldsymbol{\theta}) = \sum_{i=1}^{n} \log f_{U,\mathbf{Y}}(u_i, \mathbf{y}_i) = \sum_{i=1}^{n} \log f_{\mathbf{Y}|U}(\mathbf{y}_i \mid u_i) + \sum_{i=1}^{n} \log f_U(u_i),$$

which can be rewritten as

$$\ell_c(\boldsymbol{\theta}) = \sum_{u} \sum_{\mathbf{y}} a_{u\mathbf{y}} \log f_{\mathbf{Y}|U}(\mathbf{y} \mid u) + \sum_{u} b_u \log f_U(u),$$

where

$$a_{u\mathbf{y}} = \sum_{i=1}^{n} I(u_i = u, \mathbf{y}_i = \mathbf{y}) \text{ and } b_u = \sum_{i=1}^{n} I(u_i = u).$$

To split the conditional probability, we introduce a new counting number,

$$a_{ju\mathbf{y}} = \sum_{i=1}^{n} I(u_i = u, y_{ij} = y).$$

By conditional independence, we get

$$\ell_c(\boldsymbol{\theta}) = \sum_{j=1}^{r}\sum_{u=1}^{k}\sum_{y=0}^{c-1} a_{juy}\log P_{jy|u} + \sum_{u=1}^{k} b_u \log \pi_u. \tag{2.12}$$

Because of the constraints, Equation (2.12) is equivalent to

$$\ell_c(\boldsymbol{\theta}) = \sum_{u=1}^{k}\sum_{y=0}^{c-1}\left(\sum_{j=1}^{r-1} a_{juy}\log P_{jy|u} + a_{ruy}\log\left(1 - \sum_{j=1}^{r-1} P_{jy|u}\right)\right)$$
$$+ \sum_{u=1}^{k-1} b_u \log \pi_u + \left(n - \sum_{u=1}^{k-1} b_u\right)\log\left(1 - \sum_{u=1}^{k-1} \pi_u\right). \tag{2.13}$$

The EM algorithm iterates between the E-step and the M-step until convergence. In the E-step, we calculate the expectation of the complete log-likelihood $\ell_c(\boldsymbol{\theta})$ conditional on the observed data and current parameter updates, which can be simplified to the computation of conditional expectations of (a_{juy}, b_u). The formula is as follows:

$$\hat{a}_{juy} = \sum_{i=1}^{n} I\left(y_{ij} = y\right)\hat{f}_{U|\boldsymbol{Y}}\left(u \mid \boldsymbol{y}_i\right),$$
$$\hat{b}_u = \sum_{i=1}^{n} \hat{f}_{U|\boldsymbol{Y}}\left(u \mid \boldsymbol{y}_i\right). \tag{2.14}$$

Here we use the probability of the latent class conditional on observations. We express the posterior probability using the parameters in this model:

$$f_{U|\boldsymbol{Y}}(u \mid \boldsymbol{y}) = \frac{\left(\prod_{j=1}^{r} P_{jy|u}\right)\pi_u}{P_{\boldsymbol{Y}}(\boldsymbol{y})} = \frac{\left(\prod_{j=1}^{r} P_{jy|u}\right)\pi_u}{\sum_u \left(\prod_{j=1}^{r} P_{jy|u}\right)\pi_u}.$$

In the M-step, we maximize the expected value of $\ell_c(\boldsymbol{\theta})$ obtained by substituting a_{juy} with $\hat{a}_{juy}$ and b_u with $\hat{b}_u$. The first-order conditions of Equation (2.13) is

$$\frac{\partial \ell_c(\boldsymbol{\theta})}{\partial \pi_u} = \frac{b_u}{\pi_u} - \frac{n - \sum_{l=1}^{k-1} b_l}{1 - \sum_{l=1}^{k-1} \pi_l} = 0,$$

$$\frac{\partial \ell_c(\boldsymbol{\theta})}{\partial P_{jy|u}} = \frac{a_{juy}}{P_{jy|u}} - \frac{a_{ruy}\left(1 - \sum_{l=1}^{r-1} P_{jy|u}\right)}{1 - \sum_{l=1}^{r-1} P_{jy|u}} = 0.$$

Then, the solutions of the above equations are

$$\hat{\pi}_u = \frac{\hat{b}_u}{n}, \quad u = 1, \ldots, k,$$

$$\hat{P}_{jy|u} = \frac{\hat{a}_{juy}}{\hat{b}_u}, \quad j = 1, \ldots, r, u = 1, \ldots, k, y = 0, \ldots, c-1.$$

For how to set the initial value, please see Bartolucci et al. (2012).

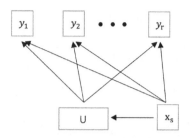

FIGURE 2.3
LCM with covariates.

2.4.3   Latent class with covariates

Sometimes we also have another type of manifest observation, which does not depend on the latent class but can still be observed. We call these observations covariates, denoted by X, where each variable has S possible outcomes. For instance, suppose we want to investigate the relationship between people's reactions to different diagnostic tests and their health status. It is reasonable to suspect that health status is a latent class variable: perhaps healthy people have a random response to different diagnostic tests, while unhealthy people will react positively to all diagnostic tests. However, we may also observe the heights and weights of the people, which could affect their health status and their reactions to the diagnostic tests. The relationship between these variables can be illustrated in Figure 2.3. This type of model is called LC regression model, which is proposed by Formann (1988); the concept of covariates in LC model first appears in Dayton and Macready (1988).

2.4.3.1   Model

To incorporate covariates into the latent class models, we assume that the conditional probability distribution function of Y given X is

$$f_{Y|X}(y \mid x) = \sum_u f_{Y|U,X}(y \mid u, x) f_{U|X}(u \mid x).$$

The model log-likelihood has the expression:

$$\ell(\theta) = \sum_{i=1}^{n} \log f_{Y|X}(y_i \mid x_i) = \sum_x \sum_y n_{xy} f_{Y|X}(y \mid x),$$

where $n_{xy} = \sum_{i=1}^{n} I(x_i = x, y_i = y)$.

Sometimes, we also impose additional assumptions, among which the two useful ones are:

Assumption 1: $f_{Y|U,X}(y \mid u, x) = f_{Y|U}(y \mid u)$.

Assumption 2: $f_{U|X}(u \mid x) = f_U(u)$.

The first assumption suggests that the covariates are conditionally independent of the response given the latent variable. On the other hand, the second assumption indicates that the covariates are independent of latent classes.

2.4.3.2 Parameter estimation

We can still use the EM algorithm to estimate the parameters. The complete likelihood function is

$$\ell_c(\boldsymbol{\theta}) = \sum_u \sum_x \sum_y a_{uxy} \log f_{Y|U,X}(y \mid u, x) + \sum_u \sum_x b_{ux} \log f_{U|X}(u \mid x),$$

where

$$a_{uxy} = \sum_{i=1}^{n} I(u_i = u, x_i = x, y_i = y), \quad b_{ux} = \sum_{i=1}^{n} I(u_i = u, x_i = x).$$

We first assume the covariate $\boldsymbol{X}$ is one-dimensional with S possible outcomes. Define the parameters as

$$P_{jy|ux} = f_{Y_j|U,X}(y \mid u, x) \text{ and } \pi_{u|x} = f_{U|X}(u \mid x),$$

where $j = 1, \ldots, r, u = 1, \ldots, k, y = 0, \ldots, c-1, x = 1, \ldots, S$. Then the log-likelihood function can be expressed as

$$\ell_c(\boldsymbol{\theta}) = \sum_{j=1}^{r} \sum_{u=1}^{k} \sum_{s=1}^{S} \sum_{y=0}^{c-1} a_{juxy} \log P_{jy|ux} + \sum_{u=1}^{k} \sum_{x=1}^{S} b_{ux} \log \pi_{u|x}, \qquad (2.15)$$

where

$$a_{juxy} = \sum_{i=1}^{n} I(u_i = u, y_{ij} = y, x_i = x).$$

The degrees of freedom of this model is $kSr(c-1) + S(k-1)$.

The EM algorithm contains the E-step and the M-step. In the E-step, we compute

$$\hat{a}_{juxy} = \sum_{i=1}^{n} I(y_{ij} = y, x_i = x) \hat{f}_{U|Y,X}(u \mid y_i, x),$$

$$\hat{b}_{ux} = \sum_{i=1}^{n} I(x_i = x) \hat{f}_{U|X}(u \mid x),$$

where

$$\begin{aligned} f_{U|X,Y}(u \mid x, y) &= \frac{f_{Y|U,X}(y \mid u, x) f_{u|x}(u \mid x)}{f_{Y|X}(y \mid x)} \\ &= \frac{f_{Y|U,X}(y \mid u, x) f_{u|x}(u \mid x)}{\sum_{l=1}^{k} f_{Y|U,X}(y \mid l, x) f_{u|x}(l \mid x)} \\ &= \frac{\prod_{j=1}^{r} P_{jy|ux} \pi_{u|x}}{\sum_{l=1}^{k} \left(\prod_{j=1}^{r} P_{jy|ux} \pi_{l|x} \right)}. \end{aligned}$$

In the M-step, maximizing (2.15) provides the following results:

$$\hat{\pi}_{u|x} = \frac{\hat{b}_{ux}}{\sum_{i=1}^{N} I\left(x_i = x\right)},$$

$$\hat{P}_{jy|ux} = \frac{\hat{a}_{juxy}}{\hat{b}_{ux}}.$$

2.4.4   Latent class regression

In the previous section, we presented how to estimate the entire model simultaneously. The method to estimate all parameters together is also called *One-step method*. The One-step method is a full information method, representing unbiased estimations of the relationships among all variables. However, this approach can become computationally intensive as the number of covariates or categories increases, since the number of parameters grows quickly. To address this issue, Bolck et al. (2004) proposed the *three-step method*, which has fast computation and does not need to specify a priori the complete measurement and structural model. However, the estimated coefficients are biased. Vermunt (2010) proposed two improved Three-step methods that have relaxed assumptions, but are still biased.

Another alternative approach is to regress π_U on X(Hagenaars and McCutcheon, 2002). One way to do this is to first model the relationship between Y and U (either using random assignment or modal class assignment), then assign each observation y_i to a latent class u, followed by building a regression model between u and X. However, this approach may introduce bias because neither method can surely assign the true unobserved class. Another method is to use the logit link function in the regression model on π_U. Formann (1992) introduced the linear logistic LCM, while Agresti (2003) further developed an LC regression model using the logit function. This model has been implemented in the R package poLCA (Linzer and Lewis, 2011).

2.4.4.1   Model

First, we consider a special case to illustrate the idea of the LC regression model before presenting the general LC regression model. We assume that for every category latent class u we have one covariate $\mathbf{v} = \left(v_1, \ldots, v_p\right)^{\top}$, and every category outcome in different latent class u, we have a covariate $\mathbf{q} = (q_1, \ldots, q_t)^{\top}$.

Under this model, we need to rewrite the response variable using the approach described above. Recall that we have r polytomous variables y_j, each with c categories denoted as $K_{j0}, \ldots, K_{jh}, \ldots, K_{jc-1}$. The response of subject i to variable y_j is coded by a selection vector $\mathbf{A}_{ij} = (A_{ij0}, \ldots, A_{ijc-1})$, where $A_{ijh} = 1$ if person i responds to variable y_j in category K_{jh}, and $A_{ijh} = 0$ otherwise. Let $P_{jh|u} = f_{Y_j|U}(h \mid u) = P(A_{ijh} = 1 \mid u)$ be the probability of observing category K_{jh} for variable y_j in latent class u. Assuming local independence, the probability of the response vector $\mathbf{A}_i = (\mathbf{A}_{i1}, \ldots, \mathbf{A}_{ir})$ is given by

$$f(\mathbf{A}_i) = \sum_{u=1}^{k} \pi_u f(\mathbf{A}_i \mid u) = \sum_{u=1}^{k} \pi_u \prod_{j=1}^{r} \prod_{h=0}^{c-1} P_{jh|u}^{A_{ijh}},$$

where π_u is the proportion of subjects in latent class u.

We introduce the logistic representations:

$$\pi_u = \exp(\alpha_u) / \left\{ \sum_{l=1}^{k} \exp(\alpha_l) \right\},$$

$$P_{jh|u} = \exp(\omega_{ujh}) / \left\{ \sum_{l=0}^{c-1} \exp(\omega_{ujl}) \right\},$$

where we restrict π_u and $P_{jh|u}$ to the admissible interval $(0,1)$, and α, ω are unrestricted $\in R$. We denote the fixed effect of $\mathbf{v}$ by $\boldsymbol{\eta} = (\eta_1, \ldots, \eta_p)^\top$ and that of $\boldsymbol{q}$ by $\boldsymbol{\lambda} = (\lambda_1, \ldots, \lambda_t)^\top$. Following the tradition of linear logistic models, we assume

$$\alpha_u = \mathbf{v}_u^\top \boldsymbol{\eta} \text{ and } \omega_{ujh} = \mathbf{q}_{ujh}^\top \boldsymbol{\lambda}.$$

The log-likelihood is given by

$$\ell(\boldsymbol{\theta}) = \sum_{d=1}^{D} n_d \log f_d(\boldsymbol{\eta}, \boldsymbol{\lambda}),$$

where n_d represents the observed frequency of the response pattern $\mathbf{A}_d = (A_{d10}, \ldots, A_{drc-1})$, with $D = r \times c$ being the total number of response patterns. The function $f_d(\boldsymbol{\eta}, \boldsymbol{\lambda})$ is the unconditional probability of the response pattern, expressed as a function of the unknown parameters $\boldsymbol{\eta}$ and $\boldsymbol{\lambda}$, and defined as follows:

$$f_d(\boldsymbol{\eta}, \boldsymbol{\lambda}) = \sum_{u=1}^{k} \frac{\exp\left(\mathbf{v}_u^\top \boldsymbol{\eta}\right)}{\sum_{l=1}^{k} \exp\left(\mathbf{v}_u^\top \boldsymbol{\eta}\right)} \prod_{j=1}^{r} \prod_{h=0}^{c-1} \left(\frac{\exp\left(\mathbf{q}_{ujh}^\top \boldsymbol{\lambda}\right)}{\sum_{l=0}^{c-1} \exp\left(\mathbf{q}_{ujl}^\top \boldsymbol{\lambda}\right)} \right)^{A_{ijh}}.$$

2.4.4.2 Parameter estimation

With respect to LCA, the observed data are the frequencies n_d of the response patterns $\mathbf{A}_d = (A_{d10}, \ldots, A_{dr(c-1)})$, and the unobserved data are the corresponding frequencies n_{du} per class u. The log-likelihood of the complete data, $\ell_c(\boldsymbol{\theta})$, is just the same as Equation (2.12):

$$\ell_c(\boldsymbol{\theta}) = \log \prod_{u=1}^{k} \pi_u^{b_u} \prod_{j=1}^{r} \prod_{y=0}^{c-1} P_{jy|u}^{a_{juh}} = \sum_{j=1}^{r} \sum_{u=1}^{k} \sum_{y=0}^{c-1} a_{juy} \log P_{jy|u} + \sum_{u=1}^{k} b_u \log \pi_u.$$

$$(2.16)$$

For the EM algorithm, in the E-step, each response pattern $\mathbf{A}_d = \left(A_{d10}, \ldots, A_{dr(c-1)}\right)$ is mapped to a corresponding pattern of $\mathbf{Y}$ as described in Equation (2.14) in Section 2.4.2. The main difference in this step from the previous section is that the parameters being considered are now $\boldsymbol{\eta}$ and $\boldsymbol{\lambda}$, which will be estimated in the M-step. In the M-step, to maximize the expected complete log-likelihood function, we take the partial derivative of Equation (2.16) with respect to each variable in $\boldsymbol{\eta}$ and $\boldsymbol{\lambda}$. Specifically, we first compute $\partial \pi_u / \partial \eta_g$ using the expression:

$$\frac{\partial \pi_u}{\partial \eta_g} = \frac{v_{ug} \exp\left(\mathbf{v}_u^\top \boldsymbol{\eta}\right)\left(\sum_{l=1}^k \exp\left(\mathbf{v}_l^\top \boldsymbol{\eta}\right)\right) - \exp\left(\mathbf{v}_u^\top \boldsymbol{\eta}\right)\left(\sum_{l=1}^k v_{lg} \exp\left(\mathbf{v}_l^\top \boldsymbol{\eta}\right)\right)}{\left(\sum_{l=1}^k \exp\left(\mathbf{v}_l^\top \boldsymbol{\eta}\right)\right)^2}$$

$$= v_{ug}\pi_u - \pi_u \frac{\sum_{l=1}^k v_{lg} \exp\left(\mathbf{v}_l^\top \boldsymbol{\eta}\right)}{\sum_{l=1}^k \exp\left(\mathbf{v}_l^\top \boldsymbol{\eta}\right)} = v_{lg}\pi_u - \pi_u \left(\sum_{l=1}^k v_{lg}\pi_l\right),$$

and similar for $\partial P_{jy|u} / \partial \lambda_z$. Then the partial derivatives are

$$\frac{\partial \ell_c(\boldsymbol{\theta})}{\partial \eta_g} = \sum_{u=1}^k b_u \frac{1}{\pi_u}\frac{\partial \pi_u}{\partial \eta_g} = \sum_{u=1}^k b_u \left(v_{ug} - \sum_{l=1}^k v_{lg}\pi_l\right), \quad g = 1, \ldots, p$$

and

$$\frac{\partial \ell_c(\boldsymbol{\theta})}{\partial \lambda_z} = \sum_{u=1}^k \sum_{j=1}^r \sum_{h=0}^{c-1} a_{juy}\left(q_{ujhz} - \sum_{l=0}^{c-1} P_{jl|u}q_{ujlz}\right), \quad z = 1, \ldots, t.$$

Because the above derivatives do not have explicit solutions, they have to be solved iteratively. The matrix of second-order partial derivatives of $\ell_c(\boldsymbol{\theta})$ has elements:

$$\frac{\partial^2 \ell_c(\boldsymbol{\theta})}{\partial \eta_g \partial \lambda_z} = 0,$$

$$\frac{\partial^2 \ell_c(\boldsymbol{\theta})}{\partial \eta_g \partial \eta_\gamma} = -N\left\{\sum_u \pi_u v_{ug} v_{u\gamma} - \left(\sum_u \pi_u v_{ug}\right)\left(\sum_u \pi_u v_{u\gamma}\right)\right\},$$

$$\frac{\partial^2 \ell_c(\boldsymbol{\theta})}{\partial \lambda_z \partial \lambda_\delta} = -\sum_u b_u \sum_j \left\{\sum_h q_{ujhz}q_{ujh\delta}P_{jy|h} - \left(\sum_l q_{ujlz}P_{jy|l}\right)\left(\sum_l q_{ujl\delta}P_{jy|l}\right)\right\},$$

where $g, \gamma = 1, \ldots, u$, $z, \delta = 1, \ldots, t$. We can then solve the problem iteratively, for example, by the Newton-Raphson procedure.

2.4.4.3 General LC regression model

We assume that there are S covariates, and therefore the covariate vector $\mathbf{x}$ has a length of S. With S covariates, the coefficients β_u have a length of $S+1$, where $\beta_1 = 0$ is fixed by definition. Recall that $\pi_{u|x} = f_{U|X}(u \mid x)$. Based on the assumptions:

$$\log\left(\frac{\pi_{u|x}}{\pi_{1|x}}\right) = x^\top \beta_u, \ u = 2, \ldots, k,$$

we can obtain the following general result:

$$\pi_{u|i} = \frac{e^{x_i^T \beta_u}}{\sum_{l=1}^{k} e^{x_i^T \beta_l}}, \ u = 2, \ldots, k.$$

Given estimates $\hat{\beta}_u$ and $\hat{P}_{jy|u}$, we can obtain the posterior class membership probabilities in the latent class regression model using

$$\hat{f}(u \mid x_i; y_i) = \frac{\hat{\pi}_{u|x_i} \hat{f}_{Y|U}(y_i \mid u)}{\sum_{l=1}^{k} \hat{\pi}_{l|x_i} \hat{f}_{Y|U}(y_i \mid u_l)}.$$

The latent class regression model estimates $kr(c-1) + (S+1)(k-1)$ parameters. This number does not depend on the number of patterns for the covariates, therefore as S increases, the number of parameters grows linearly, rather than exponentially as in the method described in Section 2.4.3.

By the assumption of local independence, and recalling the selection vector $\mathbf{A}_{ij} = (A_{ij0}, \ldots, A_{ijc-1})$, we can express the log-likelihood function as

$$\ell(\boldsymbol{\theta}) = \sum_{i=1}^{n} \log \sum_{u=1}^{k} \pi_{u|x_i} \prod_{j=1}^{r} \prod_{y=0}^{c-1} \left(P_{jy|u}\right)^{A_{ijy}}.$$

The coefficients are updated by the Newton-Raphson algorithm:

$$\hat{\beta}^{\text{new}} = \hat{\beta}^{\text{old}} - \mathbf{H}_\beta^{-1} \boldsymbol{\nabla}_\beta.$$

Here $\boldsymbol{\nabla}_\beta$ represents the gradient and $\mathbf{H}_\beta$ is the Hessian matrix of the log-likelihood function with respect to β. For more details about the Newton-Raphson process, please refer to Linzer and Lewis (2011), and Bandeen-Roche et al. (1997).

2.4.5 Latent class model with random effect

In some cases, the assumption of local independence might not be realistic in practice. For example, when studying the dependence between problematic behaviors exhibited by teenagers, such as lying to parents, fighting with others, or stealing from a store, the latent class may be based on the age of the teenager, as a 16-year-old may be more aggressive than a 10-year-old (Li et al., 2018). However, even after conditioning on age, problematic behaviors may still be correlated. For instance, teenagers who have fought with others may also be more likely to lie to their parents. To address this issue, various models have been proposed to relax the conditional independence assumption. Harper (1972) proposed a local dependence model that includes direct effects for different clusters, while Qu et al. (1996) proposed a latent class model with random effects to model the conditional dependence among multiple diagnostic tests. Beath and Heller (2009) further extended this model to include random effects for multilevel samples, which are particularly useful for longitudinal data where the LCM may be correlated but not identical at every time point. Their model includes random effects on both time points and latent class. Recently, Beath (2017) developed an R package, `randomLCA`, for random effect LC models.

2.4.5.1   Model

The model presented here assumes binary outcomes ($C = 2$) for the manifest variables. The model can be written as

$$f_{\boldsymbol{Y}|U}(\boldsymbol{y} \mid u) = \prod_{j=1}^{r} P_{jy_j|u} = \prod_{j=1}^{r} P_{j1|u}^{y_j}(1 - P_{j1|u})^{1-y_j},$$

where $\boldsymbol{y} = (y_1, \ldots, y_r)^{\top}$ and $P_{jy_j|u} = P(Y_j = y_j \mid u)$.

To model the weights of the latent classes and the conditional probabilities, the logistic transformation is used. Specifically, the weight of each latent class u is modeled as

$$\pi_u = \exp\left(\alpha_u\right) / \left\{ \sum_{l=1}^{k} \exp\left(\alpha_l\right) \right\},$$

and the conditional probability of manifest variable Y_j given latent class u is modeled as

$$f_{Y_j|u}(1 \mid u) = P_{j1|u} = \exp\left(\omega_{uj}\right) / \left\{ 1 + \exp\left(\omega_{uj}\right) \right\}.$$

The posterior probability can be written as

$$f_{U|\boldsymbol{Y}}(u \mid \boldsymbol{y}) = \frac{\left(\prod_{j=1}^{r} P_{jy_j|u}\right) \pi_u}{f_{\boldsymbol{Y}}(\boldsymbol{y})} = \frac{\pi_u \prod_{j=1}^{r} P_{j1|u}^{y_j} \left(1 - P_{j1|u}\right)^{1-y_j}}{\sum_{l=1}^{k} \pi_l \prod_{j=1}^{r} P_{j1|u}^{y_j} \left(1 - P_{j1|u}\right)^{1-y_j}}.$$

Incorporating a random effect λ in the LCM allows for modeling the conditional dependence among manifest variables within the same latent class, even after conditioning on the latent class. The assumption of local independence means that the manifest variables y_j are independent of each other, conditional on both the latent class and the random effect λ_i. Now, the conditional probability can be written as

$$f_{\boldsymbol{Y}|U,\lambda}(\boldsymbol{y}_i \mid u, \lambda_i) = \prod_{j=1}^{r} \left\{ f_{\boldsymbol{Y}|U,\lambda}(1 \mid u, \lambda_i) \right\}^{y_{ij}} \left\{ 1 - f_{\boldsymbol{Y}|U,\lambda}(1 \mid u, \lambda_i) \right\}^{1-y_{ij}}.$$

For the probit scaling, the conditional probability can be written as

$$f_{Y|U,\lambda}(y_j = 1 \mid u, \lambda_i) = \Phi\left(a_{uj} + b_{uj}\lambda\right).$$

For the logistic scaling, the conditional probability can be written as

$$f_{Y|U,\lambda}(y_j = 1 \mid u, \lambda_i) = \frac{\exp\left(a_{uj} + b_{uj}\lambda\right)}{1 + \exp\left(a_{uj} + b_{uj}\lambda\right)}. \tag{2.17}$$

The loadings can be constrained to be the same for each outcome ($b_{uj} = b$) or independent for each outcome ($b_{uj} = b_j$) case by case.

Consider the probit scaling case and assume λ has distribution $\Phi(\cdot)$. We have

$$P_{j1|u} = f_{Y|U}(y_j = 1 \mid u) = \int_{-\infty}^{+\infty} \Phi\left(a_{uj} + b_{uj}\lambda\right) d\Phi(\lambda) = \Phi\left(\frac{a_{uj}}{\sqrt{1 + b_{uj}^2}}\right),$$

and

$$f_{Y|U}(y_i \mid u) = \int_{-\infty}^{+\infty} \prod_{j=1}^{r} \Phi\left(a_{uj} + b_{uj}\lambda\right)^{y_{ij}} \left(1 - \Phi\left(a_{uj} + b_{uj}\lambda\right)\right)^{1-y_{ij}} d\Phi(\lambda). \tag{2.18}$$

In practice, to compute the integration in Equation (2.18), we use Gauss-Hermite quadrature. This involves splitting the integration into a summation over a set of mass points $(\lambda_1, \ldots, \lambda_M)$, where each point is assigned a weight w_m. Using these mass points, we can express the conditional probability for a subject as a sum

$$f_{Y|U,\lambda}(y_i \mid u, \lambda) = \sum_{m=1}^{M} w_m \prod_{j=1}^{r} \Phi\left(a_{uj} + b_{uj}\lambda_m\right)^{y_{ij}} \left(1 - \Phi\left(a_{uj} + b_{uj}\lambda_m\right)\right)^{1-y_{ij}}.$$

Using this summation, we can compute the joint posterior probability for the latent variable and the random effect variable given the observations:

$$f_{U,\lambda|Y}(u, \lambda_m \mid y) = \frac{\pi_u w_m f_{Y|U,\lambda}(y \mid u, \lambda_m)}{\sum_{l=1}^{k} \sum_{j=1}^{M} \pi_l w_j f_{Y|U,\lambda}(y \mid l, \lambda_j)}.$$

2.4.5.2 Parameter Estimation

Let $\beta = (a_{11}, \ldots, a_{k,r}, b_{1,1}, \ldots, b_{k,r})^{\top}$. The parameter vector to be estimated can be expressed as $\theta = \left(\beta^{\top}, \pi_1, \ldots, \pi_k\right)^{\top}$. The complete likelihood function is

$$\ell^*(\theta) = \sum_u \sum_\lambda \sum_y o_{u\lambda y} \log f_{Y,U,\lambda}(y, u, \lambda)$$

$$= \sum_u \sum_\lambda v_{u\lambda} \log \pi_u w_m + \sum_u \sum_\lambda \sum_y o_{u\lambda y} \log f_{Y|U,\lambda}(y \mid u, \lambda),$$

where $o_{u\lambda y} = \sum_{i=1}^{n} I\left(u_i = u, \lambda_i = \lambda, y_i = y\right)$ and $v_{u\lambda} = \sum_{i=1}^{n} I\left(u_i = u, \lambda_i = \lambda\right)$.

We can use the EM algorithm for estimating parameters. The expectations of $v_{u\lambda}$ and $o_{u\lambda y}$ are

$$E[v_{u\lambda}] = \hat{v}_{u\lambda} = \sum_{i=1}^{n} \hat{f}_{U,\lambda|Y}(u,\lambda \mid \boldsymbol{y}_i),$$

$$E[o_{u\lambda y}] = \hat{o}_{u\lambda y} = \sum_{i=1}^{n} I(\boldsymbol{y}_i = \boldsymbol{y}) \hat{f}_{U,\lambda|Y}(u,\lambda \mid \boldsymbol{y}_i).$$

The M-step involves taking the derivative of the expected complete log-likelihood function and solving for the parameters. First denote $\mu_{jum} = \phi(a_{uj} + b_{uj}\lambda_m)$ and $\mu_{um} = (\mu_{1um}, \ldots, \mu_{rum})$. We obtain the estimate for the mixing proportion as

$$\hat{\pi}_u = \frac{\sum_{m=1}^{M} \hat{v}_{u\lambda_m}}{n}.$$

Next, for estimating $f_{\boldsymbol{Y}|U,\lambda}(\boldsymbol{y} \mid u, \lambda)$, it satisfies the following equation derived by Qu et al. (1996):

$$\sum_{i=1}^{n} \sum_{u=1}^{k} \sum_{m=1}^{M} \hat{f}_{\boldsymbol{Y}|U,\lambda}(\boldsymbol{y} \mid u, \lambda) \left(\frac{\partial \mu_{um}}{\partial \boldsymbol{\beta}}\right)^{\top} V_{um}^{-1} (\boldsymbol{y}_i - \mu_{um}) = 0,$$

where V_{um} is the covariance matrix of $\boldsymbol{y}_i$ conditional on u and λ, and it is assumed to be a diagonal matrix under the local independence assumption. The diagonal elements of V_{um} are equal to $\mu_{jum}(1 - \mu_{jum})$. This equation can be solved iteratively (McCullagh and Nelder, 2019).

2.4.6 Bayes latent class analysis

One of the challenges in latent class models is computing standard errors when parameter estimates are close to the boundary of the parameter space. To address this issue, researchers have explored various methods, including Bayesian maximum a posteriori (MAP) estimation (Galindo Garre and Vermunt, 2006) and the use of different Bayesian prior distributions, such as the multinormal distribution, to relax the assumption of local independence (Asparouhov and Muthén, 2011). Next, we introduce the basic Bayes LCM, which employs Beta and Dirichlet distributions.

2.4.6.1 Model

We still assume the manifest variables are binary. Recall

$$f_{U,Y}(u,y) = \sum_{l=1}^{k} \pi_l \prod_{j=1}^{r} P_{j1|u}^{y_j} \left(1 - P_{j1|u}\right)^{1-y_j}.$$

Now, we transform the indicator of latent class, u_{il}, as an indicator vector c, where $c_{i,l}$ takes on the value of 1 if person i belongs to latent class l and 0 otherwise. Using this notation, we have

$$f\left(y_i, u_{il} \mid \pi_u, P_{j1|l}\right) = \prod_{l=1}^{k} \left\{\pi_l f(y_i|P_{j1|l})\right\}^{c_{il}}. \tag{2.19}$$

An important idea in Bayesian methods is to consider the parameters as random variables. The joint posterior distribution of the parameters π_u and $P_{j1|u}$ given y_i and c_{iu} is given by

$$f(\pi_u, P_{j1|u} \mid y_i, c_{iu}) = \frac{f(y_i, c_{iu} \mid \pi_u, P_{j1|u})f(\pi_u)f(P_{j1|u})}{f(y_i, c_{iu})}.$$

The denominator $f(y_i, c_{iu})$ is a constant, as its value is not affected by π_u nor $P_{j1|u}$. The numerator consists of three terms: the likelihood function and two priors, $f(\pi_u)$ (prior for the latent class distribution) and $f(P_{j1|u})$ (prior for the response probabilities). Thus, the joint posterior distribution is proportional to the numerator, which is denoted by the symbol $\propto$:

$$f(\pi_u, P_{j1|u} \mid y_i, c_{iu}) \propto f(y_i, c_{iu} \mid \pi_u, P_{j1|u})f(\pi_u)f(P_{j1|u}). \tag{2.20}$$

The following equations describe the conjugate priors for the proportion parameters $P_{j1|u}$ and the categorical variable π_u. Since $P_{j1|u}$ are proportions derived from Bernoulli trials, their conjugate priors follow a Beta distribution with two parameters α and β. On the other hand, since π_u is a categorical variable, its conjugate prior follows the Dirichlet distribution with parameters ν_u. The equations for the conjugate priors are presented below:

$$f\left(\pi_u\right) \sim \text{Dirichlet}\left(\pi_u; \nu_u\right) = \frac{\Gamma\left(\nu_1 + \cdots + \nu_u\right)}{\Gamma\left(\nu_1\right)\cdots\Gamma\left(\nu_u\right)}\pi_1^{\nu_1 - 1}\cdots\pi_u^{\nu_u - 1} \propto \prod_{u=1}^{k}\pi_u^{\nu_u - 1},$$

$$P_{j1|u} \sim \text{Beta}\left(P_{j1|u}; \alpha, \beta\right) = \frac{\Gamma(\alpha + \beta)}{\Gamma(\alpha)\Gamma(\beta)}P_{j1|u}^{\alpha - 1}\left(1 - P_{j1|u}\right)^{\beta - 1}$$

$$\propto P_{j1|u}^{\alpha - 1}\left(1 - P_{j1|u}\right)^{\beta - 1}.$$

Incorporating the conjugate priors for $P_{j1|u}$ and π_u into Equation (2.20) results in a specified joint posterior distribution:

$$f(\pi_u, P_{j1|u} \mid y_i, c_{iu})$$
$$\propto f(y_i, c_{iu} \mid \pi_u, P_{j1|u})f(\pi_u)f(P_{j1|u})$$
$$= \underbrace{\prod_{i=1}^{N}\prod_{l=1}^{k}\left[\pi_l f\left(y_i \mid P_{j1|l}\right)\right]^{c_{il}}}_{\text{Equation (2.19)}} \underbrace{\prod_{l=1}^{k}\pi_l^{\nu_l - 1}}_{\text{Dirichlet prior}} \underbrace{\prod_{j=1}^{r}P_{j1|u}^{\alpha_{uj} - 1}\left(1 - P_{j1|u}\right)^{\beta_{uj} - 1}}_{\text{Beta priors}}$$

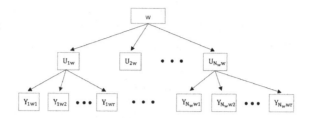

FIGURE 2.4
Multi-level LCM.

$$= \prod_{i=1}^{N} \underbrace{\prod_{u=1}^{k} \pi_u^{c_{iu}+\nu_u-1}}_{\text{Dirichelet posterior}} \underbrace{\prod_{j=1}^{r} P_{j1|u}^{y_{ij}c_{iu}+\alpha_{uj}-1} \left(1-P_{j1|u}\right)^{(1-y_{ij})c_{iu}+\beta_{uj}-1}}_{\text{Beta posteriors}}.$$

2.4.6.2 Gibbs sampling

The essence of Gibbs sampling is to estimate properties of the joint posterior distribution by simulation. We provide an algorithm for Gibbs sampling adapted from White and Murphy (2014):

$$\pi_u^{(t+1)} \sim \text{Dirichlet}\left(\pi_u^{(t)}; c_{iu}^{(t)} + \nu_u^{(t)}\right)$$

$$= \text{Dirichlet}\left(\sum_{i=1}^{n} c_{i1}^{(t)} + \nu_1^{(t)}, \dots, \sum_{i=1}^{n} c_{ik}^{(t)} + \nu_k^{(t)}\right),$$

$$P_{j1|u}^{(t+1)} \sim \text{Beta}\left(P_{j1|u}; y_i c_{iu}^{(t)} + \alpha_{uj} - 1, (1-y_i) c_{iu}^{(t)} + \beta_{uj} - 1\right)$$

$$= \text{Beta}\left(\sum_{i=1}^{n} y_i c_{iu}^{(t)} + \alpha_{uj} - 1, \sum_{i=1}^{n} (1-y_i) c_{iu}^{(t)} + \beta_{uj} - 1\right),$$

$$c_i^{(t+1)} \sim \text{Multinomial}$$

$$\times \left(1, \frac{\pi_1^{(t+1)} f_{Y|u}(\boldsymbol{y} \mid 1)}{\sum_{l=1}^{k} \pi_l^{(t+1)} f_{Y|u}(\boldsymbol{y} \mid 1)}, \dots, \frac{\pi_k^{(t+1)} f_{Y|u}(\boldsymbol{y} \mid k)}{\sum_{l=1}^{k} \pi_l^{(t+1)} f_{Y|u}(\boldsymbol{y} \mid k)}\right).$$

The R package BayesLCA provides several methods for estimating parameters, including the EM algorithm, empirical bootstrap, and Gibbs sampling. For more details about this package, see White and Murphy (2014).

2.4.7 Multi-level latent class model

Consider a more complex case where the sample has a multilevel structure, as illustrated in Figure 2.4.

The sample is composed of different groups, where the group variable w is observable. For instance, in the case of problematic behaviors among teenagers, it is reasonable to suspect that teenagers whose parents divorced might exhibit different behaviors compared to those whose parents did not divorce. In this case, the parents' marital status can be the group variable, and different forms of the latent class model (LCM) may be required for each group.

In the multi-level LCM (MLCM), we assign a level-2 latent variable Z_w to each group w, which ranges from 1 to T. Under every group w, we assign a level-1 latent variable U_{iw}, ranging from 1 to k, to each manifest subject y_{iw}. The number of manifest subjects in group w is denoted by N_w. This multilevel LCM was first proposed by Vermunt (2003). Henry and Muthén (2010) applied MLCM to adolescent smoking typologies using both parametric and nonparametric models, as well as MLCM with random effects. Finch and French (2014) conducted a simulation study under MLCM using a Monte Carlo method.

The conditional probability of the manifest variable Y_{iw}, given the level-2 latent variable $Z_w = t$, is given by

$$f\left(Y_{iw} \mid Z_w = t\right) = \sum_{u=1}^{k} f\left(U_{iw} = u \mid Z_w = t\right) \times \prod_{j=1}^{r} f\left(Y_{iwj} \mid U_{iw} = u\right),$$

where $Y_{iw} = (Y_{iw1}, Y_{iw2}, \ldots, Y_{iwr})^{\top}$. The joint probability distribution for the entire group w is

$$f\left(Y_w\right) = \sum_{t=1}^{T} f\left(Z_w = t\right) \prod_{i=1}^{N_w} f\left(Y_{iw} \mid Z_w = t\right),$$

where $Y_w = (Y_{1w}, \ldots, Y_{N_w w})^{\top}$. Expanding the joint probability distribution yields

$$f\left(Y_w\right) = \sum_{t=1}^{T} \left[f(Z_w = t) \prod_{i=1}^{N_w} \left\{ \sum_{l=1}^{k} f\left(U_{iw} = u \mid Z_w = t\right) \times \prod_{j=1}^{r} f\left(Y_{iwj} \mid U_{iw} = u\right) \right\} \right].$$

The model parameters can be estimated by the EM algorithm (Vermunt, 2003).

2.4.8   Latent transition analysis

Latent transition analysis (LTA) has been studied since the 1980s and is designed for longitudinal data (Collins and Wugalter, 1992; Velicer et al., 1996). LTA can be considered as another version of Multi-level latent class model (MLCM), where the group variable in the multi-level LCM is treated as a time variable. The basic idea of LTA is to introduce a transition matrix between different models to model the transition of latent variables into different stages. Detailed reviews of LTA and its applications were provided by Lanza et al. (2013) and Collins and Lanza (2009). Recent developments in LTA include the proposal of LTA with random intercepts by Muthén and Asparouhov (2022), which allows for variation across subjects by random intercepts.

One special case of Latent Transition Analysis is the Latent Markov Model, where the formulation is time-homogeneous and also has Markov properties. This model is widely used in LTA, and we will primarily focus on this model.

2.4.8.1   Latent Markov model

Let $\boldsymbol{U} = \left(U^{(1)}, \ldots, U^{(T)}\right)$ denote the latent process, which is assumed to follow a first-order Markov chain with state space $1, \ldots, k$. For $t = 2, \ldots, T$, the latent variable $U^{(t)}$ is conditionally independent of $U^{(1)}, \ldots, U^{(t-2)}$ given $U^{(t-1)}$. The conditional response probabilities are given by

$$\boldsymbol{P}_{\boldsymbol{y}|u} = f_{\boldsymbol{Y}^{(t)}|U^{(t)}}(\boldsymbol{y} \mid u), \ \ t=1,\ldots,T, u=1,\ldots,k, y=0,\ldots,c-1,$$

$$P_{jy|u} = f_{Y_j^{(t)}|U^{(t)}}(y \mid u), \ \ t=1,\ldots,T, u=1,\ldots,k, j=1,\ldots,r, y=0,\ldots,c-1,$$

where we assume that every time the conditional probability stays the same. The initial probabilities follow

$$\pi_u = f_{U^{(1)}}(u), \quad u = 1, \ldots, k,$$

and the transition probabilities follow

$$\pi_{u|\bar{u}}^{(t)} = f_{U^{(t)}|U^{(t-1)}}(u \mid \bar{u}), \quad t = 2, \ldots, T, \bar{u}, u = 1, \ldots, k.$$

Assuming that the transition probabilities are constant over time, the model can be considered a hidden Markov model (HMM) (Rabiner, 1989). The number of free parameters in this case is given by

$$\#\mathrm{par} = \underbrace{k \sum_{j=1}^{r}(c_j - 1)}_{\boldsymbol{P}_{y|u}} + \underbrace{k - 1}_{\pi_u} + \underbrace{(T - 1)k(k - 1)}_{\pi_{u|\bar{u}}^{(t)}}.$$

In the case of binary response variables, this expression simplifies to

$$\#\,\mathrm{par} = (r + 1)k + (T - 1)k(k - 1) - 1.$$

Furthermore, regarding the conditional distribution of $\tilde{\boldsymbol{Y}}$ given $\boldsymbol{U}$, we have

$$f_{\tilde{\boldsymbol{Y}}|\boldsymbol{U}}(\tilde{\boldsymbol{y}} \mid \boldsymbol{u}) = \prod_{t=1}^{T} \boldsymbol{P}_{\boldsymbol{y}^{(t)}|u^{(t)}} = \prod_{i=1}^{T} \prod_{j=1}^{r} P_{y_j^{(t)}|u^{(t)}}.$$

For the manifest distribution of $\tilde{\boldsymbol{Y}} = \left(\boldsymbol{Y}^{(1)}, \ldots, \boldsymbol{Y}^{(T)}\right)$, we have

$$f_{\tilde{\boldsymbol{Y}}}(\tilde{\boldsymbol{y}}) = \sum_{\boldsymbol{u}} \pi_{u^{(1)}} \pi_{u^{(2)}|u^{(1)}}^{(2)} \cdots \pi_{u^{(T)}|u^{(T-1)}}^{(T)} \boldsymbol{P}_{\boldsymbol{y}^{(1)}|u^{(1)}} \cdots \boldsymbol{P}_{\boldsymbol{y}^{(T)}|u^{(T)}}.$$

To estimate the parameters using the EM algorithm, we use a forward-backward recursion (Baum et al., 1970). The log-likelihood of the LMM can be expressed as

$$\ell(\boldsymbol{\theta}) = \sum_{i=1}^{n} \log f_{\tilde{\boldsymbol{Y}}}\left(\tilde{\boldsymbol{y}}_i\right) = \sum_{\tilde{\boldsymbol{y}}} n_{\tilde{\boldsymbol{y}}} \log f_{\tilde{\boldsymbol{y}}}(\tilde{\boldsymbol{y}}).$$

We have the following expression for the complete data log-likelihood:

$$\ell_c(\boldsymbol{\theta}) = \sum_{t=1}^{T} \sum_{u=1}^{k} \sum_{j=1}^{r} \sum_{y=0}^{c-1} a_{juy}^{(t)} \log P_{jy|u}$$

$$+ \sum_{u=1}^{k} b_u^{(1)} \log \pi_u + \sum_{t=2}^{T} \sum_{\bar{u}=1}^{k} \sum_{u=1}^{k} b_{\bar{u}u}^{(t)} \log \pi_{u|\bar{u}}^{(t)},$$

where

$$a_{juy}^{(t)} = \sum_{i=1}^{n} I\left(u_i^{(t)} = u, y_{ij}^{(t)} = y\right),$$

$$b_u^{(t)} = \sum_{i=1}^{n} I\left(u_i^{(t)} = u\right),$$

$$b_{\bar{u}u}^{(t)} = \sum_{i=1}^{n} I\left(u_i^{(t-1)} = \bar{u}, u_i^{(t)} = u\right).$$

The EM algorithm proceeds by alternating the following two steps until convergence:

E-step: the expected values of the frequencies $a_{uy}^{(t)}$, $b_u^{(t)}$, and $b_{\bar{u}u}^{(t)}$ can be obtained as

$$\hat{a}_{juy}^{(t)} = \sum_{i=1}^{n} f_{U^{(t)}|\tilde{\boldsymbol{Y}}}\left(u \mid \tilde{\boldsymbol{y}}_i\right) I\left(y_{ij}^{(t)} = y\right),$$

$$\hat{b}_u^{(t)} = \sum_{i=1}^{n} f_{U^{(t)}|\tilde{\boldsymbol{y}}}\left(u \mid \tilde{\boldsymbol{y}}_i\right) = \sum_{\tilde{\boldsymbol{y}}} n_{\tilde{\boldsymbol{y}}} f_{U^{(t)}|\tilde{\boldsymbol{Y}}}(u \mid \tilde{\boldsymbol{y}}),$$

$$\hat{b}_{\bar{u}u}^{(t)} = \sum_{i=1}^{n} f_{U^{(t-1)},U^{(t)}|\tilde{\boldsymbol{Y}}}\left(\bar{u}, u \mid \tilde{\boldsymbol{y}}_i\right) = \sum_{\tilde{\boldsymbol{y}}} n_{\tilde{\boldsymbol{y}}} f_{U^{(t-1)},U^{(t)}|\tilde{\boldsymbol{Y}}}(\bar{u}, u \mid \tilde{\boldsymbol{y}}).$$

M-step: Update parameter estimates:

$$P_{jy|u} = \frac{\sum_{t=1}^{T} \hat{a}_{juy}^{(t)}}{\sum_{t=1}^{T} \hat{b}_u^{(t)}}, \quad u = 1, \ldots, k, \ y = 0, \ldots, c-1,$$

$$\pi_u = \frac{\hat{b}_u^{(1)}}{n}, \quad u = 1, \ldots, k,$$

$$\pi_{u|\bar{u}}^{(t)} = \frac{\hat{b}_{\bar{u}u}^{(t)}}{\hat{b}_{\bar{u}}^{(t-1)}}, \quad t = 2, \ldots, T, \ \bar{u}, u = 1, \ldots, k.$$

TABLE 2.5

AIC values for each model

# of components	2	3	4	5	6	7	8
LCM	38788	37882	37659	37658	37661	37668	37664
Random LCM	37712	37660	37649	37647	37648	37655	37657
Multilevel LCM	37712	36846	36697	36513	36487	36477	36472

TABLE 2.6

BIC values for each model

# of components	2	3	4	5	6	7	8
LCM	38876	38019	37843	37888	37939	37993	38037
Random LCM	37806	37803	37839	37884	37933	37988	38037
Multilevel LCM	37833	37049	36980	36877	36933	37004	37080

For selecting initial values, latent Markov model with covariates, and a more detailed explanation, please see Bartolucci et al. (2012).

2.4.9 Case study

In this section, we apply latent class models (LCM) to a large data set, namely the publicly accessible Wave 1 data of the National Longitudinal Study of Adolescent to Adult Health (Add Health), 1994-2008 [Public Use] (Harris and Udry, 2018), which is available at www.icpsr.umich.edu/icpsrweb/ICPSR/studies/21600. The data used in this study was extracted from six survey items that measured problematic behaviors of teenagers, including (1) Lie to parents; (2) Loud/Rowdy in a public place; (3) Mos-damage property; (4) Shoplift; (5) Steal Worth; (6) Take part in a group fight; as well as one item measuring their grades.

The study first built two LCMs with one and two classes, respectively. When the number of classes is one, it means that the model assumes there are no latent layers. The AIC values of the one-class and two-class models were 44100.30 and 38788.10, respectively, while the corresponding BIC values were 44140.98 and 38876.24. These results suggest that including latent layers in the model could lead to a better model fit.

Next, the study examined which type of latent class the responses were following, including the latent class model, Random LCA model, and Multilevel model. The number of latent classes was chosen to be between 2 and 8, and the AIC and BIC values of each model are presented in Table 2.5 and 2.6, respectively. The rationale behind examining these models was to determine the best fit for the data and to identify the underlying patterns of problematic behaviors among teenagers.

The results indicate that the Multi-level LCM provides the best fit for the sample, with AIC and BIC suggesting that there are likely to be 4 or 5

TABLE 2.7
AIC values for each model

# of components	2	3	4	5	6	7	8
LC Regression	21148	20481	22486	22476	23369	23402	22230
Multi-level Regression	20534	19881	19661	19649	19643	19637	19633

TABLE 2.8
BIC values for each model

# of components	2	3	4	5	6	7	8
LC Regression	21223	20603	22656	22693	23634	23714	22589
Multilevel Regression	20642	20070	19931	20000	20075	20150	20228

classes of latent groups. These groups may correspond to differences in location, personality, or other properties. Furthermore, the proportions of these latent variables differ for teenagers in different grades.

We then attempted to construct an LC regression model, using (1) Lie to parents and (2) Loud/Rowdy in a public place as explanatory variables, and the remaining four variables as response variables. The AIC and BIC results are presented in Tables 2.7 and 2.8. The results indicate that the multilevel LC regression outperforms the LC regression. The model with a latent variable consisting of 4 classes has the lowest BIC, and for models with latent variables consisting of more than 4 classes, the AICs do not exhibit large differences. Therefore, the multilevel LC regression with 4 classes is chosen.

2.4.10 Further reading

In this section, we provided an introduction to some topics in LCA. For recent work on the practical applications of LCA, please refer to Lanza and Cooper (2016), Petersen et al. (2019), and Zhou et al. (2018). The concept of LCA is also applied in other models, such as Growth Mixture Modeling (GMM) (Jung and Wickrama, 2008). GMM splits a sample into groups and assumes that within each group, subjects follow the same intercept and slope but can have different variances. In contrast, Latent Class Growth Analysis (LCGA) assumes that within each class, the variance and covariance are zero, which is a special case of GMM. The lcmm R package is designed for the LCGA model (Proust-Lima et al., 2015).

For other LC models, the Latent Class Tree (LCT) model is proposed for dealing with models with a large number of classes (van den Bergh et al., 2017). The LCT model is a recursive process that sequentially splits parent classes into two children classes. Van Den Bergh and Vermunt (2018) and Van Den Bergh and Vermunt (2019) provided more detailed steps and evaluation methods for the LCT model and LCT model with covariates.

Another advanced version of Latent Transition Analysis is the Dynamic Latent Class Analysis (DLCA) proposed by Asparouhov et al. (2017). This model is designed for time-intensive longitudinal latent class models. DLCA splits the observed vector of measurements for individual i at time t into a within-vector and a random effect vector:

$$Y_i^{(t)} = Y_{1,i}^{(t)} + Y_{2,i},$$

where the conditional distribution of $Y_{1,i}^{(t)}$ based on the latent variable is modeled with covariates and autoregressive effects. Moreover, the latent class changes across time following a Markov process. The model is estimated by Markov Chain Monte Carlo (MCMC), which can deal with a large sample that has an unlimited number of random effects.

Kelava and Brandt (2019) proposed a nonlinear DLCA, which separates the random effect into person-specific and time-specific random effects and provides a nonlinear model for within-between equations.

2.5   Mixture models for mixed data

Let us consider the case of mixed data, which contains both continuous and categorical observations. Typically, we assume that $\mathbf{x}_i = (x_{i1}, \ldots, x_{ip_0}, x_{i,p_0+1}, \ldots, x_{ip})^{\top}$, where the first p_0 variables are categorical, taking values from $(1, \ldots, G_j)$ for $j = 1, \ldots, p_0$, and the last $p - p_0$ variables are continuous. Early approaches, like Everitt (1984), modeled such mixed data with a latent class model. It assumes that, given the component membership, the p variables are conditionally independent. The categorical variables are mostly assumed to have multinomial distributions, and the continuous variables are normally distributed. The density of the most general model is:

$$f(\mathbf{x}_i) = \sum_{c=1}^{C} \pi_c \left[\prod_{j=1}^{p_0} \prod_{g=1}^{G_j} \theta_{cjg}^{I(x_{ij}=g)} \prod_{j=p_0+1}^{p} \phi(x_{ij}; \mu_{cj}, \sigma_{cj}^2) \right].$$

2.5.1   Location mixture model

A location model, which allows for arbitrary dependence between categorical variables, is then introduced by Hunt and Jorgensen (1999, 2003); Lawrence and Krzanowski (1996). To start with, among the p variables, suppose p_0 of them are categorical. Assume that, for $j = 1, \ldots, p_0$, the j-th categorical variable has G_j distinct categories. Then, these p_0 categorical variables result in a total of $G = \prod_{j=1}^{p_0} G_j$ distinct combinations. In a location model, the p_0 distinct categorical variables are replaced by a single multinomial random variable $\mathbf{X}^{(1)}$ with G distinct values. Then, conditional on the j-th pattern of

$\mathbf{X}^{(1)}$ and the membership of the c-th component being true, the distribution of the rest $p - p_0$ continuous variables, denoted by $\mathbf{X}^{(2)}$, follows a normal distribution with mean $\boldsymbol{\mu}_{cj}$ and covariance $\boldsymbol{\Sigma}_c$, $j = 1, \ldots, G$.

To be more specific, the location model assumes that

$$f(\mathbf{x}_i) = \sum_{c=1}^{C} \pi_c g(\mathbf{x}_i; \boldsymbol{\theta}_c, \boldsymbol{\mu}_c, \boldsymbol{\Sigma}_c),$$

where

$$\begin{aligned}
g(\mathbf{x}_i; \boldsymbol{\theta}_c, \boldsymbol{\mu}_c, \boldsymbol{\Sigma}_c) &= g(x_{i1}, \ldots, x_{i,p_0}|\boldsymbol{\theta}_c)g(x_{i,p_0+1}, \ldots, x_{i,p}|x_{i1}, \ldots, x_{i,p_0}, \boldsymbol{\mu}_c, \boldsymbol{\Sigma}_c) \\
&= g(\mathbf{x}_i^{(1)}|\boldsymbol{\theta}_c)g(\mathbf{x}_i^{(2)}|x_i^{(1)}, \boldsymbol{\mu}_c, \boldsymbol{\Sigma}_c) \\
&= \prod_{g=1}^{G} \theta_{cg}^{I(\mathbf{x}_i^{(1)}=g)} \times \prod_{g=1}^{G} \phi(\mathbf{x}_i^{(2)}; \boldsymbol{\mu}_{cg}, \boldsymbol{\Sigma}_c)^{I(\mathbf{x}_i^{(1)}=g)} \\
&= \prod_{g=1}^{G} [\theta_{cg}\phi(\mathbf{x}_i^{(2)}; \boldsymbol{\mu}_{cg}, \boldsymbol{\Sigma}_c)]^{I(\mathbf{x}_i^{(1)}=g)},
\end{aligned}$$

where $\boldsymbol{\theta}_c = (\theta_{c1}, \ldots, \theta_{cG})$, $\boldsymbol{\mu}_c = (\boldsymbol{\mu}_{c1}, \ldots, \boldsymbol{\mu}_{cG})$, and $\boldsymbol{\Sigma}_c$ are parameters of the location model for each component. Based on these model assumptions, the maximum likelihood estimator of the model parameters can be readily estimated by the EM algorithm. Let $\delta_{ij} = 1$ if the categorical value of the i-th observation belongs to the j-th category, $i = 1, \ldots, n, j = 1, \ldots, G$.

At the $(t+1)$-th iteration, the E-step needs to calculate

$$p_{icj}^{(t)} = \frac{\pi_c^{(t)} \tau_{cj}^{(t)} \phi(\mathbf{x}_i^{(2)}; \boldsymbol{\mu}_{cj}^{(t)}, \boldsymbol{\Sigma}_c^{(t)})}{\sum_{c'=1}^{C} \pi_{c'}^{(t)} \tau_{c'j}^{(t)} \phi(\mathbf{x}_i^{(2)}; \boldsymbol{\mu}_{c'j}^{(t)}, \boldsymbol{\Sigma}_{c'}^{(t)})}, i = 1, \ldots, n, c = 1, \ldots, C, j = 1, \ldots, G.$$

In the M-step, update the estimators by

$$\pi_c^{(t+1)} = \sum_{j=1}^{G} \sum_{i=1}^{n} \delta_{ij} p_{icj}^{(t)}/n,$$

$$\tau_{cj}^{(t+1)} = \sum_{i=1}^{n} \delta_{ij} p_{icj}^{(t)} / \sum_{j=1}^{G} \sum_{i=1}^{n} \delta_{ij} p_{icj}^{(t)},$$

$$\boldsymbol{\mu}_{cj}^{(t+1)} = \sum_{i=1}^{n} \delta_{ij} p_{icj}^{(t)} \mathbf{x}_i^{(2)} / \sum_{i=1}^{n} \delta_{ij} p_{icj}^{(t)},$$

$$\boldsymbol{\Sigma}_c^{(t+1)} = \sum_{j=1}^{G} \sum_{i=1}^{n} \delta_{ij} p_{icj}^{(t)} (\mathbf{x}_i^{(2)} - \boldsymbol{\mu}_{cj}^{(t+1)})(\mathbf{x}_i^{(2)} - \boldsymbol{\mu}_{cj}^{(t+1)})^{\top} / \sum_{j=1}^{G} \sum_{i=1}^{n} \delta_{ij} p_{icj}^{(t)},$$

where $c = 1, \ldots, C$ and $j = 1, \ldots, G$.

The nature of the location model makes it clear that it is only suitable when the number of categorical variables p_0 and the total number of categories G are not too large. To mitigate this issue, Hunt and Jorgensen (1999) proposed imposing conditional independence among the categorical variables and dependence between a small subset of the categorical variables and the remaining continuous variables, conditional on knowing the component labels. For example, suppose that conditional on component labels, $x_{i2}, \ldots, x_{i,p_0}$ are independent of each other and other variables, but that x_{i1} is not independent of the remaining continuous variables. Then, the c-th component is assumed to have the following density within each component

$$g(\mathbf{x}_i; \boldsymbol{\theta}_c, \boldsymbol{\mu}_c, \boldsymbol{\Sigma}_c) = g(x_{i1}, \ldots, x_{i,p_0} | \boldsymbol{\theta}_c) g(x_{i,p_0+1}, \ldots, x_{i,p} | x_{i1}, \ldots, x_{i,p_0}, \boldsymbol{\mu}_c, \boldsymbol{\Sigma}_c)$$

$$= \prod_{j=1}^{p_0} g_j(x_{ij}; \boldsymbol{\theta}_{cj}) \times \prod_{g=1}^{G_1} \phi(\mathbf{x}_i^{(2)}; \boldsymbol{\mu}_{cg}, \boldsymbol{\Sigma}_c)^{I(x_{i1}=g)},$$

where $\boldsymbol{\theta}_c = (\boldsymbol{\theta}_{c1}, \ldots, \boldsymbol{\theta}_{cp_0})$, $g_j(\cdot)$ is the probability mass function of the jth categorical variable x_{ij}, and $\boldsymbol{\mu}_c = (\boldsymbol{\mu}_{c1}, \ldots, \boldsymbol{\mu}_{cG_1})$ and $\boldsymbol{\Sigma}_c$ are parameters of the location model for each component.

2.5.2 Mixture of latent variable models

Latent variable models are a class of models that includes latent class analysis, latent trait analysis, factor analysis, and latent factor models. In a latent variable model, the density function of a p-dimensional random vector $\mathbf{X}_i$ is assumed to follow

$$f(\mathbf{x}_i) = \int g(\mathbf{x}_i; \mathbf{y}, \boldsymbol{\theta}) h(\mathbf{y}) d\mathbf{y} = \int \prod_{j=1}^{p} g_j(x_{ij}; \mathbf{y}, \boldsymbol{\theta}_j) h(\mathbf{y}) d\mathbf{y},$$

where $g_j(x_{ij}; \mathbf{y}, \boldsymbol{\theta}_j)$ is the conditional distribution of X_{ij} given $\mathbf{Y} = \mathbf{y}$ and $h(\mathbf{y})$ is the marginal distribution of $\mathbf{Y}$, which is a q-dimensional random vector ($q < p$). If X_{ij} is categorical with levels $0, 1, \ldots, G_j - 1$, then given $\mathbf{y}, \boldsymbol{\theta}_j$, X_{ij} is assumed to follow a multinomial distribution. On the other hand, if X_{ij} is continuous, then it is assumed to have a Gaussian conditional distribution.

Recently, an increasing number of research articles have focused on applying latent variable models to cluster mixed data. Cai et al. (2011) developed a mixture of generalized latent variable models to handle mixed types of heterogeneous data, and applied identity link, cumulative probit link, logit or log link, and multinomial logit link functions to model continuous, ordinal, counts and nominal variables, respectively. A Bayesian approach, together with the Markov chain Monte Carlo (MCMC) method, is used to conduct the analysis.

Browne and McNicholas (2012) developed a mixture of latent variable models suitable for data with mixed types as follows:

$$f(\mathbf{x}_i) = \sum_{c=1}^{C} \pi_c \left[\int \prod_{j=1}^{p} g_j(x_{ij}; \mathbf{y}, \boldsymbol{\theta}_{cj}) h(\mathbf{y}) d\mathbf{y} \right].$$

The variables are allowed to be dependent by using latent normal random variables within each component. An EM algorithm is carried out to facilitate the maximum likelihood estimation, where the missing data, in this case, includes both the group membership and the latent variables. At the $(t+1)$-th iteration, the E-step needs to calculate

$$\mathrm{E}[Z_{ic}|\mathbf{x}_i] = \frac{\pi_c g(\mathbf{x}_i; \boldsymbol{\theta}_c)}{\sum_{h=1}^{C} \pi_h g(\mathbf{x}_i; \boldsymbol{\theta}_h)} = p_{ic},$$

and the conditional expectation of any function $a(\mathbf{y})$ as

$$\mathrm{E}[a(\mathbf{Y}_i)|\mathbf{x}_i, Z_{ic} = 1] = \int a(\mathbf{y}_i) \frac{g(\mathbf{x}_i; \mathbf{y}_i, \boldsymbol{\theta}_c) h(\mathbf{y}_i)}{f(\mathbf{x}_i)} d\mathbf{y}_i.$$

In the M-step, the mixing proportion is updated by $\pi_c^{(t+1)} = \sum_{i=1}^{n} p_{ic}/n$. Depending on whether the variable $X_j, j = 1, \ldots, p$ is continuous or categorical, the parameter updates are constructed accordingly. See Browne and McNicholas (2012) for more details.

For data mixed with binary and continuous variables, Morlini (2012) used the scores of the latent variables in place of the observed original binary attributes and then specified a mixture of continuous distributions. This method is particularly useful in economics as well as marketing, since it can be applied to a large number of binary variables and the latent variables may be interpreted as utility functions. As the authors declared, by imposing different types of restrictions on the model parameters to obtain sparsity, this methodology is suitable for a large number of variables, and local dependencies can be specified between continuous variables, original categorical variables, and a continuous variable and a categorical variable. More specifically, for the i-th object $(i = 1, \ldots, n)$, let $\mathbf{x}_i^{(1)} = [x_{i1}, \ldots, x_{i,p_0}]$ contain the values of the binary variables and $\mathbf{x}_i^{(2)} = [x_{i,p_0+1}, \ldots, x_{i,p}]$ contain the values of the continuous variables. The method assumes that the binary values are generated from latent continuous variables $\xi_j, j = 1, \ldots, p_0$, which are assumed to be hidden by being normally distributed, as follows:

$$x_{ij} = \begin{cases} 1, & \text{if } \xi_{ij} \geq T_j, \\ 0, & \text{otherwise.} \end{cases}$$

Based on the data, one can obtain the score estimator, indicated by following the steps below:

TABLE 2.9
Contingency table

	$x_l = 0$	$x_l = 1$	Total
$x_j = 0$	a_{jl}	b_{jl}	$a_{jl} + b_{jl}$
$x_j = 1$	c_{jl}	d_{jl}	$c_{jl} + d_{jl}$
Total	$a_{jl} + c_{jl}$	$b_{jl} + d_{jl}$	n

1. Estimate the threshold T_j of each latent variable. Construct the contingency Table 2.9 for $l, j = 1, \ldots, p_0$. Then, the thresholding value T_l of x_l is estimated by satisfying $\Phi(T_l) = (a_{jl} + c_{jl})/n$, and the thresholding value T_j of x_j is estimated by satisfying $\Phi(T_j) = (a_{jl} + b_{jl})/n$. In addition, the correlation coefficient r_{jl} of ξ_j and ξ_l is also estimated from the above table.

2. Perform a principal component analysis on the matrix of r_{jl}, and obtain the eigenvectors and eigenvalues.

3. Estimate the score of each principal component for each object, given the eigenvectors and eigenvalues.

4. Estimate ξ_{ij} given the scores of the principal components, denoted by $\tilde{\xi}_i = [\tilde{\xi}_{i1}, \ldots, \tilde{\xi}_{i,p_0}]$.

Given $\tilde{\xi}_i$, one can then apply a Gaussian model or a mixture of p-dimensional continuous distributions to fit the data. However, none of these can analyze the combination of continuous, nominal, and ordinal variables without transforming the original variables.

2.5.3   The MFA-MD model

McParland et al. (2014, 2017) developed a mixture of factor analyzer models for mixed data (MFA-MD) suitable for data mixed with ordinal, nominal, and continuous random variables, to quantify household socio-economic status in rural population living in northeast South Africa and analyze for both phenotypic and genetic factors when studying the metabolic syndrome. To be more specific, within each component, an item response theory (IRT) model for ordinal data and a factor analytic model for nominal data are combined together and then extended to the MFA-MD model. In these models, the ordinal variables are modeled by item response theory models with a latent Gaussian random variable, the nominal variables are modeled by factor analytic models with $G_j - 1$-dimensional latent normal random variable, and each continuous random variable is modeled as a rescaled version of a latent normal random variable. Then, the p variables in the original data are modeled as a manifestation of a $D = \sum_{j=1}^{p_0}(G_j - 1) + (p - p_0)$-dimensional latent random vector, which follows a mixture of multivariate normals.

An IRT model for ordinal data

Suppose item $j(j = 1, \ldots, J)$ is ordinal and the corresponding response is denoted by $\{1, 2, \ldots, K_j\}$ where K_j denotes the number of response levels to item j. An IRT model assumes that, for respondent i, a latent Gaussian variable z_{ij} corresponds to each categorical response y_{ij}. It is assumed that the mean of the conditional distribution of z_{ij} depends on a q-dimensional-respondent specific, latent variable $\boldsymbol{\theta}_i$, and on some item-specific parameters as

$$z_{ij}|\boldsymbol{\theta}_i \sim N(\mu_j + \boldsymbol{\Lambda}_j\boldsymbol{\theta}_i, 1).$$

A factor analytic model for nominal data

On the other hand, nominal response data is more difficult to model compared to ordinal data, since the set of possible responses is not ordered. Similar to the above notations, we still use $\{1, 2, \ldots, K_j\}$ to denote the response choices, but there is no specific ordering among the choices. In the factor analytic model for nominal data, a $K_j - 1$-dimensional latent variable is required for each observed nominal response, denoted by

$$\mathbf{z}_{ij} = (z_{ij}^1, \ldots, z_{ij}^{K_j-1}),$$

which is assumed to follow a multivariate normal distribution as

$$\mathbf{z}_{ij}|\boldsymbol{\theta}_i \sim MVN_{K_j-1}(\mu_j + \boldsymbol{\Lambda}_j\boldsymbol{\theta}_i, \mathbf{I}),$$

where $\boldsymbol{\Lambda}_j$ is a $(K_j - 1) \times q$ loading matrix. The observed nominal response is then assumed to be a manifestation of the values of the elements of $\mathbf{z}_{ij}$ relative to each other and to cut off points such as

$$y_{ij} = \begin{cases} 1, & \text{if } max_k\{z_{ij}^k\} < 0; \\ k, & \text{if } z_{ij}^{k-1} = max_k\{z_{ij}^k\} \text{ and } z_{ij}^{k-1} > 0 \text{ for } k = 2, \ldots, K_j. \end{cases}$$

A factor analysis model for mixed data

It can be seen that an IRT model and a factor analytic model are similar in structure. Hence, when there are both binary/ordinal and nominal data in the dataset, a hybrid factor analysis model is used. Suppose there are O binary or ordinal items, and J nominal items, and then let

$$\mathbf{z}_i = (z_{i1}, \ldots, z_{iO}, z_{i,O+1}^1, \ldots, z_{i,O+1}^{K_{O+1}-1}, \ldots, z_{iJ}^1, \ldots, z_{iJ}^{K_J-1}),$$

to be the vector of latent variables corresponding to these items. Clearly, the dimension of $\mathbf{z}_i$ is $D = O + \sum_{j=O+1}^{J}(K_j - 1)$. Then,

$$\mathbf{z}_i|\boldsymbol{\theta}_i \sim MVN_D(\boldsymbol{\mu} + \boldsymbol{\Lambda}\boldsymbol{\theta}_i, \mathbf{I}), \tag{2.21}$$

and marginally, the latent vector is distributed as

$$\mathbf{z}_i \sim MVN_D(\boldsymbol{\mu}, \boldsymbol{\Lambda}\boldsymbol{\Lambda}^\top + \mathbf{I}).$$

A mixture of factor analysis modeling for mixed data

By applying a mixture modeling framework to the latent variable level, an MFA-MD (McParland et al., 2014, 2017) assumes that the latent data $\mathbf{z}_i$ is modeled as a mixture of C Gaussian densities

$$\mathbf{z}_i \sim \sum_{c=1}^{C} \pi_c MVN_D(\boldsymbol{\mu}_c, \boldsymbol{\Lambda}_c \boldsymbol{\Lambda}_c^\top + \mathbf{I}). \tag{2.22}$$

While this model can explicitly model the inherent nature of each variable type directly, it can be computationally expensive.

2.5.4   The clustMD model

McParland and Gormley (2016) developed a clustMD model for observed data of any combination of continuous, binary, ordinal, or nominal variables. It is a mixture of latent Gaussian distributions and provides a parsimonious and computationally efficient approach to cluster mixed data. The clustMD model makes the following assumptions for different types of data:

1. If variable j is continuous, then it follows a Gaussian distribution $x_{ij} = \xi_{ij} \sim N(\mu_j, \sigma_j^2)$.

2. If the j-th variable is ordinal with G_j levels, assuming the latent variable ξ_{ij} follows a Gaussian distribution, then the probability of obtaining an observed ordinal response g is

$$P(x_{ij} = g) = \Phi\left(\frac{\gamma_{jg} - \mu_j}{\sigma_j}\right) - \Phi\left(\frac{\gamma_{j,g-1} - \mu_j}{\sigma_j}\right),$$

   where $\boldsymbol{\gamma}_j = (\gamma_{j,0}, \ldots, \gamma_{j,G_j})$ denotes a vector of length $G_j + 1$ that partitions the real line. For identifiability and efficiency reasons, $\gamma_{j,g}$ is set to be $\gamma_{j,g} = \Phi^{-1}(\delta_g)$, where δ_g is the proportion of the observed values being less than or equal to level g.

3. If the j-th variable is nominal with G_j possible responses, then the observed y_{ij} is a manifestation of the latent $\boldsymbol{\xi}_{ij} = (\xi_{ij}^1, \ldots, \xi_{ij}^{G_j - 1})$ through

$$x_{ij} = \begin{cases} 1, & \text{if } \max_s\{\xi_{ij}^s\} < 0, \\ k, & \text{if } \xi_{ij}^{k-1} = \max_s\{\xi_{ij}^s\} \text{ and } \xi_{ij}^{k-1} > 0, \end{cases}$$

   where the underlying continuous vector is assumed to follow a $G_j - 1$-dimensional multivariate normal distribution: $\boldsymbol{\xi}_{ij} \sim N_{G_j-1}(\boldsymbol{\mu}_j, \boldsymbol{\Sigma}_j)$.

4. If the variable is binary, then it can be either considered as nominal with two unordered responses, or ordinal variables with two levels.

If in the original data, p_1 of them are continuous, p_2 of them are ordinal or binary, and $p - p_1 - p_2$ of them are nominal, then based on the above modeling procedures, the joint vector of the observed and latent continuous variables are assumed to follow a C-component mixture of multivariate Gaussian distributions with dimension $P = p_1 + p_2 + \sum_{j=p_1+p_2+1}^{p}(G_j - 1)$. Employing a parsimonious covariance structure for the latent variables gives rise to a group of 6 clustMD models with varying levels of parsimony.

2.6   Fitting mixture models for discrete data using R

In Section 2.4, latent class analysis (LCA) is introduced as a popular technique for analyzing multiple categorical outcomes. The sBIC and poLCA packages are available for fitting latent class analysis, where poLCA is useful for fitting LCA models with two states, and sBIC is capable of handling LCA with more than two states. The following codes show how the poLCA package can be applied to simulated data.

```
library(poLCA)
data(gss82)
f <- cbind(PURPOSE,ACCURACY,UNDERSTA,COOPERAT)~1
gss.lc2 <- poLCA(f,gss82,nclass=2)
```

The sBIC package is designed to compute the sBIC for various singular model collections. The following is an example showing how SBIC can be computed on a list of models.

```
library(sBIC)
lcas = LCAs(maxNumClasses = 6, numVariables = 8,
            numStatesForVariables = 2)
results = sBIC(X, lcas)
```

In Section 2.5, McParland et al. (2014) and McParland and Gormley (2017) developed a mixture of factor analyzers models for mixed data, suitable for data mixed with ordinal, nominal, and continuous random variables. The clustMD package is suitable for carrying out the procedure. We apply the methodology to the data set "Byar," which consists of mixed-type variables measured on a group of prostate cancer patients. Patients have either stage 3 or stage 4 prostate cancer. The classification result can be obtained through the following codes.

```
library(clustMD)
data(Byar, package = "clustMD")
# Reorder variables (8 continuous, 4 categorical)
```

```
Y = as.matrix(Byar[, c(1:2, 5:6, 8:11, 3:4, 12, 7)])
# Transform skewed variables
Y[, 6] = sqrt(Y[, 6]) # Size.of.primary.tumour
Y[, 8] = log(Y[, 8])  # Serum.prostatic.acid.phosphatase
# Standardize continuous variables
Y[, 1:8] = scale(Y[, 1:8])
# Start categorical variables at 1 rather than 0
Y[, 9:12] = Y[, 9:12] + 1
# Merge categories of EKG variable for efficiency
new.levels <- c(1, 2, 2, 2, 3, 3, 3, 3)
Y[, 12] <- new.levels[factor(Y[, 12])]
res = clustMD(X = Y, G = 3, CnsIndx = 8, OrdIndx = 11,
              Nnorms = 20000, MaxIter = 500, model = "EVI",
              store.params = FALSE, scale = TRUE,
              startCL = "kmeans", autoStop = TRUE,
              ma.band = 30, stop.tol = 0.0001)
res$cl
```

3

Mixture regression models

3.1 Mixtures of linear regression models

Multiple linear regression models are commonly used to explore the relationship between a response variable $Y \in \mathbb{R}$ and covariates $\mathbf{x} \in \mathbb{R}^p$, where p is the number of predictors and $\mathbb{R}$ is the set of real numbers. However, in many applications, the assumption that the regression relationship is homogeneous among the whole population does not hold. Rather, the population may consist of several distinct clusters, indicating mixed relationships between the responses and the predictors. Consider a study of the effect of prenatal care timing on birth weight, where there are essentially two kinds of pregnancies, "complicated" and "normal" ones. Complicated pregnancies typically entail a large amount of prenatal care, resulting in poorer outcomes. Combining these pregnancies with normally progressing pregnancies could cause prenatal care to appear ineffective. Such heterogeneity can be more appropriately modeled by a *finite mixture regression model*, consisting of C homogeneous linear regression components. Finite mixtures of linear regression (FMR) models are very flexible statistical tools in studying the relationship among some interesting variables coming from several unknown latent components/groups and have greatly enriched the toolkit of regression analysis due to their simplicity. Since the FMR model was first introduced by Goldfeld and Quandt (1973), it has been widely used in many areas such as business, marketing, social sciences, etc; see in literature, e.g., Jiang and Tanner (1999), Böhning (1999), Wedel and Kamakura (2000), Hennig (2000), McLachlan and Peel (2000), Skrondal and Rabe-Hesketh (2004), and Frühwirth-Schnatter (2006).

The unknown mixture regression parameters are usually estimated by maximum likelihood estimation (MLE) using the EM algorithm based on the normality assumption of component error density. Let Z be a latent class variable such that given $Z = c$, the continuous response y depends on the $p-$dimensional predictor $\mathbf{x}$ in a linear way

$$y = \mathbf{x}^\top \boldsymbol{\beta}_c + \epsilon_c, c = 1, 2, \cdots, C,$$

where $\boldsymbol{\beta}_c \in \mathbb{R}^p$ is a fixed and unknown coefficient vector, C is the number of components in the population, and $\epsilon_c \sim N(0, \sigma_c^2)$ is independent of $\mathbf{x}$. Note that the intercept term can be included by setting the first element of $\mathbf{x}$ vector

DOI: 10.1201/9781003038511-3

as one. Suppose $P(Z = c) = \pi_c, c = 1, 2, \cdots, C$, and Z is independent of $\mathbf{x}$, then the conditional density of Y given $\mathbf{x}$, without observing Z, is

$$f(y|\mathbf{x}, \boldsymbol{\theta}) = \sum_{c=1}^{C} \pi_c \phi(y; \mathbf{x}^\top \boldsymbol{\beta}_c, \sigma_c^2), \tag{3.1}$$

where $\boldsymbol{\theta} = (\pi_1, \boldsymbol{\beta}_1^\top, \sigma_1, \ldots, \pi_C, \boldsymbol{\beta}_C^\top, \sigma_C)^\top$ is the unknown parameter vector, π_cs are the mixing proportions with $\pi_c \geq 0$ and $\sum_{c=1}^{C} \pi_c = 1$, and $\phi(\cdot\,; \mu, \sigma^2)$ denotes the probability density function (pdf) of the normal distribution $N(\mu, \sigma^2)$. The above model is the so-called *finite mixtures of linear regression models* (FMR). Hennig (2000) proved the identifiability of the model (3.1) under some general conditions of the covariates. In general, the model (3.1) is identifiable if the number of components, C, is smaller than the number of distinct $(p-1)$-dimensional hyperplanes that one needs to cover the covariates of each cluster. The above conditions are usually satisfied if the domain of $\mathbf{x}$ contains an open set in $\mathbb{R}^p$

Let $\{(\mathbf{x}_i, y_i), i = 1, \ldots, n\}$ be the collected observations from the mixture model (3.1). Maximum likelihood estimation (MLE) is commonly used to estimate $\boldsymbol{\theta}$ in (3.1) for the Gaussian FMR model by maximizing the following log-likelihood function

$$\ell_n(\boldsymbol{\theta}) = \sum_{i=1}^{n} \log \left[\sum_{c=1}^{C} \pi_c \phi(y_i; \mathbf{x}_i^\top \boldsymbol{\beta}_c, \sigma_c^2) \right]. \tag{3.2}$$

The MLE does not have an explicit form and is usually obtained by the Expectation-Maximization (EM) algorithm (Dempster et al., 1977b).

Define a latent variable z such that $z_{ic} = 1$ if the ith observation is from the cth component and 0 otherwise, and let $\mathbf{z}_i = (z_{i1}, z_{i2}, \ldots, z_{iC})^\top$. Then, the complete log-likelihood function for the complete data set $\{(\mathbf{x}_i, \mathbf{z}_i, y_i) : i = 1, 2, \ldots, n\}$ is

$$\ell_n^c(\boldsymbol{\theta}) = \sum_{i=1}^{n} \sum_{c=1}^{C} z_{ic} \log \left\{ \pi_c \phi(y_i; \mathbf{x}_i^\top \boldsymbol{\beta}_c, \sigma_c^2) \right\},$$

$$\propto \sum_{i=1}^{n} \sum_{c=1}^{C} z_{ic} \left\{ \log(\pi_c) - \frac{1}{2} \log(\sigma_c^2) \right\} - \sum_{i=1}^{n} \sum_{c=1}^{C} z_{ic} \frac{(y_i - \mathbf{x}_i^\top \boldsymbol{\beta}_c)^2}{2\sigma_c^2}.$$

Based on Section 1.5, the E step of EM algorithm computes the conditional expectation of the complete log-likelihood $\mathrm{E}[\ell_n^c(\boldsymbol{\theta})|\mathbf{y}, \mathbf{X}, \boldsymbol{\theta}^{(k)}]$ and the M step updates $\boldsymbol{\theta}$ by maximizing $\mathrm{E}[\ell_n^c(\boldsymbol{\theta})|\mathbf{y}, \mathbf{X}, \boldsymbol{\theta}^{(k)}]$, where $\mathbf{y} = (y_1, \ldots, y_n)^\top$, $\mathbf{X} = (\mathbf{x}_1, \ldots, \mathbf{x}_n)^\top$, and $\boldsymbol{\theta}^{(k)}$ is the update of $\boldsymbol{\theta}$ at the kth iteration of EM algorithm.

Specifically, the EM algorithm to maximize (3.2) can be summarized in Algorithm 3.1.1. If we assume the variance is equal, i.e., $\sigma_1 = \sigma_2 = \cdots = \sigma_C = \sigma$, then in the M-step σ is updated by

$$\sigma^{2(k+1)} = \frac{1}{n} \sum_{i=1}^{n} \sum_{c=1}^{C} p_{ic}^{(k+1)} (y_i - \mathbf{x}_i^\top \boldsymbol{\beta}_c^{(k+1)})^2.$$

Algorithm 3.1.1 EM algorithm to maximize (3.2)

Given the initial parameter estimate $\boldsymbol{\theta}^{(0)}$, at the $(k+1)th$ step, the EM algorithm iterates the following E-step and M-step until convergence:

E-step: Calculate the classification probabilities

$$p_{ic}^{(k+1)} = \frac{\pi_c^{(k)} \phi(y_i; \mathbf{x}_i^\top \boldsymbol{\beta}_c^{(k)}, \sigma_c^{2(k)})}{\sum_{l=1}^{C} \pi_l^{(k)} \phi(y_i; \mathbf{x}_i^\top \boldsymbol{\beta}_l^{(k)}, \sigma_l^{2(k)})}, \quad i = 1, \ldots, n; c = 1, \ldots, C.$$

M-step: Update parameter estimations

$$\boldsymbol{\beta}_c^{(k+1)} = \arg\min_{\boldsymbol{\beta}_c} \sum_{i=1}^{n} p_{ic}^{(k+1)} (y_i - \mathbf{x}_i^\top \boldsymbol{\beta}_c)^2$$

$$= \left(\sum_{i=1}^{n} p_{ic}^{(k+1)} \mathbf{x}_i \mathbf{x}_i^\top \right)^{-1} \sum_{i=1}^{n} p_{ic}^{(k+1)} \mathbf{x}_i y_i$$

$$= (\mathbf{X}^\top \mathbf{W}_c^{(k+1)} \mathbf{X})^{-1} \mathbf{X}^\top \mathbf{W}_c^{(k+1)} \mathbf{y},$$

$$\pi_c^{(k+1)} = \frac{1}{n} \sum_{i=1}^{n} p_{ic}^{(k+1)},$$

$$\sigma_c^{2(k+1)} = \frac{1}{\sum_{i=1}^{n} p_{ic}^{(k+1)}} \sum_{i=1}^{n} p_{ic}^{(k+1)} (y_i - \mathbf{x}_i^\top \boldsymbol{\beta}_c^{(k+1)})^2,$$

where $\mathbf{W}_c^{(k+1)}$ is a $n \times n$ diagonal matrix with diagonal elements $\{p_{ic}^{(k+1)}, i = 1, \ldots, n\}$, and $c = 1, \ldots, C$.

3.2 Unboundedness of mixture regression likelihood

For the mixture regression log-likelihood (3.2), we have

$$\ell_n(\boldsymbol{\theta}) = \log \left\{ \sum_{c=1}^{C} \pi_c \phi(y_1; \mathbf{x}_1^\top \boldsymbol{\beta}_c, \sigma_c^2) \right\} + \sum_{i=2}^{n} \log \left\{ \sum_{c=1}^{C} \pi_c \phi(y_i; \mathbf{x}_i^\top \boldsymbol{\beta}_c, \sigma_c^2) \right\}$$

$$\geq \log \left[\frac{\pi_1}{\sqrt{2\pi}\sigma_1} \exp \left\{ -\frac{(y_1 - \mathbf{x}_1^\top \boldsymbol{\beta}_1)^2}{2\sigma_1^2} \right\} \right] + \sum_{i=2}^{n} \log \left\{ \sum_{c=2}^{C} \pi_c \phi(y_i; \mathbf{x}_i^\top \boldsymbol{\beta}_c, \sigma_c^2) \right\}.$$

$$(3.3)$$

When $y_1 = \mathbf{x}_1^\top \boldsymbol{\beta}_1$, the first term of (3.3) tends to ∞ as σ_1 goes to zero while the second term is finite for any other fixed parameters. Therefore, the mixture regression log-likelihood $\ell_n(\boldsymbol{\theta})$ in (3.2) is also unbounded, similar to normal mixture models introduced in Section 1.9, and technically, the global MLE is not well defined for (3.2).

When running the EM algorithm, some initial values may converge to boundary points with small variances and very large log-likelihood values. In such situations, our objective is to find a local maximum of (3.2) in the interior of the parameter space (Kiefer, 1978; Peters and Walker, 1978). However, the challenge is finding this interior local maximum. Hathaway (1985, 1986) proposed adding some constraints on the parameter space such that the component variance has some low limit. Yao (2010) proposed a profile likelihood method to locate the interior local maximum by reducing the dimension of the mixture likelihood function without the requirement of choosing a tuning parameter. Practically, the interior local maximum can usually be found by starting from some "good" initial values such as the K-means (MacQueen et al., 1967) and the moment method estimator (Lindsay and Basak, 1993). Chen et al. (2008) also proposed using a penalized likelihood method to avoid the unboundedness of mixture likelihood. Please refer to Section 1.9 for more details.

3.3   Mixture of experts model

Since the introduction of the mixture of experts (ME) model (Jacobs et al., 1991), it has been widely used in numerous regression, classification, and fusion applications in healthcare, finance, surveillance, and recognition. The ME model is very useful for regression problems with nonstationary, piecewise continuous data, and for nonlinear classification problems with data that contains naturally distinctive subsets of patterns.

The ME model allows the component label z to depend on the predictors by modeling the dependence through the logistic regression such that

$$p(z_{ic} = 1 \mid \mathbf{x}) = \pi_c(\mathbf{x}; \boldsymbol{\alpha}) = \frac{\exp(\gamma_c)}{1 + \exp(\gamma_1) + \cdots + \exp(\gamma_{C-1})},$$

where $\gamma_c = \mathbf{z}^\top \boldsymbol{\alpha}_c$ and $\boldsymbol{\alpha} = (\boldsymbol{\alpha}_1, \ldots, \boldsymbol{\alpha}_{C-1})$. Then, the conditional density of Y given $\mathbf{x}$ is

$$f(y \mid \mathbf{x}, \boldsymbol{\theta}) = \sum_{c=1}^{C} \pi_c(\mathbf{x}) \phi(y; \mathbf{x}^\top \boldsymbol{\beta}_c, \sigma_c^2). \qquad (3.4)$$

In (3.4), we can also allow the covariates in $\pi_c(\cdot)$ and component regression functions to be different. All the discussions in this section follow similarly. The log-likelihood function for the model (3.4) is

$$\ell(\boldsymbol{\theta}) = \sum_{i=1}^{n} \log \left\{ \sum_{c=1}^{C} \pi_c(\mathbf{x}_i) \phi(y_i; \mathbf{x}_i^\top \boldsymbol{\beta}_c, \sigma_c^2) \right\},$$

where $\boldsymbol{\theta}$ collects all unknown parameters. The MLE can be found by the well-known EM algorithm as described in Algorithm 3.3.1.

Algorithm 3.3.1 EM algorithm for mixture of experts model

Given an initial parameter $\boldsymbol{\theta}^{(0)}$ repeat the following steps:

E-Step: Calculate the classification probabilities

$$p_{ic}^{(k+1)} = \frac{\pi_c(\mathbf{x}_i; \boldsymbol{\alpha}^{(k)})\phi(y_i; \mathbf{x}_i^\top \boldsymbol{\beta}_c^{(k)}, \sigma_c^{2(k)})}{\sum_{l=1}^{C} \pi_l(\mathbf{x}_i; \boldsymbol{\alpha}^{(k)})\phi(y_i; \mathbf{x}_i^\top \boldsymbol{\beta}_l^{(k)}, \sigma_l^{2(k)})}, \quad i = 1, \dots, n; c = 1, \dots, C.$$

M-Step: Update the parameters

$$\boldsymbol{\beta}_c^{(k+1)} = \arg\min_{\boldsymbol{\beta}_c} \sum_{i=1}^{n} p_{ic}^{(k+1)}(y_i - \mathbf{x}_i^\top \boldsymbol{\beta}_c)^2$$

$$= \left(\sum_{i=1}^{n} p_{ic}^{(k+1)} \mathbf{x}_i \mathbf{x}_i^\top \right)^{-1} \sum_{i=1}^{n} p_{ic}^{(k+1)} \mathbf{x}_i y_i$$

$$= (\mathbf{X}^\top \mathbf{W}_c^{k+1} \mathbf{X})^{-1} \mathbf{X}^\top \mathbf{W}_c^{(k+1)} \mathbf{y},$$

$$\sigma_c^{2(k+1)} = \frac{1}{\sum_{i=1}^{n} p_{ic}^{(k+1)}} \sum_{i=1}^{n} p_{ic}^{(k+1)}(y_i - \mathbf{x}_i^\top \boldsymbol{\beta}_c^{(k+1)})^2,$$

$$\boldsymbol{\alpha}^{(k+1)} = \arg\max_{\boldsymbol{\alpha}} \sum_{i=1}^{n} \sum_{c=1}^{C} p_{ic}^{(k+1)} \log \pi_c(\mathbf{x}_i; \boldsymbol{\alpha}),$$

where $c = 1, \dots, C$, $\mathbf{X} = (\mathbf{x}_1, \mathbf{x}_2, \dots, \mathbf{x}_n)^\top$, $\mathbf{y} = (y_1, \dots, y_n)^\top$, and $\mathbf{W}_c^{(k+1)}$ is a $n \times n$ diagonal matrix with diagonal elements $\{p_{ic}^{(k+1)}, i = 1, \dots, n\}$.

Note that the update of $\boldsymbol{\alpha}$ in the M-step does not have an explicit formula and needs to be done by some optimization algorithm such as Newton-Raphson algorithm.

3.4 Mixture of generalized linear models

3.4.1 Generalized linear models

3.4.1.1 Introduction

For discrete response variables, such as binary and count responses, the traditional linear regression $y = \mathbf{x}^\top \boldsymbol{\beta} + \epsilon$ is not suitable partly because the predicted values may not be in range and the normality assumption of the error term does not hold. For this reason, generalized linear models are popularly used for discrete responses.

It is common and convenient to assume that the random discrete variable Y has a density from the exponential family:

$$f_Y(y; \theta, \phi) = \exp[\{y\theta - b(\theta)\}/\phi - c(y, \phi)],$$

for some specific functions $b(\cdot)$ and $c(\cdot)$, where ϕ is a dispersion parameter, and θ is called a *canonical/natural parameter*. The above model is called the exponential dispersion family. If ϕ is known, it is called natural exponential family.

Theorem 3.4.1. For the above exponential dispersion family density:

$$\mu = \mathrm{E}_\theta(Y) = b'(\theta)$$

and

$$\mathrm{Var}_\theta(Y) = \phi b''(\theta) = \phi \frac{\partial}{\partial \theta} \mathrm{E}_\theta(Y) = \phi v(\mu),$$

where $v(\mu) = b''(\theta)$ is often called the variance function because it indicates how the variance of Y depends on its mean.

Proof *Note that $\int f_Y(y; \theta, \phi) dy = 1$ for any θ. Hence*

$$0 = \int \frac{\partial f_Y(y; \theta, \phi)}{\partial \theta} dy = \int \frac{y - b'(\theta)}{\phi} f_Y(y; \theta, \phi) dy = E(Y) - b'(\theta).$$

Hence $\mu = \mathrm{E}_\theta(Y) = b'(\theta)$.   □

In addition,

$$0 = \int \frac{\partial^2 f_Y(y; \theta, \phi)}{\partial \theta^2} dy = \int f_Y(y; \theta, \phi) \frac{\{y - b'(\theta)\}^2}{\phi^2} + f_Y(y; \theta, \phi) \frac{-b''(\theta)}{\phi} dy.$$

Hence $\mathrm{Var}(Y) = \phi b''(\theta)$.

The *link function* is commonly used to link the mean $\mu = \mathrm{E}(Y)$ with predictors $\mathbf{x}$:

$$g(\mu) = \mathbf{x}^\top \boldsymbol{\beta},$$

where $g(\mu)$ is a known function, called the link function. One common choice is the canonical link $\theta = g(\mu)$, i.e., $g(\cdot) = b'^{-1}(\cdot)$. Then

$$f(y \mid \mathbf{x}) = \exp[\{y\mathbf{x}^\top \boldsymbol{\beta} - b(\mathbf{x}^\top \boldsymbol{\beta})\}/\phi - c(y, \phi)].$$

One nice property of using the canonical link is that there exists a sufficient statistic equal to the dimension of $\boldsymbol{\beta}$ and the sufficient statistic given $\{(y_1, \mathbf{x}_1), \ldots, (y_n, \mathbf{x}_n)\}$ is $\sum_{i=1}^n \mathbf{x}_i y_i$. In addition, the log-likelihood is a concave function and easy to maximize.

Example 3.4.1. For the normal model $N(\mu, \sigma^2)$, we have

$$\log f(\mu, \sigma^2) = \log\left[\frac{1}{\sqrt{2\pi\sigma^2}}\exp\{-\frac{(x-\mu)^2}{2\sigma^2}\}\right] \propto \frac{x\mu - \mu^2/2}{\sigma^2} - \frac{1}{2}\log\sigma^2 - \frac{x^2}{2\sigma^2},$$

hence the normal model is in the exponential family with a canonical parameter $\theta = \mu$, dispersion parameter $\phi = \sigma^2$, and $b(\theta) = \theta^2/2$ and $c(x, \phi) = -\frac{1}{2}\log\phi - \frac{1}{2}x^2/\phi$. In addition, it can be seen that the canonical link is $g(\mu) = \mu$, an identity link. □

Example 3.4.2. For a Bernoulli random variable with a success probability π, we have

$$\log\{f(y)\} = \log\{\pi^y(1-\pi)^{1-y}\} = y\log\{\pi/(1-\pi)\} + \log(1-\pi).$$

So $\theta = \log\{\pi/(1-\pi)\}$ is the canonical parameter and the dispersion parameter $\phi = 1$. Therefore, the canonical link is $g(\pi) = \log(\pi/(1-\pi))$, which is the log-odds or the logit of π. The logistic regression models are also called logit models.

For scalar x,

$$\pi(x) = \frac{\exp\{\alpha + \beta x\}}{1 + \exp\{\alpha + \beta x\}},$$

and $\pi(x)$ has a nonlinear but monotone relationship with x. As x increases, $\pi(x)$ increases when $\beta > 0$ and decreases when $\beta < 0$. In addition, a fixed change in x often has less impact when $\pi(x)$ is near 0 or 1 than when $\pi(x)$ is near 0.5 because $\pi'(x) = \beta\pi(x)(1 - \pi(x))$. □

Example 3.4.3. For the Poisson model with mean λ, we have

$$\log f(x; \lambda) = \log\left\{\frac{\lambda^x e^{-\lambda}}{x!}\right\} = x\log\lambda - \lambda - \log x!, x = 0, 1, 2, \ldots,$$

so we have a canonical parameter $\theta = \log\lambda$, dispersion parameter $\phi = 1$ and $b(\theta) = \lambda = e^\theta$. Therefore, the canonical link is $g(\lambda) = \log(\lambda)$ and $\lambda = \exp(\mathbf{x}^\top\boldsymbol{\beta})$, which is called a Poisson log-linear model. □

3.4.1.2 Maximum likelihood estimation

The log-likelihood function for GLM given observations $\{(\mathbf{x}_1, y_1), \ldots, (\mathbf{x}_n, y_n)\}$ is

$$\ell(\boldsymbol{\beta}) = \sum_{i=1}^{n}\log f(y_i; \mu_i) = \sum_{i}\{y_i\theta_i - b(\theta_i)\}/\phi - \sum_{i=1}^{n}c(y_i, \phi), \qquad (3.5)$$

where the link function $g(\mu) = \mathbf{x}^\top\boldsymbol{\beta} \triangleq \eta$ and $\mu = b'(\theta)$. MLE tries to maximize the log-likelihood $\ell(\boldsymbol{\beta})$. However, in general, there is no closed-form solution for maximizing (3.5).

Newton-Raphson algorithm or Fisher scoring algorithm is usually used. Note that MLE tries to maximize log-likelihood $\ell(\beta)$, i.e., solve the score equation $S(\beta) = \partial\ell(\beta)/\partial\beta = 0$. Newton-Raphson algorithm approximates the log-likelihood by a local quadratic function and then maximizes it. Taking the Taylor series of the log-likelihood function around the starting value β_0, we have the following result:

$$\frac{\partial\ell(\beta)}{\partial\beta} \approx \frac{\partial\ell(\beta_0)}{\partial\beta} + \frac{\partial^2\ell(\beta_0)}{\partial\beta\partial\beta^\top}(\hat{\beta} - \beta_0).$$

Setting $\frac{\partial\ell(\beta)}{\partial\beta} = 0$, we have

$$\hat{\beta} = \beta_0 - \left\{\frac{\partial^2\ell(\beta_0)}{\partial\beta\partial\beta^\top}\right\}^{-1}\frac{\partial\ell(\beta_0)}{\partial\beta}.$$

The Newton-Raphson algorithm iterates the above procedure until convergence and, in general, has the following iteration formula:

$$\beta_{m+1} = \beta_m - \left\{\frac{\partial^2\ell(\beta_m)}{\partial\beta\partial\beta^\top}\right\}^{-1}\frac{\partial\ell(\beta_m)}{\partial\beta},$$

where β_m is the update of β at mth iteration. The above algorithm is called *Newton-Raphson algorithm*.

If we replace the Hessian matrix $\frac{\partial^2\ell(\beta_0)}{\partial\beta\partial\beta^\top}$ with its expectation, then we get the Fisher scoring algorithm, i.e., the *Fisher scoring method* has the following update formula:

$$\beta_{m+1} = \beta_m + \{\mathcal{I}(\beta_m)\}^{-1}\frac{\partial\ell(\beta_m)}{\partial\beta},$$

where

$$\mathcal{I}(\beta) = -\mathrm{E}\left\{\frac{\partial^2\ell(\beta)}{\partial\beta\partial\beta^\top}\right\} = \mathrm{Var}\left\{\frac{\partial\ell(\beta)}{\partial\beta}\right\}.$$

In many cases, $\mathcal{I}(\beta)$ is easier to calculate, and it generally stabilizes the algorithm.

For the generalized linear model, we have

$$\frac{\partial\log f(y;\mu)}{\partial\beta} = \frac{\partial\log f(y;\mu)}{\partial\theta}\frac{\partial\theta}{\partial\mu}\frac{\partial\mu}{\partial\beta},$$

$$\frac{\partial\log f(y;\mu)}{\partial\theta} = \frac{y-b'(\theta)}{\phi} = \frac{y-\mu}{\phi},$$

$$\frac{\partial\theta}{\partial\mu} = \left(\frac{\partial\mu}{\partial\theta}\right)^{-1} = \left(\frac{\partial^2 b(\theta)}{\partial\theta^2}\right)^{-1} = \{b''(\theta)\}^{-1} = \frac{1}{v(\mu)},$$

$$\frac{\partial\mu}{\partial\beta} = \left(\frac{\partial\mu}{\partial g(\mu)}\right)\left(\frac{\partial g(\mu)}{\partial\beta}\right) = \left(\frac{\partial g(\mu)}{\partial\mu}\right)^{-1}\left(\frac{\partial\mathbf{x}^\top\beta}{\partial\beta}\right) = \frac{1}{g'(\mu_i)}\mathbf{x}.$$

Therefore,

$$\frac{\partial \log f(y; \mu)}{\partial \beta} = \frac{y - \mu}{\phi v(\mu)} \frac{1}{g'(\mu_i)} \mathbf{x},$$

and the likelihood equation is

$$S(\beta) = \frac{\partial \ell(\beta)}{\partial \beta} = \sum_{i=1}^{n} \frac{y_i - \mu_i}{\phi v(\mu_i)} \frac{1}{g'(\mu_i)} \mathbf{x}_i = \phi^{-1} \sum (y_i - \mu_i) w_i g'(\mu) \mathbf{x}_i^{\top} = 0,$$

where $w_i = [v(\mu_i)g'(\mu_i)^2]^{-1}$. We can rewrite the above equation in a matrix notation as

$$S(\beta) = \phi^{-1} \mathbf{X}^{\top} \mathbf{W} \mathbf{\Delta}(\mathbf{y} - \boldsymbol{\mu}),$$

where $\mathbf{X} = (\mathbf{x}_1, \mathbf{x}_2, \ldots, \mathbf{x}_n)^{\top}$, $\mathbf{W} = \text{diag}\{w_i\}$, and $\mathbf{\Delta} = \text{diag}\{g'(\mu)\}$.

The second derivative of log-likelihood is

$$\frac{\partial^2 \ell(\beta)}{\partial \beta \partial \beta^{\top}} = -\phi^{-1} \mathbf{X}^{\top} \mathbf{W} \mathbf{\Delta} \frac{\partial \mu}{\partial \beta^{\top}} + \phi^{-1} \mathbf{X}^{\top} \frac{\partial \mathbf{W} \mathbf{\Delta}}{\partial \beta^{\top}} (\mathbf{y} - \boldsymbol{\mu}),$$

so that

$$I(\beta) = -\mathrm{E}\left[\frac{\partial^2 \ell(\beta)}{\partial \beta \partial \beta^{\top}}\right] = \phi^{-1} \mathbf{X}^{\top} \mathbf{W} \mathbf{\Delta} \frac{\partial \mu}{\partial \beta^{\top}} + 0 = \phi^{-1} \mathbf{X}^{\top} \mathbf{W} \mathbf{X}.$$

In addition, note that

$$-\mathrm{E}\left[\frac{\partial^2 \ell(\beta)}{\partial \beta \partial \phi}\right] = -\mathrm{E}\left[\frac{\partial}{\partial \phi} \phi^{-1} \mathbf{X}^{\top} \mathbf{W} \mathbf{\Delta}(\mathbf{y} - \boldsymbol{\mu})\right] = \phi^{-2} \mathbf{X}^{\top} \mathbf{W} \mathbf{\Delta} \mathrm{E}(\mathbf{y} - \boldsymbol{\mu}) = 0.$$

So the estimation of ϕ doesn't affect the large sample variance of $\hat{\beta}$. Therefore, the asymptotic variance of $\hat{\beta}$ is

$$\text{Var}_{\infty}(\hat{\beta}) = \phi^{-1}(\mathbf{X}^{\top} \mathbf{W} \mathbf{X})^{-1},$$

where Var_{∞} indicates the limiting or asymptotic variance.

If the canonical link is used, $g(\mu) = \theta = \eta$, $\partial \mu_i / \partial \eta_i = \partial b'(\theta) / \partial \theta_i = b''(\theta)$. Then

$$\frac{\partial \log f(y; \mu)}{\partial \beta} = \frac{y - \mu}{\phi v(\mu)} b''(\theta) \mathbf{x} = \frac{y - \mu}{\phi} \mathbf{x},$$

$$\frac{\partial^2 \log f(y; \mu)}{\partial \beta \beta^{\top}} = -\frac{\partial \mu}{\partial \beta^{\top}} \mathbf{x}/\phi = -b''(\theta) \frac{\partial \theta}{\partial \beta^{\top}} \mathbf{x}/\phi = -b''(\theta) \mathbf{x} \mathbf{x}^{\top}/\phi = -v(\mu) \mathbf{x} \mathbf{x}^{\top}.$$

Therefore, the second derivative of the log-likelihood is

$$\frac{\partial^2 \ell(\boldsymbol{\beta})}{\partial \boldsymbol{\beta} \partial \boldsymbol{\beta}^\top} = \sum_{i=1}^{n} -v(\mu_i) \mathbf{x}_i \mathbf{x}_i^\top < 0.$$

The log-likelihood is a concave function of $\boldsymbol{\beta}$ and the solution is unique.

Based on the above derivations, Fisher scoring for the generalized linear model is

$$\boldsymbol{\beta}_{m+1} = \boldsymbol{\beta}_m + (\mathbf{X}^\top \mathbf{W} \mathbf{X})^{-1} \mathbf{X}^\top \mathbf{W} \boldsymbol{\Delta} (\mathbf{y} - \boldsymbol{\mu}).$$

Note that

$$\boldsymbol{\beta}_{m+1} = (\mathbf{X}^\top \mathbf{W} \mathbf{X})^{-1} \mathbf{X}^\top \mathbf{W} \{ \mathbf{X} \boldsymbol{\beta}_m + \boldsymbol{\Delta} (\mathbf{y} - \boldsymbol{\mu}) \} = (\mathbf{X}^\top \mathbf{W} \mathbf{X})^{-1} \mathbf{X}^\top \mathbf{W} \mathbf{y}_m^*,$$

where the new response $\mathbf{y}_m^* = \mathbf{X} \boldsymbol{\beta}_m + \boldsymbol{\Delta} (\mathbf{y} - \boldsymbol{\mu})$ with the predictor $\mathbf{X}$ yields the generalized least square estimate with weight $\mathbf{W}$.

3.4.2 Mixtures of generalized linear models

To account for heterogeneity in regression coefficients, mixtures of generalized linear models (Grün and Leisch, 2008) are proposed to combine the flexibility of mixture models and generalized linear models.

The density of a mixture of C-component GLMs is given by

$$f(y_i; \boldsymbol{\Psi}) = \sum_{c=1}^{C} \pi_c f(y_i; \theta_{ic}, \phi_c), \qquad (3.6)$$

where $\boldsymbol{\Psi}$ collects all unknown parameters and

$$\log f(y_i; \theta_{ic}, \phi_c) = \{ y_i \theta_{ic} - b(\theta_{ic}) \} / \phi_c - c(y_i, \phi_c).$$

For the cth component, the mean is μ_{ic}, and $g_c(\mu_{ic}) = \mathbf{x}_i^\top \boldsymbol{\beta}_c$ is the linear predictor, where $g_c(\cdot)$ is the link function of the cth component. Please refer to Grün and Leisch (2008) about the identifiability results and real data applications of the model (3.6).

3.4.2.1 Maximum likelihood estimation

Similar to other mixture models, the mixture of GLMs model (3.6) can be estimated through the EM algorithm of Algorithm 3.4.1. The parameter updates in M step can be done using the methods introduced in Section 3.4.1.2.

Algorithm 3.4.1 EM algorithm for Mixtures of GLM

Given the initial parameter estimate $\boldsymbol{\Psi}^{(0)}$, at $(k+1)th$ step, the EM algorithm iterates the following E-step and M-step until convergence.

E-step: Calculate the classification probabilities:

$$p_{ic}^{(k+1)} = \frac{\pi_c^{(k)} f(y_i; \theta_{ic}, \phi_c)}{\sum_{l=1}^{C} \pi_l^{(k)} f(y_i; \theta_{il}, \phi_l)}, \quad i = 1, \ldots, n; c = 1, \ldots, C.$$

M-step: Update parameter estimates:

$$\pi_c^{(k+1)} = \frac{\sum_{i=1}^{n} p_{ic}^{(k+1)}}{n},$$

Update the estimator of $\boldsymbol{\beta}_c$ by solving the following equation:

$$\sum_{i=1}^{n} p_{ic}^{(k+1)} w(\mu_{ic})(y_i - \mu_{ic}) g_c'(\mu_{ic}) \mathbf{x}_i = \mathbf{0},$$

where $g_c(\mu_{ic}) = \mathbf{x}_i^\top \boldsymbol{\beta}_c$ and $w(\mu_{ic}) = 1/[\{g_c'(\mu_{ic})\}^2 V(\mu_{ic})]$ with $V(\mu_{ic}) = \phi_c v(\mu_{ic})$ being the variance of Y_i.

3.5 Hidden Markov model regression

Hidden Markov Model Regression (HMMR) proposed by Fridman (1993) is a natural extension of the Hidden Markov Model (HMM) to regression analysis. The parameters of the regression model are assumed to be determined by the outcome of a finite-state Markov chain and the error terms; the error terms are conditionally independent and normally distributed, with mean zero and state-dependent variance.

3.5.1 Model setting

Let us consider a scenario in which a stochastic triplet sequence $\{s_t, \mathbf{x}_t, y_t\}_0^\infty$ is a Markov process and s_t is a latent, unobserved, homogeneous, and irreducible Markov chain with a finite state space $\{1, \ldots, S\}$. The transition matrix of s_t is represented by Γ, which is an $S \times S$ matrix with elements $\gamma_{jk} = P(s_{t+1} = k | s_t = j)$ denoting the transition probability from state j to state k and satisfying $\sum_{k=1}^{S} \gamma_{jk} = 1$ for all j. Additionally, let $\boldsymbol{\pi} = (\pi_1, \ldots, \pi_S)$ be the initial probabilities of the states satisfying $\sum_{k=1}^{S} \pi_k = 1$. All initial probabilities and transition probabilities are positive.

The HMMR assumes that the response variable y_t follows a normal regression model with mean $\mathbf{x}_t^\top \boldsymbol{\beta}_k$ and variance σ_k^2 given covariates $\mathbf{x}_t$ and state $s_t = k$, and is independent of $(s_{t-1}, \mathbf{x}_{t-1}, y_{t-1})$. Thus, we have

$$p(s_{t+1}|s_t, \mathbf{x}_t, y_t) = p(s_{t+1}|s_t),$$
$$Y_t|\mathbf{x}_t, s_t = k \sim N\{\mathbf{x}_t^\top \boldsymbol{\beta}_k, \sigma_k^2\},$$

where $N(\mu, \sigma^2)$ is the normal distribution with mean μ and variance σ^2 and has the density function $\phi(\cdot|\mu, \sigma^2)$. Let $p_k(\mathbf{x}, y) = P(y_t = y|\mathbf{x}_t = \mathbf{x}, s_t = k) = \phi\{y|\mathbf{x}^\top \boldsymbol{\beta}_k, \sigma_k^2\}$ for $k = 1, \ldots, S$, where $P(\mathbf{x}, y) = \text{diag}\{p_1(\mathbf{x}, y), \ldots, p_S(\mathbf{x}, y)\}$ is an $S \times S$ diagonal matrix.

3.5.2   Estimation algorithm

Define $\{(\mathbf{x}_t, y_t), t = 1, \ldots, T\}$ to be a finite realization of $\{\mathbf{X}_t, Y_t\}_0^\infty$. Then, the likelihood function for the observed data is

$$\ell(\boldsymbol{\pi}, \Gamma, \boldsymbol{\theta}) = \boldsymbol{\pi} P(\mathbf{x}_1, y_1) \Gamma P(\mathbf{x}_2, y_2) \Gamma P(\mathbf{x}_3, y_3) \ldots \Gamma P(\mathbf{x}_T, y_T) \mathbf{1}^\top, \qquad (3.7)$$

where $\boldsymbol{\theta} = \{\boldsymbol{\beta}_k, \sigma_k^2, k = 1, \ldots, S\}$. Define the latent state variable $\mathbf{z}_t = (z_{t1}, \ldots, z_{tS})$, where z_{tk} is the associated indicator of S_t. That is

$$z_{tk} = \begin{cases} 1, & \text{if } s_t = k, \\ 0, & \text{otherwise.} \end{cases}$$

Then, the complete log-likelihood function for $\{(x_t, y_t, \mathbf{z}_t), t = 1, \ldots, T\}$ is

$$\ell_c = \sum_{k=1}^{S} z_{1k} \log(\pi_k) + \sum_{t=2}^{T} \sum_{j=1}^{S} \sum_{k=1}^{S} z_{t-1,j} z_{tk} \log(\gamma_{jk}) + \sum_{t=1}^{T} \sum_{k=1}^{S} z_{tk} \log \phi\{y_t|\mathbf{x}_t^\top \boldsymbol{\beta}_k, \sigma_k^2\}.$$
$$(3.8)$$

Given the entire sequence of $(\mathbf{x}_t, y_t)$ and the estimator of $(\boldsymbol{\pi}, \Gamma, \boldsymbol{\theta})$, in the E-step, the expectation of the complete log-likelihood (3.8) is calculated. This is equivalent to calculating the conditional expectation:

$$E(z_{tk}|(\mathbf{x}_1, y_1, \ldots, \mathbf{x}_T, y_T)) \triangleq r_{tk},$$
$$E(z_{t-1,j} z_{tk}|(\mathbf{x}_1, y_1, \ldots, \mathbf{x}_T, y_T)) \triangleq h_{tjk}.$$

Note that r_{tk} denotes the conditional probability of being in state k at time t given the entire observed sequence, and h_{tjk} is the conditional probability of being in state k at time t and in state j at time $t-1$ given the entire observed sequence. In order to calculate r_{tk} and h_{tjk}, define two $1 \times S$ vectors of forward and backward probabilities, respectively, as

$$
\boldsymbol{\alpha}_t = (\alpha_{t1}, \dots, \alpha_{tS})^\top = \boldsymbol{\pi} P(\mathbf{x}_1, y_1) \Gamma P(\mathbf{x}_2, y_2) \dots, \Gamma P(\mathbf{x}_t, y_t),
$$
$$
\boldsymbol{\beta}_t = (\beta_{t1}, \dots, \beta_{tS})^\top = \Gamma P(\mathbf{x}_{t+1}, y_{t+1}) \Gamma P(\mathbf{x}_{t+2}, y_{t+2}) \dots, \Gamma P(\mathbf{x}_T, y_T) \mathbf{1}^\top,
$$

for $t = 1, \dots, T$. Then, the conditional probabilities r_{tk} and h_{tjk} can be calculated as

$$
r_{tk} = \alpha_{tk} \beta_{tk} / \ell(\boldsymbol{\pi}, \Gamma, \boldsymbol{\theta}),
$$
$$
h_{tjk} = \alpha_{t-1,j} \gamma_{jk} p_k(x_t, y_t) \beta_{tk} / \ell(\boldsymbol{\pi}, \Gamma, \boldsymbol{\theta}),
$$

where $\ell(\boldsymbol{\pi}, \Gamma, \boldsymbol{\theta})$ is defined in (3.7). In the M step, the conditional expectation of the complete likelihood (3.8) is maximized. Note that the conditional expectation consists of three separate parts, namely,

$$
\mathrm{E}(\ell_c | (\mathbf{x}_1, y_1, \dots, \mathbf{x}_T, y_T)) = L_1(\boldsymbol{\pi}) + L_2(\Gamma) + L_3(\boldsymbol{\theta}),
$$

where $L_1(\boldsymbol{\pi}) = \sum_{k=1}^{S} r_{1k} \log(\boldsymbol{\pi})$, $L_2(\Gamma) = \sum_{t=2}^{T} \sum_{j=1}^{S} \sum_{k=1}^{S} h_{tjk} \log(\gamma_{jk})$, and

$$
L_3(\boldsymbol{\theta}) = \sum_{t=1}^{T} \sum_{k=1}^{S} r_{tk} \log \phi\{y_t | \mathbf{x}_t^\top \boldsymbol{\beta}_k, \sigma_k^2\}.
$$

Hence, the maximization of $\mathrm{E}(\ell_c | (\mathbf{x}_1, y_1, \dots, \mathbf{x}_T, y_T))$ can be done by optimizing one part at a time. It is easy to see that given $\sum_{k=1}^{S} \pi_k = 1$, the optimization result for maximizing $L_1(\boldsymbol{\pi})$ is $\pi_k = r_{1k}$. In addition, for $j = 1, \dots, S$, we can maximize $\sum_{t=2}^{T} \sum_{k=1}^{S} h_{tjk} \log(\gamma_{jk})$ with constraints $\sum_{k=1}^{S} \gamma_{jk} = 1$ and $\sum_{k=1}^{S} h_{tjk} = r_{t-1,j}$, which gives

$$
\gamma_{jk} = \frac{\sum_{t=2}^{T} h_{tjk}}{\sum_{t=2}^{T} r_{t-1,j}}.
$$

$L_3(\boldsymbol{\theta})$ can be maximized similarly to the regular mixture of linear regressions model:

$$
\boldsymbol{\beta}_k = \left(\sum_{t=1}^{T} r_{tk} \mathbf{x}_t \mathbf{x}_t^\top \right)^{-1} \sum_{t=1}^{T} r_{tk} \mathbf{x}_t y_t,
$$
$$
\sigma_k^2 = \frac{1}{\sum_{t=1}^{T} r_{tk}} \sum_{t=1}^{T} r_{tk} (y_t - \mathbf{x}_t^\top \boldsymbol{\beta}_t)^2,
$$

for $k = 1, \dots, S$.

3.6   Mixtures of linear mixed-effects models

Linear mixed models are widely used for longitudinal/panel data when multiple correlated measurements are made on each unit of interest. However, if there is heterogeneity in the data, that is, units come from several distinct clusters, then finite mixtures of linear mixed-effects models (Yau et al., 2003; Celeux et al., 2005; Bai et al., 2016) are more appropriate.

Consider I subjects, where n_i repeated measurements are taken on the ith subject. Then, for each $i = 1, \ldots, I$, let z_i be a latent variable with $P(z_i = c) = \pi_c$, $c = 1, \ldots, C$. Given $z_i = c$, it is assumed that the response $\mathbf{y}_i \in \mathbb{R}^{n_i}$ follows a linear mixed effects model as

$$\mathbf{y}_i = \mathbf{X}_i \boldsymbol{\beta}_c + \mathbf{U}_i \mathbf{b}_{ic} + \mathbf{e}_{ic},$$

where $\mathbf{X}_i$ is an $n_i \times p$ covariate matrix for fixed effects, $\boldsymbol{\beta}_c$ is a $p \times 1$ vector of fixed-effect coefficients, $\mathbf{U}_i$ is a $n_i \times q$ covariate matrix for random effects, $\mathbf{b}_{ic} \sim N_q(0, \boldsymbol{\Psi}_c)$ is a $q \times 1$ vector of random effect coefficients, $\mathbf{e}_{ic} \sim N_{n_i}(\mathbf{0}, \boldsymbol{\Lambda}_{ic})$ is an $n_i \times 1$ vector of random errors, and all the $\mathbf{b}_{ic}$s and $\mathbf{e}_{ic}$s are independent of each other. Then, conditional on $z_i = c$, the joint distribution of $(\mathbf{y}_i, \mathbf{b}_{ic})$ is

$$\begin{pmatrix} \mathbf{y}_i \\ \mathbf{b}_{ic} \end{pmatrix} \Big| z_i = c \sim N_{n_i+q} \left(\begin{pmatrix} \mathbf{X}_i \boldsymbol{\beta}_c \\ \mathbf{0} \end{pmatrix}, \begin{pmatrix} \mathbf{U}_i \boldsymbol{\Psi}_c \mathbf{U}_i^\top + \boldsymbol{\Lambda}_{ic} & \mathbf{U}_i \boldsymbol{\Psi}_c \\ \boldsymbol{\Psi}_c \mathbf{U}_i^\top & \boldsymbol{\Psi}_c \end{pmatrix} \right),$$

and without observing z_i, the mixture distribution of $\mathbf{y}_i$ is

$$\mathbf{y}_i \sim \sum_{c=1}^{C} \pi_c N_{n_i} \left(\mathbf{X}_i \boldsymbol{\beta}_c, \mathbf{U}_i \boldsymbol{\Psi}_c \mathbf{U}_i^\top + \boldsymbol{\Lambda}_{ic} \right). \tag{3.9}$$

In the following, the error covariance matrices are assumed as $\boldsymbol{\Lambda}_{ic} = \sigma_c^2 \mathbf{R}_i$ for $i = 1, \ldots, I$ and $c = 1, \ldots C$, where $\mathbf{R}_i$ are identity matrices.

The ECM algorithm can be used to estimate model (3.9) and is summarized as below:

Algorithm 3.6.1 ECM algorithm for model (3.9)

Given the initial parameter estimate $\boldsymbol{\theta}^{(0)}$, at the $(k+1)th$ step, the ECM algorithm iterates the following E-step and CM-step until convergence.

E-step: Calculate the classification probabilities:

$$
p_{ic}^{(k+1)} = \frac{\pi_c^{(k)} \phi(\mathbf{y}_i; \mathbf{X}_i^\top \boldsymbol{\beta}_c^{(k)}, \mathbf{U}_i \boldsymbol{\Psi}_c^{(k)} \mathbf{U}_i^\top + \boldsymbol{\Lambda}_{ic}^{(k)})}{\sum_{l=1}^{C} \pi_l^{(k)} \phi(\mathbf{y}_i; \mathbf{X}_i^\top \boldsymbol{\beta}_l^{(k)}, \mathbf{U}_i \boldsymbol{\Psi}_l^{(k)} \mathbf{U}_i^\top + \boldsymbol{\Lambda}_{il}^{(k)})},
$$
$$
i = 1, \ldots, I; c = 1, \ldots, C.
$$

CM-step: Update parameter estimates:

M-0: Obtain $\pi_c^{(k+1)}, c = 1, \ldots, C$, as

$$
\pi_c^{(k+1)} = \frac{1}{I} \sum_{i=1}^{I} p_{ic}^{(k+1)}.
$$

M-1: Given $\sigma_c^2 = \sigma_c^{2(k)}$, obtain $\boldsymbol{\beta}_c^{(k+1)}, c = 1, \ldots, C$, as

$$
\boldsymbol{\beta}_c^{(k+1)} = \left\{ \sum_{i=1}^{I} p_{ic}^{(k+1)} \mathbf{X}_i \mathbf{X}_i^\top / \sigma_c^{2(k)} \right\}^{-1}
$$
$$
\times \left\{ \sum_{i=1}^{I} p_{ic}^{(k+1)} \mathbf{X}_i^\top \mathbf{R}_i^{-1} (\mathbf{y}_i - \mathbf{U}_i b_{ic}^{(k+1)}) / \sigma_c^{(k+1)} \right\}.
$$

M-2: Given $\boldsymbol{\beta}_c = \boldsymbol{\beta}_c^{(k+1)}$, obtain $\sigma_c^{2(k+1)}, c = 1, \ldots, C$:

$$
\sigma_c^{2(k+1)} = \frac{\sum_{i=1}^{I} p_{ic}^{(k+1)} \left\{ \mathbf{E}_{ic}^\top \mathbf{R}_i^{-1} \mathbf{E}_{ic} + \mathrm{tr}(\boldsymbol{\Omega}_{ic}^{(k+1)} \mathbf{U}_i^\top \mathbf{R}_i^{-1} \mathbf{U}_i) \right\}}{\sum_{i=1}^{I} p_{ic}^{(k+1)} n_i},
$$

where $\mathbf{E}_{ic} = \mathbf{y}_i - \mathbf{X}_c \boldsymbol{\beta}_c - \mathbf{U}_i \mathbf{b}_{ic}$ and $\boldsymbol{\Omega}_{ic} = \left(\boldsymbol{\Psi}_c^{-1} + \frac{1}{\sigma_c^2} \mathbf{U}_i^\top \mathbf{R}_i^{-1} \mathbf{U}_i \right)^{-1}$.

M-3 : Obtain $\boldsymbol{\Psi}_c^{(k+1)}, c = 1, \ldots, C$, as

$$
\boldsymbol{\Psi}_c^{(k+1)} = \frac{\sum_{i=1}^{I} p_{ic}^{(k+1)} \left\{ \mathbf{b}_{ic}^{(k+1)} \mathbf{b}_{ic}^{(k+1)\top} + \boldsymbol{\Omega}_{ic}^{(k+1)} \right\}}{\sum_{i=1}^{I} p_{ic}^{(k+1)}}.
$$

3.7   Mixtures of multivariate regressions

3.7.1   Multivariate regressions with normal mixture errors

Multivariate regression analysis has been widely used in many branches where there are $d \geq 1$ response variables and $p \geq 1$ independent variables. The model is generally assumed as:

$$\mathbf{y}_i = \mathbf{B}^\top \mathbf{x}_i + \boldsymbol{\epsilon}_i,$$

where $\mathbf{y}_i = (y_{i1}, \dots, y_{id})^\top$ is a d-dimensional vector of response variables for the i-th sample unit, $\mathbf{x}_i$ is a p-dimensional vector of predictor variables, with the first element 1 for an intercept term, $\mathbf{B}$ is a matrix of dimension $p \times d$ whose (j, k)th element is the regression coefficient of the jth predictor on the kth response, and $\boldsymbol{\epsilon}_i$ is a d-dimensional random vector of the error terms corresponding to the ith observation. A classical multivariate linear regression model assumes that $\boldsymbol{\epsilon}_i$s are independent and identically distributed random vectors whose distribution is assumed to be a multivariate Gaussian with a d-dimensional vector of zeros as mean and a $d \times d$ positive-definite covariance matrix $\boldsymbol{\Sigma}$ as covariance matrix, that is $\boldsymbol{\epsilon}_i \sim N_d(\mathbf{0}, \boldsymbol{\Sigma})$.

To relax the normality assumption of the error terms, a mixture of multivariate normal distributions can be used by assuming that

$$\boldsymbol{\epsilon}_i \sim \sum_{c=1}^{C} \pi_c N(\boldsymbol{\lambda}_c, \boldsymbol{\Sigma}_c),$$

where π_cs are positive weights that sum to 1, and the $\boldsymbol{\lambda}_c$s are d-dimensional mean vectors satisfying $\sum_{c=1}^{C} \pi_c \boldsymbol{\lambda}_c = \mathbf{0}$. As a result, the probability density function of $\mathbf{y}_i$ becomes

$$\mathbf{y}_i \sim \sum_{c=1}^{C} \pi_c \phi_d(\mathbf{y}; \boldsymbol{\mu}_{ic}, \boldsymbol{\Sigma}_c), \tag{3.10}$$

where $\boldsymbol{\mu}_{ic} = \boldsymbol{\lambda}_c + \mathbf{B}^\top \mathbf{x}_i$, and $\phi_d(\mathbf{y}; \boldsymbol{\mu}, \boldsymbol{\Sigma})$ is the density of the d-dimensional Gaussian distribution $MVN(\boldsymbol{\mu}, \boldsymbol{\Sigma})$, i.e.,

$$\phi_d(\mathbf{y}; \boldsymbol{\mu}, \boldsymbol{\Sigma}) = (2\pi)^{-\frac{d}{2}} |\boldsymbol{\Sigma}|^{-\frac{1}{2}} \exp\left\{ -\frac{1}{2}(\mathbf{y} - \boldsymbol{\mu})^\top \boldsymbol{\Sigma}^{-1}(\mathbf{y} - \boldsymbol{\mu}) \right\}.$$

Therefore, $\mathbf{y}_i$ has a mixture distribution with heterogeneous intercepts and the same covariate coefficients across different components. In Section 3.7.3, we will introduce more general cases when both the intercepts and covariate coefficients vary across mixture components.

3.7.2 Parameter estimation

Given a random sample of n observations, the log-likelihood of the proposed model is

$$\ell(\boldsymbol{\theta}) = \sum_{i=1}^{n} \log \left(\sum_{c=1}^{C} \pi_c \phi_d(\mathbf{y}_i; \boldsymbol{\mu}_{ic}, \boldsymbol{\Sigma}_c) \right), \tag{3.11}$$

and the complete log-likelihood of the model is

$$\ell_c(\boldsymbol{\theta}) = \sum_{i=1}^{n} \sum_{c=1}^{C} [\log \pi_c + \log \phi_d(\mathbf{y}_i; \boldsymbol{\mu}_{ic}, \boldsymbol{\Sigma}_c)],$$

which can be divided into two parts $\ell_c = \ell_{c1} + \ell_{c2}$ up to a constant, where

$$\ell_{c1} = \sum_{c=1}^{C} z_{.c} \log \pi_c,$$

$$\ell_{c2} = -\frac{1}{2} \sum_{c=1}^{C} z_{.c} \log|\boldsymbol{\Sigma}_c| - \frac{1}{2} \sum_{i=1}^{n} \sum_{c=1}^{C} z_{ic}(\mathbf{y}_i - \boldsymbol{\mu}_{ic})^{\top} \boldsymbol{\Sigma}_c^{-1}(\mathbf{y}_i - \boldsymbol{\mu}_{ic}),$$

where $z_{.c} = \sum_{i=1}^{n} z_{ic}$, and $|\mathbf{A}|$ denotes the determinant of a matrix $\mathbf{A}$. Note that ℓ_{c1} can be simply maximized by $\hat{\pi}_c = z_{.c}/n$.

In order to maximize ℓ_{c2}, define $\boldsymbol{\Gamma}$ as a matrix of dimensions $(C + p) \times d$, which is built by combining $\boldsymbol{\lambda}_1^{\top}, \ldots, \boldsymbol{\lambda}_C^{\top}$ and matrix $\mathbf{B}$ by rows. Similarly, define $\boldsymbol{\mu}_c$ as an $n \times d$ matrix whose rows are $\boldsymbol{\mu}_{1c}^{\top}, \ldots, \boldsymbol{\mu}_{nc}^{\top}$. Since $\boldsymbol{\mu}_{ic} = \boldsymbol{\lambda}_c + \mathbf{B}^{\top}\mathbf{x}_i$, it can be seen that $\boldsymbol{\mu}_c = \mathbf{X}_c \boldsymbol{\Gamma}$, where $\mathbf{X}_c = (\mathbf{0}_c \ \mathbf{X})$, $\mathbf{X} = (\mathbf{x}_1, \ldots, \mathbf{x}_n)^{\top}$, and $\mathbf{0}_c$ is a $n \times C$-dimensional matrix whose c-th column elements are all equal to 1 and all the other elements are equal to 0. In addition, let $\mathbf{z}_{+c} = (z_{1c}, \ldots, z_{nc})^{\top}$ and $\mathbf{Y} = (\mathbf{y}_1, \ldots, \mathbf{y}_n)^{\top}$, then ℓ_{c2} can be written as

$$\ell_{c2} = -\frac{1}{2} \sum_{c=1}^{C} z_{.c} \log|\boldsymbol{\Sigma}_c| - \frac{1}{2} \sum_{c=1}^{C} \text{tr}(\boldsymbol{\Sigma}_c^{-1}\mathbf{D}_c), \tag{3.12}$$

where $\mathbf{D}_c = (\mathbf{Y} - \mathbf{X}_c\boldsymbol{\Gamma})^{\top} \text{diag}(\mathbf{z}_{+c})(\mathbf{Y} - \mathbf{X}_c\boldsymbol{\Gamma})$, and $\text{diag}(\mathbf{z}_{+c})$ is an $n \times n$ diagonal matrix whose elements are $\mathbf{z}_{+c}$. Then, ℓ_{c2} of (3.12) can be maximized by taking a first-order derivative and setting the vector to equal to $\mathbf{0}$. If $\mathbf{M} = \sum_{c=1}^{C} \boldsymbol{\Sigma}_c^{-1} \otimes [\mathbf{X}_c^{\top} \text{diag}(\mathbf{z}_{+c})\mathbf{X}_c]$ is nonsingular, where $\otimes$ is the Kronecker product operator, the solutions for maximizing (3.12) are

$$vec(\hat{\boldsymbol{\Gamma}}) = \mathbf{M}^{-1}\mathbf{N}vec(\mathbf{Y}) \text{ and } \hat{\boldsymbol{\Sigma}}_c = z_{.c}^{-1}\mathbf{D}_c, c = 1, \ldots, C,$$

where $\mathbf{N} = \sum_{c=1}^{C} \boldsymbol{\Sigma}_c^{-1} \otimes [\mathbf{X}_c^{\top} \text{diag}(\mathbf{z}_{+c})]$ and $vec(\mathbf{A})$ is the vector formed by stacking columns of the matrix $\mathbf{A}$ one underneath the other. Once $\hat{\boldsymbol{\Gamma}}$ is calculated, $\hat{\boldsymbol{\lambda}}_1, \ldots, \hat{\boldsymbol{\lambda}}_C$ and $\hat{\mathbf{B}}$ can be obtained directly, followed by $\hat{\boldsymbol{\mu}}_{ic}$ being computed as $\hat{\boldsymbol{\mu}}_{ic} = \hat{\boldsymbol{\lambda}}_c + \hat{\mathbf{B}}^{\top}\mathbf{x}_i$ for $i = 1, \ldots, n$ and $c = 1, \ldots, C$.

To sum up, the EM algorithm for maximizing (3.11) can be summarized as follows.

Algorithm 3.7.1 EM algorithm for maximizing (3.11)

Given an initial parameter estimate $\boldsymbol{\theta}^{(0)}$, at the $(k+1)th$ step, the EM algorithm iterates the following E-step and M-step until convergence.

E-step: Calculate the classification probabilities:

$$p_{ic}^{(k+1)} = \frac{\pi_c^{(k)}\phi_d(\mathbf{y}_i; \hat{\boldsymbol{\mu}}_{ic}^{(k)}, \hat{\boldsymbol{\Sigma}}_c^{(k)})}{\sum_{h=1}^{C}\pi_h^{(k)}\phi_d(\mathbf{y}_i; \hat{\boldsymbol{\mu}}_{ih}^{(k)}, \hat{\boldsymbol{\Sigma}}_h^{(k)})}, i = 1,\dots,n; c = 1,\dots,C. \quad (3.13)$$

M-step: Update parameter estimates:

$$\pi_c^{(k+1)} = \frac{1}{n}\sum_{i=1}^{n}p_{ic}^{(k+1)}, \ c = 1,\dots,C.$$

Let

$$\hat{\boldsymbol{\Sigma}}_{c0}^{(k+1)} = \hat{\boldsymbol{\Sigma}}_c^{(k)},$$

$$\mathbf{M}_0^{(k+1)} = \sum_{c=1}^{C}(\hat{\boldsymbol{\Sigma}}_{c0}^{(k+1)})^{-1} \otimes [\mathbf{X}_c^\top \operatorname{diag}(\mathbf{p}_{+c}^{(k)})\mathbf{X}_c],$$

$$\mathbf{N}_0^{(k+1)} = \sum_{c=1}^{C}(\hat{\boldsymbol{\Sigma}}_{c0}^{(k+1)})^{-1} \otimes [\mathbf{X}_c^\top \operatorname{diag}(\mathbf{p}_{+c}^{(k)})].$$

Then for $j = 0,\dots$, repeat the following steps until convergence:

$$vec(\hat{\boldsymbol{\Gamma}}_{j+1}^{(k+1)}) = (\mathbf{M}_j^{(k+1)})^{-1}\mathbf{N}_j^{(k+1)}vec(\mathbf{Y}),$$

$$\hat{\boldsymbol{\Sigma}}_{c,j+1}^{(k+1)} = (p_{.c}^{(k)})^{-1}\mathbf{R}_{j+1}^{(k+1)\top}\operatorname{diag}(\mathbf{p}_{+c}^{(k)})\mathbf{R}_{j+1}^{(k+1)},$$

$$\mathbf{M}_{j+1}^{(k+1)} = \sum_{c=1}^{C}(\hat{\boldsymbol{\Sigma}}_{c,j}^{(k+1)})^{-1} \otimes [\mathbf{X}_c^\top \operatorname{diag}(\mathbf{p}_{+c}^{(k)})\mathbf{X}_c],$$

$$\mathbf{N}_{j+1}^{(k+1)} = \sum_{c=1}^{C}(\hat{\boldsymbol{\Sigma}}_{c,j}^{(k+1)})^{-1} \otimes [\mathbf{X}_c^\top \operatorname{diag}(\mathbf{p}_{+c}^{(k)})],$$

where $\mathbf{R}_{j+1}^{(k+1)} = \mathbf{Y} - \mathbf{X}_c\hat{\boldsymbol{\Gamma}}_{j+1}^{(k+1)}, p_{.c} = \sum_{i=1}^{n}p_{ic}$, and $\mathbf{p}_{+c} = (p_{1c},\dots,p_{nc})^\top$.

3.7.3 Mixtures of multivariate regressions

In model (3.10), more generally, we can also allow $\mathbf{B}$ to vary over different components, i.e.,

$$\mathbf{y}_i \sim \sum_{c=1}^{C} \pi_c \phi_d(\mathbf{y}; \boldsymbol{\mu}_{ic}, \boldsymbol{\Sigma}_c),$$

where $\boldsymbol{\mu}_{ic} = \mathbf{B}_c^{\top} \mathbf{x}_i$. Note that the intercept term is absorbed in the notation of $\mathbf{B}_c$. The above model is called *mixtures of multivariate regressions*. The log-likelihood of this C-component mixture of multivariate regressions is

$$\ell(\boldsymbol{\theta}) = \sum_{i=1}^{n} \log \left[\sum_{c=1}^{C} \pi_c \phi_d(\mathbf{y}_i; \mathbf{B}_c^{\top} \mathbf{x}_i, \boldsymbol{\Sigma}_c) \right], \tag{3.14}$$

where $\boldsymbol{\theta}$ is a collection of the unknown parameters.

To maximize (3.14), we can employ the EM algorithm. At the $(k+1)$th step, the E step can be calculated similarly to (3.13), and the M step updates π_c by

$$\pi_c^{(k+1)} = \frac{1}{n} \sum_{i=1}^{n} p_{ic}^{(k+1)}, \quad c = 1, \ldots, C,$$

and updates $\mathbf{B}_c$ and $\boldsymbol{\Sigma}_c$ by iterating the following two steps until convergence:

$$\mathbf{B}_c^{(k+1)} = (\mathbf{X}^{\top} W_c^{(k)} \mathbf{X})^{-1} \mathbf{X}^{\top} W_c^{(k)} \mathbf{Y},$$

$$\boldsymbol{\Sigma}_c^{(k+1)} = \frac{\sum_{i=1}^{n} p_{ic}^{(k+1)} (\mathbf{y}_i - \mathbf{B}_c^{(k+1)\top} \mathbf{x}_i)(\mathbf{y}_i - \mathbf{B}_c^{(k+1)\top} \mathbf{x}_i)^{\top}}{\sum_{i=1}^{n} p_{ic}^{(k+1)}},$$

where $\mathbf{X} = (\mathbf{x}_1, \ldots, \mathbf{x}_n)^{\top}$, $\mathbf{Y} = (\mathbf{y}_1, \ldots, \mathbf{y}_n)^{\top}$, and

$$W_c^{(k)} = \mathrm{diag}\{p_{1c}^{(k+1)}, \ldots, p_{nc}^{(k+1)}\}.$$

Additionally, if $\boldsymbol{\Sigma}_1 = \boldsymbol{\Sigma}_2 = \cdots = \boldsymbol{\Sigma}_C = \boldsymbol{\Sigma}$, the update of the common variance-covariance matrix $\boldsymbol{\Sigma}$ is

$$\boldsymbol{\Sigma}^{(k+1)} = \frac{\sum_{i=1}^{n} \sum_{c=1}^{C} p_{ic}^{(k+1)} (\mathbf{y}_i - \mathbf{B}_c^{(k+1)\top} \mathbf{x}_i)(\mathbf{y}_i - \mathbf{B}_c^{(k+1)\top} \mathbf{x}_i)^{\top}}{n}.$$

3.8 Seemingly unrelated clusterwise linear regression

In addition to finite mixtures of Gaussian linear regression models, Galimberti and Soffritti (2020) introduced a more flexible class of finite mixtures of multivariate Gaussian linear regression models in which different regressors can be used for different dependent variables, as in the seemingly unrelated regression context. The application of such a model can be found in multivariate economic data, where different economic variables are expected to be relevant in the prediction of different aspects of economic behaviors. The idea of applying the seemingly unrelated approach to multivariate regression can also be found in medicine (Keshavarzi et al., 2012), and food quality (Cadavez and Henningse, 2012), among others.

3.8.1 Mixtures of Gaussian seemingly unrelated linear regression models

A typical seemingly unrelated Gaussian linear regression model is defined as follows. Let $\mathbf{Y}_i = (Y_{i1}, \ldots, Y_{id})^\top$ be the vector of d-dimension dependent variables for the ith observation, $i = 1, \ldots, n$, and let $\mathbf{x}_i = (\mathbf{x}_{i1}^\top, \ldots, \mathbf{x}_{id}^\top)$ be the vector of p-dimension independent variables. Suppose that only p_j independent variables $(p_j \leq p)$, $\mathbf{x}_{ij}$, are relevant for the prediction of the jth dependent variable Y_{ij}. Then, a seemingly unrelated linear regression model for the conditional distribution of $Y_{ij}|\mathbf{X}_{ij} = \mathbf{x}_{ij}, j = 1, \ldots, d$ is defined as

$$
\begin{cases}
Y_{i1} = \lambda_1 + \mathbf{x}_{i1}^\top \tilde{\boldsymbol{\beta}}_1 + \epsilon_{i1}, \\
\vdots \\
Y_{ij} = \lambda_j + \mathbf{x}_{ij}^\top \tilde{\boldsymbol{\beta}}_j + \epsilon_{ij}, \quad i = 1, \ldots, n, \\
\vdots \\
Y_{id} = \lambda_d + \mathbf{x}_{id}^\top \tilde{\boldsymbol{\beta}}_d + \epsilon_{id},
\end{cases}
\tag{3.15}
$$

where $\lambda_j, \tilde{\boldsymbol{\beta}}_j$, and ϵ_{ij} are the intercept, regression coefficients, and the error term, respectively, for the ith observation of the jth dependent variable. In a classical model of this type, it is assumed that $\boldsymbol{\epsilon}_i = (\epsilon_{i1}, \ldots, \epsilon_{id})^\top$ are independent and identically distributed, and $\boldsymbol{\epsilon}_i \sim MVN_d(\mathbf{0}, \boldsymbol{\Sigma})$. In a matrix form, equation (3.15) can also be formulated as

$$
\mathbf{Y}_i = \boldsymbol{\lambda} + \tilde{\mathbf{X}}_i^\top \boldsymbol{\beta} + \boldsymbol{\epsilon}_i, \quad i = 1, \ldots, n,
$$

where $\boldsymbol{\lambda} = (\lambda_1, \ldots, \lambda_d)^\top, \boldsymbol{\beta} = (\tilde{\boldsymbol{\beta}}_1^\top, \ldots, \tilde{\boldsymbol{\beta}}_d^\top)^\top$, and $\tilde{\mathbf{X}}_i$ is a $g \times d$ matrix defined as

$$
\tilde{\mathbf{X}}_i = \begin{bmatrix}
\mathbf{x}_{i1} & \mathbf{0}_{p_1} & \cdots & \mathbf{0}_{p_1} \\
\mathbf{0}_{p_2} & \mathbf{x}_{i2} & \cdots & \mathbf{0}_{p_2} \\
\vdots & \vdots & & \vdots \\
\mathbf{0}_{p_d} & \mathbf{0}_{p_d} & \cdots & \mathbf{x}_{id}
\end{bmatrix}
$$

with $g = \sum_{j=1}^{d} p_j$. Since the d equations in (3.15) are correlated, they have to be considered jointly. This model is different from the classical multivariate linear regression, in that different vectors of regressors can be used for different dependent variables.

Given the above definitions, we now formally define the mixture of Gaussian seemingly unrelated linear regression models. Let

$$
\mathbf{Y}_i = \begin{cases}
\boldsymbol{\lambda}_1 + \tilde{\mathbf{X}}_i^\top \boldsymbol{\beta}_1 + \boldsymbol{\epsilon}_i, \boldsymbol{\epsilon}_i \sim N_d\left(\mathbf{0}, \boldsymbol{\Sigma}_1\right), & \text{with probability } \pi_1, \\
\quad\vdots & \\
\boldsymbol{\lambda}_c + \tilde{\mathbf{X}}_i^\top \boldsymbol{\beta}_c + \boldsymbol{\epsilon}_i, \boldsymbol{\epsilon}_i \sim N_d\left(\mathbf{0}, \boldsymbol{\Sigma}_c\right), & \text{with probability } \pi_c, \\
\quad\vdots & \\
\boldsymbol{\lambda}_C + \tilde{\mathbf{X}}_i^\top \boldsymbol{\beta}_C + \boldsymbol{\epsilon}_i, \boldsymbol{\epsilon}_i \sim N_d\left(\mathbf{0}, \boldsymbol{\Sigma}_C\right), & \text{with probability } \pi_C,
\end{cases}
$$

where $\sum_{c=1}^{C} \pi_c = 1$. If all the covariance matrices $\boldsymbol{\Sigma}_c, c = 1, \ldots, C$ are positive definite, then the conditional distribution of $\mathbf{Y}_i | \mathbf{X}_i = \mathbf{x}_i$ can be written as a weighted average of C Gaussian seemingly unrelated linear regression models:

$$
f\left(\mathbf{y}_i \mid \mathbf{x}_i; \boldsymbol{\theta}\right) = \sum_{c=1}^{C} \pi_c \phi\left(\mathbf{y}_i; \boldsymbol{\mu}_{ic}, \boldsymbol{\Sigma}_c\right), \mathbf{y}_i \in \mathbb{R}^d, \tag{3.16}
$$

where

$$
\boldsymbol{\mu}_{ic} = \boldsymbol{\lambda}_c + \tilde{\mathbf{X}}_i^\top \boldsymbol{\beta}_c,
$$

$\boldsymbol{\theta} = (\boldsymbol{\pi}^\top, \boldsymbol{\theta}_1^\top, \ldots, \boldsymbol{\theta}_C^\top)^\top$, $\boldsymbol{\theta}_c = (\boldsymbol{\lambda}_c^\top, \boldsymbol{\beta}_c^\top, v(\boldsymbol{\Sigma}_c)^\top)^\top$, and $v(\mathbf{A})$ contains only the distinct elements of $\mathbf{A}$.

3.8.2 Maximum likelihood estimation

Similar to other mixture models, the log-likelihood of the model (3.16) is

$$
\ell_c(\boldsymbol{\theta}) = \sum_{i=1}^{n} \log \left(\sum_{c=1}^{C} \pi_c \phi\left(\mathbf{y}_i; \boldsymbol{\lambda}_c + \tilde{\mathbf{X}}_i^\top \boldsymbol{\beta}_c, \boldsymbol{\Sigma}_c\right) \right),
$$

which is hard to be maximized.

The EM algorithm is summarized in Algorithm 3.8.1.

Algorithm 3.8.1 EM algorithm for (3.16)

Given an initial value of the parameter $\boldsymbol{\theta}^{(0)}$, update the following steps until convergence:

E-step: Calculate the classification probabilities:

$$p_{ic}^{(k+1)} = \frac{\hat{\pi}_c^{(k)}\phi\left(\mathbf{y}_i;\hat{\boldsymbol{\mu}}_{ic}^{(k)},\hat{\boldsymbol{\Sigma}}_c^{(k)}\right)}{\sum_{h=1}^C \hat{\pi}_h^{(k)}\phi\left(\mathbf{y}_i;\hat{\boldsymbol{\mu}}_{ih}^{(k)},\hat{\boldsymbol{\Sigma}}_h^{(k)}\right)}, i=1,\dots,n; c=1,\dots,C.$$

M-step: Update parameter estimates:

$$\hat{\pi}_c^{(k+1)} = \frac{1}{n}\sum_{i=1}^n p_{ic}^{(k)}, c=1,\dots,C.$$

for $c = 1,\dots,C$. In addition within the $(k+1)$th iteration, let $\tilde{\boldsymbol{\lambda}}_c^{(0)} = \hat{\boldsymbol{\lambda}}_c^{(k)}, \tilde{\boldsymbol{\beta}}_c^{(0)} = \hat{\boldsymbol{\beta}}_c^{(k)}$ and $\tilde{\boldsymbol{\Sigma}}_c^{(0)} = \hat{\boldsymbol{\Sigma}}_c^{(k)}$ be the starting values. Then, update the following until convergence:

$$\tilde{\boldsymbol{\lambda}}_c^{(t+1)} = \frac{\sum_{i=1}^n p_{ic}^{(k+1)}\left(\mathbf{y}_i - \tilde{\mathbf{X}}_i^\top\tilde{\boldsymbol{\beta}}_c^{(t)}\right)}{\sum_{i=1}^n p_{ic}^{(k+1)}},$$

$$\tilde{\boldsymbol{\beta}}_c^{(t+1)} = \left(\sum_{i=1}^n p_{ic}^{(k+1)}\tilde{\mathbf{X}}_i\left(\tilde{\boldsymbol{\Sigma}}_c^{(t)}\right)^{-1}\tilde{\mathbf{X}}_i^\top\right)^{-1}\sum_{i=1}^n p_{ic}^{(k+1)}\tilde{\mathbf{X}}_i\left(\tilde{\boldsymbol{\Sigma}}_c^{(t)}\right)^{-1}\left(\mathbf{y}_i - \tilde{\boldsymbol{\lambda}}_c^{(t+1)}\right),$$

$$\tilde{\boldsymbol{\Sigma}}_c^{(t+1)} = \frac{\sum_{i=1}^n p_{ic}^{(k+1)}\left(\mathbf{y}_i - \tilde{\boldsymbol{\lambda}}_c^{(t+1)} - \tilde{\mathbf{X}}_i^\top\tilde{\boldsymbol{\beta}}_c^{(t+1)}\right)\left(\mathbf{y}_i - \tilde{\boldsymbol{\lambda}}_c^{(t+1)} - \tilde{\mathbf{X}}_i^\top\tilde{\boldsymbol{\beta}}_c^{(t+1)}\right)^\top}{\sum_{i=1}^n p_{ic}^{(k+1)}}.$$

Parsimonious models are also available, where parsimony is attained by constraining the component covariance matrices based on their spectral decomposition. Please see Section 1.7.2 for more details about how to create parsimonious mixture models based on the spectral decomposition of covariance matrices.

3.9   Fitting mixture regression models using R

We will use the R package `MixSemiRob` to analyze mixture regression models. For example, there is `ethanol`, which gives the equivalence ratios and peak nitrogen oxide emissions in a study using pure ethanol as a spark-ignition engine fuel. To fit the `ethanol` with a finite mixture of linear regression models, one can run

```
set.seed(123)
library(MixSemiRob)
data(ethanol, package = "MixSemiRob")
em.out = mixreg(ethanol$NO, ethanol$Equivalence)
```

Results show that there exist 2 components with mixing proportions 0.47 and 0.53, regression coefficients $(0.56, 0.08)$ and $(1.25, -0.08)$, and standard deviations 0.0012 and 0.0012.

In addition, `mixtools` could be used to fit the mixture of experts. The following code fit the `ethanol` with a mixture of experts:

```
set.seed(123)
library(mixtools)
em.out = regmixEM(ethanol$Equivalence, ethanol$NO)
hme.out = hmeEM(ethanol$Equivalence, ethanol$NO,
                beta = em.out$beta)
```

Based on a mixture of experts model, the estimated regression coefficients are $(0.56, 0.09)$ and $(1.25, -0.08)$, with variances 0.0020 and 0.0005. The mixing proportions depend on the independent variables $\mathbf{x}$.

Furthermore, `mixtools` is also useful for fitting mixtures of generalized linear models as follows:

```
# Mixture of logistic regressions
res = logisregmixEM(y, x, verb = TRUE, epsilon = 1e-01)
# Mixture of multinomials
res = multmixEM(y)
# Mixture of Poisson regressions
res = poisregmixEM(y, x)
```

The `depmixS4` package is useful to fit the hidden Markov model. The `speed` dataset is a bivariate series of response times and accuracy scores of a single participant switching between slow/accurate responding and fast guessing on a lexical decision task. It contains three-time series with three variables, namely response time, accuracy, and a covariate, which defines the relative pay-off for speeded versus accurate responding. The following code fits the `speed` data with a hidden Markov model.

```
library(depmixS4)
data(speed, package = "depmixS4")
set.seed(123)
msp = depmix(response = rt ~ 1, nstates = 2, data = speed)
fmsp = fit(msp, emc = em.control(rand = FALSE))
```

Results show that a two-state model fits the data better, with one having slow responses and the other having fast responses. The probability of remaining in either state 1 or state 2 is approximately 0.9, indicating that the states are stable.

The `flexmix` package is useful to fit mixtures of multivariate and seemingly unrelated Gaussian regression models with diagonal component covariance matrices. The `NPreg` dataset contains artificial values generated from a 2-component mixture model, where one independent variable was generated uniformly from $[0, 10]$, and the other three dependent variables were generated from the Gaussian, Poisson, and Binomial distribution.

```
library(flexmix)
data(NPreg, package = "flexmix")
```

First, we can fit the data set with a two-component Gaussian mixture model:

```
ex1 = flexmix(yn ~ x + I(x^2), data = NPreg, k = 2,
              control = list(verb = 5, iter = 100))
```

The results show that mixing proportions are 0.51 and 0.49, with 100 observations in each group. Now, if we switch gears and fit a mixture of one Gaussian response and one Poisson response:

```
ex2 = flexmix(yn ~ x, data = NPreg, k = 2,
              model = list(FLXMRglm(yn ~ . + I(x^2)),
              FLXMRglm(yp ~ ., family = "poisson")))
```

The mixing proportions become 0.49 and 0.51, with 96 observations from the 1st component, and 104 from the second. In addition, regression coefficients for the Gaussian and Poisson component can be obtained through:

```
parameters(ex2, component = 1, model = 1) # Gaussian
parameters(ex2, component = 1, model = 2) # Poisson
```

4

Bayesian mixture models

4.1 Introduction

A Bayesian approach to mixture analysis is usually implemented by simulating posterior distributions with Markov Chain Monte Carlo (MCMC) methods (Tanner and Wong, 1987; Gelfand and Smith, 1990). Important initial papers on Bayesian analysis of mixtures using MCMC methods include Diebolt and Robert (1994) and Escobar and West (1995). One main advantage of the Bayesian approach is that all statistical inferences about unknown parameters can be easily performed via the posterior distribution. When the posterior distribution of the unknown parameters is available, the Bayesian method can yield statistical inference without relying on the asymptotic normality. This is one of the main benefits of the Bayesian method when the sample size n is small. In contrast, the asymptotic theory of the MLE can be applied only when the sample size n is very large (McLachlan and Peel, 2000, p.68). However, there are some challenges for Bayesian mixture models. One main difficulty is that improper priors yield improper posterior distributions. One possible solution is to use "partially proper prior" (Roeder and Wasserman, 1997), which does not require subjective input for model parameters. The second difficulty is that when the number of components C is unknown, the parameter space is not well defined with infinite dimension. A possible solution is to fit mixture models for a sequence of C values followed by choosing the best C value based on some model selection criteria, such as the information criterion. In addition, many methods have been proposed for Bayesian mixture analysis to choose the number of components and estimate the mixture models simultaneously, such as the methods of Dirichlet process mixtures by Escobar and West (1995), distributional distances by Mengersen (1996), the reversible jump MCMC by Richardson and Green (1997), and the birth-and-death MCMC by Stephens (2000a). The third difficulty for Bayesian mixture models is the effect of label switching, which arises from the invariance of the likelihood under relabelling of the mixture components if there is no prior information to distinguish components. Please see Chapter 5 for more details on how to solve label-switching issues for Bayesian mixture models.

DOI: 10.1201/9781003038511-4

4.2   Markov chain Monte Carlo methods

4.2.1   Hastings-Metropolis algorithm

For the simplicity of explanation, we use discrete distribution to illustrate the idea of Markov chain Monte Carlo (MCMC) methods. However, all the explanations and results can easily be generalized to continuous distributions. We first provide a theorem that is important for Markov chain Monte Carlo methods.

Theorem 4.2.1. Let $\{X_i, i \geq 1\}$ be an irreducible Markov chain with stationary probabilities

$$\pi_j = \lim_{i \to \infty} P(X_i = j), j \geq 0,$$

and let h be a bounded function. Then,

$$\lim_{n \to \infty} \frac{1}{n} \sum_{i=1}^{n} h(X_i) = \sum_{j=0}^{\infty} h(j) \pi_j.$$

In general, let $\mathbf{X}$ be a discrete random vector whose set of possible values is $\mathbf{x}_j, j \geq 1$. Suppose that we are interested in calculating

$$\theta = \mathrm{E}[h(\mathbf{X})] = \sum_{j=1}^{\infty} h(\mathbf{x}_j) P(\mathbf{X} = \mathbf{x}_j),$$

for a specified function h. Sometimes, it is difficult to compute θ, for example, if $p(\mathbf{x}_j) = P(\mathbf{X} = \mathbf{x}_j) = cb_j$, where b_js are specified, but c is unknown and it is not computationally feasible to sum the b_j to determine c. Based on Theorem 4.2.1, one possible solution is to generate a sequence of successive states of a vector-valued Markov chain $\mathbf{X}_1, \mathbf{X}_2, \ldots$, whose stationary probabilities are $P(\mathbf{X} = \mathbf{x}_i)$, then θ can be estimated by $n^{-1} \sum_{i=1}^{n} h(\mathbf{X}_i)$.

Next, we introduce the Hastings-Metropolis algorithm, a general tool to generate a Markov chain $\{X_n, n \geq 0\}$ with stationary probabilities $\pi(j) = cb_j$ where $c^{-1} = \sum_{j=1}^{\infty} b_j$.

Let Q be any specified irreducible Markov transition probability matrix on the integers, with $q(i, j)$ representing the (i, j)th element of Q. Let

$$\alpha(i, j) = \min \left(\frac{\pi(j) q(j, i)}{\pi(i) q(i, j)}, 1 \right).$$

Define a Markov chain $\{X_n, n \geq 0\}$ as follows.

Step 1: When $X_n = i$, generate a random variable Y such that $P(Y = j) = q(i, j)$.

Step 2: If $Y = j$, then set X_{n+1} as follows

$$X_{n+1} = \begin{cases} j, & \text{with probability } \alpha(i,j); \\ i, & \text{with probability } 1 - \alpha(i,j). \end{cases}$$

The Markov chain $\{X_n, n \geq 0\}$ has transition probabilities

$$P_{i,j} = q(i,j)\alpha(i,j), \text{ if } j \neq i;$$
$$P_{i,i} = q(i,i) + \sum_{k \neq i} q(i,k)(1 - \alpha(i,k)), \text{otherwise};$$

with stationary probabilities $\pi(j)$.

Since $\pi(j) = cb_j$,

$$\alpha(i,j) = \min \left(\frac{b_j q(j,i)}{b_i q(i,j)}, 1 \right),$$

which shows that the value of c is not needed to define the Markov chain.

4.2.2   Gibbs sampler

Gibbs sampler is one of the most important Hastings-Metropolis algorithms to generate MCMC samples for Bayesian data analysis. Let $\mathbf{X} = (\mathbf{X}_1, \ldots, \mathbf{X}_d)$ and suppose we want to generate a random vector having mass function $p(\mathbf{x}) = cg(\mathbf{x})$, where $g(\mathbf{x})$ is known, but c is not. Gibbs sampler simulates directly from the conditional distribution of a sub-vector of $\mathbf{X}$ given all other parameters in $\mathbf{X}$ and assumes that we can generate a random variable $\mathbf{X}$ from the conditional distribution $P(\mathbf{X}_i = \mathbf{x} \mid \mathbf{X}_j, j \neq i)$. Gibbs sampler cycles through all the parameters iteratively until we have enough draws after a sufficiently long burn-in.

Step 1: Suppose the present state is $\mathbf{x}_n = (\mathbf{x}_1, \ldots, \mathbf{x}_d)$, a coordinate that is equally likely to be any of $1, \ldots, d$ is chosen. If coordinate i is chosen, then a random variable $\mathbf{X}_i$ with probability mass function $P(\mathbf{X}_i = \mathbf{x} \mid \mathbf{X}_j, j \neq i)$ is generated.

Step 2: If $\mathbf{X}_i = \mathbf{x}$, then the next state is $\mathbf{x}_{n+1} = (\mathbf{x}_1, \ldots, \mathbf{x}_{i-1}, \mathbf{x}, \mathbf{x}_{i+1}, \ldots, \mathbf{x}_d)$. The Markov chain $\{\mathbf{x}_n, n \geq 0\}$ has stationary probabilities $p(\mathbf{x})$.

The above Gibbs sampler is a special case of the Hastings-Metropolis algorithm. After step 1, let $\mathbf{y} = (\mathbf{x}_1, \ldots, \mathbf{x}_{i-1}, \mathbf{x}, \mathbf{x}_{i+1}, \ldots, \mathbf{x}_d)$ be the candidate of next state with

$$q(\mathbf{x}, \mathbf{y}) = \frac{1}{d} P(\mathbf{X}_i = \mathbf{x} \mid \mathbf{X}_j = \mathbf{x}_j, j \neq i) = \frac{1}{d} \frac{p(\mathbf{y})}{P(\mathbf{X}_j = \mathbf{x}_j, j \neq i)}.$$

Using the result of the Hastings-Metropolis algorithm,

$$\alpha(\mathbf{x},\mathbf{y}) = \min\left(\frac{\pi(\mathbf{y})q(\mathbf{y},\mathbf{x})}{\pi(\mathbf{x})q(\mathbf{x},\mathbf{y})},1\right) = \min\left(\frac{\pi(\mathbf{y})p(\mathbf{x})}{\pi(\mathbf{x})p(\mathbf{y})},1\right) = 1,$$

where the second equation can be derived based on the result $q(\mathbf{y},\mathbf{x})/q(\mathbf{x},\mathbf{y}) = p(\mathbf{x})/p(\mathbf{y})$. Hence utilizing the Gibbs sampler, the candidate state is always accepted as the next state of the chain. Although the theory for the Hastings-Metropolis algorithm and Gibbs sampler is developed for discrete random variables, it also holds for continuous random variables.

Example 4.2.1. Let $X_i, i = 1,\ldots,d$ be independent exponential random variables with respective rate $\lambda_i, i = 1,\ldots,d$. Let $S = \sum_{i=1}^{d} X_i$, and suppose we want to generate the random vector $\mathbf{x} = (x_1,\ldots,x_d)$, conditioning on the event that $S > c$ for some large positive constant c. That is, we want to generate the value of a random vector whose density function is

$$f(x_1,\ldots,x_d) = \frac{1}{P(S > c)} \prod_{i=1}^{d} \lambda_i e^{-\lambda_i x_i}, \quad \sum_{i=1}^{d} x_i > c.$$

In order to use the Gibbs sampler, we only need to compute $P(X = x) = P(X_i = x \mid X_j, j \neq i)$, which calls for generating an exponential random variable conditioning on $x + \sum_{j \neq i} x_j > c$ and is distributed as the constant $(c - \sum_{j \neq i} x_j)_+$ plus the exponential. □

4.3　Bayesian approach to mixture analysis

Given $(\mathbf{x}_1,\ldots,\mathbf{x}_n)$ from a C-component mixture density:

$$f(\mathbf{x};\boldsymbol{\theta}) = \sum_{c=1}^{C} \pi_c g(\mathbf{x};\boldsymbol{\lambda}_c),$$

its likelihood function is $L(\boldsymbol{\theta}) = \prod_{i=1}^{n} f(\mathbf{x}_i;\boldsymbol{\theta})$. Let

$$z_{ij} = \begin{cases} 1, & \text{if } i\text{th observation } \mathbf{x}_i \text{ is from the } j\text{th component;} \\ 0, & \text{otherwise.} \end{cases}$$

$\mathbf{z}_i = (z_{i1},\ldots,z_{iC})^\top$, and $\mathbf{z} = (\mathbf{z}_1,\ldots,\mathbf{z}_n)$. Let $p(\boldsymbol{\theta})$ be the prior distribution for $\boldsymbol{\theta}$ and $\mathbf{X} = (\mathbf{x}_1,\ldots,\mathbf{x}_n)^\top$, then the posterior distribution for $\boldsymbol{\theta}$ given $\mathbf{X}$ is

$$p(\boldsymbol{\theta} \mid \mathbf{X}) = p(\boldsymbol{\theta})L(\boldsymbol{\theta})/f(\mathbf{X}) = \sum_{\mathbf{z}} p(\boldsymbol{\theta})L_c(\boldsymbol{\theta})/f(\mathbf{X}), \tag{4.1}$$

where $f(\mathbf{X}) = \int p(\boldsymbol{\theta})L(\boldsymbol{\theta})d\boldsymbol{\theta}$, and

$$L_c(\boldsymbol{\theta}) = \prod_{i=1}^{n}\prod_{c=1}^{C}\{\pi_c g(\mathbf{x}_i;\boldsymbol{\lambda}_c)\}^{z_{ij}}$$

denotes the complete likelihood formed based on both $\mathbf{X}$ and their component-indicator vectors $\mathbf{z}$. Usually, the posterior $p(\boldsymbol{\theta} \mid \mathbf{X})$ does not have an explicit form and it is difficult to compute $f(\mathbf{X})$ due to the multivariate dimensional integration; the Bayesian inference is then based on simulated samples from the posterior distribution using Markov chain Monte Carlo (MCMC) methods. More specifically, we first draw a sufficient number of samples from the posterior $p(\boldsymbol{\theta} \mid \mathbf{X})$, using some MCMC methods such as Gibbs sampling. Then we discard some initial burn-in samples to ensure that the distribution of the remaining samples converges to the posterior distribution and can provide an accurate approximation. Please refer to Cowles and Carlin (1996) and Brooks and Roberts (1998) for a review of convergence diagnostics and how to determine the number of samples required for initial burn-in. Robert (1996) reported that 5000 draws, after the initial burn-in, were adequate for univariate and bivariate mixture models. Let $(\boldsymbol{\theta}_1, \boldsymbol{\theta}_2, \ldots, \boldsymbol{\theta}_N)$ be the MCMC samples after a sufficiently long burn-in, the posterior quantity, such as $\mathrm{E}(T(\boldsymbol{\theta}) \mid \mathbf{X})$, is usually estimated by the ergodic average

$$\mathrm{E}(T(\boldsymbol{\theta}) \mid \mathbf{X}) \approx \frac{1}{N}\sum_{i=1}^{N}T(\boldsymbol{\theta}_i).$$

Another commonly used Bayesian point estimator of $\boldsymbol{\theta}$ is the maximum a posterior (MAP) estimator, which is the mode of the posterior $p(\boldsymbol{\theta} \mid \mathbf{X})$. Let $\hat{\boldsymbol{\theta}}$ be the MAP estimator of $\boldsymbol{\theta}$. Then, $T(\hat{\boldsymbol{\theta}})$ is the MAP estimator of $T(\boldsymbol{\theta})$. A $(1-\alpha)100\%$ confidence interval estimate for $T(\boldsymbol{\theta})$ can be obtained by ordering the sample values of $\{T(\boldsymbol{\theta}_i), i = 1, \ldots, N\}$ and finding the $\alpha/2$ and $(1 - \alpha/2)$ sample quantiles.

4.4 Conjugate priors for Bayesian mixture models

If a conjugate prior is specified, then the posterior $p(\boldsymbol{\theta} \mid \mathbf{X})$ can be written in a closed form. However, we shall demonstrate that the exact calculation is most commonly feasible with small sample sizes only. For moderate to large sample sizes, the Bayesian inference needs to be performed based on Markov chain Monte Carlo (MCMC) samples. Let's consider the model setting when the component densities $g(\mathbf{x};\boldsymbol{\lambda}_c)$ belong to the same exponential family, i.e.,

$$g(\mathbf{x};\boldsymbol{\lambda}) = \exp\{\mathbf{x}^{\top}\boldsymbol{\lambda} - b(\boldsymbol{\lambda}) - c(\mathbf{x})\},$$

for some specific functions $b(\cdot)$ and $c(\cdot)$. Then the conjugate prior for $\boldsymbol{\lambda}_c$ has the form:

$$p(\boldsymbol{\lambda}_c; \boldsymbol{\gamma}_c, \phi_c) \propto \exp\{\boldsymbol{\lambda}_c^\top \boldsymbol{\gamma}_c - \phi_c c(\mathbf{x})\}, \tag{4.2}$$

where $\boldsymbol{\gamma}_c$ is a real-value vector of constants and ϕ_c is a scalar constant ($c = 1, \ldots, C$). A conjugate prior for the component proportions $\boldsymbol{\pi} = (\pi_1, \ldots, \pi_C)$ is the Dirichlet distribution $\mathcal{D}(\delta_1, \delta_2, \cdots, \delta_C)$, which is a generalization of the beta distribution $\mathrm{Beta}(\delta_1, \delta_2)$ and has the density

$$p_{\mathcal{D}}(\boldsymbol{\pi}; \boldsymbol{\delta}) = \Gamma(\sum_{c=1}^C \delta_c - C) \prod_{c=1}^C \frac{\pi_c^{\delta_c - 1}}{\Gamma(\delta_c)}, \tag{4.3}$$

where $\boldsymbol{\delta} = (\delta_1, \delta_2, \cdots, \delta_C)$.

Based on the above priors, we can find the posterior density:

$$p(\boldsymbol{\theta} \mid \mathbf{X}) \propto \sum_{\mathbf{z}} p_{\mathcal{D}}(\boldsymbol{\pi}; \delta_1 + n_1, \cdots, \delta_C + n_C) \prod_{c=1}^C p(\boldsymbol{\lambda}_c; \boldsymbol{\gamma}_c + n_c \bar{\mathbf{x}}_c, \phi_c + n_c), \tag{4.4}$$

where $n_c = \sum_{i=1}^n z_{ic}$ and $\bar{\mathbf{x}}_c = \sum_{i=1}^n z_{ic}\mathbf{x}_i / n_c$. Note that the first summation is over all possible combinations of $\mathbf{z}$, which contains C^n terms. Therefore, the calculation of (4.4) is very large even for moderate sample sizes.

Based on the above conjugate priors (4.2) and (4.3), the Gibbs sampler to simulate MCMC samples can be implemented based on the following steps.

Step 1. Simulate

$$\boldsymbol{\pi} \mid \cdots \sim \mathcal{D}(\delta_1 + n_1, \ldots, \delta_C + n_C),$$

and

$$\boldsymbol{\lambda}_c \sim p(\boldsymbol{\lambda}; \boldsymbol{\gamma}_c + n_c \bar{\mathbf{x}}_c, \phi_c + n_c), \quad c = 1, \ldots, C,$$

where $p(\boldsymbol{\lambda}; \boldsymbol{\gamma}, \phi)$ is defined in (4.2) and " $\mid \cdots$ " denotes "conditional on all other parameters."

Step 2. Simulate

$$\mathbf{z}_i \sim \mathrm{Mult}(1, \mathbf{p}_i), \quad i = 1, \ldots, n,$$

which is a multinomial distribution consisting of C categories, where $\mathbf{p}_i = (p_{i1}, \ldots, p_{iC})^\top$ and

$$p_{ic} = \frac{\pi_c g(\mathbf{x}; \lambda_c)}{\sum_{l=1}^C \pi_l g(\mathbf{x}; \lambda_l)}.$$

Step 3. Update $n_c = \sum_{i=1}^n z_{ic}$ and $\bar{\mathbf{x}}_c = \sum_{i=1}^n z_{ic}\mathbf{x}_i / n_c$, $c = 1, \ldots, C$.

4.5 Bayesian normal mixture models

4.5.1 Bayesian univariate normal mixture models

Consider a C-component normal mixture distribution,

$$p(x; \boldsymbol{\pi}, \boldsymbol{\mu}, \boldsymbol{\sigma}^2) = \pi_1 N(x; \mu_1, \sigma_1^2) + \dots, + \pi_C N(x; \mu_C, \sigma_C^2). \qquad (4.5)$$

Diebolt and Robert (1994) and Bensmail et al. (1997) considered the following conjugate priors:

$$\boldsymbol{\pi} \sim \mathcal{D}(\delta_1, \delta_2, \cdots, \delta_C), \; \mu_c \sim N(\xi_c, \sigma_c^2/\gamma_c), \; \sigma_c^{-2} \sim \Gamma(\alpha_c, \beta_c),$$

where $\mathcal{D}(\cdot)$ is Dirichlet distribution. Bensmail et al. (1997) recommended taking $\gamma_c = 1, \xi_c = \bar{x}, \alpha_c = 2.5, \beta_c = 0.5 S_x^2$. The amount of information in this prior is similar to that in a typical, single observation.

Let

$$Z_{ic} = \begin{cases} 1, & \text{if } x_i \text{ comes from the } c^{th} \text{ component;} \\ 0, & \text{otherwise.} \end{cases}$$

Let $n_c = \sum_{i=1}^n Z_{ic}, \bar{x}_c = \sum_{i=1}^n Z_{ic} x_i, c = 1, \dots, C$. Then, Gibbs sampling can be implemented based on the following conditional posterior distributions:

$$\boldsymbol{\pi} \mid \cdots \sim \mathcal{D}(\delta_1 + n_1, \dots, \delta_C + n_C),$$

$$\mu_c \mid \cdots \sim N\left(\frac{n_c \bar{x}_c + \gamma_j \xi_c}{n_c + \gamma_c}, \frac{\sigma^2}{n_c + \gamma_c}\right),$$

$$\sigma_c^{-2} \mid \cdots \sim \Gamma\left(0.5 + 0.5 n_c + \alpha_c, \; \beta_c + 0.5 \gamma_c (\mu_c - \xi_c)^2 + 0.5 \sum_{i=1}^n Z_{ic}(x_i - \mu_c)^2\right),$$

$$P(Z_{ic} = 1 \mid \cdots) = \frac{\pi_c \phi(x_i; \mu_c, \sigma_c^2)}{\sum_{c=1}^C \pi_c \phi(x_i; \mu_c, \sigma_c^2)},$$

where $\phi(x; \mu, \sigma^2)$ is the density function for $N(\mu, \sigma^2)$. $\{Z_{i1}, \dots, Z_{iC}\}$ can be generated based on multinomial distribution. Phillips and Smith (1996) and Richardson and Green (1997) suggested using the following conjugate priors to estimate the unknown parameters in (4.5):

$$\boldsymbol{\pi} \sim \mathcal{D}(\delta, \delta, \cdots, \delta), \; \mu_c \sim N(\xi, \kappa^{-1}), \; \sigma_c^{-2} \sim \Gamma(\alpha, \beta), \; c = 1, \dots, C. \quad (4.6)$$

Richardson and Green (1997) recommended that $\delta = 1, \xi$ equals to the sample mean of the observations, $\kappa = 1/R^2, \alpha = 2$, and $\beta = R^2/200$, where R is the range of the observations. Then, Gibbs sampling can be implemented based on the following conditional posterior distributions:

$$\boldsymbol{\pi} \mid \cdots \sim \mathcal{D}(\delta + n_1, \ldots, \delta + n_C),$$

$$\mu_c \mid \cdots \sim N\left(\frac{\sigma_c^{-2} n_c \bar{x}_c + \kappa \xi}{\sigma_c^{-2} n_c + \kappa}, \frac{1}{n_c \sigma_c^{-2} + \kappa}\right),$$

$$\sigma_c^{-2} \mid \cdots \sim \Gamma\left(0.5 n_c + \alpha, \ \beta + 0.5 \sum_{i=1}^{n} Z_{ic}(x_i - \mu_c)^2\right),$$

$$P(Z_{ic} = 1 \mid \cdots) \propto \pi_c \phi(x_i; \mu_c, \sigma_c^2), \ c = 1, \ldots, C.$$

4.5.2   Bayesian multivariate normal mixture models

For a p-dimensional multivariate normal mixture,

$$p(\mathbf{x}; \boldsymbol{\theta}) = \pi_1 N(\mathbf{x}; \boldsymbol{\mu}_1, \Sigma_1) + \ldots + \pi_C N(\mathbf{x}; \boldsymbol{\mu}_C, \Sigma_C),$$

Diebolt and Robert (1994) and Bensmail et al. (1997) considered the following conjugate priors:

$$\boldsymbol{\pi} \sim \mathcal{D}(\delta_1, \delta_2, \cdots, \delta_C), \ \mu_c \sim N(\boldsymbol{\xi}_c, \Sigma_c / \gamma_c), \ \Sigma_c^{-1} \sim W(\alpha_c, \mathbf{V}_c),$$

where $W(\alpha, \mathbf{V})$ denotes a Wishart distribution with the following density:

$$p(\mathbf{X}) = \frac{|\mathbf{X}|^{0.5(\alpha - p - 1)} \exp\{-0.5 \mathrm{tr}(\mathbf{X}\mathbf{V}^{-1})\}}{2^{0.5 \alpha p} \pi^{p(p-1)/4} |\mathbf{V}|^{0.5\alpha} \prod_{j=1}^{p} \Gamma(0.5(\alpha + 1 - j))}.$$

When $p = 1$, the Wishart distribution reduces to the gamma distribution $(0.5\alpha, 0.5V^{-1})$. Then, Gibbs sampling can be implemented based on the following conditional posterior distributions:

$$\boldsymbol{\pi} \mid \cdots \sim \mathcal{D}(\delta_1 + n_1, \ldots, \delta_C + n_C),$$

$$\mu_c \mid \cdots \sim N\left(\frac{n_c \bar{\mathbf{x}}_c + \gamma_c \boldsymbol{\xi}_c}{n_c + \gamma_c}, \frac{\Sigma_c}{n_c + \gamma_c}\right),$$

$$\Sigma_c^{-1} \mid \cdots \sim W\left(n_c + \alpha_c, \left\{\mathbf{V}_c^{-1} + n_c S_c + \frac{n_c \alpha_c}{n_c + \alpha_c}(\bar{\mathbf{x}}_c - \boldsymbol{\xi}_c)(\bar{\mathbf{x}}_c - \boldsymbol{\xi}_c)^{\top}\right\}^{-1}\right),$$

$$P(Z_{ic} = 1 \mid \cdots) \propto \pi_c \phi(\mathbf{x}_i; \boldsymbol{\mu}_c, \Sigma_c),$$

where $n_c = \sum_{i=1}^{n} Z_{ic}, \bar{\mathbf{x}}_c = \sum_{i=1}^{n} Z_{ic} \mathbf{x}_i$, and

$$S_c = n_c^{-1} \sum_{i=1}^{n} z_{ic}(\mathbf{x}_i - \bar{\mathbf{x}}_c)(\mathbf{x}_i - \bar{\mathbf{x}}_c)^{\top}, c = 1, \ldots, C.$$

Bensmail et al. (1997) recommended that $\gamma_c = 1, \boldsymbol{\xi}_c = \bar{\mathbf{x}}, \alpha_c = 5$ and $\mathbf{V}_c^{-1} = S_c, c = 1, \ldots, C$, noting that the amount of information in this prior is similar to that contained in a typical, single observation.

4.6 Improper priors

In Bayesian data analysis, it is common to use some noninformative priors and it is well known that impropriety of the prior distribution does not automatically lead to the impropriety of the posterior distribution. However, we shall show that fully noninformative priors (improper priors) will result in the impropriety of the posterior for mixture models (Berger, 2013).

Let $\boldsymbol{\theta} = (\theta_1, \ldots, \theta_C)$ and θ_c be the cth component parameters. Suppose independent improper priors are used, i.e., $p(\boldsymbol{\theta}) = \prod_{c=1}^{C} p(\theta_c)$, where $\int p(\theta_c) d\theta_c = \infty$. Let $\mathbf{z}_0$ be a partition of the data such that the first component is empty, i.e., no observations are allocated to the first component. Then, $p(\mathbf{X}, \mathbf{z}_0; \boldsymbol{\theta}) = p(\mathbf{X}, \mathbf{z}_0; \boldsymbol{\theta}_{-1})$, where $\boldsymbol{\theta}_{-1} = (\theta_2, \ldots, \theta_C)$, and the marginal likelihood $f(\mathbf{X})$ can be written as

$$f(\mathbf{X}) = \int p(\boldsymbol{\theta}) L(\boldsymbol{\theta}) d\boldsymbol{\theta} = \int \sum_{\mathbf{z}} p(\boldsymbol{\theta}) p(\mathbf{X}, \mathbf{z}; \boldsymbol{\theta}) d\boldsymbol{\theta}$$

$$\geq \int p(\boldsymbol{\theta}) p(\mathbf{X}, \mathbf{z}_0; \boldsymbol{\theta}) d\boldsymbol{\theta}$$

$$= \int p(\mathbf{X}, \mathbf{z}_0; \boldsymbol{\theta}_{-1}) \prod_{c=1}^{C} p(\theta_c) d\boldsymbol{\theta}$$

$$= \int p(\theta_1) d\theta_1 \int p(\mathbf{X}, \mathbf{z}_0; \boldsymbol{\theta}_{-1}) \prod_{c=2}^{C} p(\theta_c) d\boldsymbol{\theta}$$

$$= \infty.$$

Note that the above issue occurs mainly because when no observation is allocated to one or more components in (4.1), the data are uninformative about those component parameters. To solve this issue of improper noninformative priors, Diebolt and Robert (1994) and Wasserman (2000) proposed changing the calculation of the posterior (4.1) by restricting $\mathbf{z}$ to those that have a minimal number of samples, say n_0, for each component. Wasserman (2000) showed that such a change is equivalent to imposing a "data-dependent" prior.

In Bayesian probability, Jeffreys prior (Jeffreys, 1998) is the well-known default noninformative prior. Grazian and Robert (2018) and Rubio and Steel (2014) proved that Jeffreys priors also lead to improper posterior distribution in the content of regular location-scale mixture models.

4.7 Bayesian mixture models with unknown numbers of components

In previous sections, we have assumed that the number of components C is a fixed known constant. When it is unknown and treated as an unknown parameter, Phillips and Smith (1996) and Richardson and Green (1997) recommended using a Poisson distribution truncated at the origin as the prior for C:

$$p(c) = \frac{\lambda^c}{\{\exp(\lambda) - 1\}c!}, \; c = 1, 2, \ldots.$$

Richardson and Green (1997) also considered a uniform distribution between 1 and some prespecified upper bound integer c_u as the prior for C.

Bayesian inference is performed based on simulated MCMC samples from the joint posterior distribution for C and $\boldsymbol{\theta}$. Many Bayesian methods have been proposed to handle discrete transitions or jump between mixture models with different numbers of components. For example, Phillips and Smith (1996) proposed an iterative jump-diffusion sampling algorithm; Richardson and Green (1997) proposed a reversible jump MCMC Metropolis-Hastings algorithm to undertake split or combine components; Stephens (2000a) proposed a birth-and-death MCMC procedure. Based on the priors (4.6), Richardson and Green (1997) noticed that the posterior distribution of C is very sensitive to the choice of the hyperparameter β. Richardson and Green (1997) demonstrated that adding a hyperprior to the hyperparameter β using another gamma distribution, say Gamma(a, b), can greatly enhance the stability of the posterior distribution of C. Based on their empirical experience, they recommended $\beta_1 = 2, a = 0.2$, and $b = 100(a/\beta_1)/R^2$, where R is the range of the observations. Gruet et al. (1999) extended the reversible jump MCMC technique of Richardson and Green (1997) to an exponential mixture model, and Robert et al. (2000) extended the reversible jump MCMC technique to a normal hidden Markov chain model. In Section 9.7, we will introduce Bayesian nonparametric mixture models, which treat the mixing distribution completely unspecified.

4.8 Fitting Bayesian mixture models using R

To apply the Bayesian methodology (Hurn et al., 2003), `mixtools` offered some solutions. The `ethanol` data from the `MixSemiRob` package contains the equivalence ratios and peak nitrogen oxide emissions in a study using pure ethanol as a spark-ignition engine fuel. To fit a mixture of linear regressions through the Bayesian approach, the `regmixMH` function could be applied where a proper prior has to be assumed.

```
set.seed(123)
library(MixSemiRob)
data(ethanol, package = "MixSemiRob")
beta = matrix(c(1.3, -0.1, 0.6, 0.1), 2, 2)
sigma = c(.02, .05)

library(mixtools)
MH.out = regmixMH(ethanol$Equivalence, ethanol$NO,
                  beta = beta, s = sigma,
                  sampsize = 2500, omega = .0013)
summary(MH.out$theta)
```

The `regcr` functions can then add a confidence region or Bayesian credible region for regression lines to a scatterplot:

```
beta.c1 = MH.out$theta[2400:2499, c("beta0.1", "beta1.1")]
beta.c2 = MH.out$theta[2400:2499, c("beta0.2", "beta1.2")]
plot(ethanol$NO, ethanol$Equivalence)
regcr(beta.c1, x = ethanol$NO, nonparametric = TRUE,
      plot = TRUE, col = 2)
regcr(beta.c2, x = ethanol$NO, nonparametric = TRUE,
      plot = TRUE, col = 3)
```

The results are shown in Figure 4.1.

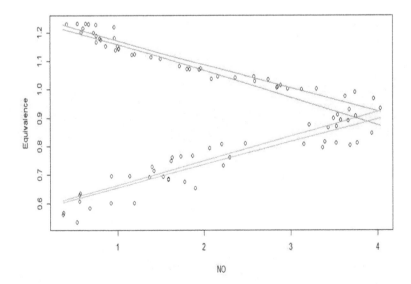

FIGURE 4.1
The NO data fitted with a two-component mixture of linear regressions.

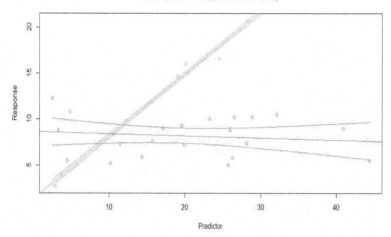

FIGURE 4.2
The CO2 data along with the fitted two-component mixture of linear regressions.

For another example, the CO2 data gives the gross national product (GNP) per capita in 1996 for various countries as well as their estimated carbon dioxide (CO2) emission per capita for the same year. Similarly, a two-component mixture of linear regressions model can be fitted through the regmixEM function:

```
data(CO2data, package = "mixtools")
CO2.out = regmixEM(CO2data$CO2, CO2data$GNP,
                   beta = matrix(c(8, -1, 1, 1), 2, 2),
                   sigma = c(2, 1), lambda = c(1, 3)/4)
summary(CO2.out)
```

The result shows the mixing proportions to be 0.75 and 0.25, with betas being $(8.68, 1.42)$ and $(-0.05, 0.68)$. In addition, the fitting results can be plotted as in Figure 4.2, by the following code.

```
plot(CO2.out, density = TRUE, alpha = 0.01)
```

5

Label switching for mixture models

Label switching has long been a challenging problem for both Bayesian and Frequentist finite mixtures. It arises due to the invariance of the mixture likelihood to the relabeling of the mixture components. For Bayesian mixtures, if the prior is symmetric for the component parameters, the posterior distribution is invariant under the relabeling of the mixture components. Without solving the label switching, the ergodic averages (based on posterior or bootstrap samples) of component-specific quantities will be identical, becoming useless for inference relating to individual components, such as the component-specific parameters, predictive component densities, and marginal classification probabilities. In this chapter, we will introduce some commonly used methods to solve the label-switching issue for mixture models.

5.1   Introduction of label switching

Supposing that $\mathbf{x} = (x_1, \ldots, x_n)$ is a random sample from a C-component mixture density,

$$f(\boldsymbol{\theta}; x) = \sum_{c=1}^{C} \pi_c g(x; \lambda_c),$$

the likelihood for $\mathbf{x}$ is

$$L(\boldsymbol{\theta}; \mathbf{x}) = \prod_{i=1}^{n} \{\pi_1 g(x_i; \lambda_1) + \pi_2 g(x_i; \lambda_2) + \cdots + \pi_C g(x_i; \lambda_C)\}, \qquad (5.1)$$

where

$$\boldsymbol{\theta} = \left[\begin{pmatrix} \pi_1 \\ \lambda_1 \end{pmatrix}, \cdots, \begin{pmatrix} \pi_C \\ \lambda_C \end{pmatrix} \right].$$

For any permutation $\boldsymbol{\omega} = \{\omega(1), \ldots, \omega(C)\}$ of the integers $\{1, \ldots, C\}$, let

$$\boldsymbol{\theta}^{\boldsymbol{\omega}} = \left[\begin{pmatrix} \pi_1^{\boldsymbol{\omega}} \\ \lambda_1^{\boldsymbol{\omega}} \end{pmatrix}, \cdots, \begin{pmatrix} \pi_C^{\boldsymbol{\omega}} \\ \lambda_C^{\boldsymbol{\omega}} \end{pmatrix} \right] = \left[\begin{pmatrix} \pi_{\boldsymbol{\omega}(1)} \\ \lambda_{\boldsymbol{\omega}(1)} \end{pmatrix}, \cdots, \begin{pmatrix} \pi_{\boldsymbol{\omega}(C)} \\ \lambda_{\boldsymbol{\omega}(C)} \end{pmatrix} \right]$$

be the corresponding permutation of the parameter vector $\boldsymbol{\theta}$. One interesting property of finite mixture models is that for any permutation $\boldsymbol{\omega} =$

DOI: 10.1201/9781003038511-5

$\{\boldsymbol{\omega}(1), \ldots, \boldsymbol{\omega}(C)\}$, $L(\boldsymbol{\theta}^{\boldsymbol{\omega}}; \mathbf{x}) = L(\boldsymbol{\theta}; \mathbf{x})$ for any $\mathbf{x}$. Therefore, if $\hat{\boldsymbol{\theta}}$ is the maximum likelihood estimator (MLE), $\hat{\boldsymbol{\theta}}^{\boldsymbol{\omega}}$ is also the MLE for any permutation $\boldsymbol{\omega}$. This is called *label switching*. Due to label switching, the labeling subscripts, $(1, \ldots, C)$, we assign to any MLE and component parameters in (5.1) are unidentifiable, becoming meaningless unless we put additional restrictions on the model. For example, considering a univariate two-component normal mixture with known equal component variance, let $\boldsymbol{\theta} = (\pi_1, \pi_2, \mu_1, \mu_2)$ be the true parameter value and $\hat{\boldsymbol{\theta}}_j = (\hat{\pi}_{j1}, \hat{\pi}_{j2}, \hat{\mu}_{j1}, \hat{\mu}_{j2})$ be the original parameter estimate based on jth simulated or bootstrap sample, where $j = 1, \ldots, N$. The label in $(\hat{\pi}_{j1}, \hat{\mu}_{j1})$ is meaningless, and $(\hat{\pi}_{j1}, \hat{\mu}_{j1})$ can be the estimate of either (π_1, μ_1) or (π_2, μ_2). The true labels of $(\hat{\pi}_{j1}, \hat{\mu}_{j1})$s might be different for different js. Therefore, one cannot directly use $\{(\hat{\pi}_{j1}, \hat{\mu}_{j1}), j = 1, \ldots, N\}$ to estimate the bias and the variation of the MLE for (π_1, μ_1). Similar arguments hold for any other component-specific parameter.

Label switching also exists for Bayesian mixtures. Let $p(\boldsymbol{\theta})$ be the prior for the parameters. The Bayesian inference is based on the posterior distribution of $\boldsymbol{\theta}$: $p(\boldsymbol{\theta} \mid \mathbf{x}) = p(\boldsymbol{\theta}) L(\boldsymbol{\theta}; \mathbf{x}) / p(\mathbf{x})$, where $p(\mathbf{x})$ is the marginal density for $\mathbf{x}$ and $L(\boldsymbol{\theta}; \mathbf{x})$ is defined in (5.1). Usually, the posterior $p(\boldsymbol{\theta} \mid \mathbf{x})$ does not have an explicit form due to the intractable $p(\mathbf{x})$, and the Bayesian inference is based on simulated samples from the posterior distribution using Markov chain Monte Carlo (MCMC) methods. If there is no prior information available to distinguish components of mixture models, a symmetric prior is usually used, i.e., $p(\boldsymbol{\theta}) = p(\boldsymbol{\theta}^{\boldsymbol{\omega}})$ for any permutation $\boldsymbol{\omega}$. Then, we have $p(\boldsymbol{\theta} \mid \mathbf{x}) = p(\boldsymbol{\theta}^{\boldsymbol{\omega}} \mid \mathbf{x})$ for any permutation $\boldsymbol{\omega}$, which creates the label-switching problem. Due to the label switching, marginal posterior distributions for the component parameters will be identical across different mixture components. In addition, the posterior means of component-specific parameters are the same across mixture components and are thus poor estimates of these parameters. A similar issue occurs when estimating quantities relating to individual mixture components such as predictive component densities and marginal classification probabilities. It is then meaningless to draw inferences, relating to individual components, directly from Markov chain Monte Carlo (MCMC) samples using ergodic averaging before solving the label-switching problem.

Next, we illustrate the label-switching phenomena using the galaxy data (Roeder, 1990). The data set consists of the velocities (in thousands of kilometers per second) of 82 distant galaxies diverging from our galaxy. They are sampled from six well-separated conic sections of the Corona Borealis. A histogram of the 82 data points is given in Figure 5.1.

A 6-component normal mixture is fitted to the galaxy data as follows:

$$p(x; \boldsymbol{\pi}, \boldsymbol{\mu}, \boldsymbol{\sigma}^2) = \pi_1 N(x; \mu_1, \sigma_1^2) + \ldots, + \pi_6 N(x; \mu_6, \sigma_6^2). \qquad (5.2)$$

To estimate the unknown parameters in (5.2), Gibbs sampling is applied based on the following conjugate priors (Richardson and Green, 1997):

$$\boldsymbol{\pi} \sim D(\delta, \delta, \cdots, \delta), \ \mu_c \sim N(\xi, \kappa^{-1}), \ \sigma_c^{-2} \sim \Gamma(\alpha, \beta), c = 1, \ldots, 6,$$

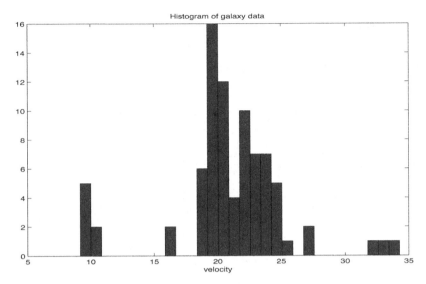

FIGURE 5.1
Histogram plot of galaxy data. The number of bins used is 30.

where $D(\cdot)$ is the Dirichlet distribution. Based on Richardson and Green (1997), we let $\delta = 1, \xi$ equal the sample mean of the observations, κ equal $1/R^2$, $\alpha = 2$, and $\beta = R^2/200$, where R is the range of the observations.

Figure 5.2 is the plot of $10,000$ original Gibbs samples for 6 component means. From the plot, we can see that there are distinct jumps in the traces of the means as the Gibbs samples move between permutation symmetric regions. Figure 5.3 is the plot of estimated marginal posterior densities based on the original Gibbs samples. There are multiple modes for each marginal density. Theoretically, if we run the MCMC scheme sufficiently long, the Gibbs samples should visit all the 6! modal regions and the marginal posterior densities of component parameters would be exactly the same across different components. That is how label switching happens for Bayesian analysis.

Note that the posterior $p(\boldsymbol{\theta} \mid \mathbf{x})$ has $C!$ permutation symmetric maximal modes, each of which is associated with one of $C!$ permutation symmetric modal regions. Therefore, each modal region can be considered as one well-labeled region for the whole parameter space and can be used for valid statistical inference. Practically, given generated MCMC samples $(\boldsymbol{\theta}_1, \ldots, \boldsymbol{\theta}_N)$ from the posterior, we solve the label switching by finding the labels $(\boldsymbol{\omega}_1, \ldots, \boldsymbol{\omega}_N)$ such that $\boldsymbol{\theta}_1^{\boldsymbol{\omega}_1}, \ldots, \boldsymbol{\theta}_N^{\boldsymbol{\omega}_N}$ are all in the same modal region therefore having the same label meaning. Only after we solve the label-switching problem, can we then use the labeled samples to do component-specific statistical inference. Since each of $C!$ symmetric modal regions is well labeled, there are essentially $C!$ sets of latent "true" labels, which, however, are identifiable up to the same

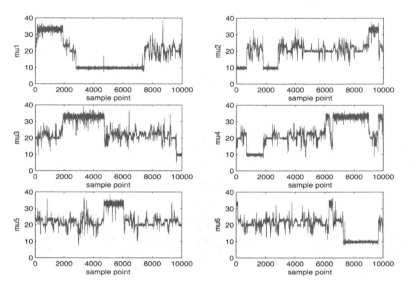

FIGURE 5.2
Plots of Gibbs samples of the component means for Galaxy data.

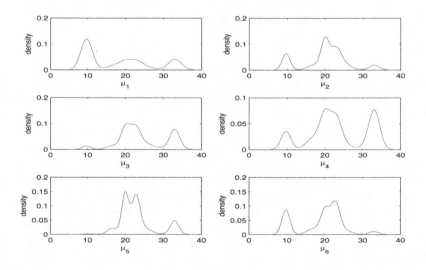

FIGURE 5.3
Plot of marginal posterior densities of the component means based on original
Gibbs samples.

permutation. The label switching can be solved if we recover one of the labeled modal regions.

Many methods have been proposed to solve the label switching in Bayesian analysis. The easiest way to solve the label switching is to use an explicit parameter constraint(Diebolt and Robert, 1994; Dellaportas et al., 1996; Richardson and Green, 1997) so that only one permutation can satisfy it. A generally used constraint is $\lambda_1 < \lambda_2 < \ldots < \lambda_C$ if the component parameter λ is univariate. Similarly, we can also put left-right ordering on π, $\pi_1 < \pi_2 < \ldots < \pi_C$. Stephens (1997) proved that inference using an identifiability constraint can be performed after the constraint is imposed after the MCMC run, which is also equivalent to changing the prior distribution. Frühwirth-Schnatter (2001) proposed using an identifiability constraint on the MCMC output from the *random permutation sampler*. For the random permutation sampler, at every iteration of an MCMC sampler, a Metropolis-Hastings move is used to propose a new permutation of the labels, which ensures the sampler visits all $C!$ symmetric regions.

Celeux (1998); Celeux et al. (2000); Stephens (1997, 2000b) all expressed their concerns about the parameter constraint labeling. One problem with parameter constraint labeling is the choice of constraint, especially for multivariate data. For example, with p-dimensional multivariate normal mixtures, the order constraint can be based on any dimension of the component means, component variances, or component proportions. Different order constraints may generate markedly different results as demonstrated by Celeux et al. (2000) and Yao and Lindsay (2009). In addition, it is difficult to anticipate the overall effect, which depends on the true mixture models to be estimated as well as the sample size. Furthermore, many choices of order constraints fail to remove the symmetry of the posterior distribution. As a result, the label-switching problem may remain after imposing an identifiability constraint. See the example by Stephens (2000b).

Celeux (1998) and Stephens (2000b) proposed a relabeling algorithm by minimizing a Monte Carlo risk. Stephens (2000b) suggested a particular choice of loss function based on the Kullback-Leibler (KL) divergence. We will refer to this particular relabeling algorithm as the *KL algorithm*. However, such risk-based relabeling algorithms are sensitive to the choice of starting labels and require computing $C!$ permutations in each iteration. Celeux et al. (2000) and Hurn et al. (2003) proposed a decision-theoretic procedure by defining a *label invariant loss function* for each quantity we want to estimate. Both the relabeling algorithm and the labeling method using label invariant loss functions require batch processing of the Bayesian samples, which can be computationally demanding on storage when the sample size and/or the number of components is large. Marin et al. (2005) proposed a method related to relabeling algorithm. They found the MAP (maximum a posterior) estimate of the parameters based on all the MCMC samples and then labeled the samples by minimizing their distance from the MAP estimate. Chung et al. (2004) proposed using *data-dependent priors* to solve the labeling problems. They proposed to break the symmetry of the prior and thus the posterior distribution by assigning one observation to each component. Yao and Lindsay (2009) proposed labeling the samples based on the posterior modes they are associated with when they are used as the starting points for an ascending algorithm of the posterior. Their method is an online algorithm and does not require one to compare $C!$ permutations when doing the labeling, which makes it much faster than some other relabeling algorithms. Sperrin et al. (2010) proposed a soft labeling method by assigning labeling probabilities for all possible labels to account for the uncertainty in the relabeling process. However, their method assigns the labels to the allocation vectors directly instead of the MCMC samples, which makes the method depend on the assumption that there is no label switching between the allocation vectors and the corresponding MCMC samples, which is not necessarily true for all the samples. Yao (2012b) proposed finding soft labels by estimating the posterior using a mixture distribution with $C!$ symmetric components. Papastamoulis and Iliopoulos (2010) proposed an artificial allocations-based solution to the label-switching problem. One of the advantages of their method is that it requires a small computational effort compared to many other sophisticated solutions. However, similar to Sperrin et al. (2010), their method also assigns the labels to the allocation vectors directly. Rodríguez and Walker (2012) proposed using the latent allocations to find the labels by minimizing a k-means type of diverging measure. Other labeling methods include, for example, Celeux et al. (2000), Frühwirth-Schnatter (2001), Hurn et al. (2003), Chung et al. (2004), Marin et al. (2005), Geweke (2007), Grün and Leisch (2009), Cron and West (2011), Yao (2012a), Yao (2013b), Yao and Li (2014), Wu and Yao (2017). Jasra et al. (2005) provided a selective overview of existing methods to solve the label-switching problem in Bayesian mixture modeling.

5.2   Loss functions-based relabeling methods

The first type of labeling method we consider is to use batch processing. After the samples $(\boldsymbol{\theta}_1, \ldots, \boldsymbol{\theta}_N)$ have been collected, one goes back and assigns labels/permutations so that the labeled $\boldsymbol{\theta}$ values $(\boldsymbol{\theta}_1^{\boldsymbol{\omega}_1}, \ldots, \boldsymbol{\theta}_N^{\boldsymbol{\omega}_N})$ are clustered tightly together. Stephens (2000b) and Celeux (1998) developed such a relabeling algorithm. For a given loss function $L : A \times \boldsymbol{\theta} \to [0, \infty)$ such that

$$L(a, \boldsymbol{\theta}) = \min_{\boldsymbol{\omega}} L_0(a, \boldsymbol{\theta}^{\boldsymbol{\omega}})$$

for some $L_0 : A \times \boldsymbol{\theta} \to [0, \infty)$, the risk function is: $R(a) = \int_{\boldsymbol{\theta}} L(a, \boldsymbol{\theta}) p(\boldsymbol{\theta}|\mathbf{x}) d\boldsymbol{\theta}$, where $p(\boldsymbol{\theta} \mid \mathbf{x})$ is the posterior distribution of $\boldsymbol{\theta}$. We aim to find a which minimizes $R(a)$. It is hard to compute $R(a)$ directly, and so instead, we find labels by minimizing its Monte Carlo (MC) estimator

$$\hat{R}(a) = \sum_{t=1}^{N} L(a, \boldsymbol{\theta}_t). \tag{5.3}$$

The general algorithm to find labels based on minimizing (5.3) is as follows: Starting with some initial values for $(\boldsymbol{\omega}_1, \ldots, \boldsymbol{\omega}_N)$ (setting them all to the identity permutation, for example), iterate the following steps until a fixed point is reached.

Step 1: choose $\hat{a}$ to minimize $\sum_{t=1}^{N} L_0(\hat{a}, \boldsymbol{\theta}_t^{\boldsymbol{\omega}_t})$.

Step 2: for $t = 1, \ldots, N$ choose $\boldsymbol{\omega}_t$ to minimize $L_0(\hat{a}, \boldsymbol{\theta}_t^{\boldsymbol{\omega}_t})$.

Notice that the above algorithm requires $C!$ comparisons for each t in Step 2. Hence the computation speed is slow when C is large.

5.2.1 KL algorithm

Stephens (2000b) suggests a particular choice of L_0 as

$$L_0(Q; \boldsymbol{\theta}) = \sum_{i=1}^{n} \sum_{c=1}^{C} p_{ic}(\boldsymbol{\theta}^{\boldsymbol{\omega}}) \log \left[\frac{p_{ic}(\boldsymbol{\theta}^{\boldsymbol{\omega}})}{q_{ic}} \right]$$

with the corresponding Monte Carlo estimate of the risk function

$$L(Q; \boldsymbol{\theta}) = \sum_{t=1}^{N} \sum_{i=1}^{n} \sum_{c=1}^{C} p_{ic}(\boldsymbol{\theta}_t^{\boldsymbol{\omega}_t}) \log \left[\frac{p_{ic}(\boldsymbol{\theta}_t^{\boldsymbol{\omega}_t})}{q_{ic}} \right].$$

Here Stephens (2000b) used the Kullback-Leibler (KL) divergence to measure the loss of reporting $Q = \{q_{ic}\}_{1 \leq i \leq n, 1 \leq c \leq C}$ when the true classification probability is $\{p_{ic}(\boldsymbol{\theta})\}_{1 \leq i \leq n, 1 \leq c \leq C}$. The above risk function can be minimized by the following algorithm, referred to as the *KL algorithm*.

Starting with some initial values for $(\boldsymbol{\omega}_1, \ldots, \boldsymbol{\omega}_N)$ (the identity permutation for example), iterate the following two steps until convergence.
Step 1: Choose $\hat{Q} = (\hat{q}_{ic})_{1 \leq i \leq n, 1 \leq c \leq C}$ to minimize

$$\sum_{t=1}^{N} \sum_{i=1}^{n} \sum_{c=1}^{C} p_{ic}(\boldsymbol{\theta}_t^{\boldsymbol{\omega}_t}) \log \left[\frac{p_{ic}(\boldsymbol{\theta}_t^{\boldsymbol{\omega}_t})}{\hat{q}_{ic}} \right],$$

which gives

$$\hat{q}_{ic} = \frac{1}{N} \sum_{t=1}^{N} p_{ic}(\boldsymbol{\theta}_t^{\boldsymbol{\omega}_t}).$$

Step 2: For $t = 1, \ldots, N$, choose $\boldsymbol{\omega}_t$ to minimize

$$\sum_{i=1}^{n} \sum_{c=1}^{C} p_{ic}(\boldsymbol{\theta}^{\boldsymbol{\omega}_t}) \log \left[\frac{p_{ic}(\boldsymbol{\theta}_t^{\boldsymbol{\omega}_t})}{\hat{q}_{ic}} \right],$$

where $p_{ic}(\boldsymbol{\theta}^{\boldsymbol{\omega}}) = \dfrac{\pi_c^{\boldsymbol{\omega}} g(y_i; \lambda_c^{\boldsymbol{\omega}})}{\sum_{l=1}^{C} \pi_l^{\boldsymbol{\omega}} g(y_i; \lambda_l^{\boldsymbol{\omega}})}$ is the classification probability that the observation y_i comes from the component c based on the permuted parameter $\boldsymbol{\theta}^{\boldsymbol{\omega}}$.

5.2.2 The K-means method

We will first argue that the labeling problem can be transferred to a clustering problem. Let $\boldsymbol{\Delta} = \{\boldsymbol{\theta}_t^{\boldsymbol{\omega}_{(j)}}, t = 1, \ldots, N, j = 1, \ldots, C!\}$, where $\{\boldsymbol{\omega}_{(1)}, \ldots, \boldsymbol{\omega}_{(C!)}\}$ are the $C!$ permutations of $(1, \ldots, C)$. Note that this $\boldsymbol{\Delta}$ includes both of the original samples and all of their permutations. Suppose one can find $C!$ tight clusters for $\boldsymbol{\Delta}$, each containing exactly one permutation of each sample element $\boldsymbol{\theta}$. One can then choose any one of these tight clusters to be the newly labeled sample set and assume they are in the same modal region. So, the labeling problem can be transferred to the clustering problem if only one permutation of each sample element $\boldsymbol{\theta}$ is allowed in each cluster.

Different definitions of the size of a cluster lead to different labeling methods. Yao (2012a) defined the size of the cluster consisting of $(\boldsymbol{\theta}_1, \ldots, \boldsymbol{\theta}_N)$ by the trace of their covariance matrix and proposed finding labels by minimizing the following criterion

$$L(\boldsymbol{\theta}_c, \boldsymbol{\Omega}) = \text{Tr} \left(\sum_{t=1}^{N} (\boldsymbol{\theta}_t^{\boldsymbol{\omega}_t} - \boldsymbol{\theta}_c)(\boldsymbol{\theta}_t^{\boldsymbol{\omega}_t} - \boldsymbol{\theta}_c)^\top \right) = \sum_{t=1}^{N} (\boldsymbol{\theta}_t^{\boldsymbol{\omega}_t} - \boldsymbol{\theta}_c)^\top (\boldsymbol{\theta}_t^{\boldsymbol{\omega}_t} - \boldsymbol{\theta}_c),$$

(5.4)

where $\boldsymbol{\Omega} = (\boldsymbol{\omega}_1, \ldots, \boldsymbol{\omega}_N)$, and $\text{Tr}(A)$ is the trace of the matrix A. When the labels $(\boldsymbol{\omega}_t, t = 1, \ldots, N)$ are fixed, the minimum over $\boldsymbol{\theta}_c$ of (5.4) occurs at the sample mean of $\{\boldsymbol{\theta}_1^{\boldsymbol{\omega}_1}, \ldots, \boldsymbol{\theta}_N^{\boldsymbol{\omega}_N}\}$. When $\boldsymbol{\theta}_c$ is fixed, the optimum over $\boldsymbol{\omega}_t, t = 1, \ldots, N$ can be done independently for all t. The algorithm to minimize (5.4) is listed as follows.

Algorithm 5.2.1. Labeling by trace of covariance (TRCOV)
Starting with some initial values for $(\boldsymbol{\omega}_1, \ldots, \boldsymbol{\omega}_N)$ (based on the order constraint for example), iterate the following steps until a fixed point is reached.
Step 1: Update $\boldsymbol{\theta}_c$ by the sample mean based on the current labels $\{\boldsymbol{\omega}_1, \ldots, \boldsymbol{\omega}_N\}$,

$$\boldsymbol{\theta}_c = \frac{1}{N} \sum_{t=1}^{N} \boldsymbol{\theta}_t^{\boldsymbol{\omega}_t}.$$

Step 2: Given the currently estimated value $\boldsymbol{\theta}_c$, $\{\boldsymbol{\omega}_1, \ldots, \boldsymbol{\omega}_N\}$ are updated by

$$\boldsymbol{\omega}_t = \arg \min_{\boldsymbol{\omega}} (\boldsymbol{\theta}_t^{\boldsymbol{\omega}} - \boldsymbol{\theta}_c)^\top (\boldsymbol{\theta}_t^{\boldsymbol{\omega}} - \boldsymbol{\theta}_c).$$

The loss function $L(\boldsymbol{\theta}_c, \boldsymbol{\Omega})$ defined in (5.4) decreases after each of the above two steps. So this algorithm must converge. Since this algorithm may converge to a local minimum, we recommend running this algorithm starting from several initial values. In Step 2, we can also update $\boldsymbol{\theta}_c$ after each change of $\boldsymbol{\omega}_t$, which may increase the speed of convergence.

It is important to note that when one component-specific parameter dominates the others in the information contained within $\boldsymbol{\theta}$, the labeling method discussed here will yield results that closely resemble those obtained through order constraint labeling applied solely to this dominant parameter. To elaborate further, if $\boldsymbol{\theta}$ comprises only C parameters, each representing a component, such as the C component means for univariate data, then the labeling method in question will be identical to the order constraint labeling applied specifically to the component means. Consequently, the order constraint labeling can be regarded as a special case of the TRCOV method. However, unlike the order constraint labeling approach, the TRCOV method possesses the capability to simultaneously incorporate information from various component parameters, making it highly adaptable to the high-dimensional scenario.

5.2.3 The determinant-based loss

A drawback of the TRCOV method is that the objective function (5.4) is not invariant to the scale transformation of the parameters. For example, if the scale of one component-specific parameter is too large compared to other parameters, the labeling will be mainly dominated by this component parameter and nearly close to the order constraint labeling based on this parameter.

To solve the scale effect of the parameters, Yao (2012a) proposed defining the size of a cluster consisting of $(\boldsymbol{\theta}_1, \ldots, \boldsymbol{\theta}_N)$ by the determinant of their covariance matrix

$$L(\boldsymbol{\theta}_c, \boldsymbol{\Omega}) = \det\left(\sum_{t=1}^{N}(\boldsymbol{\theta}_t^{\boldsymbol{\omega}_t} - \boldsymbol{\theta}_c)(\boldsymbol{\theta}_t^{\boldsymbol{\omega}_t} - \boldsymbol{\theta}_c)^{\top}\right), \qquad (5.5)$$

where $\boldsymbol{\Omega} = (\boldsymbol{\omega}_1, \ldots, \boldsymbol{\omega}_N)$ and $\det(A)$ is the determinant of matrix A. Yao (2012a) proved that this determinant criterion is invariant to all permutation invariant linear transformations of the parameters (changing both component means by a scale factor, both variances by a different one, for example).

Based on the next theorem, we know when $(\boldsymbol{\omega}_1, \ldots, \boldsymbol{\omega}_N)$ are fixed, the minimum of (5.5) over $\boldsymbol{\theta}_c$ occurs at the sample mean of $\{\boldsymbol{\theta}_1, \ldots, \boldsymbol{\theta}_N\}$.

Theorem 5.2.1. When $(\boldsymbol{\omega}_1, \ldots, \boldsymbol{\omega}_N)$ are fixed, let $\bar{\boldsymbol{\theta}}$ be the sample mean of $\{\boldsymbol{\theta}_1^{\boldsymbol{\omega}_1}, \ldots, \boldsymbol{\theta}_N^{\boldsymbol{\omega}_N}\}$. Then $\bar{\boldsymbol{\theta}}$ minimizes (5.5) over $\boldsymbol{\theta}_c$.

Proof: Notice that

$$\det\left(\sum_{t=1}^{N}(\boldsymbol{\theta}_t^{\boldsymbol{\omega}_t} - \boldsymbol{\theta}_c)(\boldsymbol{\theta}_t^{\boldsymbol{\omega}_t} - \boldsymbol{\theta}_c)^{\top}\right)$$

$$= \det\left(\sum_{t=1}^{N}(\boldsymbol{\theta}_t^{\boldsymbol{\omega}_t} - \bar{\boldsymbol{\theta}})(\boldsymbol{\theta}_t^{\boldsymbol{\omega}_t} - \bar{\boldsymbol{\theta}})^{\top} + N(\bar{\boldsymbol{\theta}} - \boldsymbol{\theta}_c)(\bar{\boldsymbol{\theta}} - \boldsymbol{\theta}_c)^{\top}\right)$$

$$\geq \det\left(\sum_{t=1}^{N}(\boldsymbol{\theta}_t^{\boldsymbol{\omega}_t} - \bar{\boldsymbol{\theta}})(\boldsymbol{\theta}_t^{\boldsymbol{\omega}_t} - \bar{\boldsymbol{\theta}})^{\top}\right) .$$

So the minimum of (5.5) over $\boldsymbol{\theta}_c$ occurs at the sample mean of $\{\boldsymbol{\theta}_1^{\boldsymbol{\omega}_1}, \ldots, \boldsymbol{\theta}_N^{\boldsymbol{\omega}_N}\}$.

Unlike the trace of covariance case, the optimum over $\boldsymbol{\omega}_t, t = 1, \ldots, N$ can not truly be done independently for all t. Rather we need to optimize over $\boldsymbol{\omega}_t$ one t at a time while holding all the other fixed. Let

$$\Sigma_{<t>} = \sum_{l \neq t}(\boldsymbol{\theta}_l^{\boldsymbol{\omega}_l} - \boldsymbol{\theta}_c)(\boldsymbol{\theta}_l^{\boldsymbol{\omega}_l} - \boldsymbol{\theta}_c)^{\top}.$$

It can be proved that

$$L(\boldsymbol{\theta}_c, \boldsymbol{\Omega}) = \det(\Sigma_{<t>})\left[1 + (\boldsymbol{\theta}_t^{\boldsymbol{\omega}_t} - \boldsymbol{\theta}_c)^{\top}\Sigma_{<t>}^{-1}(\boldsymbol{\theta}_t^{\boldsymbol{\omega}_t} - \boldsymbol{\theta}_c)\right] .$$

Thus to optimize over $\boldsymbol{\omega}_t$ for a particular t, other terms fixed, we just minimize

$$(\boldsymbol{\theta}_t^{\boldsymbol{\omega}_t} - \boldsymbol{\theta}_c)^{\top}\Sigma_{<t>}^{-1}(\boldsymbol{\theta}_t^{\boldsymbol{\omega}_t} - \boldsymbol{\theta}_c), \tag{5.6}$$

which is a weighted distance between $\boldsymbol{\theta}_t^{\boldsymbol{\omega}_t}$ and $\boldsymbol{\theta}_c$. The leave-one-out weight matrix $\Sigma_{<t>}^{-1}$ makes this labeling method invariant to the affine transformation of the component parameters.

In summary, the algorithm to minimize (5.5) will be as follows.

Algorithm 5.2.1 Labeling by determinant of covariance (DETCOV)

Starting with some initial values for $(\boldsymbol{\omega}_1, \ldots, \boldsymbol{\omega}_N)$ (the order constraint or trace of covariance loss labeling, for example), iterate the following two steps until a fixed point is reached.

Step 1: Update $\boldsymbol{\theta}_c$ by the sample mean based on the current values $\{\boldsymbol{\omega}_1, \ldots, \boldsymbol{\omega}_N\}$,

$$\boldsymbol{\theta}_c = \frac{1}{N}\sum_{t=1}^{N}\boldsymbol{\theta}_t^{\boldsymbol{\omega}_t} .$$

Step 2: For $t = 1, \ldots, N$, given the current estimated value $\boldsymbol{\theta}_c$, and $\{\boldsymbol{\omega}_l, l \neq t\}$, $\boldsymbol{\omega}_t$ are updated by

$$\boldsymbol{\omega}_t = \arg\min_{\boldsymbol{\omega}}(\boldsymbol{\theta}_t^{\boldsymbol{\omega}} - \boldsymbol{\theta}_c)^{\top}\Sigma_{<t>}^{-1}(\boldsymbol{\theta}_t^{\boldsymbol{\omega}} - \boldsymbol{\theta}_c) .$$

Let $\Sigma = \sum_{t=1}^{N} (\boldsymbol{\theta}_t^{\boldsymbol{\omega}_t} - \boldsymbol{\theta}_c)(\boldsymbol{\theta}_t^{\boldsymbol{\omega}_t} - \boldsymbol{\theta}_c)^{\top}$ and $u_t = \Sigma^{-1/2}(\boldsymbol{\theta}_t^{\boldsymbol{\omega}_t} - \boldsymbol{\theta}_c)$. Yao (2012a) proved that

$$\Sigma_{<t>}^{-1} = \Sigma^{-1/2}\left(I + \frac{1}{1 - u_t^{\top} u_t} u_t u_t^{\top}\right)\Sigma^{-1/2}, \qquad (5.7)$$

and

$$\Sigma^{-1} = \Sigma_{<t>}^{-1/2}\left(I - \frac{1}{1 + v_t^{\top} v_t} v_t v_t^{\top}\right)\Sigma_{<t>}^{-1/2}.$$

Based on (5.7), in order to calculate $\Sigma_{<t>}^{-1}$, we only need to find $\Sigma^{-1/2}$. Since $\Sigma^{-1/2}$ only needs to be updated after some label $\boldsymbol{\omega}_t$ changes, the computation of (5.7) is much less than updating $\Sigma_{<t>}^{-1}$ for each t.

Using the above technique, we can use the inverse of the initial matrix Σ to find all other matrix inversions. Suppose that we have optimized over $t = 1$, and changed the permutation involved. Then, move to $t = 2$. However, technically the covariance matrix for the change in $t = 1$ needs to be updated if we want to guarantee that the objective function is monotonically decreasing. We can do so in a way that we do not need to recalculate inverses. First, we can remove the original $t = 1$ term as above:

$$\Sigma_{<1>}^{-1} = \Sigma^{-1/2}(I + \frac{1}{1 - u^{\top} u} u u^{\top})\Sigma^{-1/2},$$

where $u = \Sigma^{-1/2}(\boldsymbol{\theta}_1^{\text{old}} - \boldsymbol{\theta}_c)$. Then when the new $(\boldsymbol{\theta}_1^{\text{new}} - \boldsymbol{\theta}_c)(\boldsymbol{\theta}_1^{\text{new}} - \boldsymbol{\theta}_c)^{\top}$ is added back in, let $w = \Sigma_{<1>}^{-1/2}(\boldsymbol{\theta}_1^{\text{new}} - \boldsymbol{\theta}_c)$,

$$\Sigma_{\text{new}} = \Sigma_{<1>} + (\boldsymbol{\theta}_1^{\text{new}} - \boldsymbol{\theta}_c)(\boldsymbol{\theta}_1^{\text{new}} - \boldsymbol{\theta}_c)^{\top} = \Sigma_{<1>}^{1/2}(I + w w^{\top})\Sigma_{<1>}^{1/2},$$

so

$$\Sigma_{\text{new}}^{-1} = \Sigma_{<1>}^{-1/2}\left(I - \frac{1}{1 + w^{\top} w} w w^{\top}\right)\Sigma_{<1>}^{-1/2}.$$

Thus we can carry out all these calculations with only the initial matrix inversion of Σ.

5.2.4 Asymptotic normal likelihood method

Based on the asymptotic theory for the posterior distribution, see Walker (1969) and Frühwirth-Schnatter (2006)[Sec 1.3 & 3.3] for example, when the sample size is large, the "correctly" labeled MCMC samples, i.e., the samples from the same modal region, should approximately follow a normal distribution. Yao and Lindsay (2009) proposed labeling the samples by maximizing the following normal likelihood

$$LR(\boldsymbol{\theta}_c, \boldsymbol{\Sigma}, \boldsymbol{\Omega}) = |\boldsymbol{\Sigma}|^{-N/2} \prod_{t=1}^{N} \exp\{-\frac{1}{2}(\boldsymbol{\theta}_t^{\boldsymbol{\omega}_t} - \boldsymbol{\theta}_c)^{\top}\boldsymbol{\Sigma}^{-1}(\boldsymbol{\theta}_t^{\boldsymbol{\omega}_t} - \boldsymbol{\theta}_c)\},$$

or, equivalently, minimizing the following negative log normal likelihood over $(\boldsymbol{\theta}_c, \boldsymbol{\Sigma}, \boldsymbol{\omega})$

$$L(\boldsymbol{\theta}_c, \boldsymbol{\Sigma}, \boldsymbol{\Omega}) = N \log(|\boldsymbol{\Sigma}|) + \sum_{t=1}^{N} (\boldsymbol{\theta}_t^{\boldsymbol{\omega}_t} - \boldsymbol{\theta}_c)^{\top} \boldsymbol{\Sigma}^{-1} (\boldsymbol{\theta}_t^{\boldsymbol{\omega}_t} - \boldsymbol{\theta}_c), \qquad (5.8)$$

where $\boldsymbol{\theta}_c$ is the center for the normal distribution, $\boldsymbol{\Sigma}$ is the covariance structure, and $\boldsymbol{\Omega} = (\boldsymbol{\omega}_1, \ldots, \boldsymbol{\omega}_N)$.

By maximizing (5.8), in fact, we are choosing the labels $\boldsymbol{\omega}$ so that the maximum of the normal likelihood is greatest. This method is affine transformation invariant, as the determinant loss, but is much easier to calculate. Given labels $\boldsymbol{\omega}$, minimizing (5.8) is equivalent to finding the MLE of a regular normal likelihood, i.e.,

$$\boldsymbol{\theta}_c = \frac{1}{N} \sum_{t=1}^{N} \boldsymbol{\theta}_t^{\boldsymbol{\omega}_t},$$

$$\boldsymbol{\Sigma} = \frac{1}{N} \sum_{t=1}^{N} (\boldsymbol{\theta}_t^{\boldsymbol{\omega}_t} - \boldsymbol{\theta}_c)(\boldsymbol{\theta}_t^{\boldsymbol{\omega}_t} - \boldsymbol{\theta}_c)^{\top}.$$

Given the center $\boldsymbol{\theta}_c$ and covariance structure $\boldsymbol{\Sigma}$ of the normal distribution, the label $\boldsymbol{\omega}_t$ is chosen by minimizing

$$(\boldsymbol{\theta}_t^{\boldsymbol{\omega}} - \boldsymbol{\theta}_c)^{\top} \boldsymbol{\Sigma}^{-1} (\boldsymbol{\theta}_t^{\boldsymbol{\omega}} - \boldsymbol{\theta}_c),$$

which is a weighted distance between $\boldsymbol{\theta}_t^{\boldsymbol{\omega}}$ and $\boldsymbol{\theta}_c$. The weight $\boldsymbol{\Sigma}$ will make this labeling method invariant to the affine transformation of the component parameters. So the algorithm to find the labels by maximizing (5.8) is as follows.

Algorithm 5.2.2 Labeling by normal likelihood (NORMLH)

Starting with some initial values for $(\boldsymbol{\omega}_1, \ldots, \boldsymbol{\omega}_N)$ (setting them based on the order constraint or trace of covariance loss labeling, for example), iterate the following two steps until a fixed point is reached.

Step 1: Update $\boldsymbol{\theta}_c$ and $\boldsymbol{\Sigma}$ based on current labels $\{\boldsymbol{\omega}_1, \ldots, \boldsymbol{\omega}_N\}$,

$$\boldsymbol{\theta}_c = \frac{1}{N} \sum_{t=1}^{N} \boldsymbol{\theta}_t^{\boldsymbol{\omega}_t},$$

$$\boldsymbol{\Sigma} = \frac{1}{N} \sum_{t=1}^{N} (\boldsymbol{\theta}_t^{\boldsymbol{\omega}_t} - \boldsymbol{\theta}_c)(\boldsymbol{\theta}_t^{\boldsymbol{\omega}_t} - \boldsymbol{\theta}_c)^{\top}.$$

Step 2: For $t = 1, \ldots, N$, update $\boldsymbol{\omega}_t$ by

$$\boldsymbol{\omega}_t = \arg \min_{\boldsymbol{\omega}} (\boldsymbol{\theta}_t^{\boldsymbol{\omega}} - \boldsymbol{\theta}_c)^{\top} \boldsymbol{\Sigma}^{-1} (\boldsymbol{\theta}_t^{\boldsymbol{\omega}} - \boldsymbol{\theta}_c).$$

If N is big enough, which is generally the case for MCMC samples, $\Sigma_{<t>}^{-1}$ in (5.6) is approximately equal to Σ^{-1}. Hence, generally, the NORMLH labeling method produces similar labeling results to the DETCOV method. However, NORMLH is much faster than DETCOV and provides a nice explanation based on the normal likelihood and the asymptotic labeling.

5.3 Modal relabeling methods

As explained in Section 5.1, the posterior of Bayesian mixtures has $C!$ symmetric major modes each of which is associated with one modal region. To solve the label switching, we need to find the labels, $\{\omega_1, \ldots, \omega_N\}$, such that the labeled samples $\{\theta_1^{\omega_1}, \ldots, \theta_N^{\omega_N}\}$ are from the same modal region. Yao and Lindsay (2009) proposed a modal relabeling method called PM(ECM) that uses each MCMC sample as the starting value for an ascending algorithm such as the ECM algorithm (Meng and Rubin, 1993), and then relabels the samples according to the posterior modes to which they converge. Yao and Lindsay (2009) argued that the PM(ECM) automatically matches the labels in the highest posterior density (HPD) region, $\{\theta : p(\theta|\mathbf{x}) > c\}$, where $p(\theta|\mathbf{x})$ is the posterior of θ given observation $\mathbf{x} = (x_1, \ldots, x_n)$. In addition, PM(ECM) is an online algorithm, which does not require batch processing, reducing the amounts of storage. Furthermore, many existing labeling methods depend on the choice of the initial labels for the samples. In contrast, the PM(ECM) method does not require initial labels, which can save considerable computation time and provide more stable results.

5.3.1 Ideal labeling based on the highest posterior density region

Suppose that the parameter space is the product space

$$\Omega = \{(\pi_1, \ldots, \pi_C, \lambda_1, \ldots, \lambda_C) : \sum_{c=1}^{C} \pi_c = 1, 0 < \pi_c < 1, \lambda_c \neq \lambda_{c'}, 1 \leq c \neq c' \leq C\}.$$

For any point $\theta \in \Omega$, define an equivalence class of points, call it $EC(\theta)$, that contains all parameter θ' generating the same posterior as θ, i.e.,

$$EC(\theta) = \{\theta' : p(\theta'|\mathbf{x}) = p(\theta|\mathbf{x})\}.$$

This equivalent class of nonidentifiability arises due to label switching. For each θ, $EC(\theta)$ will consist of $C!$ points θ^ω that correspond to $C!$ possible relabelings.

Let us say that a subset S of Ω is an *identifiable subset* if, for every $\theta \in S$, we have $EC(\theta) \cap S = \{\theta\}$. If we restrict the parameters to lie in an identifiable subset, then all the parameters are identifiable and have unique labels, as there is a one-to-one map from parameters to distributions. If we have an identifiable subset S, then we can create mirrored sets by permutation: $S^{\omega} = \{\theta^{\omega} : \theta \in S\}$. Each of the mirrored sets is also identifiable and they are disjoint from each other. One can find a small neighborhood around every $\theta \in \Omega$ that is identifiable, making the points in Ω *locally identifiable*.

In a standard mixture problem, one says that an estimator $\hat{\theta}_n$ is *consistent* for true value θ_0 if there exists a sequence of labels ω_n such that $\hat{\theta}_n^{\omega_n} \xrightarrow{p} \theta_0$. That is, given any ball around θ_0, the probability that one element of $EC(\hat{\theta}_n)$ is in that ball goes to one. It will be *consistent and asymptotically normal* if $\sqrt{n}(\hat{\theta}_n^{\omega_n} - \theta_0) \xrightarrow{L} N(0, \sigma^2)$. In a standard mixture problem, these results hold for the regular MLE sequence and are clearly related to the fact that the parameters are locally identifiable. Hence, when the sample size is large enough, we can meaningfully assign labels, such that all the labeled θ values are close to the same labeled true value.

Suppose that the posterior density $p(\theta|\mathbf{x})$ attains its maximum at a regular point $\hat{\theta}$, with value $p(\hat{\theta}|\mathbf{x}) = d$. It will also attain this maximum at all points in $EC(\hat{\theta})$. More generally, if there are other local maxima, they must occur in entire sets of equivalence classes. Let us suppose that our goal is to build credible regions for the parameters, for any fixed credibility level $1 - \alpha$, using regions of the highest posterior density (HPD). Such credible regions have the theoretical justification of being the smallest volume of credible regions at a fixed level. To be specific, let the regions have the form $\psi_c = \{\theta : p(\theta|\mathbf{x}) \geq c\}$, where $c = c_\alpha$ is chosen to give the target credibility level. From the proceeding asymptotic theory, one can say that for any fixed α, for n sufficiently large the HPD region will consist of $C!$ disjoint sets $\{S_j, j = 1, \ldots, C!\}$. Each S_j will be an identifiable set clustered around θ_0^{ω}, for some ω. And each set S_j must be a permutation image S_1^{ω} of S_1 for some ω. In such a setting, one can use the identifiable set S_1 to describe the HPD region, as all others are just permuted images of S_1. The parameters have unique labels in S_1, so the HPD region can be described using labeled parameters. It is also reasonable to consider using asymptotic normality for the labeled parameters of S_1. In this sense, we can consider the problem to be well-labeled at infinity.

Figure 5.4 shows the plot of the two component means (μ_1, μ_2) of 1000 Gibbs samples, assuming equal variances, and their permutation (μ_2, μ_1), based on the 500 observations generated from $0.5N(0, 1) + 0.5N(2, 1)$. From Figure 5.4, we can see that either one of the clusters can be used to assign labels to each estimate as there appears to be no uncertainty about which cluster a sample is in.

Figure 5.5 is the plot of the two component means of 1000 Gibbs samples, assuming equal variances, and their permutations based on the 100 observations generated from $0.5N(0, 1) + 0.5N(2, 1)$, which is the same as above. Note that compared to 500, 100 is not so large, and based on such a sample size, the two permuted regions are connected together. Clearly, there are no

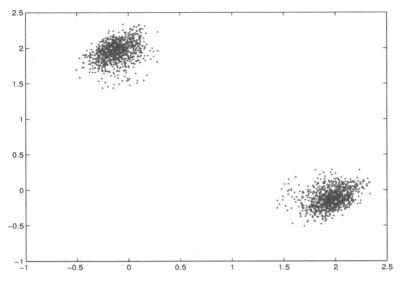

FIGURE 5.4
Plot of the two component means of 1000 Gibbs samples for the observations
with a sample size of 500 generated from 0.5N(0,1)+0.5N(2,1).

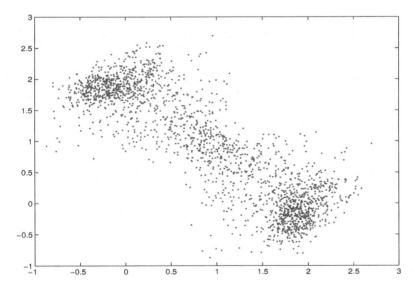

FIGURE 5.5
Plot of the two component means of 1000 Gibbs samples for the observations
with a sample size of 100 generated from $0.5N(0,1) + 0.5N(2,1)$.

well-separated clusters anymore, and one can define a boundary line to separate the regions, but the choice is quite arbitrary.

We need to transfer the asymptotic labeling theory to finite n. For a given maximal mode $\hat{\boldsymbol{\theta}}$, we define $S_c(\hat{\boldsymbol{\theta}})$ to be the maximal connected subset of the HPD region ψ_c that contains $\hat{\boldsymbol{\theta}}$, where $\psi_c = \{\boldsymbol{\theta} : p(\boldsymbol{\theta}|\mathbf{x}) \geq c\}$, with $c = c_\alpha$ chosen to give the target credibility level. We will call $S_c(\hat{\boldsymbol{\theta}})$ the *modal region* defined by c and $\hat{\boldsymbol{\theta}}$. When $c = p(\hat{\boldsymbol{\theta}}|\mathbf{x})$, ψ_c equals $EC(\hat{\boldsymbol{\theta}})$, and $S_c(\hat{\boldsymbol{\theta}})$ is the single point $\{\hat{\boldsymbol{\theta}}\}$. As c decreases, the size of the set $S_c(\hat{\boldsymbol{\theta}})$ increases. Note also that $S_c^{\boldsymbol{\omega}}(\hat{\boldsymbol{\theta}})$ is automatically the maximal connected subset that contains $\hat{\boldsymbol{\theta}}^{\boldsymbol{\omega}}$, i.e., $S_c^{\boldsymbol{\omega}}(\hat{\boldsymbol{\theta}}) = S_c(\hat{\boldsymbol{\theta}}^{\boldsymbol{\omega}})$. In a smooth problem, as long as c is sufficiently large, the set ψ_c will be the union of disjoint $S_c^{\boldsymbol{\omega}}(\hat{\boldsymbol{\theta}})$ over all permutations $\boldsymbol{\omega}$ and each of them is an identifiable subset. Note that none of such identifiable sets, $S_c^{\boldsymbol{\omega}}(\hat{\boldsymbol{\theta}})$, contains any degenerate point, which corresponds to mixture models with less than C components. (Otherwise, they won't be disjoint).

If the modal region $S_c(\hat{\boldsymbol{\theta}})$ is an identifiable subset, we can mimic the asymptotic theory and use $S_c(\hat{\boldsymbol{\theta}})$, with its labels, to describe the HPD region of the posterior distribution, as the rest of the region is just permuted copies of $S_c(\hat{\boldsymbol{\theta}})$. In this case, the identifiable HPD region provides $\boldsymbol{\theta}$ values it contains a natural labeling from the local identifiability property. We will call these labels the *HPD labels*, and consider them to be the *ideal labels*, when they exist. In fact, these identifiable sets will eventually be disjoint for any c when n is large enough in accordance with the asymptotic identifiability of the labels.

Unfortunately, for finite sample size n, there will always exist a value of c such that the HPD region around one mode intersects the HPD region around another mode, thereby creating an HPD region that is not identifiable. Let c_0 be the supremum of such c values. For any point $\boldsymbol{\theta}$ with $p(\boldsymbol{\theta}|\mathbf{x}) \leq c_0$, we will say that its labeling is HPD ambiguous and denote its proportion by $\alpha_0 = \Pr(p(\boldsymbol{\theta}|\mathbf{x}) \leq c_0)$. We will say that any point $\boldsymbol{\theta}$ with $p(\boldsymbol{\theta}|\mathbf{x}) > c_0$ has a well-defined ideal HPD label, and we will call its proportion $(1 - \alpha_0)$ the *labeling credibility*. In Section 5.3.4, we will consider the calculation of c_0 and $(1 - \alpha_0)$ as a way of determining which points have HPD labels, and what fraction of points could be ideally labeled.

5.3.2   Introduction of the HPD modal labeling method

In this section, we introduce an HPD modal labeling method proposed by Yao and Lindsay (2009) to explore the geometry of the mixture posterior by using each MCMC draw as a starting point for an ascent algorithm and labeling the samples based on the modes of the posterior to which they converge. An algorithm is called an "ascent algorithm" if it nondecreases the posterior after each iteration (see Section 5.3.3 for more detail). The main property/idea of the HPD modal labeling method is that *the samples converged to the same mode have the same labels*. If we start at a point with an HPD label, each iteration of this type of algorithm will stay in the same identifiable region while climbing up the posterior until it reaches the mode within that region. Thus, if you label points by the mode they reach, these labels will exactly reconstruct the HPD labels for all $\boldsymbol{\theta}$ with $p(\boldsymbol{\theta}|\mathbf{x}) > c_0$, where c_0 is defined at the end of the previous section. Thus the HPD modal labeling method can reproduce the ideal HPD labels when they exist, which is a primary motivation and essentially unique benefit of labeling based on an ascending algorithm, and hence can be considered the best method for describing the HPD region. However, different ascent algorithms will potentially differ on points with $p(\boldsymbol{\theta}|\mathbf{x}) \leq c_0$, the HPD ambiguous points.

More specifically, if the posterior density has a maximal/major mode at $\tilde{\boldsymbol{\theta}}$, it also has maximal modes at all permutations of $\tilde{\boldsymbol{\theta}}$. We can pick one such mode to be our reference mode (and the corresponding modal region), such as by order constraint labeling on some parameter. Denote by $\hat{\boldsymbol{\theta}}$ the chosen reference maximal mode. The objective of labeling is to find labels $\{\boldsymbol{\omega}_1, \ldots, \boldsymbol{\omega}_N\}$ such that $\{\boldsymbol{\theta}_1^{\boldsymbol{\omega}_1}, \ldots, \boldsymbol{\theta}_N^{\boldsymbol{\omega}_N}\}$ have the same label meaning (i.e., stay in the same modal region) as the reference mode $\hat{\boldsymbol{\theta}}$. If a sampled $\boldsymbol{\theta}$ converges to the reference maximal mode $\hat{\boldsymbol{\theta}}$, then $\boldsymbol{\theta}$ is said to have the same label as $\hat{\boldsymbol{\theta}}$ and hence has the identity label. If $\boldsymbol{\theta}$ converges to another maximal mode $\hat{\boldsymbol{\theta}}_*$, a permuted version of the reference mode, such that $\hat{\boldsymbol{\theta}}_*^{\boldsymbol{\omega}} = \hat{\boldsymbol{\theta}}$, then the label of $\boldsymbol{\theta}$ is also $\boldsymbol{\omega}$ and $\boldsymbol{\theta}^{\boldsymbol{\omega}}$ would ascend to $\hat{\boldsymbol{\theta}}_*^{\boldsymbol{\omega}} = \hat{\boldsymbol{\theta}}$. If $\boldsymbol{\theta}$ converges to a minor mode, say $\boldsymbol{\theta}_*$, $\boldsymbol{\theta}$ and $\boldsymbol{\theta}_*$ have the same label (i.e., $\boldsymbol{\theta}$ and $\boldsymbol{\theta}_*$ are in the same modal region), and the label can be found by making $\boldsymbol{\theta}_*^{\boldsymbol{\omega}}$ match the label of the reference mode $\hat{\boldsymbol{\theta}}$ using a risk-based criterion, such as the Euclidean distance.

In Algorithm 5.3.1 below, we summarize the above HPD modal labeling method (Yao and Lindsay, 2009) when given an ascent algorithm (ALG) (see Section 5.3.3 for more detail) and a reference maximal mode $\hat{\boldsymbol{\theta}}$, say by Algorithm 5.3.2. Yao and Lindsay (2009) also called this algorithm *Labeling based on Posterior Modes and an Ascent Algorithm* (PM(ALG)).

Algorithm 5.3.1 HPD Modal Labeling (HPDML)

Step 1: Take each MCMC sample $\{\boldsymbol{\theta}_t, t = 1, \ldots, N\}$ as the initial value, and find the corresponding converged mode $\{m_t, t = 1, \ldots, N\}$ using any ascent algorithm.

Step 2: Label $\{\boldsymbol{\theta}_t, t = 1, \ldots, N\}$ by the associated converged mode $\{m_t, t = 1, \ldots, N\}$, i.e.

$$\boldsymbol{\omega}_t = \arg\min_{\boldsymbol{\omega}} (m_t^{\boldsymbol{\omega}} - \hat{\boldsymbol{\theta}})^\top (m_t^{\boldsymbol{\omega}} - \hat{\boldsymbol{\theta}}) . \qquad (5.9)$$

In our experience, most of the samples will converge to one of the $C!$ maximal modes. If m_t is one of the maximal modes i.e., there exists $\boldsymbol{\omega}_t$ such that $m_t^{\boldsymbol{\omega}_t} = \hat{\boldsymbol{\theta}}$. So the Step 2 in Algorithm 5.3.1 is to find the label $\boldsymbol{\omega}_t$ that makes $m_t^{\boldsymbol{\omega}_t}$ the same as the referenced major mode $\hat{\boldsymbol{\theta}}$. Using the objective function in Step 2 to determine the labels of m_t, we can relax the stopping rule for the ascent algorithm when finding $\{m_t, t = 1, \ldots, N\}$.

Note that practically, we do not need to compute all $C!$ permutations in (5.9) for all samples. Instead, we can apply to m_t the same order constraint as the one used to define $\hat{\boldsymbol{\theta}}$, denoted by $\boldsymbol{\omega}_t^*$. To illustrate the advantage of the HPD modal labeling method over other risk-based relabeling algorithms, let's consider the case of a univariate normal mixture. Suppose we have a reference maximal mode denoted as $\hat{\boldsymbol{\theta}}$, which is ordered based on the component means, given by:

$$\hat{\boldsymbol{\theta}} = (\hat{\pi}_1, \ldots, \hat{\pi}_C, \hat{\mu}_1, \ldots, \hat{\mu}_C, \hat{\sigma}_1, \ldots, \hat{\sigma}_C),$$

where $\hat{\mu}_1 < \hat{\mu}_2 < \ldots < \hat{\mu}_C$. Now, let $\mathbf{m}_t$ represent one of the $C!$ maximal modes, and our goal is to determine the labeling $\boldsymbol{\omega}_t$ such that $\mathbf{m}_t^{\boldsymbol{\omega}_t} = \hat{\boldsymbol{\theta}}$. If we identify the label $\boldsymbol{\omega}_0$ such that $m_t^{\boldsymbol{\omega}_0}$ is also ordered by the component means μ's, we can conclude that $\boldsymbol{\omega}_0 = \boldsymbol{\omega}_t$. Consequently, when dealing with maximal modes, the labeling process becomes as straightforward as the order constraint labeling approach. This characteristic makes the HPD modal labeling significantly faster, particularly in scenarios with a large number of components, in contrast to other risk-based relabeling algorithms that necessitate $C!$ comparisons during each iteration. Another main advantage of HPDML is that it is an online algorithm and can do labeling along the process of MCMC sampling, thereby greatly reducing storage requirements.

When dealing with minor modes, we employ the distance criteria (5.9) to determine the labeling $\boldsymbol{\omega}_t$ that minimizes the distance between $\mathbf{m}_t^{\boldsymbol{\omega}_t}$ and $\hat{\boldsymbol{\theta}}$. Alternatively, several other existing labeling methods can also be applied to label the minor modes. For instance, similar to the KL algorithm, we can utilize the Kullback-Leibler divergence from the distribution of clusterings based on the reference mode $\hat{\boldsymbol{\theta}}$ to the distribution of clusterings based on $\mathbf{m}_t^{\boldsymbol{\omega}}$. Consequently, the criteria (5.9) can be replaced by the following expression:

$$\boldsymbol{\omega}_t = \arg\max_{\boldsymbol{\omega}} \sum_{i=1}^{n} \sum_{c=1}^{C} p_{ic}(\hat{\boldsymbol{\theta}}) \log(p_{ic}(m_t^{\boldsymbol{\omega}})),$$

where, $p_{ic}(\boldsymbol{\theta}) = \pi_c g(x_i; \lambda_c) / \sum_{l=1}^{C} \{\pi_l g(x_i; \lambda_l)\}$ represents the classification probability of x_i belonging to the cth component based on the parameter $\boldsymbol{\theta}$. Notably, this criterion possesses the advantageous characteristic of being invariant to the scale effect of parameters.

Note that the number of labels we need to determine equals the number of different genuine modes of $(m_1, \ldots, m_N)$, which is quite small compared to the number of Gibbs samples. The genuine modes are the modes that do not include the permuted modes. Hence, the labeling process is very fast after we get the modes sequence $(m_1, \ldots, m_N)$.

5.3.3 ECM algorithm

To locate the posterior modes, one common method is to apply a gradient ascent algorithm, which, unfortunately, is not easy for Bayesian mixtures, due to the intractable posterior. Yao and Lindsay (2009) proposed an ECM algorithm to find the posterior modes of Bayesian mixtures by combining the properties of the Gibbs sampler and the ideas of ECM (Meng and Rubin, 1993).

Define the latent variables

$$
z_{ic} = \begin{cases} 1, & \text{if the } i\text{th observation is from } c\text{th component;} \\ 0, & \text{otherwise.} \end{cases},
$$

where $i = 1, \ldots, n, c = 1, \ldots, C$. Let $\mathbf{z} = \{z_{ij}, 1 \le i \le n, 1 \le c \le C\}$. Then the complete likelihood for $(\mathbf{x}, \mathbf{z})$ is

$$
L_c(\boldsymbol{\theta}; \mathbf{x}, \mathbf{z}) = \prod_{i=1}^{n} \prod_{c=1}^{C} [\pi_c g(x_i; \lambda_c)]^{z_{ic}},
$$

and the complete posterior distribution is

$$
p(\boldsymbol{\theta}, \mathbf{z} \mid \mathbf{x}) = \frac{1}{p(\mathbf{x})} p(\boldsymbol{\theta}) L_c(\boldsymbol{\theta}; \mathbf{x}, \mathbf{z}),
$$

where $p(\mathbf{x})$ is the marginal density for $\mathbf{x} = \{x_1 \cdots, x_n\}$. Suppose that all the prior parameters are fixed and we are in a setting such that we can use Gibbs sampler to get the MCMC samples, i.e., there exists a partition of $\boldsymbol{\theta} = \{\boldsymbol{\theta}_{(1)}, \ldots, \boldsymbol{\theta}_{(p)}\}$ such that all the conditional complete posterior distributions $\{p(\boldsymbol{\theta}_{(i)} \mid \ldots), 1 \le i \le p\}$ have an explicit formula, where $\boldsymbol{\theta}_{(i)}$ can be scalar or vector and $|\ldots$ denotes conditioning on all other parameters and the latent variable $\mathbf{z}$. Yao and Lindsay (2009) proposed the following ECM algorithm to find the posterior modes of Bayesian mixtures.

Algorithm 5.3.2 ECM algorithm for Bayesian mixtures (ECM(BM))

Starting with the initial value of $\boldsymbol{\theta}$, iterate the following two steps until a fixed point is reached.

E-step: Find the conditional expectation of the latent variable $\mathbf{z}$, i.e., the classification probability for each observation:

$$
p_{ic} = E(z_{ic} \mid \mathbf{x}, \boldsymbol{\theta}) = \frac{\pi_c g(x_i; \lambda_c)}{\sum_{l=1}^{C} \pi_l g(x_i; \lambda_l)}, i = 1, \ldots, n, c = 1, \ldots, C.
$$

M-step: Update $\boldsymbol{\theta}$ by maximizing the conditional complete posterior distribution $p(\boldsymbol{\theta}_{(i)} \mid \ldots), 1 \le i \le p$ sequentially with the latent variable z_{ic} replaced by the classification probability p_{ic}.

Based on the theory of Expectation-Conditional Maximization (ECM), the posterior distribution $p(\boldsymbol{\theta})$ increases after each iteration. Additionally, the algorithm possesses a notable property of natural equivalence. Specifically, if $\boldsymbol{\theta}$ converges to $\boldsymbol{\theta}_*$, then $\boldsymbol{\theta}^\omega$ converges to $\boldsymbol{\theta}_*^\omega$. In other words, if the algorithm forms a modal cluster, any fixed permutation of its elements will also form a cluster that corresponds to the permuted mode.

Let us complete an example to demonstrate how ECM(BM) algorithm works.

Example 5.3.1. Suppose $\{x_1, \ldots, x_n\}$ come from a C-component normal mixture:

$$f(x; \boldsymbol{\theta}) = \pi_1 N(x; \mu_1, \sigma_1^2) + \pi_2 N(x; \mu_2, \sigma_2^2) + \cdots + \pi_C N(x; \mu_C, \sigma_C^2).$$

Conjugate priors for normal densities have been considered by Diebolt and Robert (1994) and Bensmail et al. (1997), and are given by

$$\pi \sim D(\delta_1, \delta_2, \cdots, \delta_C), \ \mu_c \sim N(\xi_c, \sigma_c^2/\gamma_c), \ \sigma_c^{-2} \sim \Gamma(\alpha_c, \beta_c), c = 1, \ldots, C,$$

where $D(\cdot)$ is a Dirichlet distribution. Then we have the following conditional posterior distributions:

$$\pi \mid \cdots \sim D(\delta_1 + n_1, \ldots, \delta_C + n_C),$$

$$\mu_c \mid \cdots \sim N\left(\frac{n_c \bar{x}_c + \gamma_c \xi_c}{n_c + \gamma_c}, \frac{\sigma_c^2}{n_c + \gamma_c}\right),$$

$$\sigma_c^{-2} \mid \cdots \sim \Gamma\left(0.5 + 0.5 n_c + \alpha_c, \ \beta_c + 0.5\gamma_c(\mu_c - \xi_c)^2 + 0.5\sum_{i=1}^{n} z_{ic}(x_i - \mu_c)^2\right),$$

$$P(z_{ic} = 1 \mid \cdots) \propto \pi_c \phi(x_i; \mu_c, \sigma_c^2), \qquad (5.10)$$

where $n_c = \sum_{i=1}^{n} z_{ic}$, $\bar{x}_c = \sum_{i=1}^{n} z_{ic} x_i$, $i = 1, \ldots, n, c = 1, \ldots, C$, and "$\mid \cdots$" denotes conditioning on all other parameters.

Based on the above conditional posterior distributions, the $(k+1)$th step of the ECM(BM) algorithm can be performed as follows.

E-step: Find the conditional expectation of $\mathbf{z}$:

$$p_{ic}^{(k+1)} = E(Z_{ic} \mid \cdots) = \frac{\pi_c^{(k)} \phi(x_i; \mu_c^{(k)}, \sigma_c^{2(k)})}{\sum_{l=1}^{C} \pi_l^{(k)} \phi(x_i; \mu_l^{(k)}, \sigma_l^{2(k)})}.$$

Denote $n_c^{(k+1)} = \sum_{i=1}^{n} p_{ic}^{(k+1)}$, $\bar{x}_c^{(k+1)} = \sum_{i=1}^{n} p_{ic}^{(k+1)} x_i$.

M-step: Update the parameters by maximizing the conditional posterior distribution (5.10) with z_{ic} replaced by $p_{ic}^{(k+1)}$:

$$\pi_c^{(k+1)} = \frac{n_c^{(k+1)} - 1 + \delta_c}{n - C + \sum_{l=1}^{C} \delta_l},$$

$$\mu_c^{(k+1)} = \frac{n_c \bar{x}_c^{(k+1)} + \gamma_c \xi_c}{n_c + \gamma_c},$$

$$\sigma_c^{2(k+1)} = \frac{\beta_c + 0.5\gamma_c(\mu_c^{(k+1)} - \xi_c)^2 + 0.5\sum_{i=1}^{n} p_{ic}^{(k+1)}(x_i - \mu_c^{(k+1)})^2}{0.5 n_c^{(k+1)} + \alpha_c - 0.5},$$

where $c = 1, \ldots, C$. $\square$

Other popular conjugate priors for normal densities have been considered by Richardson and Green (1997) and are given by

$$\pi \sim D(\delta, \delta, \cdots, \delta), \ \mu_c \sim N(\xi, \kappa^{-1}), \ \sigma_c^{-2} \sim \Gamma(\alpha, \beta), c = 1, \ldots, C.$$

Then we have the following conditional posterior distributions:

$$\pi \mid \cdots \sim D(\delta + n_1, \ldots, \delta + n_C),$$

$$\mu_c \mid \cdots \sim N \left(\frac{\sigma_c^{-2} n_c \bar{x}_c + \kappa \xi}{\sigma_c^{-2} n_c + \kappa}, \frac{1}{n_c \omega_c^{-2} + \kappa} \right),$$

$$\sigma_c^{-2} \mid \cdots \sim \Gamma \left(0.5 n_c + \alpha, \ \beta + 0.5 \sum_{i=1}^{n} z_{ic}(x_i - \mu_c)^2 \right),$$

$$P(z_{ic} = 1 \mid \cdots) \propto \pi_c \phi(x_i; \mu_c, \sigma_c^2), c = 1, \ldots, C.$$

Based on the above conditional posterior distributions, the $(k+1)$th step of the ECM(BM) algorithm is as follows.

E step: Find the conditional expectation of $\mathbf{z}$:

$$p_{ic}^{(k+1)} = E(z_{ic} \mid \cdots) = \frac{\pi_c^{(k)} \phi(x_i; \mu_c^{(k)}, \sigma_c^{2(k)})}{\sum_{l=1}^{C} \pi_l^{(k)} \phi(x_i; \mu_l^{(k)}, \sigma_l^{2(k)})}, i = 1, \ldots, n, c = 1, \ldots, C.$$

M step: Update the parameters estimates:

$$\pi_c^{(k+1)} = \frac{n_c^{(k+1)} - 1 + \delta}{n - C + C\delta},$$

$$\mu_c^{(k+1)} = \frac{\sigma_c^{-2(k+1)} n_c \bar{x}_c^{(k+1)} + \kappa \xi}{\sigma_c^{-2(k+1)} n_c + \kappa},$$

$$\sigma_c^{2(k+1)} = \frac{\beta + 0.5 \sum_{i=1}^{n} p_{ic}^{(k+1)} (x_i - \mu_c^{(k+1)})^2}{0.5 n_c^{(k+1)} + \alpha - 1},$$

where $n_c^{(k+1)} = \sum_{i=1}^{n} p_{ic}^{(k+1)}, \bar{x}_c^{(k+1)} = \sum_{i=1}^{n} p_{ic}^{(k+1)} x_i$, and $c = 1, \ldots, C$.

Besides the above ECM(BM) algorithm, there are several other methods to estimate the maximal posterior modes before applying HDPML. If the sample size is large enough and the MLE is not difficult to calculate (by the EM algorithm), we can simply use the MLE as the estimator of one of the maximum posterior modes. We can also calculate the posterior likelihood for all the (burn-in) Bayesian samples, and then choose the one with the highest value as the estimate of the major posterior mode.

5.3.4 HPD modal labeling credibility

Let c_* be the maximum posterior value among all the degenerate points that correspond to mixture models with less than C components. Yao and Lindsay

(2009) argued that c_* is a lower bound and a good approximation to c_0 and then $\Pr(p(\boldsymbol{\theta}|\mathbf{x}) > c_*) = 1 - \alpha_* \geq \Pr(p(\boldsymbol{\theta}|\mathbf{x}) > c_0)$ provides an upper bound to, and a good approximation to the labeling credibility $1 - \alpha_0$, where $\alpha_* = \Pr(p(\boldsymbol{\theta}|\mathbf{x}) \leq c_*)$ is an approximation of the proportion of HPD ambiguous. As such, α_* is a good measure of how difficult the labeling problem is, while also indicating the level of arbitrariness involved in assigning labels to all sample points.

When $c < c_*$, the modal region $S_c(\hat{\boldsymbol{\theta}})$ will contain one or more degenerate points and thus is unidentifiable. Therefore, $c_0 \geq c_*$ and this lower bound becomes a tight one if $S_c(\hat{\boldsymbol{\theta}})$ and $S_c(\hat{\boldsymbol{\theta}}^{\boldsymbol{\omega}})$ first connect at a degenerate point because when $c > c_*$, $S_c(\hat{\boldsymbol{\theta}})$ and $S_c(\hat{\boldsymbol{\theta}}^{\boldsymbol{\omega}})$ are not connected and do not contain any degenerate points. Unfortunately, it is difficult to verify whether this property holds in general, or even in a specific data analysis. When c reaches a sufficiently small level, the maximal connected subsets S_c and $S_c^{\boldsymbol{\omega}}$ will intersect each other. In this case, the corresponding modes can themselves be connected by a continuous path. When the component level parameter is univariate, every such path must pass through a degenerate parameter value, so in these cases, c_* and c_0 are necessarily the same. Yao and Lindsay (2009) provided some graphical checking methods and empirically demonstrated that the result of $c_* = c_0$ holds in general. Therefore, we can also conclude that sample points with posterior greater than c_* are HPD labeled.

The value c_* and hence the labeling credibility level $1 - \alpha_*$ can be easily estimated based on the ECM(BM) algorithm. Note that for the ECM(BM), the updated point after each iteration from a degenerate point will be also a degenerate point. Therefore, we can find the c^* value by running the ECM(BM) algorithm starting from multiple degenerate points and choosing the converged degenerate mode with the largest posterior. Denote the estimate of c_* by $\hat{c}^*$. Then the corresponding label credibility $1 - \alpha_*$ can be estimated by the proportion of MCMC samples with a posterior larger than $\hat{c}^*$.

If the labeling credibility is very high, say 95%, then one may feel comfortable saying that the modal labeling method is providing a close description of the HPD region, concluding that it is "valid." If the labeling credibility is low, say 50%, then one may think that any labeling algorithm is somewhat ad hoc in nature, and sensible methods can reasonably disagree. The artificial separation by any labeling method will cause the bias effect of the parameter estimates. In this situation, the soft labeling, which assigns a probability of labeling for each Gibbs sample, may help reduce the bias effect. Please see Section 5.4 for an introduction to the soft labeling.

5.4 Soft probabilistic relabeling methods

All the labeling methods introduced so far are deterministic labeling methods, which assign a unique label for each sample. In this section, we introduce two probabilistic relabeling methods that can provide probabilities for each label, thus acknowledging the uncertainty in relabeling the samples.

5.4.1 Model-based labeling

As explained in Section 5.1, due to the label switching, the posterior $p(\boldsymbol{\theta} \mid \mathbf{x})$ has $C!$ permutation symmetric modal regions, each of which can be considered as a well-labeled region. Therefore, we can solve the label-switching issue if we can recover the distribution of one of the symmetric modal regions. Yao (2012b) proposed finding soft labels by estimating the posterior using a mixture distribution with $C!$ symmetric components, each component corresponding to one of the $C!$ symmetric modal regions. Compared to existing labeling methods, the method of Yao (2012b) aims to approximate the posterior directly and provide the labeling probabilities for all possible labels.

Let us first consider the situation when $C = 2$ and there will be two symmetric modal regions of the posterior, each of which can be considered as the region for the "true" labeled parameter space. The target of labeling is to recover one of the modal regions, which will be called the reference modal region. Suppose that the reference modal region has the well-labeled posterior density $g(\boldsymbol{\theta} \mid \mathbf{x})$. Let $\boldsymbol{\omega}_0 = (2, 1)$. If $\boldsymbol{\theta}$ comes from the reference modal region, i.e., $\boldsymbol{\theta}$ has an identical label $(1, 2)$, it has the density $g(\boldsymbol{\theta} \mid \mathbf{x})$; if $\boldsymbol{\theta}$ comes from the other modal region, i.e., $\boldsymbol{\theta}^{\boldsymbol{\omega}_0}$ comes from the reference modal region, then $\boldsymbol{\theta}$ has the density $g(\boldsymbol{\theta}^{\boldsymbol{\omega}_0} \mid \mathbf{x})$. Let $\boldsymbol{\omega}_t$ be the true latent label for $\boldsymbol{\theta}_t$, where $(\boldsymbol{\theta}_1, \ldots, \boldsymbol{\theta}_N)$ are N MCMC samples. We have

$$p\{\boldsymbol{\theta}_t \mid \mathbf{x}, \boldsymbol{\omega}_t = (1, 2)\} = g(\boldsymbol{\theta}_t \mid \mathbf{x}),$$
$$p\{\boldsymbol{\theta}_t \mid \mathbf{x}, \boldsymbol{\omega}_t = (2, 1)\} = g(\boldsymbol{\theta}_t^{\boldsymbol{\omega}_0} \mid \mathbf{x}),$$
$$P\{\boldsymbol{\omega}_t = (1, 2)\} = P\{\boldsymbol{\omega}_t = (2, 1)\} = 1/2.$$

Therefore, the original full posterior density for $\boldsymbol{\theta}_t$, without knowing its true label $\boldsymbol{\omega}_t$, has a mixture form

$$p(\boldsymbol{\theta}_t \mid \mathbf{x}) = \frac{1}{2}g(\boldsymbol{\theta}_t \mid \mathbf{x}) + \frac{1}{2}g(\boldsymbol{\theta}_t^{\boldsymbol{\omega}_0} \mid \mathbf{x}). \tag{5.11}$$

The model (5.11) is called the symmetric mixture model by Yao (2012b), due to the permutation symmetry of the mixture components. The label switching can be solved if we can estimate $g(\boldsymbol{\theta} \mid \mathbf{x})$ and then the Bayesian inference can be performed based on the well-labeled posterior $g(\boldsymbol{\theta} \mid \mathbf{x})$ instead of the original unlabeled posterior $p(\boldsymbol{\theta} \mid \mathbf{x})$.

Let $\hat{g}(\boldsymbol{\theta} \mid \mathbf{x})$ be the resulting estimate of $g(\boldsymbol{\theta} \mid \mathbf{x})$. We can then find the classification/labeling probabilities $(\hat{p}_{t1}, \hat{p}_{t2})$ for each $\boldsymbol{\theta}_t$, where

$$\hat{p}_{t1} = \frac{\hat{g}(\boldsymbol{\theta}_t \mid \mathbf{x})}{\hat{g}(\boldsymbol{\theta}_t \mid \mathbf{x}) + \hat{g}(\boldsymbol{\theta}_t^{\boldsymbol{\omega}_0} \mid \mathbf{x})},$$

$$\hat{p}_{t2} = 1 - \hat{p}_{t1}, \ t = 1, \dots, N.$$

The estimated probabilities, $\hat{p}_{t1}$ and $\hat{p}_{t2}$, can be interpreted as the probabilities that $\boldsymbol{\theta}_t$ belongs to the reference modal region (i.e., the label of $\boldsymbol{\theta}_t$ is $\boldsymbol{\omega}_t = (1, 2)$) and that $\boldsymbol{\theta}_t^{\boldsymbol{\omega}_0}$ (permuted according to $\boldsymbol{\omega}_0 = (2, 1)$) belongs to the reference modal region, respectively. To determine the labeling $\boldsymbol{\omega}_t$, we can maximize the labeling probabilities $\{\hat{p}_{t1}, \hat{p}_{t2}\}$, thereby assigning the identity permutation label $\hat{\boldsymbol{\omega}}_t = (1, 2)$ if $\hat{p}_{t1} > \hat{p}_{t2}$, or assigning the permutation label $\hat{\boldsymbol{\omega}}_t = (2, 1)$ if $\hat{p}_{t1} < \hat{p}_{t2}$. Subsequently, Bayesian inference can be performed based on the labeled samples $\{\boldsymbol{\theta}_t^{\hat{\boldsymbol{\omega}}_t}, t = 1, \dots, N\}$.

Next, we introduce both a parametric method and a semi-parametric method proposed by Yao (2012b) to estimate the symmetric mixture model (5.11) given the MCMC samples $(\boldsymbol{\theta}_1, \dots, \boldsymbol{\theta}_N)$.

5.4.1.1   Parametric soft labeling

Based on the asymptotic theory for the posterior distribution (Walker, 1969; Frühwirth-Schnatter, 2006), when the sample size is large, the labeled MCMC samples should approximately follow a normal distribution. In other words, there exist permutations $\{\boldsymbol{\omega}_1, \dots, \boldsymbol{\omega}_N\}$ such that $\{\boldsymbol{\theta}_1^{\boldsymbol{\omega}_1}, \dots, \boldsymbol{\theta}_N^{\boldsymbol{\omega}_N}\}$ approximately follows a normal distribution. Consequently, the distribution $g(\boldsymbol{\theta} \mid \mathbf{x})$ in (5.11) can be approximated by a normal density as follows:

$$p(\boldsymbol{\theta} \mid \mathbf{x}) \approx \frac{1}{2}\phi(\boldsymbol{\theta}; \boldsymbol{\mu}, \boldsymbol{\Sigma}) + \frac{1}{2}\phi(\boldsymbol{\theta}^{\boldsymbol{\omega}_0}; \boldsymbol{\mu}, \boldsymbol{\Sigma}). \tag{5.12}$$

Here $\boldsymbol{\omega}_0 = (2, 1)$, $\boldsymbol{\mu}$ and $\boldsymbol{\Sigma}$ denote the center and covariance matrix for the reference modal region, and $\phi(\boldsymbol{\theta}; \boldsymbol{\mu}, \boldsymbol{\Sigma})$ corresponds to the density function of the multivariate normal distribution $N(\boldsymbol{\mu}, \boldsymbol{\Sigma})$. In practice, some transformation of the MCMC samples may help make the normality approximation work better for the labeled samples. For example, for the standard error parameters, one might take the log transformation. For the mixing proportion parameters, one might take the log odds transformation.

Yao (2012b) provided the EM algorithm in Algorithm 5.4.1 to estimate the symmetric normal mixture model (5.12). As Yao (2012b) explained if we use the classification EM, i.e., assign hard labels in E-step, the above algorithm will be the same as the NORMLH method explained in Algorithm 5.2.2 (Yao and Lindsay, 2009).

5.4.1.2   Semi-parametric soft labeling

When the sample size is small, the labeled samples may not be approximated well by a normal distribution. Yao (2012b) further proposed a semi-parametric

---

Algorithm 5.4.1 Model-based labeling by symmetric normal mixture model (MBLSNM)

Starting with some initial values $\boldsymbol{\mu}^{(0)}$ and $\boldsymbol{\Sigma}^{(0)}$, in the $(k+1)$th step,

E step: Compute the labeling/classification probabilities:

$$p_{t1}^{(k+1)} = \frac{\phi(\boldsymbol{\theta}_t; \boldsymbol{\mu}^{(k)}, \boldsymbol{\Sigma}^{(k)})}{\phi(\boldsymbol{\theta}_t; \boldsymbol{\mu}^{(k)}, \boldsymbol{\Sigma}^{(k)}) + \phi(\boldsymbol{\theta}_t^{\boldsymbol{\omega}_0}; \boldsymbol{\mu}^{(k)}, \boldsymbol{\Sigma}^{(k)})},$$

$$p_{t2}^{(k+1)} = 1 - p_{t1}^{(k+1)}, \quad t = 1, \ldots, N.$$

M step: Update $\boldsymbol{\mu}$ and $\boldsymbol{\Sigma}$:

$$\boldsymbol{\mu}^{(k+1)} = \frac{1}{N} \sum_{t=1}^{N} \left\{ p_{t1}^{(k+1)} \boldsymbol{\theta}_t + p_{t2}^{(k+1)} \boldsymbol{\theta}_t^{\boldsymbol{\omega}_0} \right\},$$

$$\boldsymbol{\Sigma}^{(k+1)} = \frac{1}{N} \sum_{t=1}^{N} \left\{ p_{t1}^{(k+1)} (\boldsymbol{\theta}_t - \boldsymbol{\mu}^{(k+1)})(\boldsymbol{\theta}_t - \boldsymbol{\mu}^{(k+1)})^{\top} \right.$$
$$\left. + p_{t2}^{(k+1)} (\boldsymbol{\theta}_t^{\boldsymbol{\omega}_0} - \boldsymbol{\mu}^{(k+1)})(\boldsymbol{\theta}_t^{\boldsymbol{\omega}_0} - \boldsymbol{\mu}^{(k+1)})^{\top} \right\}.$$

---

method to estimate the mixture model (5.11) without any parametric assumption about $g(\boldsymbol{\theta} \mid \mathbf{x})$ based on the semi-parametric EM-type as described in Algorithm 5.4.2.

To use the above semi-parametric labeling, we need to choose the bandwidth matrix $\mathbf{H}$ first. A good rule of thumb is to use a bandwidth matrix proportional to $\hat{\boldsymbol{\Sigma}}^{1/2}$, i.e., $\mathbf{H} = h\hat{\boldsymbol{\Sigma}}^{1/2}$, where $\hat{\boldsymbol{\Sigma}}$ is the estimated covariance matrix based on the initial labeled samples. By assuming a multivariate normal distribution for the labeled samples, we can derive the rule of thumb for the bandwidth matrix $\mathbf{H}$

$$\hat{\mathbf{H}} = N^{-1/(p+4)} \hat{\boldsymbol{\Sigma}}^{1/2},$$

where p is the dimension of $\boldsymbol{\theta}$ (Scott, 1992, Page 152).

5.4.1.3 More than two components

When there are C components, there will be $C!$ symmetric modal regions, and, the posterior distribution $p(\boldsymbol{\theta} \mid x)$ is a mixture with $C!$ permutation symmetric components:

$$p(\boldsymbol{\theta} \mid \mathbf{x}) = \frac{1}{C!} \sum_{j=1}^{C!} g(\boldsymbol{\theta}^{\boldsymbol{\omega}_{(j)}} \mid \mathbf{x}),$$

Algorithm 5.4.2 Model-based labeling by semi-parametric mixture models (MBLSP)

Starting with the initial density estimate $g^{(0)}(\boldsymbol{\theta} \mid \mathbf{x})$, in the $(k+1)$th step,

E step: Compute the labeling probabilities:

$$
\begin{aligned}
p_{t1}^{(k+1)} &= \frac{g^{(k)}(\boldsymbol{\theta}_t \mid \mathbf{x})}{g^{(k)}(\boldsymbol{\theta}_t \mid \mathbf{x}) + g^{(k)}(\boldsymbol{\theta}_t^{\boldsymbol{\omega}_0} \mid \mathbf{x})}, \\
p_{t2}^{(k+1)} &= 1 - p_{t1}^{(k+1)}, \quad t = 1, \ldots, N.
\end{aligned}
$$

Nonparametric step: Update $g(\boldsymbol{\theta} \mid \mathbf{x})$ by a kernel density estimator:

$$
g^{(k+1)}(\boldsymbol{\theta} \mid \mathbf{x}) = \frac{1}{N} \sum_{t=1}^{N} \left\{ p_{t1}^{(k+1)} K_{\mathbf{H}}(\boldsymbol{\theta}_t - \boldsymbol{\theta}) + p_{t2}^{(k+1)} K_{\mathbf{H}}(\boldsymbol{\theta}_t^{\boldsymbol{\omega}_0} - \boldsymbol{\theta}) \right\},
$$

where $K_{\mathbf{H}}(\boldsymbol{\theta}) = \det(\mathbf{H})^{-1} K(\mathbf{H}^{-1}\boldsymbol{\theta})$, $\mathbf{H}$ is a bandwidth matrix, and $K(\boldsymbol{\theta})$ is a multivariate kernel density such as multivariate Gaussian density.

---

where $g(\boldsymbol{\theta} \mid \mathbf{x})$ is the posterior density for a labeled modal region, and $\{\boldsymbol{\omega}_{(1)}, \ldots, \boldsymbol{\omega}_{(C!)}\}$ are the $C!$ permutations of $(1, \ldots, C)$.

The model-based labeling algorithms introduced in the previous two sections can be easily extended to the situation when the number of components is larger than two. The extensions of E steps are trivial. For the symmetric normal mixture model, the M step is now

$$
\boldsymbol{\mu}^{(k+1)} = \frac{1}{N} \sum_{t=1}^{N} \sum_{j=1}^{C!} p_{tj}^{(k+1)} \boldsymbol{\theta}_t^{\boldsymbol{\omega}_{(j)}},
$$

$$
\boldsymbol{\Sigma}^{(k+1)} = \frac{1}{N} \sum_{t=1}^{N} \sum_{j=1}^{C!} p_{tj}^{(k+1)} (\boldsymbol{\theta}_t^{\boldsymbol{\omega}_{(j)}} - \boldsymbol{\mu}^{(k+1)})(\boldsymbol{\theta}_t^{\boldsymbol{\omega}_{(j)}} - \boldsymbol{\mu}^{(k+1)})^{\top}. \tag{5.13}
$$

For the symmetric semi-parametric mixture model, the *nonparametric step* is

$$
g^{(k+1)}(\boldsymbol{\theta} \mid \mathbf{x}) = \frac{1}{N} \sum_{t=1}^{N} \sum_{j=1}^{C!} p_{tj}^{(k+1)} K_{\mathbf{H}}(\boldsymbol{\theta}_t^{\boldsymbol{\omega}_{(j)}} - \boldsymbol{\theta}).
$$

Based on the labeling probabilities $\{\hat{p}_{tj}, t = 1, \ldots, N, j = 1, \ldots, C!\}$, we can then choose the label $\boldsymbol{\omega}_t$ for $\boldsymbol{\theta}_t$ by maximizing $\{\hat{p}_{t1}, \ldots, \hat{p}_{tC!}\}$. For example, if $\hat{p}_{tk}$ maximizes $\{\hat{p}_{t1}, \ldots, \hat{p}_{tC!}\}$ for some k, then $\hat{\boldsymbol{\omega}}_t = \boldsymbol{\omega}_{(k)}$.

Given the labeling probabilities, one may also use them to do weighted averaging when doing Bayesian inference. For example, the posterior mean of

θ can be estimated by

$$\hat{\theta} = \frac{1}{N} \sum_{t=1}^{N} \sum_{j=1}^{C!} \hat{p}_{tj} \theta_t^{\boldsymbol{\omega}_{(j)}} . \qquad (5.14)$$

Note that the weighted average (5.14) mimics the idea of the M step in (5.13).

5.4.2 Probabilistic relabeling strategies

Sperrin et al. (2010) developed a probabilistic relabeling method by extending the probabilistic relabeling of Jasra et al. (2005). Sperrin et al. (2010) proposed using the current allocation vector $\mathbf{z}$ to perform a probabilistic relabeling for the MCMC samples. Let $p(\boldsymbol{\omega}; \mathbf{z}, \boldsymbol{\theta})$ be the probability distribution for the label $\boldsymbol{\omega}$ of $\mathbf{z}$ given the current parameter estimate $\boldsymbol{\theta}$. Sperrin et al. (2010) proposed estimating $p(\boldsymbol{\omega}; \mathbf{z}, \boldsymbol{\theta})$ and $\boldsymbol{\theta}$ in an iterative fashion similar to an EM-type approach. Given $\boldsymbol{\theta}$, $p(\boldsymbol{\omega}; \mathbf{z}, \boldsymbol{\theta})$ is estimated based on the complete likelihood function

$$\hat{p}(\boldsymbol{\omega}; \mathbf{z}, \boldsymbol{\theta}) \propto \prod_{i=1}^{n} \prod_{j=1}^{C} \left\{ \pi_{\boldsymbol{\omega}(j)} f(x_i; \lambda_{\boldsymbol{\omega}(j)}) \right\}^{z_{ij}} .$$

Given $p(\boldsymbol{\omega}; \mathbf{z}, \boldsymbol{\theta})$, we can update $\boldsymbol{\theta}$ based on the weighted average (5.14)

$$\boldsymbol{\theta} \leftarrow \frac{1}{N} \sum_{t=1}^{N} \sum_{j=1}^{C!} \hat{p}(\boldsymbol{\omega}_{(j)}; \mathbf{z}_t, \boldsymbol{\theta}_t) \theta_t^{\boldsymbol{\omega}_{(j)}} .$$

However, the method of Sperrin et al. (2010) assigns the labels to the allocation vectors directly instead of the MCMC samples. This makes the method valid if there is no label switching between the allocation vectors and the corresponding MCMC samples, which is not necessarily true for all the samples.

5.5 Label switching for frequentist mixture models

Most of the existing labeling methods are proposed for Bayesian mixture models partly because solving the label-switching issue is more crucial for Bayesian mixture models. In contrast, there has been much less attention on the label-switching issue for frequentist mixture models. For frequentist mixture models, in order to compare different estimation methods or perform statistical inference for some estimators, we might want to compute their bias, variance, mean square errors, or other criteria, using a simulation study or bootstrap methods.

For a simulation study or a parametric bootstrap, when generating a random sample $(x_1, \ldots, x_n)$ from a mixture model, $f(x) = \sum_{c=1}^{C} \pi_c g(x; \lambda_c)$, we can first generate a multinomial random variable C_i with $P(C_i = c) = \pi_c, c = 1, 2, \ldots, C$, and then generate x_i from the component density $g(x; \lambda_{c_i})$. Marginally, x_i will have the required mixture density. Based on $(x_1, \ldots, x_n)$, one can then find the MLE of $\boldsymbol{\theta}$ without using the latent labels $(c_1, \ldots, c_n)$. By repeating the above procedures, one can get a sequence of simulated/bootstrapped estimates, say $(\hat{\boldsymbol{\theta}}_1, \ldots, \hat{\boldsymbol{\theta}}_N)$, which do not have meaningful labels. Before performing statistical inference, one must find their corresponding labels $(\boldsymbol{\omega}_1, \ldots, \boldsymbol{\omega}_N)$ such that $(\hat{\boldsymbol{\theta}}_1^{\boldsymbol{\omega}_1}, \ldots, \hat{\boldsymbol{\theta}}_N^{\boldsymbol{\omega}_N})$ have the same label meaning.

Compared to Bayesian mixture models, one distinct feature of labeling for frequentist mixture models in simulation studies or parametric bootstrap is the availability of true component labels, $(c_1, \ldots, c_n)$, of generated observations. Although these labels are not used for estimation, they can provide valuable information when solving label-switching issues for parameter estimates $\{\hat{\boldsymbol{\theta}}_j, j = 1, \ldots, N\}$.

Let $\mathbf{z} = \{z_{ic}, i = 1, \ldots, n, c = 1, \ldots, C\}$, where

$$z_{ic} = \begin{cases} 1, & \text{if } c_i = c, \\ 0, & \text{otherwise.} \end{cases}$$

Yao (2015) proposed finding the label $\boldsymbol{\omega}$ for $\hat{\boldsymbol{\theta}}$ by maximizing

$$L(\boldsymbol{\omega}; \hat{\boldsymbol{\theta}}, \mathbf{x}, \mathbf{z}) = \sum_{i=1}^{n} \sum_{c=1}^{C} z_{ic} \log p_{ic}(\hat{\boldsymbol{\theta}}^{\boldsymbol{\omega}}), \tag{5.15}$$

where

$$p_{ic}(\boldsymbol{\theta}^{\boldsymbol{\omega}}) = \frac{\pi_c^{\boldsymbol{\omega}} g(x_i; \lambda_c^{\boldsymbol{\omega}})}{f(x_i; \boldsymbol{\theta}^{\boldsymbol{\omega}})} \equiv p_{i\boldsymbol{\omega}(c)}(\boldsymbol{\theta})$$

is the classification probability, and $f(x_i; \boldsymbol{\theta}^{\boldsymbol{\omega}}) = \sum_{c=1}^{C} \pi_c^{\boldsymbol{\omega}} g(x_i; \lambda_c^{\boldsymbol{\omega}})$. Note that maximizing (5.15) is equivalent to minimizing the Kullback-Leibler divergence between the true classification probability z_{ij} and the estimated classification probability $p_{ij}(\hat{\boldsymbol{\theta}})$. Since the classification probabilities $p_{ij}(\boldsymbol{\theta})$ is a byproduct of an EM algorithm, the computation of (5.15) is usually very fast.

Yao (2015) proved that maximizing (5.15) is equivalent to maximizing the complete likelihood of $(\mathbf{x}, \mathbf{z})$ over $\boldsymbol{\omega}$

$$L(\hat{\boldsymbol{\theta}}^{\boldsymbol{\omega}}; \mathbf{x}, \mathbf{z}) = \prod_{i=1}^{n} \prod_{c=1}^{C} \{\hat{\pi}_c^{\boldsymbol{\omega}} g(x_i; \hat{\lambda}_c^{\boldsymbol{\omega}})\}^{z_{ic}},$$

by noting that

$$\log\{L(\hat{\boldsymbol{\theta}}^{\boldsymbol{\omega}}; \mathbf{x}, \mathbf{z})\} = \sum_{i=1}^{n} \sum_{c=1}^{C} \left[z_{ic} \log \left\{ \frac{\hat{\pi}_c^{\boldsymbol{\omega}} g(x_i; \hat{\lambda}_c^{\boldsymbol{\omega}})}{f(x_i; \hat{\boldsymbol{\theta}}^{\boldsymbol{\omega}})} \right\} \right]$$
$$+ \sum_{i=1}^{n} \sum_{c=1}^{C} \left[z_{ic} \log\{f(x_i; \hat{\boldsymbol{\theta}}^{\boldsymbol{\omega}})\} \right]$$
$$= \sum_{i=1}^{n} \sum_{c=1}^{C} z_{ic} \log p_{ic}(\hat{\boldsymbol{\theta}}^{\boldsymbol{\omega}}) + \sum_{i=1}^{n} \log f(x_i; \hat{\boldsymbol{\theta}}^{\boldsymbol{\omega}}), \quad (5.16)$$

where the second term of (5.16) is a log mixture likelihood and is invariant to the permutation of component labels of $\hat{\boldsymbol{\theta}}$. For this reason, Yao (2015) called the above labeling method *complete likelihood-based labeling*.

More generally, one could find labels by minimizing

$$\sum_{i=1}^{n} \sum_{c=1}^{C} \rho(z_{ij}, p_{ij}(\hat{\boldsymbol{\theta}}^{\boldsymbol{\omega}})),$$

where ρ is a loss function. It is expected that any reasonable choice of loss function ρ should give similar relabeling results. The (5.15) can be considered as a special case in which the Kullback-Leibler divergence is used. Another natural choice of ρ is the L_2 loss $\rho(t) = t^2$ leading to the *Distance-based labeling (DISTLAT)* proposed by Yao (2015).

5.6 Solving label switching for mixture models using R

Yao (2015) proposed two labeling methods for frequentist mixture models, namely the complete likelihood-based labeling (COMPLH) and the distance-based labeling (DISTLAT). In addition, order constraint labeling based on component means (OC-μ), and order constraint labeling based on component proportions (OC-π) are compared.

In Example 2 of Yao (2015), we consider the multivariate normal mixture model

$$\pi_1 N \left(\begin{pmatrix} \mu_{11} \\ \mu_{12} \end{pmatrix}, \begin{pmatrix} 1 & 0 \\ 0 & 1 \end{pmatrix} \right) + (1 - \pi_1) N \left(\begin{pmatrix} \mu_{21} \\ \mu_{22} \end{pmatrix}, \begin{pmatrix} 1 & 0 \\ 0 & 1 \end{pmatrix} \right),$$

where $\pi_1 = 0.3, \mu_{11} = 0, \mu_{12} = 0, \mu_{21} = 0.5, \mu_{22} = 0.5$. The following code can be used to apply the aforementioned labeling methods.

```
set.seed(123)
library(MixSemiRob)
```

```
library(mvtnorm)
mu1 = 0.5; mu2 = 0.5; prop = 0.3
n = 100; n1 = rbinom(1, n, prop)
pm = c(2, 1, 4, 3, 6, 5)
sigma = diag(c(1, 1))
mu = matrix(c(0, mu1, 0, mu2), ncol = 2)
pi = c(prop, 1 - prop)
ini = list(sigma = sigma, mu = mu, pi = pi)
x1 = rmvnorm(n1, c(0, 0), ini$sigma)
x2 = rmvnorm(n - n1, c(mu1, mu2), ini$sigma)
x = rbind(x1, x2)

out = mixnorm(x, C = 2)
lat = rbind(rep(c(1, 0), times = c(n1, n - n1)),
            rep(c(0, 1), times = c(n1, n - n1)))
clhest = complh(out, lat)
distlatest = distlat(out, lat)

COMPLH = c(clhest$mu[, 1], clhest$mu[, 2], clhest$pi)
DISTLAT = c(distlatest$mu[, 1], distlatest$mu[, 2],
            distlatest$pi)
NORMLH = c(out$mu[, 1], out$mu[, 2], out$pi)

if(out$mu[1, 1] < out$mu[2, 1]){ # if mu11 < mu12
  OCmu1 = NORMLH[1:6]
} else {
  OCmu1 = NORMLH[pm]
}
if(out$mu[1, 2] < out$mu[2, 2]){ # if mu21 < mu22
  OCmu2 = NORMLH[1:6]
} else {
  OCmu2 = NORMLH[pm]
}
if(out$pi[1] < out$pi[2]){        # if pi1 < pi2
  OCpi = NORMLH[1:6]
} else {
  OCpi = NORMLH[pm]
}

# TABLE 5.1
rbind(COMPLH, DISTLAT, NORMLH, OCmu1, OCmu2, OCpi)
```

TABLE 5.1

One example of Case IV of Example 2 in Yao (2015)

method	μ_{11}	μ_{12}	μ_{21}	μ_{22}	π
COMPLH	-0.16	2.95	0.16	-0.10	0.60
DISTLAT	-0.16	2.95	0.16	-0.10	0.60
NORMLH	-0.16	2.95	0.16	-0.10	0.60
OC$-\mu_1$	-0.16	2.95	0.16	-0.10	0.60
OC$-\mu_2$	2.95	-0.16	-0.10	0.16	0.40
OC$-\pi$	2.95	-0.16	-0.10	0.16	0.40

Table 5.1 shows a sample output of the above example when $n = 100$.

Yao (2015) provides a more detailed simulation study, which demonstrates that both COMPLH and DISTLAT work well, providing better results than the rule-of-thumb methods of order constraint labeling. Please also see the R package MixSemiRob for other labeling methods.

6

Hypothesis testing and model selection for mixture models

The Gaussian mixture models play important roles in statistical analysis. In fact, it is well known that any continuous distribution can be approximated well by a finite mixture of normal densities (Lindsay, 1995; McLachlan and Peel, 2000). When the number of components is unknown, one could use the nonparametric maximum likelihood estimate (NPMLE) introduced in Section 1.13 to estimate the mixing distribution nonparametrically. However, when the data has an underlying clustering structure, the NPMLE tends to overestimate the true number of components. If too many components are used, the mixture model may overfit the data and produce poor interpretations. On the contrary, if too few components are considered, the mixture may not be flexible enough to approximate the true underlying data structure. Therefore, selecting the number of components is not only of theoretical interest but also for practical applications.

6.1 Likelihood ratio tests for mixture models

For a general parametric model $f(x; \boldsymbol{\theta}), \boldsymbol{\theta} \in \Theta$, to test $H_0 : \boldsymbol{\theta} \in \Theta_0$ versus $H_a : \boldsymbol{\theta} \in \Theta_a$ given $\mathbf{x} = \{x_1, x_2, ..., x_n\}$, where $\Theta_0 \cap \Theta_a = \phi$ and $\Theta_0 \cup \Theta_a = \Theta$, the likelihood ratio test (LRT) statistic is commonly used and has the following form:

$$\Lambda = \frac{\max_{\boldsymbol{\theta} \in \Theta_0} \mathcal{L}(\boldsymbol{\theta}; \mathbf{x})}{\max_{\boldsymbol{\theta} \in \Theta} \mathcal{L}(\boldsymbol{\theta}; \mathbf{x})},$$

where $\mathcal{L}(\boldsymbol{\theta}; \mathbf{x}) = \prod_{i=1}^{n} f(x_i; \boldsymbol{\theta})$ is the likelihood function. Under some regularity conditions, the LRT statistic $-2 \log(\Lambda)$ has an asymptotic χ^2 distribution with a degree of freedom $df = dim(\Theta) - dim(\Theta_0)$, where $dim(\Theta)$ means the number of unknown parameters in the parameter space of Θ.

For finite mixture models $f(x; \boldsymbol{\theta}) = \sum_{c=1}^{C} \pi_c g(x; \boldsymbol{\lambda}_c)$, we are usually interested in testing the number of components, i.e., $H_0 : C = C_0$ vs $H_a : C = C_0 + 1$, which simplifies to the homogeneity test when $C_0 = 1$. It is clear that under the null hypothesis, the mixture model parameters are not identifiable. For example, when testing $H_0 : C = 1$ v.s. $H_a : C = 2$,

DOI: 10.1201/9781003038511-6

the model $f(x; \boldsymbol{\theta}) = \pi_1 g(x; \boldsymbol{\lambda}_1) + \pi_2 g(x; \boldsymbol{\lambda}_2)$ will be reduced to the null model ($C = 1$) if $\pi_1 = 0$ or $\pi_2 = 0$ or $\boldsymbol{\lambda}_1 = \boldsymbol{\lambda}_2$. Therefore, a surprising but well-known result is that the traditional asymptotic χ^2 distribution of the LRT does not hold for mixture models when testing for the number of components due to the violations of regularity conditions. Besides the aforementioned nonidentifiability issue, there is also the unboundedness issue of the likelihood function as explained in Section 1.9.

Hartigan (1985) and Bickel and Chernoff (1993) proved that the LRT is unbounded for finite normal mixture models when testing the number of components. Many researchers have proposed adding some restrictions on the parameter space so that the limit distribution of the LRT can be derived. For example, Dacunha-Castelle and Gassiat (1999) addressed the problem of testing a hypothesis using likelihood ratio statistics in nonidentifiable models and applied it to model selection when the smaller model is nonidentifiable due to the parameterization of the larger model. With the restriction of compact parameter space, a local conic parameterization was applied to this problem, and the asymptotic distribution for the (pseudo)-LRT was then achieved. Also, assuming a compact parameter space, Liu and Shao (2003) applied a general quadratic approximation of the mixture log-likelihood ratio function in a Hellinger neighborhood of the true density. Then, by maximizing the quadratic form, the asymptotic distribution of LRT is achieved, even when there is a loss of identifiability of the true distribution. Chen and Chen (2003) and Zhu and Zhang (2004) studied the asymptotic properties of the LRT for testing the homogeneity of normal mixtures and mixture regression models, respectively. Azaïs et al. (2009) worked on the LRT for the mixing measure in mixture models with or without structural parameters. Under some necessary conditions, they applied the LRT to test a single distribution against any mixture and to test the number of populations in a finite mixture with or without structural parameters. Chen and Li (2009) developed an expectation-maximization (EM) test for testing the homogeneity in normal mixture models. Chen et al. (2012) extended the result of Chen and Li (2009) and developed a likelihood-based EM test for testing the null hypothesis for an arbitrary number of components (C) in finite normal mixture models. However, their testing procedures require all component mean parameters to be different. Kasahara and Shimotsu (2015) and Shen and He (2015) further extended the EM test to normal mixture regression models with heteroscedastic components and a structured logistic normal mixture model, respectively.

Next, we use some examples to illustrate the nonregularity of the LRT for mixture models.

Example 6.1.1. Consider a mixture of $C = 2$ univariate normal densities with means $\mu_1 = 0$ and $\mu_2 = \mu$, and common unit variances. Suppose we wish to test $H_0 : f(x; \boldsymbol{\theta}) = \phi(x; 0, 1)$ versus $H_1 : f(x; \boldsymbol{\theta}) = (1 - \pi)\phi(x; 0, 1) + \pi\phi(x; \mu, 1)$.

The parameter space is given by

$$\Theta = \{\boldsymbol{\theta} = (\pi, \mu)^\top : [0,1] \times (-\infty, \infty)\}.$$

The null subspace that specifies H_0 is the entire μ-axis when $\pi = 0$ and the line segment $[0,1]$ on the π-axis when $\mu = 0$; that is

$$\Theta_0 = \{\boldsymbol{\theta} = (\pi, \mu)^\top : ([0] \times (-\infty, \infty)) \cup ([0,1] \times [0])\}.$$

Therefore, Θ_0 is on the boundary of Θ.

Given samples $x_1, x_2, \cdots, x_n$, the log-likelihood function is

$$\ell(\boldsymbol{\theta}) = \sum_{i=1}^{n} \log f(x_i; \boldsymbol{\theta}),$$

and the Fisher information is given by

$$\mathcal{I}(\boldsymbol{\theta}) = \begin{pmatrix} E_\theta \frac{(\phi(x;\mu,1) - \phi(x;0,1))^2}{f^2(x;\theta)} & E_\theta \frac{\pi\phi'(x;\mu,1)(\phi(x;\mu,1) - \phi(x;0,1))}{f^2(x;\theta)} \\ E_\theta \frac{\pi\phi'(x;\mu,1)(\phi(x;\mu,1) - \phi(x;0,1))}{f^2(x;\theta)} & E_\theta \frac{(\pi\phi'(x;\mu,1))^2}{f^2(x;\theta)} \end{pmatrix}.$$

If $\boldsymbol{\theta} = (0,0)^\top$, the Fisher information is $I(\boldsymbol{\theta}) = \begin{pmatrix} 0 & 0 \\ 0 & 0 \end{pmatrix}$. If $\boldsymbol{\theta} = (\pi, 0)^\top$ with $\pi \neq 0$, the Fisher information is

$$\mathcal{I}(\boldsymbol{\theta}) = \begin{pmatrix} 0 & 0 \\ 0 & E_\theta \frac{(\pi\phi'(x;0,1))^2}{f^2(x;\theta)} \end{pmatrix}.$$

If $\boldsymbol{\theta} = (0, \mu)$ with $\mu \neq 0$, the Fisher information is

$$\mathcal{I}(\boldsymbol{\theta}) = \begin{pmatrix} E_\theta \frac{(\phi(x;\mu,1) - \phi(x;0,1))^2}{f^2(x;\theta)} & 0 \\ 0 & 0 \end{pmatrix}.$$

Standard asymptotic inference about the MLE requires the assumption that the information matrix is positive definite, which is clearly violated based on the above arguments.

Hartigan (1985) showed that $-2\log\Lambda = 2\{\ell(\hat{\boldsymbol{\theta}}_1) - \ell(\hat{\boldsymbol{\theta}}_0)\}$ is asymptotically unbounded and converges to ∞ at a slow rate of $0.5\log(\log(n))$ when H_0 is true. Bickel and Chernoff (1993) and Liu and Shao (2003) proved that $-2\log\Lambda$ diverges to ∞ at a rate of $\log(\log(n))$ and

$$\lim_{n\to\infty} P\{-2\log\Lambda - \log(\log(n)) + \log(2\pi_1^2) \leq x\} = \exp(-e^{-x/2}).$$

$\square$

Example 6.1.2. Let $X_1, \ldots, X_n$ be a random sample from the mixture of exponentials $(1 - \pi)Exp(1) + \pi Exp(\theta)$, where $Exp(\theta)$ denotes the exponential distribution with mean θ. The score statistic for π at $\pi = 0$ and given θ has the following form:

$$S(\theta) = \sum_{i=1}^{n} \left\{ \frac{\exp(-x_i/\theta)/\theta}{\exp(-x_i)} - 1 \right\}.$$

Under the homogeneous model where $\theta = 0$, we have

$$\mathrm{E}[S^2(\theta)] = \begin{cases} \frac{n(1-\theta)^2}{\theta(2-\theta)}, & \theta < 2; \\ \infty, & \theta \geq 2. \end{cases}$$

Therefore, the Fisher information is only finite when $0 < \theta < 2$. $\quad\square$

Based on the above two examples, we can see that the Fisher information for mixture models under the null hypothesis could be either degenerate or go to infinity resulting in the violation of the regularity assumption of the positive definite. Next, we provide one special case of mixture models for which the limiting distribution of the LRT exists.

Example 6.1.3. For a mixture of two known but general univariate densities with unknown proportions, $f(x; \pi) = \pi g_0(x) + (1 - \pi)g_1(x)$, where g_0 and g_1 are known densities and $0 < \pi \leq 1$ is unknown, consider the test of $H_0 : \pi = 1$ $(C = 1)$ versus $H_1 : \pi < 1$ $(C = 2)$. Titterington et al. (1985) showed that under H_0:

$$- 2 \log \Lambda = \{\max(0, W)\}^2 + o_p(1) \sim \frac{1}{2}\chi_0^2 + \frac{1}{2}\chi_1^2, \tag{6.1}$$

where $W \sim N(0, 1)$, and χ_0^2 denotes the degenerate distribution that puts mass 1 at zero. Lindsay (1995, Section 4.2) named the above-limiting distribution (6.1) a chi-bar squared distribution, i.e., a mixture of chi-squares distributions. $\quad\square$

Dacunha-Castelle and Gassiat (1999); Liu and Shao (2003); Chen and Chen (2001) derived the limiting distribution of the LRT for the general mixture model:

$$f(x) = (1 - \pi)g(x, \theta_1) + \pi g(x, \theta_2) \tag{6.2}$$

under the null hypothesis

$$H_0 : \pi(1 - \pi)(\theta_1 - \theta_2) = 0,$$

where $\theta_i \in \Theta$. The log-likelihood function is

$$\ell(\pi, \theta_1, \theta_2) = \sum_{i=1}^{n} \log\{(1 - \pi)g(x, \theta_1) + \pi g(x, \theta_2)\}.$$

The likelihood ratio test (LRT) statistic is defined to be

$$R_n = 2\{\ell(\hat{\pi}, \hat{\theta}_1, \hat{\theta}_2) - \ell(0.5, \hat{\theta}_0, \hat{\theta}_0)\},$$

where $\{\hat{\pi}, \hat{\theta}_1, \hat{\theta}_2\}$ is the MLE for the mixture model (6.2), and $\hat{\theta}_0$ is the MLE under the null hypothesis H_0.

If the parameter space Θ is not compact, R_n may go to infinity in probability as $n \to \infty$. However, when Θ is assumed to be compact, Dacunha-Castelle and Gassiat (1999); Liu and Shao (2003); Chen and Chen (2001) derived the following limiting distribution of the LRT.

Theorem 6.1.1. If the parameter space Θ is compact and some other regularity conditions hold, then the asymptotic distribution of the LRT under the null model is that of $\sup_\theta \{W^+(\theta)\}^2$, where $W(\theta)$ is a Gaussian process with mean 0, variance 1 and the complicated autocorrelation function:

$$\rho(\theta_1, \theta_2) = \rho\{Z_i(\theta) - h(\theta)Y_i, Z_i(\theta') - h(\theta')Y_i\},$$

where

$$Y_i(\theta) = \frac{g(x_i; \theta) - g(x_i; \theta_0)}{(\theta - \theta_0)g(x_i; \theta_0)}, \theta \neq \theta_0; \quad Y_i = Y_i(\theta_0) = \frac{g'(x_i; \theta_0)}{g(x_i; \theta_0)}$$

$$Z_i(\theta) = \frac{Y_i(\theta) - Y_i(\theta_0)}{\theta - \theta_0}, \theta \neq \theta_0; \quad Z_i = Z_i(\theta_0) = \frac{g''(x_i; \theta_0)}{2g(x_i; \theta_0)},$$

and $h(\theta) = \mathrm{E}\{Y_i Z_i(\theta)\}/\mathrm{E}(Y_i^2)$.

For the model (6.2), Ghosh and Sen (1985) proved that, under the restriction that Θ is compact and $|\theta_1 - \theta_2| > \epsilon > 0$, $-2\log\lambda$ has a limit in the form of Gaussian process. Dacunha-Castelle and Gassiat (1997) and Lemdani and Pons (1999) investigated how to remove the above separation condition. Chen and Cheng (2000) placed a condition $\min(\pi_1, \pi_2) > \epsilon > 0$ and the compact Θ. They proved that $-2\log\lambda$ has $0.5\chi_0^2 + 0.5\chi_1^2$ limiting distribution. In order to test for a single normal distribution versus a mixture of two normal distributions in the presence of a structural parameter, Chen and Chen (2003) investigated the large sample performance of the LRT statistic without separation conditions on the mean parameters. The asymptotic null distribution of such an LRT statistic is shown to be the maximum of a χ^2-variable and supremum of the square of a truncated Gaussian process with mean 0 and variance 1.

Chen et al. (2001) proposed a modified likelihood ratio test (MLRT) by adding a regularity-restoring penalty function

$$pl(\pi, \theta_1, \theta_2) = \ell(\pi, \theta_1, \theta_2) + p(\pi)$$

such that $p(\pi) \to -\infty$ as $\pi \to 0$ or 1 and $p(\pi)$ achieves its maximal value at $\pi = 0.5$. The MLRT is then defined to be

$$M_n = 2\{pl(\tilde{\pi}, \tilde{\theta}_1, \tilde{\theta}_2) - pl(1/2, \tilde{\theta}_0, \tilde{\theta}_0)\}.$$

The main reason for the nonregularity of mixture models is the possibility of $\pi = 0$ or 1. Because of the penalty function, the fitted value of π under the modified likelihood is bounded away from 0 and 1.

Theorem 6.1.2. Under some regularity conditions, the asymptotic null distribution of the MLRT statistic M_n is the mixture of χ_1^2 and χ_0^2 with the same weights, i.e., $0.5\chi_0^2 + 0.5\chi_1^2$.

Chen et al. (2001) suggested the use of penalty function $p(\pi) = \alpha \log\{4\pi(1 - \pi)\}$ where $\alpha > 0$. For some mixture models such as Binomial, Poisson, and Normal in mean, Chen et al. (2001) suggested using $\alpha = \log(M)$ if the parameters θ_1 and θ_2 are restricted to $[-M, M]$. A better penalty $p(\pi) = \log(1 - |1 - 2\pi|) \leq \log(1 - |1 - 2\pi|^2) = \log\{4\pi(1 - \pi)\}$ is later proposed by Li et al. (2009).

6.2   LRT based on bootstrap

As explained in Section 6.1, the LRT test statistics will converge to ∞ without any constraint on the parameter space. However, the null distribution of the LRT test statistic does exist for finite sample size n and can be approximated by the bootstrap method (McLachlan, 1987; Feng and McCulloch, 1996).

Consider the hypotheses

$$H_0 : C = C_0 \text{ vs } H_a : C = C_1$$

for a specific value C_0 and $C_1 > C_0$. A mixture of C_0 components has to be fitted first, and bootstrap samples are then generated from the fitted null model. Then, after fitting mixture models with C_0 and C_1 components, the test statistic $-2\log \Lambda$ is computed, and the process is repeated independently for B times. This forms an estimate of the null distribution of $-2\log \Lambda$. The j-th order statistic of the bootstrap replications can be used as an estimate of the quantile of order $(3j - 1)/(3B + 1)$.

Since the estimation of a mixture model depends on the starting values and stopping rules, the success of the testing process requires that the bootstrap data be fitted by the same procedures as the original sample. It has been calculated that the size of the test, which rejects H_0 if $-2\log \Lambda$ of the original data is greater than the j-th smallest value of the B bootstrap replications, is

$$\alpha = 1 - j/(B + 1). \tag{6.3}$$

As a result, if given a significance level α, the values of j and B can be approximated through (6.3). However, it is highly recommended that a very large value of B is needed to ensure an accurate estimate of the p-value.

6.3   Information criteria in model selection

The AIC (Akaike, 1974) and the BIC (Schwarz, 1978) are two of the most popularly used information criteria for model selection problems. The AIC criterion aims to minimize the Kullback-Leibler distance between the true distribution and the distributions for the candidate models. It was proposed by Bozdogan and Sclove (1984) and Sclove (1987) in the mixture context. It takes the form:

$$AIC = -2 \log \ell(\hat{\theta}) + 2k,$$

where k is the number of parameters to be estimated and $\ell(\hat{\theta})$ is the mixture log-likelihood.

Bozdogan (1993) provided an analytic extension of AIC, without violating Akaika's principle. The new selection criterion, called consistent AIC (CAIC), is defined as

$$CAIC = -2 \log \ell(\hat{\theta}) + k(\log n + 1).$$

On the other hand, the BIC criterion tries to maximize the posterior probability in the space of all candidate models and is defined by

$$BIC = -2 \log \ell(\hat{\theta}) + k \log n,$$

where n is the number of observations. Although limited in theoretical results, AIC and BIC have been popularly used for model selection in mixture models. Leroux (1992) showed, under mild conditions, that information criteria/penalized log-likelihood criteria, including AIC and BIC, do not underestimate the true number of components, asymptotically. Therefore, AIC and BIC would suffice to choose the number of components for mixture models for the purpose of density estimation. Roeder and Wasserman (1997) also proved that the nonparametric density estimation by the normal mixture model with the number of components selected by BIC is consistent. Keribin (2000) further showed that BIC provides a consistent estimate of the number of components in mixture models under the strong assumption of compact parameter space.

As an alternative to AIC and BIC, the Hannan–Quinn information criterion (HQIC) is another criterion for model selection. It is given as

$$HQIC = -2 \log \ell(\hat{\theta}) + 2k \log(\log n).$$

However, HQIC is not as commonly used as AIC or BIC (Burnham and Anderson, 2002), and is shown to not be asymptotically efficient (Claeskens and Hjort, 2008).

In case the number of clusters in the data is different from the number of components, Biernacki et al. (2000) proposed the integrated complete likelihood (ICL) criterion as a modification to BIC, defined by

$$ICL = -2 \log \ell_c(\hat{\theta}; \mathbf{x}_c) + 2k \log(\log n),$$

where $\ell_c(\hat{\theta}; \mathbf{x}_c)$ is the complete log-likelihood. In practice, the unknown component label indicator z_{ij} is replaced with the classification probability p_{ij}.

Many attempts have been made to improve AIC by reducing its finite sample bias. Following the derivation of the original AIC, Ishiguro et al. (1997) proposed EIC, which bootstraps the bias term instead of using an asymptotic approximation. EIC is not restricted to maximum likelihood estimation, and the bootstrapping in EIC is shown to be more accurate than the asymptotic approximation in AIC. Smyth (2000) also discussed a cross-validation-based information criterion.

6.4 Cross-validation for mixture models

Cross-validation is a well-known technique in supervised learning to select a model from a family of candidate models (Hjorth, 1994), and in unsupervised learning in the context of kernel density estimation for automatically choosing smoothing parameters (Silverman, 1986). Smyth (2000) investigated cross-validated likelihood as a tool for automatically determining the appropriate number of components in mixture modeling.

Let D be the dataset, and M be the number of partitions. For the i-th partition, let S_i be the testing set, and T_i be the training set, such that $S_i \cup T_i = D, i = 1, \ldots, M$. Then, define the cross-validated estimate of the log-likelihood for the j-th model as

$$\ell_j^{CV} = \frac{1}{M} \sum_{i=1}^{M} \ell(\hat{\theta}^{(j)}(T_i)|S_i),$$

where $\hat{\theta}^{(j)}(T_i)$ denotes the estimator of a j-component mixture model based on the sample T_i, and $\ell(\hat{\theta}^{(j)}(T_i)|S_i)$ is the corresponding log-likelihood evaluated on the testing set S_i using the parameters estimated from the training set T_i.

One of the most commonly used cross-validation methods is k-fold cross-validation, in which the data set is divided into k disjoint subsets (approximately equally sized). Each time, one of the sets is used as the test set and the remaining ones are used as the training set. When $k = n$, where n is the sample size, the k-fold cross-validation simplifies to the "leave-one-out" cross-validation, in which we take turns to leave one observation out as a test observation and use the remaining $n - 1$ observations as the training data. Another method to create training and test data sets is Monte-Carlo cross-validation, which randomly generates B independent partitions of the data set into a test sample of size $n\alpha$ and a training sample size $n(1 - \alpha)$. Smyth (2000) suggests a choice of $\alpha = 0.5$ and a value of B between 20 and 50.

6.5 Penalized mixture models

Another class of methods for the model selection is based on the distance measured between the fitted model and the nonparametric estimate of the population distribution. Let

$$F(x, H) = \sum_{c=1}^{C} \pi_c G(x, \boldsymbol{\theta}_c)$$

be a finite mixture distribution family, and

$$F_n(x) = \frac{1}{n} \sum_{i=1}^{n} I(x_i \leq x). \tag{6.4}$$

be the empirical distribution function, where $G(x, \boldsymbol{\theta}_c)$ is the cdf of cth component. Chen and Kalbeisch (1996) defined a penalized distance as

$$D(F_n, F(x, H)) = d(F_n, F(x, H)) - a_n \sum_{c=1}^{C} \log \pi_c, \tag{6.5}$$

where

$$H(\theta) = \sum_{c=1}^{C} \pi_c I(\theta_c \leq \theta),$$

is the mixing distribution, $d(F_1, F_2)$ is a distance measure in the space of probability distributions, and the last term in (6.5) is the penalty function such that too small a π_c, or in other words, too big a C, is penalized. Then, the penalized minimum distance estimator of H is defined as $\hat{H} = \arg\min D(F_n, F(x, H))$ among all finite mixing distributions H. Kolmogorov-Smirnov and Cramer-von distance are recommended distances, to satisfy the consistency condition. Please also refer to Section 6.7 for more detail about the choice of $d(F_1, F_2)$.

For the normal mixtures

$$f(x; \boldsymbol{\theta}) = \sum_{c=1}^{C} \pi_c \phi(x; \mu_c, \sigma_c^2),$$

James et al. (2001) applied the Kullback-Leibler distance to model selection and define

$$\hat{f}_C = \arg\min_f \mathrm{KL}\{\tilde{f}_h, \sum_{c=1}^{C} \pi_c \phi(x; \mu_c, \sigma_c^2 + h^2)\},$$

where

$$\mathrm{KL}(g, f) = \int g(x) \log\left(\frac{g(x)}{f(x)}\right) dx$$

is the Kullback-Leibler distance between two densities g and f, and $\tilde{f}_h$ is the kernel density estimator of $f(x; \boldsymbol{\theta})$ as defined in (1.13). Through a Taylor expansion, it is shown that the maximization procedure by James et al. (2001) could be viewed as a variant of a penalized MLE, with a random penalty function. James et al. (2001) showed that their estimator of mixture complexity is consistent.

In many applications, however, it is unrealistic to expect the component densities to be normal. In addition, if one misspecifies the mixture model then, the mixture complexity estimator based on KL distance may be unstable. As a result, Woo and Sriram (2006) developed a robust estimator of mixture complexity based on Hellinger distance. When the component densities are unknown but members of some parametric family of mixtures, the estimator is consistent, and when the model is misspecified, the method is robust. The estimator is defined as

$$\hat{C}_n = \min\{C : H^2(\tilde{f}_h, \hat{f}^C) \leq H^2(\tilde{f}_h, \hat{f}^{C+1}) + \alpha_{n,C}\},$$

where

$$H^2(f_1, f_2) = \|f_1^{1/2} - f_2^{1/2}\|_2^2$$

is the Hellinger distance between two densities f_1 and f_2, and

$$\hat{f}^C = \arg\min_f H(\tilde{f}_h, f)$$

is the Hellinger distance estimator. $\alpha_{n,C}$ is the penalty term defined as

$$\alpha_{n,C} = n^{-1}b(n)[\nu(C+1) - \nu(C)],$$

where $b(n)$ depends only on n and $\nu(C)$ is the number of parameters in the C-component mixture model. $\{\alpha_{n,C}; C \geq 1\}$ are positive sequences of threshold values that converge to 0 as $n \to \infty$.

The aforementioned methods are all based on the complete model search algorithms and are computationally expensive. As a result, Chen and Khalili (2008) proposed a penalized likelihood method with the SCAD penalty for mixtures of univariate location distributions, to improve the computational efficiency. For the one-dimensional location mixture model

$$f(x; H, \sigma) = \int_\Theta g(x; \theta, \sigma)dH(\theta),$$

assume that $\theta_1 \leq \theta_2 \leq \ldots \leq \theta_C$, and define $\eta_c = \theta_{c+1} - \theta_c$. Then, define the penalized log-likelihood function as

$$\tilde{\ell}_n(H, \sigma) = \ell_n(H, \sigma) + a_C \sum_{c=1}^{C} \log \pi_c - \sum_{c=1}^{C-1} p_n(\eta_c),$$

for some $a_C > 0$, where $\ell_n(H, \sigma)$ is the log-likelihood, $p_n(\cdot)$ is some nonnegative penalty function. The first penalty term forces the estimated value of

π_c to be away from 0 to prevent overfitting. The second penalty $p_n(\cdot)$ is set to be the SCAD penalty, which shrinks an η that is close to 0 to exactly 0 with a positive probability. When $\hat{\eta}_c = 0$ for some c, then the number of components is set to be C minus the number of 0 ηs. However, the penalized likelihood method of Chen and Khalili (2008) can only be applied to a one-dimensional location mixture model. In addition, some of the components would be incorrectly eliminated if some components in the true model have the same location.

For multivariate Gaussian mixture models, Huang et al. (2017) proposed a penalized likelihood method for model selection by shrinking some component proportions to 0. The penalized log-likelihood function is

$$\ell_p(\boldsymbol{\theta}) = \ell(\boldsymbol{\theta}) - n\lambda D_f \sum_{c=1}^{C} \{\log(\epsilon + p_\lambda(\pi_c)) - \log(\epsilon)\},$$

where λ is a tuning parameter, D_f is the number of free parameters for each component, ϵ is a very small positive number, such as 10^{-6} or $o(n^{-1/2}\log^{-1} n)$, and $p_\lambda(\pi)$ is the SCAD penalty function by Fan et al. (2001). The method is shown to be statistically consistent in determining the number of components.

6.6 EM-test for finite mixture models

6.6.1 EM-test in single parameter component density

Li et al. (2009) developed an EM test for testing the homogeneity of mixture models with a single mixing parameter. Suppose $X_1, \ldots, X_n$ is a random sample from $(1 - \pi)g(x_i; \theta_1) + \pi g(x_i; \theta_2)$. We are interested in testing $H_0 : \pi(1-\pi)(\theta_1 - \theta_2) = 0$. The modified log-likelihood function is defined as

$$pl(\pi, \theta_1, \theta_2) = \sum_{i=1}^{n} \log\{(1-\pi)g(x_i; \theta_1) + \pi g(x_i; \theta_2)\} + p(\pi)$$
$$= \ell(\pi, \theta_1, \theta_2) + p(\pi),$$

where $\ell(\pi, \theta_1, \theta_2)$ is the usual log-likelihood function and $p(\pi)$ is a penalty function on π such that $p(\pi)$ achieves the maximal value at $\alpha = 0.5$.

For each fixed $\pi = \pi_0 \in (0, 0.5]$, we compute a penalized likelihood ratio test statistic of the form:

$$M(\pi_0) = 2\{pl(\pi_0, \tilde{\theta}_{01}, \tilde{\theta}_{02}) - pl(0.5, \tilde{\theta}_0, \tilde{\theta}_0)\},$$

where $\tilde{\theta}_{01}$ and $\tilde{\theta}_{02}$ are the maximizers of $pl(\pi, \theta_1, \theta_2)$ and $\tilde{\theta}_0$ is the maximizer of $pl(0.5, \theta, \theta)$. For fixed π_0, the mixture model is fully identifiable. Consequently, we can show that $M_n(\pi_0)$ has a simple χ^2-type null limiting distribution. Note that if the data is from an alternative model with π far from π_0, this test is likely to be inefficient.

To improve the power, we adopt an EM-like algorithm to iteratively update $M_n(\alpha)$, for a fixed and finite number of times, between π and (θ_1, θ_2). In addition, we choose a number of initial values of π_0 to increase the power if the data is from the alternative model. We then use the maximum value of $M_n(\pi_0)$'s as our test statistic. The above procedure is summarized in Algorithm 6.6.1.

Theorem 6.6.1. (Li et al., 2009) Assume one of α_j's is equal to 0.5. Under the null distribution $f(x; \theta_0)$ and some regularity conditions, for any fixed finite k, as $n \to \infty$,

$$EM^{(k)} \xrightarrow{L} 0.5\chi_0^2 + 0.5\chi_1^2.$$

6.6.2 EM-test in normal mixture models with equal variance

Chen and Li (2009) extended the method of Li et al. (2009) to normal mixture models with equal variance. Suppose $X_1, \ldots, X_n$ is a random sample from $(1 - \pi)N(\mu_1, \sigma^2) + \pi N(\mu_2, \sigma^2)$. We are interested in testing $H_0 : \pi(1 - \pi)(\mu_1 - \mu_2) = 0$. Chen and Chen (2003) first noted that the normal density function is not strongly identifiable, which is a consequence of

$$\frac{\partial^2 \phi(x; \mu, \sigma^2)}{\partial \mu^2}\Big|_{(\mu, \sigma^2)=(0,1)} = 2\frac{\partial \phi(x; \mu, \sigma^2)}{\partial(\sigma^2)}\Big|_{(\mu, \sigma^2)=(0,1)}.$$

The modified log-likelihood function for the EM-test is as follows:

$$pl(\pi, \mu_1, \mu_2, \sigma) = \sum_{i=1}^{n} \log\{(1 - \pi)\phi(x_i; \mu_1, \sigma^2) + \pi\phi(x_i; \mu_2, \sigma^2)\} + p_1(\sigma) + p_2(\pi),$$

where $p_1(\sigma)$ is a penalty function on σ^2 and $p_2(\pi)$ is a penalty function on π. The penalty function $p_1(\sigma)$ will be selected to prevent the underestimation of σ under the null model. Its implementation is described in Algorithm 6.6.2.

Theorem 6.6.2. (Chen and Li, 2009) Assume one of α_j's is equal to 0.5. Under the null distribution $N(\mu_0, \sigma_0^2)$ and some regularity conditions, for any fixed finite k, as $n \to \infty$,

$$P(EM^{(k)} \leq x) \xrightarrow{p} F(x - \Delta)\{0.5 + 0.5F(x)\},$$

where $F(x)$ is the cumulative density function (cdf) for χ_1^2 and $\Delta = 2\max_{\pi_j \neq 0.5}\{p_2(\pi_j) - p_2(0.5)\}$.

Algorithm 6.6.1 Algorithm to implement the EM-test (Li et al., 2009)

Step 0: Choose a number of initial π values, say $\pi_1, \pi_2, \ldots, \pi_J \in (0, 0.5]$. Compute

$$\tilde{\theta}_0 = \arg\max_{\theta} pl(0.5, \theta, \theta).$$

Let $j = 1, k = 0$.

Step 1: Let $\pi_j^{(k)} = \pi_j$ and compute

$$(\theta_{j1}^{(k)}, \theta_{j2}^{(k)}) = \arg\max_{\theta_1, \theta_2} pl(\pi_j^{(k)}, \theta_1, \theta_2)$$

and

$$M^{(k)}(\pi_j) = 2\{pl(\pi_j^{(k)}, \theta_{j1}^{(k)}, \theta_{j2}^{(k)}) - pl(0.5, \tilde{\theta}_0, \tilde{\theta}_0)\}.$$

Step 2: Use the EM algorithm to update $M^{(k)}(\pi_j)$. For $i = 1, 2, \ldots, n$, compute the weights

$$p_{ij}^{(k)} = \frac{\pi_j^{(k)} g(x_i; \theta_{j2}^{(k)})}{(1 - \pi_j^{(k)}) g(x_i; \theta_{j1}^{(k)}) + \pi_j^{(k)} g(x_i; \theta_{j2}^{(k)})}.$$

Update the parameters:

$$\pi_j^{(k+1)} = \arg\max_{\pi}\left\{(n - \sum_{i=1}^{n} p_{ij}^{(k)})\log(1 - \pi) + \sum_{i=1}^{n} p_{ij}^{(k)}\log(\pi) + p_2(\pi)\right\},$$

$$\theta_{j1}^{(k+1)} = \arg\max_{\theta_1} \sum_{i=1}^{n}(1 - p_{ij}^{(k)})\log g(x_i; \theta_1),$$

$$\theta_{j2}^{(k+1)} = \arg\max_{\theta_2} \sum_{i=1}^{n} p_{ij}^{(k)}\log g(x_i; \theta_2),$$

Compute

$$M^{(k+1)}(\pi_j) = 2\{pl(\pi_j^{(k+1)}, \theta_{j1}^{(k+1)}, \theta_{j2}^{(k+1)}) - pl(0.5, \tilde{\theta}_0, \tilde{\theta}_0)\}.$$

Let $k = k + 1$ and repeat Step 2 for a fixed number of iterations in k.

Step 3: Let $j = j + 1, k = 0$ and go to Step 1, until $j = J$.

Step 4: For each k, calculate the test statistic as

$$EM^{(k)} = \max\{M^{(k)}(\pi_j), j = 1, 2, \ldots, J\}.$$

Algorithm 6.6.2 EM-test for a normal mixture with equal variance

---

Step 0: Choose a number of initial π values, say $\pi_1, \pi_2, \ldots, \pi_J \in (0, 0.5]$. Compute

$$(\tilde{\mu}_0, \tilde{\sigma}_0) = \arg\max_{\mu, \sigma} pl(0.5, \mu, \mu, \sigma).$$

Let $j = 1, k = 0,$.

Step 1: Let $\pi_j^{(k)} = \pi_j$ and compute

$$(\mu_{j1}^{(k)}, \mu_{j2}^{(k)}, \sigma^{(k)}) = \arg\max_{\mu_1, \mu_2, \sigma} pl(\pi_j^{(k)}, \mu_1, \mu_2, \sigma)$$

and

$$M^{(k)}(\pi_j) = 2\{pl(\pi_j^{(k)}, \mu_{j1}^{(k)}, \mu_{j2}^{(k)}, \sigma^{(k)}) - pl(0.5, \tilde{\mu}_0, \tilde{\mu}_0, \tilde{\sigma}_0)\}.$$

Step 2: Use the EM algorithm to update $M^{(k)}(\pi_j)$. For $i = 1, 2, \ldots, n$, compute the weights:

$$p_{ij}^{(k)} = \frac{\pi_j^{(k)} \phi(x_i; \mu_{j2}^{(k)}, \sigma^{2(k)})}{(1 - \pi_j^{(k)})\phi(x_i; \mu_{j1}^{(k)}, \sigma^{2(k)}) + \pi_j^{(k)}\phi(x_i; \mu_{j2}^{(k)}, \sigma^{2(k)})}.$$

Update the parameters:

$$\pi_j^{(k+1)} = \arg\max_{\pi}\{(n - \sum_{i=1}^{n} p_{ij}^{(k)})\log(1-\pi) + \sum_{i=1}^{n} p_{ij}^{(k)}\log(\pi) + p_2(\pi)\},$$

$$\mu_{j1}^{(k+1)} = \sum_{i=1}^{n}(1 - p_{ij}^{(k)})x_i / \sum_{i=1}^{n}(1 - p_{ij}^{(k)}),$$

$$\mu_{j2}^{(k+1)} = \sum_{i=1}^{n} p_{ij}^{(k)} x_i / \sum_{i=1}^{n} p_{ij}^{(k)},$$

$$\sigma^{(k+1)} = \arg\max_{\sigma}\left\{ -\frac{1}{2\sigma^2}\sum_{i=1}^{n}(1 - p_{ij}^{(k)})(x_i - \mu_{j1}^{(k+1)})^2 \right.$$

$$\left. -\frac{1}{2\sigma^2}\sum_{i=1}^{n} p_{ij}^{(k)}(x_i - \mu_{j2}^{(k+1)})^2 - \frac{n}{2}\log(\sigma^2) + p_1(\sigma) \right\}.$$

Compute

$$M^{(k+1)}(\pi_j) = 2\{pl(\pi_j^{(k+1)}, \mu_{j1}^{(k+1)}, \mu_{j2}^{(k+1)}, \sigma^{(k+1)}) - pl(0.5, \tilde{\mu}_0, \tilde{\mu}_0, \tilde{\sigma}_0)\}.$$

Let $k = k + 1$ and repeat Step 2 for a fixed number of iterations in k.

Step 3: Let $j = j + 1, k = 0$ and go to Step 1, until $j = J$.

Step 4: For each k, calculate the test statistic as

$$EM^{(k)} = \max\{M^{(k)}(\pi_j), j = 1, 2, \ldots, J\}.$$

6.6.3   EM-test in normal mixture models with unequal variance

Chen and Li (2009) further studied the EM-test for normal mixture models with unequal variance. Suppose $X_1, \ldots, X_n$ is a random sample from $(1 - \pi)N(\mu_1, \sigma_1^2) + \pi N(\mu_2, \sigma_2^2)$. We are interested in testing $H_0 : \alpha = 0$ or $\alpha = 1$ or $(\mu_1, \sigma_1^2) = (\mu_2, \sigma_2^2)$.

Note that the normal kernel does not satisfy the finite Fisher information condition, i.e., under the null model $N(0, 1)$,

$$E\left\{ \frac{\phi(x; \mu, \sigma^2)}{\phi(x; 0, 1)} - 1 \right\}^2 = \infty,$$

whenever $\sigma > 2$.

We first introduce a modified log-likelihood function as follows:

$$pl(\boldsymbol{\theta}) = \sum_{i=1}^{n} \log\{(1 - \pi)\phi(x_i; \mu_1, \sigma_1^2) + \pi\phi(x_i; \mu_2, \sigma_2^2)\} + \sum_{c=1}^{2} p_1(\sigma_c) + p_2(\pi),$$

where $\boldsymbol{\theta} = (\pi, \mu_1, \mu_2, \sigma_1, \sigma_2), p_1(\sigma)$ is a penalty function on σ^2 used to prevent the fitted value of σ to be close to 0, and $p_2(\pi)$ is a penalty function on π. The algorithm to implement the EM-test for a normal mixture with unequal variance is explained in Algorithm 6.6.3.

Theorem 6.6.3. (Chen and Li, 2009) Assume one of π_cs is equal to 0.5. Under the null distribution $N(\mu_0, \sigma_0^2)$ and some regularity conditions, for any fixed finite k, as $n \to \infty$,

$$P(EM^{(k)} \leq x) \overset{L}{\to} \chi_2^2.$$

Based on the above theorem, the EM-test has a simple limiting distribution when applied to a normal mixture model with unequal variance.

6.7   Hypothesis testing based on goodness-of-fit tests

Goodness-of-fit tests have been popularly used to test whether a homogeneous population has a given parametric distribution. In other words, the GOF tests can be used to test whether the sampled data represents the data we would expect to find in the null model. However, there is little research on how existing GOF statistics perform for finite mixture models as well as their comparisons with other proposed hypothesis testing procedures that are specifically designed for mixture models.

Wichitchan et al. (2019a) proposed a simple class of testing procedures for finite mixture models based on goodness of fit test (GOF) statistics. Their

Algorithm 6.6.3 EM-test for a normal mixture with unequal variance

Step 0: Choose a number of initial π values, say $\pi_1, \pi_2, \ldots, \pi_C \in (0, 0.5]$. Compute

$$(\tilde{\mu}_0, \tilde{\sigma}_0) = \arg\max_{\mu, \sigma} pl(0.5, \mu, \mu, \sigma, \sigma).$$

Let $c = 1, k = 0$.

Step 1: Let $\pi_c^{(k)} = \pi_c$ and compute

$$(\mu_{c1}^{(k)}, \mu_{c2}^{(k)}, \sigma_{c1}^{(k)}, \sigma_{c2}^{(k)}) = \arg\max_{\mu_1, \mu_2, \sigma_1, \sigma_2} pl(\pi_c^{(k)}, \mu_1, \mu_2, \sigma_1, \sigma_2)$$

and

$$M^{(k)}(\pi_c) = 2\{pl(\pi_c^{(k)}, \mu_{c1}^{(k)}, \mu_{c2}^{(k)}, \sigma_{c1}^{(k)}, \sigma_{c2}^{(k)}) - pl(0.5, \tilde{\mu}_0, \tilde{\mu}_0, \tilde{\sigma}_0, \tilde{\sigma}_0)\}.$$

Step 2: Use the EM algorithm to update $M^{(k)}(\pi_c)$. For $i = 1, 2, \ldots, n$, compute the weights:

$$p_{ic}^{(k)} = \frac{\pi_c^{(k)} \phi(x_i; \mu_{c2}^{(k)}, \sigma_{c2}^{2(k)})}{(1 - \pi_c^{(k)})\phi(x_i; \mu_{c1}^{(k)}, \sigma_{c1}^{2(k)}) + \pi_c^{(k)}\phi(x_i; \mu_{c2}^{(k)}, \sigma_{c2}^{2(k)})}.$$

Update the parameters:

$$\pi_c^{(k+1)} = \arg\max_{\pi}\{(n - \sum_{i=1}^n p_{ic}^{(k)})\log(1-\pi) + \sum_{i=1}^n p_{ic}^{(k)}\log(\pi) + p_2(\pi)\},$$

$$\mu_{c1}^{(k+1)} = \sum_{i=1}^n (1 - p_{ic}^{(k)})x_i / \sum_{i=1}^n (1 - p_{ic}^{(k)}),$$

$$\mu_{c2}^{(k+1)} = \sum_{i=1}^n p_{ic}^{(k)}x_i / \sum_{i=1}^n p_{ic}^{(k)},$$

$$\sigma_{c1}^{(k+1)} = \arg\max_{\sigma_1}\left\{-\frac{1}{2\sigma_1^2}\sum_{i=1}^n (1 - p_{ic}^{(k)})(x_i - \mu_{c1}^{(k+1)})^2 \right.$$
$$\left. -\frac{1}{2}\sum_{i=1}^n (1 - p_{ic}^{(k)})\log(\sigma_1^2) + p_1(\sigma_1)\right\},$$

$$\sigma_{c2}^{(k+1)} = \arg\max_{\sigma_2}\left\{-\frac{1}{2\sigma_2^2}\sum_{i=1}^n p_{ic}^{(k)}(x_i - \mu_{c2}^{(k+1)})^2 - \frac{1}{2}\sum_{i=1}^n p_{ic}^{(k)}\log(\sigma_2^2) + p_1(\sigma_2)\right\}.$$

Compute

$$M^{(k+1)}(\pi_c) = 2\{pl(\pi_c^{(k+1)}, \mu_{c1}^{(k+1)}, \mu_{c2}^{(k+1)}, \sigma_{c1}^{(k+1)}, \sigma_{c2}^{(k+1)}) - pl(0.5, \tilde{\mu}_0, \tilde{\mu}_0, \tilde{\sigma}_0, \tilde{\sigma}_0)\}.$$

Let $k = k + 1$ and repeat Step 2 for a fixed number of iterations in k.

Step 3: Let $c = c + 1, k = 0$ and go to Step 1, until $c = C$.

Step 4: For each k, calculate the test statistic as

$$EM^{(k)} = \max\{M^{(k)}(\pi_c), c = 1, 2, \ldots, C\}.$$

proposed test procedure is very simple to compute and understand, and can be applied to many other mixture models. In addition, unlike the LRT, the GOF method does not need to estimate the alternative model and thus can detect any violation of the null hypothesis for the true model. The limit distribution of test statistics is simulated based on a bootstrap procedure.

We are interested in testing H_0: *Data set follows a C-component mixture distribution*:

$$f(x; \boldsymbol{\theta}) = \sum_{c=1}^{C} \pi_c g(x; \lambda_c), \qquad (6.6)$$

where $g(\cdot)$ is a parametric component density function against H_a: *Data set does not follow a C-component mixture distribution.*

Wichitchan et al. (2019a) demonstrated the performance of the GOF method for different types of H_a, such as H_a: *the number of components is $C+1$*, and H_a: *the number of components is larger than C*, which are commonly used in hypothesis testing literature of mixture models. They investigated and compared the performance of testing the above hypothesis using the following five empirical distribution function (EDF) based goodness of fit test statistics: the Kolmogorov-Smirnov (KS) statistic, the Cramér-Von Mises Statistic, the Kuiper statistic, the Watson statistic, and the Anderson-Darling statistic.

Let $F(x; \hat{\boldsymbol{\theta}})$ be the cdf of the estimated mixture model of (6.6):

$$F(x; \hat{\boldsymbol{\theta}}) = \sum_{c=1}^{C} \hat{\pi}_c G(x; \hat{\lambda}_c), \qquad (6.7)$$

where $G(x; \lambda_c)$ is the cdf of the cth component and $\hat{\boldsymbol{\theta}} = (\hat{\pi}_1, \ldots, \hat{\pi}_C, \hat{\lambda}_1, \ldots, \hat{\lambda}_C)$ is the MLE of $\boldsymbol{\theta}$.

The *Kolmogorov-Smirnov (KS)* statistic is commonly used to evaluate the goodness of fit of the empirical distribution function $F_n(x)$ to a hypothesized distribution function $F(x; \hat{\boldsymbol{\theta}})$. The KS statistic is defined as

$$D(\hat{\boldsymbol{\theta}}) = \sup_x |F_n(x) - F(x; \hat{\boldsymbol{\theta}})|,$$

or

$$D(\hat{\boldsymbol{\theta}}) = \max(D^+(\hat{\boldsymbol{\theta}}), D^-(\hat{\boldsymbol{\theta}})),$$

where

$$D^+(\boldsymbol{\theta}) = \max_{1 \leq i \leq n} [i/n - F(x_{(i)}; \boldsymbol{\theta})],$$
$$D^-(\boldsymbol{\theta}) = \max_{1 \leq i \leq n} [F(x_{(i)}; \boldsymbol{\theta}) - (i-1)/n] \qquad (6.8)$$

and $x_{(1)} \leq x_{(2)} \leq \ldots \leq x_{(n)}$ are the order statistics of $x_1, \ldots, x_n$ (Chakravarty et al., 1967). The KS statistic is a straightforward measure to compute and has greater efficacy in identifying changes in the mean of a distribution. Nevertheless, its sensitivity is relatively higher in the central part of the distribution compared to the tails.

The *Cramér-Von Mises (CVM)* test statistic (Cramér, 1928; Mises, 2013) is defined as

$$W^2(\hat{\boldsymbol{\theta}}) = n \int\limits_{-\infty}^{\infty} \psi(x)[F_n(x) - F(x_i; \hat{\boldsymbol{\theta}})]^2 dF(x),$$

where $\psi(x)$ is a weight function such as $\psi(x) = 1$. The numerical value of $W^2(\hat{\boldsymbol{\theta}})$ is found using the following representation:

$$W^2(\hat{\boldsymbol{\theta}}) = \sum_{i=1}^{n} \left[F(x_{(i)}; \hat{\boldsymbol{\theta}}) - \frac{2i-1}{2n} \right]^2 + \frac{1}{12n},$$

where $x_{(1)} \leq x_{(2)} \leq \cdots \leq x_{(n)}$. The Cramer-von Mises (CVM) test is generally considered to be more powerful than the KS test, particularly in detecting changes in the mean of the distribution.

The *Kuiper* statistic, introduced by Kuiper (1960), is a modified version of the KS test that can be used for analyzing random points on a circle or a line. Compared to the KS test, the Kuiper statistic is better suited for detecting changes in variance, but it generally has lower statistical power. The Kuiper statistic, denoted by V, can be calculated using the following expression:

$$V(\hat{\boldsymbol{\theta}}) = D^+(\hat{\boldsymbol{\theta}}) + D^-(\hat{\boldsymbol{\theta}}),$$

where $D^+(\cdot)$ and $D^-(\cdot)$ are defined in (6.8).

The *Watson* statistic, introduced by Watson (1961), is a goodness-of-fit test statistic that is related to the CVM test. While the CVM test has the form of a second moment about the origin, the Watson statistic has the form of a variance. The Watson statistic is defined as follows:

$$U^2(\hat{\boldsymbol{\theta}}) = n \int\limits_{-\infty}^{\infty} \left\{ F_n(x) - F(y_i; \hat{\boldsymbol{\theta}}) - \int\limits_{-\infty}^{\infty} [F_n(y) - F(y_i; \hat{\boldsymbol{\theta}})]dF(y) \right\}^2 dF(x),$$

and the computational form is

$$U^2(\hat{\boldsymbol{\theta}}) = W^2(\hat{\boldsymbol{\theta}}) - n(\bar{z}(\hat{\boldsymbol{\theta}}) - 1/2)^2,$$

where W^2 is the Cramér Von Mises statistic, and $\bar{z}(\boldsymbol{\theta}) = n^{-1} \sum_{i=1}^{n} F(x_i; \boldsymbol{\theta})$. Similar to the Kuiper statistic, the Watson statistic can also be applied to random points on a circle and is better suited for detecting changes in variance.

The *Anderson-Darling (AD)* statistic, first introduced by Anderson and Darling (1952) and later refined by Anderson and Darling (1954), is another alternative statistic based on the square integral. The AD statistic is defined as follows:

$$A^2 = \int\limits_{-\infty}^{\infty} \frac{1}{F(x)\{1 - F(x)\}} \{F_n(x) - F(x_i)\}^2 dF(x).$$

Unlike the CVM test statistic, the AD test uses $\{F(x)(1 - F(x))\}^{-1}$ as a weight function and its computational form is

$$A^2(\hat{\boldsymbol{\theta}}) = -n - \frac{1}{n}\sum_{i=1}^{n}(2i - 1)\left\{\ln z_i + \ln(1 - z_{n+1-i})\right\},$$

where $z_i = F(x_{(i)}; \hat{\boldsymbol{\theta}})$. The AD statistic is one of the most powerful cdf-based GOF tests but is sensitive to the tails of distributions (Stephens, 1972, 1974, 1976). It is a modification of the KS test and gives more weight to the tails of the distribution compared to the KS test.

Unfortunately, asymptotic distributions have not been developed for the above five GOF test statistics for mixture models. Even for the homogeneous model, where $C = 1$, Babu and Rao (2004) showed that the asymptotic distribution of the GOF test statistics depends on the unknown parameters in a complex way. To address this issue, Wichitchan et al. (2019a) proposed a bootstrap method to simulate the distribution of the GOF test statistics. The steps for the bootstrap method are outlined in Algorithm 6.7.1.

Algorithm 6.7.1 Bootstrap procedure of using GOF test statistics

Given $\mathbf{x} = (x_1, \ldots, x_n)$, we obtain the MLE $\hat{\boldsymbol{\theta}} = (\hat{\pi}_1, \ldots, \hat{\pi}_C, \hat{\lambda}_1, \ldots, \hat{\lambda}_C)^{\top}$ under the null model. Then we perform the following steps:

Step 1: Generate iid sample $\tilde{x}_1, \ldots, \tilde{x}_n$ from the estimated C-component mixture model $\sum_{c=1}^{C}\hat{\pi}_c g(x; \hat{\lambda}_c)$. Let $\tilde{\mathbf{x}} = (\tilde{x}_1, \ldots, \tilde{x}_n)$.

Step 2: Obtain the MLE $\tilde{\boldsymbol{\theta}}$ based on the bootstrapped sample $\tilde{\mathbf{x}}$.

Step 3: Calculate the GOF statistic (KS, CVM, Kuiper, Watson or AD) based on the empirical cdf in (6.4) for $\tilde{\mathbf{x}}$ and the estimated cdf $F(x; \tilde{\boldsymbol{\theta}})$ in (6.7).

Step 4: Repeat Step 1–Step 3, B times.

Step 5: Calculate the empirical distribution, defined in (6.4), of the B bootstrapped GOF statistics and use it to estimate the sampling distribution of GOF statistics.

Step 6: Calculate the p-value based on the estimated sampling distribution in Step 5.

Based on extensive numerical studies of Wichitchan et al. (2019a), the tests based on CVM, Watson, and AD have the overall best performance and can provide surprisingly comparable or even superior hypothesis testing performance compared to some of the existing methods, including the EM test (Chen and Li, 2009; Chen et al., 2012). In addition, based on numerical studies and real data application in Wichitchan et al. (2019a), CVM, Watson, and AD are also effective in choosing the number of components based on sequential GOF tests.

6.8 Model selection for mixture models using R

The problem of determining the number of components C is still an important issue in mixture models. Some commonly used methods include the information criteria-based methods, and the likelihood ratio test for the hypothesis: $H_0 : C = C_0$ vs $H_a : C = C_0 + 1$ for some known constant C_0. The `mixtools` package offered some solutions to solve this problem.

`mixtools` can be used to apply the information criteria, such as Akaike's information criterion (AIC), Schwartz's Bayesian information criterion (BIC), Bozdogan's consistent AIC (CAIC), and Integrated Completed Likelihood (ICL) to assess the number of components in mixture models. To be more specific, there are three functions:

(i.) `multmixmodel.sel` for mixtures of multinomials model;

(ii.) `regmixmodel.sel` for mixtures of regressions model;

(iii.) `repnormmixmodel.sel` for mixture models with normal components and repeated measures.

For example, if we apply `multimixmodel.sel` to a synthetic data:

```
library(mixtools)
x = matrix(rpois(70, 6), 10, 7)
x.new = makemultdata(x, cuts = 5)
res = multmixmodel.sel(x.new$y, comps = 1:4, epsilon = 1e-3)
```

The output gives the AIC, BIC, CAIC, and ICL values and the "winner" model. The result shows that the criteria agree on a two-component mixture model, in this case.

For another example, the `crabs` data from the `MASS` library contains five morphological measurements of 200 crabs of the species Leptograpsus variegatus, collected at Fremantle, Australia. Among all the variables, we are particularly interested in RW (rear width in mm) and CL (carapace length in mm), with the objective of distinguishing between the two crabs sexes, without the other variables. The following code performs model selection on the `crabs` data:

```
library(MASS)
data(crabs)
regmixmodel.sel(crabs$RW, crabs$CL)
```

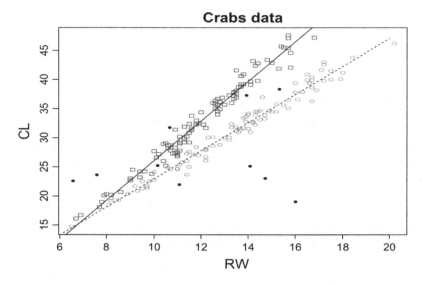

FIGURE 6.1
Crabs data: squares and circles denote the two groups of crabs based on their sexes.

Results show that the four criteria agree on a two-component mixture of linear regressions model. The estimation and clustering results are shown in Figure 6.1.

`mixtools` can also perform LRT based on bootstrap. The `boot.comp` function performs a parametric bootstrap by producing B bootstrap realizations of the likelihood ratio statistic for testing $H_0 : C = C_0$ vs $H_a : C = C_0 + 1$ for some known constant C_0. We apply the LRT on the `crabs` data again:

```
res = boot.comp(y = crabs$CL, x = crabs$RW, max.comp = 5,
                B = 200, mix.type = "regmix", epsilon = 1e-3)
```

```
res$p.values
```

Based on LRT, a three-component mixture of the regression model is recommended for the `crabs` data.

In addition, Wichitchan et al. (2019a) proposed a simple class of testing procedures for finite mixture models based on goodness of fit test (GOF) statistics. For example, consider the following four-component normal mixture model $0.25N(-2.5, 0.5^2) + 0.25N(0, 0.6^2) + 0.25N(2.5, 0.6^2) + 0.25N(5, 0.8^2)$ and four null hypotheses:

- H_0: data follows a normal distribution.

- H_0: data follows a normal mixture distribution with $C = 2$.

TABLE 6.1

GOF test of Wichitchan et al. (2019a)

	$C = 1$	$C = 2$	$C = 3$	$C = 4$
p-value	0.00	0.00	0.00	0.65

- H_0: data follows a normal mixture distribution with $C = 3$.

- H_0: data follows a normal mixture distribution with $C = 4$.

The following code performs GOF test on these hypotheses using the mixTest function of the R package `MixSemiRob`

```
n=200
numc=4
mu = c(-2.5, 0, 2.5, 5)
sd = c(0.8, 0.6, 0.6, 0.8)
w = c(0.25, 0.25, 0.25, 0.25)
cum_p=c(0.25,0.5,0.75,1);
gen_p=runif(n,0,1);
bernulli=matrix(rep(0,n*numc),nrow=n);
rand=matrix(rep(0,n*numc),nrow=n);
bernulli[,1]=(gen_p<cum_p[1]);
for(i in 2:numc){
   bernulli[,i]=(cum_p[i-1]<gen_p)*(gen_p<cum_p[i]);
}
for(i in 1:numc){
   rand[,i]=rnorm(n,mu[i],sd[i])
}
x=apply(rand*bernulli,1,sum);
out = mixTest(data, alpha = 0.10, C.max = 4, nboot = 500,
              nstart = 5)
```

The p-values for the aforementioned tests are listed in Table 6.1. Since the p-value is almost 0, the null hypotheses H_0: *data follows a normal distribution,* H_0: *data follows a normal mixture distribution with $C = 2$* and H_0: *data follows a normal mixture distribution with $C = 3$* are all rejected. In addition, based on a *p-value* of 0.65, we fail to reject the null hypothesis H_0: *data follows a normal mixture distribution with $C = 4$.* Therefore, we can conclude that the synthetic data follows a four-component normal mixture distribution.

7

Robust mixture regression models

As explained in Section 3, the unknown mixture regression parameters are usually estimated by maximum likelihood estimators (MLE) using the EM algorithm based on the normality assumption of component error density. Similar to the ordinary least-squares estimates for the linear regression, the normal mixture linear regression can be very sensitive to the presence of gross outliers or heavy-tailed error distributions, failing to accommodate for the outlying effects greatly jeopardizing both model estimation and inference. In fact, even a single atypical value may have a large effect on the parameter estimators. To overcome this problem, many robust methods for mixture regression models have been developed recently. In Section 7.1, we introduce some robust estimation methods for the traditional linear regression based on Yu and Yao (2017), and in Sections 7.3–7.9, we introduce some popular methods used to robustly estimate the mixture linear regression based on Yu et al. (2020).

7.1 Robust linear regression

Linear regression has been popularly used to study how a response variable y is related to covariates $\mathbf{x}$. Given the independent and identically distributed (iid) observations $\{(\mathbf{x}_i, y_i),\ i = 1, \ldots, n\}$, the linear regression model assumes

$$y_i = \mathbf{x}_i^\top \boldsymbol{\beta} + \varepsilon_i, \tag{7.1}$$

where $\boldsymbol{\beta}$ is an unknown $p \times 1$ vector, and the ε_is are i.i.d. with $\mathrm{E}(\varepsilon_i \mid \mathbf{x}_i) = 0$. The ordinary least-squares (OLS) estimate is commonly used to estimate $\boldsymbol{\beta}$ by minimizing the sum of squared residuals:

$$\sum_{i=1}^{n} (y_i - \mathbf{x}_i^\top \boldsymbol{\beta})^2. \tag{7.2}$$

However, it is well known that the OLS estimate is extremely sensitive to outliers.

Many robust methods have been proposed to robustly estimate $\boldsymbol{\beta}$ in (7.1). To compare different robust methods, the efficiency and breakdown point

DOI: 10.1201/9781003038511-7

(Donoho, 1982) are two commonly used criteria. The efficiency of a robust estimator is a measure of its relative performance compared to the OLS estimator under the assumption of a normal error distribution and no outliers. On the other hand, the breakdown point of an estimator indicates the proportion of outliers that it can handle before producing an infinite estimate. Both of these measures are important in assessing a robust estimator. Given a sample $\mathbf{z} = (\mathbf{z}_i, \ldots, \mathbf{z}_n)$, where $\mathbf{z}_i = (\mathbf{x}_i, y_i)$, let $T(\mathbf{z})$ be an estimator of the parameter β. Let $\mathbf{z}'$ be the corrupted sample where any m of the original points in $\mathbf{z}$ are replaced by arbitrary bad data. The finite sample breakdown point (BP) (Donoho, 1982) δ^* of $T(\mathbf{z})$ is defined as follows:

$$\delta^* (\mathbf{z}, T) = \min_{1 \leq m \leq n} \left\{ \frac{m}{n} : \sup_{\mathbf{z}'} \|T (\mathbf{z}') - T (\mathbf{z})\| = \infty \right\},$$

where $\|\cdot\|$ is the Euclidean norm.

7.1.1 M-estimators

M-estimators (Huber et al., 1964) of β replace the least squares criterion (7.2) with a robust criterion

$$\hat{\beta} = \arg \min_{\beta} \sum_{i=1}^{n} \rho \left(\frac{y_i - \mathbf{x}_i^{\top} \beta}{\hat{\sigma}} \right), \tag{7.3}$$

where $\rho(\cdot)$ is a robust loss function and $\hat{\sigma}$ is an error scale estimate. The derivative of ρ, denoted by $\psi(\cdot) = \rho'(\cdot)$, is called the influence function. In particular, the OLS estimator is the solution if $\rho(t) = \frac{1}{2}t^2$. The MLE is also a special case of (7.3) if $\rho(t)$ is the negative log density of the error distribution.

One of the most commonly used influence functions in robust statistics is Huber's ψ function (Huber, 1981), which is defined as $\psi_c(t) = \rho'(t) = \max\{-c, \min(c, t)\}$, where c is a tuning parameter. Huber (1981) suggests using $c = 1.345$ in practice, which achieves a relative efficiency of approximately 95% when the error distribution is normal. Another widely used influence function is the Tukey bisquare function, which is defined as $\psi_c(t) = t\{1 - (t/c)^2\}_+^2$, where $x_+ = \max(0, x)$. The value $c = 4.685$ is often recommended for this function, which also achieves a 95% efficiency. When $\rho(t)$ is simply the absolute value of t, the estimator based on minimizing the sum of absolute residuals is known as the *least absolute deviation* (LAD) estimator, also called median regression, defined by:

$$\hat{\beta} = \arg \min_{\beta} \sum_{i=1}^{n} |y_i - \mathbf{x}_i^{\top} \beta|.$$

The LAD is also called L_1 estimator due to the L_1 norm used. While the least absolute deviation (LAD) estimator is more robust than ordinary least squares (OLS) to atypical y values, it is still susceptible to high leverage outliers, leading to a low breakdown point (BP) of $1/n$ (Rousseeuw, 1984). Additionally, LAD estimators have low efficiency, around 0.64, when the errors follow a normal distribution. Monotone M-estimators, which are M-estimators with monotonic ψ functions, share a similar low BP of $1/n$ because of lacking of immunity to high leverage outliers (Maronna et al., 2019).

7.1.2   Generalized M-estimators (GM-estimators)

Mallows (1975) proposed Mallows GM-estimator to make M-estimator more resistant to high-leverage outliers by solving

$$\sum_{i=1}^{n} w_i \psi \left\{ \frac{r_i\left(\hat{\boldsymbol{\beta}}\right)}{\hat{\sigma}} \right\} \mathbf{x}_i = 0,$$

where $\psi(e) = \rho'(e), r_i(\boldsymbol{\beta}) = y_i - \mathbf{x}_i^\top \boldsymbol{\beta}$, and $w_i = \sqrt{1 - h_i}$ with h_i being the leverage of the ith observation. The weight assigned to each observation, w_i, is designed to reduce the influence of observations with high leverage, making them contribute less to the estimator than observations with small leverage. However, this approach may also down-weight "good" leverage points that align well with the majority of the data, leading to a reduction in efficiency.

Handschin et al. (1975) proposed Schweppe GM-estimator to improve the efficiency of Mallows GM-estimator, which is the solution of

$$\sum_{i=1}^{n} w_i \psi \left\{ \frac{r_i\left(\hat{\boldsymbol{\beta}}\right)}{w_i \hat{\sigma}} \right\} \mathbf{x}_i = 0.$$

It adjusts the leverage weights based on the size of the residual r_i. The breakdown points for both Mallows and Schweppe GM-estimators are no more than $1/(p+1)$ (Carroll and Welsh, 1988), where p is the number of unknown parameters.

By extending the original Schweppe estimator, Coakley and Hettmansperger (1993) proposed Schweppe one-step (S1S) estimator to speed up the computation. The S1S estimator is defined as

$$\hat{\boldsymbol{\beta}} = \hat{\boldsymbol{\beta}}_0 + \left[\sum_{i=1}^{n} \psi' \left(\frac{r_i\left(\hat{\boldsymbol{\beta}}_0\right)}{\hat{\sigma} w_i} \right) \mathbf{x}_i \mathbf{x}_i' \right]^{-1} \times \sum_{i=1}^{n} \hat{\sigma} w_i \psi \left(\frac{r_i\left(\hat{\boldsymbol{\beta}}_0\right)}{\hat{\sigma} w_i} \right) \mathbf{x}_i,$$

where the weight w_i is defined in the same way as Schweppe's GM-estimator. Unlike the Mallows and Schweppe GM-estimators, the S1S estimate finds the M-estimate in one step, given an initial estimator, rather than iteratively. In order to achieve a balance between robustness and efficiency, Coakley and Hettmansperger (1993) suggested using Rousseeuw's LTS method in Section 7.1.5 to obtain initial estimates of the residuals and LMS in Section 7.1.4 to obtain initial estimates of the scale. They demonstrated that the S1S estimator, which is a combination of these two methods, achieves a breakdown point of BP = 0.5 while maintaining 0.95 efficiency.

7.1.3 R-estimators

Jaeckel (1972) proposed R-estimators by minimizing the sum of some scores of the ranked residuals:

$$\min \sum_{i=1}^{n} a_n \left(R_i \right) r_i,$$

where R_i represents the rank of the ith residual r_i, and $a_n \left(\cdot \right)$ is a monotone score function that satisfies

$$\sum_{i=1}^{n} a_n \left(i \right) = 0.$$

While R-estimators are advantageous compared to M-estimators because they are scale equivalent, the optimal choice of score function for R-estimators remains unclear. Additionally, most R-estimators have a low breakdown point of BP = $1/n$. However, Naranjo and Hettmansperger (1994) proposed a bounded influence R-estimator that has a fairly high efficiency when the errors are normally distributed, but with a breakdown point of no more than 0.2.

7.1.4 LMS estimators

Siegel (1982) proposed the LMS estimator by minimizing the median of the squared residuals:

$$\hat{\beta} = \arg\min_{\beta} \text{Med}\{ \left(y_i - \mathbf{x}_i^{\top} \beta \right)^2 \}.$$

Although the LMS estimator has a high breakdown point of nearly 0.5, it has some limitations such as not having a well-defined influence function and a slow convergence rate of $n^{-\frac{1}{3}}$ with zero efficiency (Rousseeuw, 1984). However, the LMS estimator can be useful as an initial estimate for other high breakdown points and high-efficiency robust methods, which typically require an initial estimate with a high breakdown point, even if the initial estimate has low efficiency.

7.1.5 LTS estimators

The LTS estimator (Rousseeuw, 1984) is defined as

$$\hat{\beta} = \arg\min_{\beta} \sum_{i=1}^{h} r_{(i)} \left(\beta \right)^2,$$

where $|r_{(1)}(\beta)| \leq |r_{(2)}(\beta)| \leq \cdots \leq |r_{(n)}(\beta)|$ are ordered absolute values of the residuals $\{r_i = y_i - \mathbf{x}_i^\top \beta, i = 1, \ldots, n\}$, $h = [n(1-\alpha)+1]$, and α is the proportion of trimming. Therefore, instead of minimizing the residual sum of squares for all observations, LTS estimators find a subset of observations along with its associated OLS estimate such that the residual sum of squares of this subset of observations is the smallest.

Setting $h = \left(\frac{n}{2}\right) + 1$ ensures that the LTS estimator has a breakdown point of BP = 0.5 and a convergence rate of $n^{-\frac{1}{2}}$, which makes it highly resistant to outliers (Rousseeuw, 1984). However, its efficiency is quite low at around 0.08 (Stromberg et al., 2000). Despite this, the LTS estimator is often used as an initial estimate for other robust methods with higher efficiency and breakdown points.

7.1.6   S-estimators

Rousseeuw and Yohai (1984) proposed S-estimators by

$$\hat{\beta} = \arg\min_{\beta} \hat{\sigma}\left(r_1(\beta), \cdots, r_n(\beta)\right),$$

where $\hat{\sigma}\left(r_1(\beta), \cdots, r_n(\beta)\right)$ is a scale M-estimator defined as the solution of

$$\frac{1}{n}\sum_{i=1}^{n} \rho\left(\frac{r_i(\beta)}{\hat{\sigma}}\right) = \delta,$$

where $\delta = \mathrm{E}_\Phi\left[\rho(r)\right]$. When the biweight scale is used, S-estimators can attain a high breakdown point of BP = 0.5, and they also have an asymptotic efficiency of 0.29 under the assumption of normally distributed errors (Maronna et al., 2019).

7.1.7   Generalized S-estimators (GS-estimators)

In an attempt to improve the low efficiency of S-estimators, Croux et al. (1994) introduced generalized S-estimators:

$$\hat{\beta} = \arg\min_{\beta} S_n(\beta),$$

where $S_n(\beta)$ is defined as

$$S_n(\beta) = \sup\left\{S > 0; \binom{n}{2}^{-1}\sum_{i<j} \rho\left(\frac{r_i - r_j}{S}\right) \geq k_{n,p}\right\},$$

where $r_i = y_i - \mathbf{x}_i^\top \beta$, p is the number of regression parameters, and $k_{n,p}$ is a constant depending on n and p. When $\rho(x) = I(|x| \geq 1)$ and $k_{n,p} =$

$\left(\binom{n}{2} - \binom{h_p}{2} + 1\right)/\binom{n}{2}$ with $h_p = \frac{n+p+1}{2}$, the generalized S-estimator reduces to the least quartile difference (LQD) estimator:

$$\hat{\boldsymbol{\beta}} = \arg\min_{\boldsymbol{\beta}} Q_n(r_1, \ldots, r_n),$$

where $Q_n = \{|r_i - r_j|; i < j\}_{\binom{h_p}{2}}$ is the $\binom{h_p}{2}$th order statistic among the $\binom{n}{2}$ elements of the set $\{|r_i - r_j|; i < j\}$. Although the generalized S-estimators have the same high breakdown point as S-estimators, they have a higher efficiency.

7.1.8 MM-estimators

The MM-estimator, proposed by Yohai (1987), is one of the most popularly used robust regression methods due to its high breakdown point and high efficiency. The MM-estimator consists of a three-stage estimation procedure. First, we compute an initial estimate $\hat{\boldsymbol{\beta}}_0$, which is consistent and has a high breakdown point, but possibly a low efficiency. In the second stage, we compute a robust M-estimator of scale $\hat{\sigma}$ of the residuals, based on the initial estimator $\hat{\boldsymbol{\beta}}_0$. In the third stage, we obtain an M-estimator $\hat{\boldsymbol{\beta}}$ by starting at $\hat{\boldsymbol{\beta}}_0$ using the robust scale $\hat{\sigma}$.

In practice, LMS or S-estimator with Huber or bisquare functions is typically used to compute the initial estimator $\hat{\boldsymbol{\beta}}_0$ in the first stage. Let $\rho_0(r) = \rho(r/k_0)$ and $\rho_1(r) = \rho(r/k_1)$, where $k_1 \geq k_0$ and $\rho(\cdot)$ is a bounded function. Given the initial estimator $\hat{\boldsymbol{\beta}}_0$, the second stage computes the scale estimator $\hat{\sigma}$ by solving the equation

$$\frac{1}{n} \sum_{i=1}^{n} \rho_0 \left(\frac{r_i(\hat{\boldsymbol{\beta}}_0)}{\hat{\sigma}} \right) = 0.5,$$

where $r_i(\boldsymbol{\beta}) = y_i - \mathbf{x}_i^\top \boldsymbol{\beta}$. If the ρ-function is biweight, then $k_0 = 1.56$ ensures that the estimator has the asymptotic BP = 0.5. Given the scale estimator $\hat{\sigma}$ from the second stage, the final MM-estimator $\hat{\boldsymbol{\beta}}$ minimizes

$$L(\boldsymbol{\beta}) = \sum_{i=1}^{n} \rho_1 \left(\frac{r_i(\boldsymbol{\beta})}{\hat{\sigma}} \right),$$

starting from the initial estimator $\hat{\boldsymbol{\beta}}_0$.

According to Yohai (1987), if $\hat{\boldsymbol{\beta}}$ satisfies $L(\hat{\boldsymbol{\beta}}) \leq L(\hat{\boldsymbol{\beta}}_0)$, then the breakdown point of $\hat{\boldsymbol{\beta}}$ is not less than $\hat{\boldsymbol{\beta}}_0$. The MM-estimator's breakdown point depends only on k_0, and its asymptotic variance depends only on k_1. Therefore, we can select k_1 to achieve the desired normal efficiency without affecting the estimator's breakdown point. The MM-estimator can attain higher efficiency at the normal distribution with increasing k_1. The following table (Maronna et al., 2019) lists some values of k_1 with the corresponding efficiencies of the biweight ρ-function:

Efficiency	0.80	0.85	0.90	0.95
k_1	3.14	3.44	3.88	4.68

According to Yohai (1987), MM-estimators with larger values of k_1 are more susceptible to outliers compared to those with smaller values of k_1. Thus, one needs to be cautious when selecting a larger k_1 to achieve higher efficiency. In practice, it is recommended to start with a bisquare S-estimator and then use an MM-estimator with a bisquare function and an efficiency of 0.85, which corresponds to $k_1 = 3.44$. This combination strikes a balance between achieving good efficiency and maintaining robustness to outliers.

7.1.9 Robust and efficient weighted least squares estimator

A novel class of robust regression methods, named robust and efficient weighted least squares estimator (REWLSE), was introduced by Gervini et al. (2002). REWLSE achieves maximum breakdown point and full efficiency under normal errors simultaneously. This estimator is a variant of the weighted least squares estimator, where the weights are calculated in an adaptive manner using an initial robust estimator.

To detect outliers in a sample, Gervini et al. (2002) proposed using the standardized residuals defined as

$$r_i = \frac{y_i - \mathbf{x}_i^\top \hat{\boldsymbol{\beta}}_0}{\hat{\sigma}},$$

where $\hat{\boldsymbol{\beta}}_0$ and $\hat{\sigma}$ are initial robust estimators of the regression parameters and scale, respectively. A large absolute value of r_i indicates that the observation $(\mathbf{x}_i, y_i)$ may be an outlier.

To measure the proportion of outliers in the sample, Gervini et al. (2002) introduced the quantity d_n, which is defined as

$$d_n = \max_{i > i_0} \left\{ F^+(|r|_{(i)}) - \frac{(i-1)}{n} \right\}^+,$$

where $\{\cdot\}^+$ denotes the positive part, F^+ is the distribution function of $|X|$ when $X \sim F$, $|r|_{(1)} \leq \cdots \leq |r|_{(n)}$ are the order statistics of the standardized absolute residuals, and $i_0 = \max\left\{i : |r|_{(i)} < \eta\right\}$, where η is a large quantile of F^+. In practice, the cdf of a normal distribution is chosen for F, $\eta = 2.5$ (Rousseeuw and Leroy, 2005), and $\lfloor n d_n \rfloor$ observations with the largest standardized absolute residuals are considered outliers and are removed from the sample, where $\lfloor a \rfloor$ denotes the largest integer less than or equal to a.

Gervini et al. (2002) proposed the adaptive weights

$$w_i = \begin{cases} 1, & \text{if } |r_i| < t_n; \\ 0, & \text{if } |r_i| \geq t_n, \end{cases}$$

where the adaptive cut-off value is $t_n = |r|_{(i_n)}$ with $i_n = n - \lfloor nd_n \rfloor$. Let $\mathbf{W} = \mathrm{diag}(w_1, \cdots, w_n)$. Then, the REWLSE is

$$\hat{\beta} = (\mathbf{X}^\top \mathbf{W} \mathbf{X})^{-1} \mathbf{X}^\top \mathbf{W} \mathbf{y},$$

where $\mathbf{X} = (\mathbf{x}_1, \ldots, \mathbf{x}_n)^\top$ and $\mathbf{y} = (y_1, \cdots, y_n)^\top$.

When initial regression and scale estimators are chosen with a breakdown point of 0.5, the resulting REWLSE also has a breakdown point of 0.5. Moreover, the REWLSE method incorporates data adaptive cut-off values that make it asymptotically equivalent to OLS estimates under normal-error models, resulting in full asymptotic efficiency.

7.1.10 Robust regression based on regularization of case-specific parameters

The mean shift method proposed by She and Owen (2011) and Lee et al. (2012) is used to robustly estimate the regression parameters. In this model, the response vector $\mathbf{y}$ is related to the predictor matrix $\mathbf{X}$, the regression parameters β, and the mean shift parameter γ through the following equation

$$\mathbf{y} = \mathbf{X}\beta + \gamma + \varepsilon, \ \varepsilon \sim N(0, \sigma^2 I), \tag{7.4}$$

where $\varepsilon \sim N(0, \sigma^2 I)$ is the random error term. Here, $\mathbf{y} = (y_1, \cdots, y_n)^\top$, $\mathbf{X} = (\mathbf{x}_1, \ldots, \mathbf{x}_n)^\top$, and $\gamma = (\gamma_1, \ldots, \gamma_n)^\top$. The parameter γ_i is nonzero when the ith observation is an outlier and zero otherwise.

The mean shift parameter γ_i in model (7.4) is nonzero only for outliers, which implies that the majority of γ_i values are zeros, corresponding to typical observations. To exploit the sparsity of γ, She and Owen (2011) and Lee et al. (2012) proposed a penalized least squares approach to jointly estimate the regression coefficients β and the mean shift parameter γ. Specifically, they minimized the following objective function:

$$L(\beta, \gamma) = \frac{1}{2} \{\mathbf{y} - (\mathbf{X}\beta + \gamma)\}^\top \{\mathbf{y} - (\mathbf{X}\beta + \gamma)\} + \sum_{i=1}^{n} p_\lambda(|\gamma_i|), \tag{7.5}$$

where $p_\lambda(|\cdot|)$ is a penalty function that depends on a regularization parameter λ. This penalty function promotes sparsity in the estimated mean shift parameter, leading to more accurate identification of outliers.

Both She and Owen (2011) and Lee et al. (2012) studied in more detail the case when $p_\lambda(|\gamma|) = \lambda |\gamma|$, the L_1 penalty. Note that given the estimator $\hat{\gamma}$, β can be updated by the regular OLS estimator with $\mathbf{y}$ replaced by $\mathbf{y} - \hat{\gamma}$. Given the regression parameter estimator $\hat{\beta}$, there is an explicit minimizer of (7.5): $\hat{\gamma}_i = \mathrm{sgn}(r_i)(|r_i| - \lambda)_+$, where $r_i = y_i - \mathbf{x}_i^\top \hat{\beta}$ and $\mathrm{sgn}(\cdot)$ is a sign function, i.e.,

$$\mathrm{sgn}(x) = \begin{cases} 1, & \text{if } x > 0; \\ 0, & \text{if } x = 0; \\ -1, & \text{if } x < 0. \end{cases}$$

TABLE 7.1
Breakdown points and asymptotic efficiencies of various regression estimators

	Estimator	Breakdown Point	Asymptotic Efficiency
High BP	LMS	0.5	0
	LTS	0.5	0.08
	S-estimates	0.5	0.29
	GS-estimates	0.5	0.67
	MM-estimates	0.5	0.95
	GM-estimates(S1S)	0.5	0.95
	REWLSE	0.5	1.00
Low BP	GM-estimates(Mallows,Schweppe)	$1/(p+1)$	0.95
	Bounded R-estimates	< 0.2	0.90-0.95
	Monotone M-estimates	$1/n$	0.95
	LAD	$1/n$	0.64
	OLS	$1/n$	1.00

Therefore, the minimizer of (7.5) can be found by iteratively updating the two steps above. She and Owen (2011) proved that the mean shift estimator (7.5) with an L_1 penalty is equivalent to the M-estimator with the Huber's ψ function. Note, however, the monotone M-estimator is not resistant to the high-leverage outliers and has a breakdown point $1/n$.

To address this issue, She and Owen (2011) explored a generalized penalty function for $p_\lambda(|\gamma|)$ in the mean shift model. The update of β given γ remains unchanged. Once we have an estimator of the regression parameter $\hat{\beta}$, we can obtain $\hat{\gamma}$ by applying a thresholding function $\Theta(\gamma; \lambda)$ (She et al., 2009) that corresponds to the chosen penalty function. The L_1, Hard, and SCAD (Fan et al., 2001) penalty functions correspond to soft, hard, and SCAD thresholding solutions for γ, respectively. Minimizing the equation (7.5) yields a sparse $\hat{\gamma}$ for outlier detection and a robust estimate of β. She and Owen (2011) demonstrated that the mean shift estimator with hard or SCAD penalties is equivalent to the M-estimator with certain redescending ψ functions, and therefore, will be robust to high leverage outliers if high breakdown point robust estimators are used as initial values.

7.1.11 Summary

Table 7.1, (Yu and Yao, 2017) provides a summary of the breakdown points (BP) and asymptotic efficiencies for some of the estimators discussed in this section. The table shows that both MM-estimators and REWLSE have high breakdown points and high efficiency. In contrast, while LMS, LTS, S-estimators, and GS-estimators are highly resistant to outliers with high BP, their efficiencies are low. Nonetheless, these high-breakdown-point robust estimators are often used as initial estimates for other high-breakdown-point and high-efficiency robust estimators.

7.2 Robust estimator based on a modified EM algorithm

In Algorithm 3.1.1, we observe that the EM algorithm for the traditional mixture regression updates the regression parameter β, in the M step, using a weighted least squares estimate. However, the least squares criterion is well known to be sensitive to outliers and heavy-tailed error distributions, which can negatively affect the performance of the model. To solve this issue, Bai et al. (2012) proposed a modified EM algorithm to robustly estimate mixture regression parameters by replacing the least squares criterion in the M-step using a robust criterion, i.e., updating $\beta_c^{(k+1)}, c = 1, \ldots, C$ by the equation

$$\sum_{i=1}^{n} p_{ic}^{(k+1)} \mathbf{x}_i \psi \left(\frac{y_i - \mathbf{x}_i^\top \beta_c}{\sigma_c^{(k)}} \right) = 0, \tag{7.6}$$

where $\sigma^{(k)}$ is a robust scale estimator of the error terms, and $\psi(\cdot) = \rho'(t)$ is a robust influence function. One widely used choice is Huber's ψ-function, given by $\psi_a(t) = \max\{-a, \min(a, t)\}$, where $a = 1.345$ is recommended to yield a relative efficiency of approximately 95% when the error distribution is normal (Huber, 1981). Another popular option is Tukey's bisquare function $\psi_a(t) = t\{1 - (t/a)^2\}_+^2$, which downweights the tail contributions through a biweight function. A recommended value of $a = 4.685$ produces a relative efficiency of approximately 95% for normal errors. Using the ℓ_1 loss function $\rho(t) = |t|$ yields the median regression. Notably, when $\psi(\cdot) = t$, the resulting estimator is the traditional MLE based least squares estimator. For more information about choosing an appropriate $\psi(\cdot)$, see Huber (1981) and Andrews and Mallows (1974)

Let $w(t) = \psi(t)/t$. In (7.6), when β_c is in a neighborhood of $\beta_c^{(k)}$, we have the following approximation

$$\sum_{i=1}^{n} p_{ic}^{(k+1)} \mathbf{x}_i \psi \left(\frac{y_i - \mathbf{x}_i^\top \beta_c}{\sigma_c^{(k)}} \right) \approx \sum_{i=1}^{n} p_{ic}^{(k+1)} \mathbf{x}_i w \left(\frac{y_i - \mathbf{x}_i^\top \beta_c^{(k)}}{\sigma_c^{(k)}} \right) \left(\frac{y_i - \mathbf{x}_i^\top \beta_c}{\sigma_c^{(k)}} \right)$$

$$= \sum_{i=1}^{n} p_{ij}^{*(k+1)} \mathbf{x}_i \left(\frac{y_i - \mathbf{x}_i^\top \beta_c}{\sigma_c^{(k)}} \right),$$

where

$$p_{ij}^{*(k+1)} = p_{ic}^{(k+1)} w \left(\frac{y_i - \mathbf{x}_i^\top \beta_j^{(k)}}{\sigma_c^{(k)}} \right).$$

Therefore, the solution of (7.6) can be approximated by

$$\beta_c^{(k+1)} = \left(\sum_{i=1}^{n} p_{ic}^{*(k+1)} \mathbf{x}_i \mathbf{x}_i^\top\right)^{-1} \sum_{i=1}^{n} p_{ic}^{*(k+1)} \mathbf{x}_i y_i,$$

which can be considered as a weighted least squares estimator with the weights $\{p_{ij}^{*(k+1)}, i = 1, \ldots, n\}$.

Bai et al. (2012) established the consistency and the asymptotic normality of the proposed estimator. Based on their empirical studies, the method based on Tukey's bisquare has greater resistance to high-leverage outliers and has overall better performance than the method based on Huber's function. The proposed method is much more efficient than the existing MLE when there are outliers or the errors have heavy tails. In addition, the proposed robust estimation procedure has performance comparable to the MLE when there are no outliers and the error is exactly normal.

7.3 Robust mixture modeling by heavy-tailed error densities

One class of robust mixture regression modeling is to substitute the normal error densities with heavier-tailed error densities. These methods have clearly defined objective functions and their asymptotic properties have been extensively studied, as they are essentially maximum likelihood estimators based on distinct assumptions about heavy-tailed error densities. In this section, we will describe some of these techniques.

7.3.1 Robust mixture regression using t-distribution

Yao et al. (2014) proposed a robust mixture regression method by assuming a t-distribution for the component error density. More specifically, let $\mathcal{C}$ be a latent class variable such that $P(\mathcal{C} = c) = \pi_c$ and given $\mathcal{C} = c$, the response y depends on the $p-$dimensional predictor $\mathbf{x}$ in a linear way

$$y = \mathbf{x}^\top \beta_c + \epsilon_c, c = 1, 2, \cdots, C,$$

where the error density $g(\epsilon; \sigma_c, \nu_c)$ of ϵ_c is a t-distribution with a degree of freedom ν_c and scale parameter σ_c with

$$g(\epsilon; \sigma, \nu) = \frac{\Gamma(\frac{\nu+1}{2})\sigma^{-1}}{(\pi\nu)^{\frac{1}{2}}\Gamma(\frac{\nu}{2})\left\{1 + \frac{\epsilon^2}{\sigma^2\nu}\right\}^{\frac{1}{2}(\nu+1)}}. \tag{7.7}$$

The unknown parameter $\boldsymbol{\theta}$ can be estimated by maximizing the log-likelihood

$$\ell(\boldsymbol{\theta}) = \sum_{i=1}^{n} \log \left\{ \sum_{c=1}^{C} \pi_c g(y_i - \mathbf{x}_i^\top \boldsymbol{\beta}_c; \sigma_c, \nu_c) \right\}, \tag{7.8}$$

where $\boldsymbol{\theta} = (\pi_1, \boldsymbol{\beta}_1^\top, \sigma_1, \ldots, \pi_C, \boldsymbol{\beta}_C^\top, \sigma_C)^\top$ collects all unknown parameters. Note that when $\nu \to \infty$, t-distribution is reduced to a normal density. Therefore, with an adaptive choice of ν, this robust estimator can achieve the full efficiency of the normality-based MLE.

Note that the t-distribution can be considered a scale mixture of normal distributions. Let u be the latent variable such that

$$\epsilon | u \sim N(0, \sigma^2/u), \ u \sim \text{gamma}(\frac{1}{2}\nu, \frac{1}{2}\nu),$$

where $\text{gamma}(\alpha, \gamma)$ has density

$$f(u; \alpha, \gamma) = \frac{1}{\Gamma(\alpha)} \gamma^\alpha u^{\alpha-1} e^{-\gamma u}, \ u > 0.$$

Then, marginally ϵ has a $t-$distribution with a degree of freedom ν and scale parameter σ. Therefore, if the latent variable u is given, the mixture log-likelihood (7.8) is simplified to the traditional normal mixture likelihood.

Next, we introduce how to use the EM algorithm to simplify the computation. Let

$$z_{ic} = \begin{cases} 1, & \text{if } i\text{th observation is from the } c\text{th component;} \\ 0, & \text{otherwise,} \end{cases}$$

where $i = 1, \cdots, n, c = 1, \cdots, C$. Let $\mathbf{z} = (z_{11}, \ldots, z_{nC})$ and $\mathbf{u} = (u_1, \ldots, u_n)$ be the collected latent variables, $\mathbf{y} = (y_1, \ldots, y_n)$, and $\mathbf{X} = (\mathbf{x}_1, \ldots, \mathbf{x}_n)^\top$. Then the complete likelihood for $(\mathbf{y}, \mathbf{u}, \mathbf{z})$ given $\mathbf{X}$ is

$$\ell_c(\boldsymbol{\theta}; \mathbf{y}, \mathbf{u}, \mathbf{z}) = \sum_{i=1}^{n} \sum_{c=1}^{C} z_{ic} \log\{\pi_c \phi(y_i; \mathbf{x}_i^\top \boldsymbol{\beta}_c, \sigma_c^2/u_i) f(u_i; \frac{1}{2}\nu_c, \frac{1}{2}\nu_c)\}$$

$$= \sum_{i=1}^{n} \sum_{c=1}^{C} z_{ic} \log\{f(u_i; \frac{1}{2}\nu_c, \frac{1}{2}\nu_c)\} + \sum_{i=1}^{n} \sum_{c=1}^{C} z_{ic} \log(\pi_c),$$

$$+ \sum_{i=1}^{n} \sum_{c=1}^{C} z_{ic} \left\{ -\frac{1}{2}\log(2\pi\sigma_c^2) + \frac{1}{2}\log(u_i) - \frac{u_i}{2\sigma_c^2}(y_i - \mathbf{x}_i^\top \boldsymbol{\beta}_c)^2 \right\}. \tag{7.9}$$

Note that the above first term does not involve unknown parameters. E step computes $\mathrm{E}(\ell_c(\boldsymbol{\theta}; \mathbf{y}, \mathbf{u}, \mathbf{z}) \mid \mathbf{X}, \mathbf{y}, \boldsymbol{\theta}^{(k)})$, given the current estimate $\boldsymbol{\theta}^{(k)}$ at the kth step, which simplifies to the calculation of $\mathrm{E}(z_{ic} \mid \mathbf{X}, \mathbf{y}, \boldsymbol{\theta}^{(k)})$ and $\mathrm{E}(u_i \mid$

$\mathbf{X}, \mathbf{y}, \boldsymbol{\theta}^{(k)}, z_{ic} = 1$). The M step computes

$$
\mathrm{E}\{\ell_c(\boldsymbol{\theta}; \mathbf{y}, \mathbf{u}, \mathbf{z}) \mid \mathbf{X}, \mathbf{y}, \boldsymbol{\theta}^{(k)}\} \propto \sum_{i=1}^{n} \sum_{c=1}^{C} \mathrm{E}(z_{ic} \mid \mathbf{X}, \mathbf{y}, \boldsymbol{\theta}^{(k)}) \left[\log(\pi_c) \right.
$$

$$
\left. - \frac{1}{2} \log(2\pi\sigma_c^2) - \frac{\mathrm{E}(u_i \mid \mathbf{X}, \mathbf{y}, \boldsymbol{\theta}^{(k)}, z_{ic} = 1)}{2\sigma_c^2} (y_i - \mathbf{x}_i^\top \boldsymbol{\beta}_c)^2 \right],
$$

which has an explicit solution for $\boldsymbol{\theta}$. More specifically, we have the following EM algorithm to maximize (7.8).

Algorithm 7.3.1. Given the initial estimate $\boldsymbol{\theta}^{(0)}$, at $(k+1)$th iteration, we calculate the following two steps and then repeat the whole process until convergence.

E step: Calculate

$$
p_{ic}^{(k+1)} = \mathrm{E}(z_{ic} \mid \mathbf{X}, \mathbf{y}, \boldsymbol{\theta}^{(k)}) = \frac{\pi_c^{(k)} g(y_i - \mathbf{x}_i^\top \boldsymbol{\beta}_c^{(k)}; \sigma_c^{(k)}, \nu_c)}{\sum_{l=1}^{C} \pi_l^{(k)} g(y_i - \mathbf{x}_i^\top \boldsymbol{\beta}_l^{(k)}; \sigma_l^{(k)}, \nu_l)},
$$

and

$$
u_{ic}^{(k+1)} = \mathrm{E}(u_i \mid \mathbf{X}, \mathbf{y}, \boldsymbol{\theta}^{(k)}, z_{ic} = 1) = \frac{\nu + 1}{\nu + \left\{ (y_i - \mathbf{x}_i^\top \boldsymbol{\beta}_c^{(k)}) / \sigma_c^{(k)} \right\}^2}, \quad (7.10)
$$

where $g(\epsilon; \sigma, \nu)$ is defined in (7.7).

M step: Update parameter estimates:

$$
\pi_c^{(k+1)} = \sum_{i=1}^{n} p_{ic}^{(k+1)} / n,
$$

$$
\boldsymbol{\beta}_c^{(k+1)} = \left(\sum_{i=1}^{n} \mathbf{x}_i \mathbf{x}_i^\top p_{ic}^{(k+1)} u_{ic}^{(k+1)} \right)^{-1} \left(\sum_{i=1}^{n} \mathbf{x}_i y_i p_{ic}^{(k+1)} u_{ic}^{(k+1)} \right), \quad (7.11)
$$

and

$$
\sigma_c^{(k+1)} = \left\{ \frac{\sum_{i=1}^{n} p_{ic}^{(k+1)} u_{ic}^{(k+1)} (y_i - \mathbf{x}_i^\top \boldsymbol{\beta}_c^{(k+1)})^2}{\sum_{i=1}^{n} p_{ic}^{(k+1)}} \right\}^{1/2}, \quad (7.12)
$$

where $c = 1, \cdots, C$.

If we further assume that $\sigma_1 = \sigma_2 = \cdots = \sigma_C = \sigma$, then in M step, we can update σ by

$$
\sigma^{(k+1)} = \left\{ \frac{\sum_{i=1}^{n} \sum_{c=1}^{C} p_{ic}^{(k+1)} u_{ic}^{(k+1)} (y_i - \mathbf{x}_i^\top \boldsymbol{\beta}_c^{(k+1)})^2}{n} \right\}^{1/2}.
$$

One beneficial property of the above algorithm is that each iteration of the E step and M step of Algorithm 7.3.1 monotonically nondecreases the mixture log-likelihood (7.8), i.e., $\ell(\boldsymbol{\theta}^{(k+1)}) \geq \ell(\boldsymbol{\theta}^{(k)})$, for all $k \geq 0$. Note that the weight $u_{ic}^{(k+1)}$ of (7.10) decreases if the standardized residual increases. Therefore, the weights $u_{ic}^{(k+1)}$ in (7.11) and (7.12) can assist in producing robust estimators of mixture regression parameters by reducing the effects of outliers.

Just like the standard M-estimate for linear regression (Maronna et al., 2019), the mixture regression model that utilizes a t-distribution is prone to being affected by high-leverage outliers. To address this issue, Yao et al. (2014) developed a trimmed version of their method by adaptively trimming high leverage points. In contrast to TLE (Neykov et al., 2007), which uses a fixed proportion of trimming, the trimming proportion in this method is data adaptive.

The above trimming method is robust against high-leverage outliers but may lose some efficiency because some high-leverage points might have small residuals and thus can also provide valuable information to regression parameters. More research is needed on how to incorporate information from data with high leverage points and small residuals. One possible way is to borrow the ideas from GM-estimators (Krasker and Welsch, 1982; Maronna and Yohai, 1981) and one-step GM-estimators (Coakley and Hettmansperger, 1993; Simpson et al., 1998).

7.3.1.1 Adaptive choice of the degrees of freedom for the t-distribution

Up to this point, we have assumed that the degrees of freedom ν_c's for the t-distribution are known. We will now discuss how to adaptively choose ν based on the method proposed by Yao et al. (2014). First, consider the case where $\nu_1 = \nu_2 = \cdots = \nu_C = \nu$.

To adaptively choose the degrees of freedom parameter ν, we can use the profile likelihood method based on the log-likelihood defined in equation (7.8). Specifically, define the profile likelihood for ν as

$$L(\nu) = \max_{\boldsymbol{\theta}} \sum_{i=1}^{n} \log \left\{ \sum_{c=1}^{C} \pi_c f(y_i - \mathbf{x}_i^{\top} \boldsymbol{\beta}_c; \sigma, \nu) \right\}.$$

Given a fixed ν, $L(\nu)$ can be easily calculated using Algorithm 7.3.1. Then, we can estimate ν by finding the value of ν that maximizes $L(\nu)$, i.e., $\hat{\nu} = \arg\max_{\nu} L(\nu)$. Note that we can also apply the profile method when the ν_c's are different.

Alternatively, we can incorporate the estimation of ν_c's into the EM algorithm using the complete likelihood defined in equation (7.9). Specifically, in the M-step of the $(k+1)$th iteration, given the current estimate $\boldsymbol{\theta}^{(k)}$, we can update ν_c by finding the value of ν_c that maximizes the following expected

$$
\nu_c^{(k+1)} = \arg\max_{\nu_c} \mathrm{E}\left[\sum_{i=1}^{n} z_{ic} \log f(u_i; \tfrac{1}{2}\nu_c, \tfrac{1}{2}\nu_c) \mid \mathbf{X}, \mathbf{y}, \boldsymbol{\theta}^{(k)}\right],
$$

$$
= \arg\max_{\nu_c} \sum_{i=1}^{n} p_{ic}^{(k+1)} \left[-\log\Gamma(0.5\nu_c) + 0.5\nu_c \log(0.5\nu_c)\right.
$$

$$
\left. +0.5\nu_c \left\{v_{ic}^{(k+1)} - u_{ic}^{(k+1)}\right\} - v_{ic}^{(k+1)}\right], \tag{7.13}
$$

where $f(u; \nu_1, \nu_2)$ is defined in (7.9), and z_{ic} is the indicator variable that denotes whether observation i belongs to cluster c.

7.3.2 Robust mixture regression using Laplace distribution

Song et al. (2014) proposed a robust mixture regression by modeling the component error density by a Laplace/double exponential distribution. More specifically, in (7.3.1), ϵ is assumed to have a Laplace distribution with density

$$
f(\epsilon; \sigma) = \exp(-\sqrt{2}|\epsilon|/\sigma)/(\sqrt{2}\sigma). \tag{7.14}
$$

Therefore, the unknown mixture regression parameter $\boldsymbol{\theta}$ can be estimated by maximizing the log-likelihood

$$
\ell(\boldsymbol{\theta}) = \sum_{i=1}^{n} \log\left\{\sum_{c=1}^{C} \frac{\pi_c}{\sqrt{2}\sigma_c} \exp\left(-\frac{\sqrt{2}|y_i - \mathbf{x}_i^\top \boldsymbol{\beta}_c|}{\sigma_c}\right)\right\}. \tag{7.15}
$$

Based on Andrews and Mallows (1974), a Laplace distribution can be expressed as a mixture of a normal distribution and another distribution related to the exponential distribution. More specifically, let ϵ and V be two random variables such that

$$
\epsilon|V = v \sim N(0, \sigma^2/(2v^2)),
$$

and V has a density function $g(v) = v^{-3}\exp(-(2v^2)^{-1}), v > 0$. Then ϵ has a marginal Laplace distribution with the density given in (7.14). Based on this finding, we can use an EM algorithm to maximize (7.15).

Let $\mathbf{v} = (v_1, \ldots, v_n)$ be the latent scale variable. Then, the complete log-likelihood function for $(\mathbf{y}, \mathbf{v}, \mathbf{z})$ given $\mathbf{X}$ is

$$
\ell_c(\boldsymbol{\theta}; \mathbf{y}, \mathbf{v}, \mathbf{z})
$$

$$
= \sum_{i=1}^{n}\sum_{c=1}^{C} z_{ic} \log\left[\pi_c \frac{v_i}{\sqrt{\pi}\sigma_c} \exp\left\{-\frac{v_i^2(y_i - \mathbf{x}_i^\top \boldsymbol{\beta}_c)^2}{\sigma_c^2}\right\} \frac{1}{v_i^3}\exp\left(-\frac{1}{2v_i^2}\right)\right]
$$

$$
= \sum_{i=1}^{n}\sum_{c=1}^{C} z_{ic} \log\pi_c - \frac{1}{2}\sum_{i=1}^{n}\sum_{c=1}^{C} z_{ic}\log\pi\sigma_c^2 - \sum_{i=1}^{n}\sum_{c=1}^{C} \frac{z_{ic}v_i^2(y_i - \mathbf{x}_i^\top \boldsymbol{\beta}_c)^2}{\sigma_c^2}
$$

$$
- \sum_{i=1}^{n}\sum_{c=1}^{C} z_{ic}\log v_i^3 - \frac{1}{2}\sum_{i=1}^{n}\sum_{c=1}^{C} \frac{z_{ic}}{v_i^2}.
$$

Let $\boldsymbol{\theta}^{(k)}$ be the estimate of $\boldsymbol{\theta}$ at the kth step. The E-step calculates the condition expectation $\mathrm{E}[\ell_c(\boldsymbol{\theta}; \mathbf{y}, \mathbf{v}, \mathbf{z})|\mathbf{X}, \mathbf{y}, \boldsymbol{\theta}^{(k)}]$, which simplifies to the computations of $\mathrm{E}[Z_{ic}|\mathbf{X}, \mathbf{y}, \boldsymbol{\theta}^{(k)}]$ and $\mathrm{E}[V_i^2|\mathbf{X}, \mathbf{y}, \boldsymbol{\theta}^{(k)}, z_{ic}=1]$. The M-step then maximizes $\mathrm{E}[\ell_c(\boldsymbol{\theta}; \mathbf{y}, \mathbf{v}, \mathbf{z})|\mathbf{X}, \mathbf{y}, \boldsymbol{\theta}^{(k)}]$. More specifically, we could use the following EM algorithm to maximize (7.15).

Algorithm 7.3.2. Given the initial parameter estimate $\boldsymbol{\theta}^{(0)}$, we iterate the following E-step and M-step until convergence.

E-step: Calculate

$$p_{ic}^{(k+1)} = \mathrm{E}[Z_{ic}|\mathbf{X}, \mathbf{y}, \boldsymbol{\theta}^{(k)}] = \frac{\pi_c^{(k)}\sigma_c^{-1(k)}\exp(-\sqrt{2}|y_i - \mathbf{x}_i^\top\boldsymbol{\beta}_c^{(k)}|/\sigma_c^{(k)})}{\sum_{c=1}^C \pi_c^{(k)}\sigma_c^{-1(k)}\exp(-\sqrt{2}|y_i - \mathbf{x}_i^\top\boldsymbol{\beta}_c^{(k)}|/\sigma_c^{(k)})},$$

$$\delta_{ic}^{(k+1)} = \mathrm{E}[V_i^2|\mathbf{X}, \mathbf{y}, \boldsymbol{\theta}^{(k)}, z_{ic}=1] = \frac{\sigma_c^{(k)}}{\sqrt{2}|y_i - \mathbf{x}_i^\top\boldsymbol{\beta}_c^{(k)}|}.$$

M-step: Update parameters $\pi_c^{(k+1)}$, $\boldsymbol{\beta}_c^{(k+1)}$, and $\sigma_c^{2(k+1)}$ as

$$\pi_c^{(k+1)} = \sum_{i=1}^n p_{ic}^{(k+1)}/n,$$

$$\boldsymbol{\beta}_c^{(k+1)} = \left(\sum_{i=1}^n p_{ic}^{(k+1)}\delta_{ic}^{(k+1)}\mathbf{x}_i\mathbf{x}_i^\top\right)^{-1}\left(\sum_{i=1}^n p_{ic}^{(k+1)}\delta_{ic}^{(k+1)}\mathbf{x}_iy_i\right), \quad (7.16)$$

$$\sigma_c^{(k+1)} = \left\{\frac{2\sum_{i=1}^n\sum_{c=1}^C p_{ic}^{(k+1)}\delta_{ic}^{(k+1)}(y_i - \mathbf{x}_i^\top\boldsymbol{\beta}_c^{(k+1)})^2}{\sum_{i=1}^n p_{ic}^{(k+1)}}\right\}^{1/2},$$

where $c = 1, \ldots, C$.

If we further assume that $\sigma_1 = \sigma_2 = \ldots = \sigma_C = \sigma$, then σ^2 can be updated in M-step by

$$\sigma^{(k+1)} = \left\{\frac{2\sum_{i=1}^n\sum_{c=1}^C p_{ic}^{(k+1)}\delta_{ic}^{(k+1)}(y_i - \mathbf{x}_i^\top\boldsymbol{\beta}_c^{(k+1)})^2}{n}\right\}^{1/2}.$$

The robustness of the regression parameter estimates can be seen from the updates $\boldsymbol{\beta}_c^{(k+1)}$'s in (7.16), in which the weights δ_{ic}^{k+1} are reversely related to the term $|y_i - \mathbf{x}_i^\top\boldsymbol{\beta}_c^{(k)}|$, meaning that observations with larger residuals have smaller weights.

It is worth noting that the procedure based on t-distribution proposed by Yao et al. (2014) achieves both efficiency and robustness by adapting the degrees of freedom based on the data. On the other hand, the procedure based on Laplace distribution has the advantage of simpler and faster computation and does not require any tuning parameter. However, it may result in some loss of efficiency in the absence of outliers.

7.4 Scale mixtures of skew-normal distributions

Basso et al. (2010) proposed a robust mixture modeling method based on scale mixtures of skew-normal distributions. Zeller et al. (2016) extended the prior works of Basso et al. (2010) and Yao et al. (2014) by introducing a robust mixture regression model based on scale mixtures of skew-normal (SMSN) distributions. Unlike traditional regression models that assume symmetry and normality, the SMSN-based model can effectively handle asymmetric and heavy-tailed data. This allows for more accurate modeling of real-world phenomena, where deviations from normality are common.

Definition 7.4.1. We say that a random variable Y has a skew-normal distribution with a location parameter μ, dispersion parameter $\sigma^2 > 0$, and skewness parameter λ, written as $Y \sim SN(\mu, \sigma^2, \lambda)$, if its density is given by

$$f(y) = 2\phi(y; \mu, \sigma^2)\Phi(a), \; y \in \mathbb{R},$$

where $a = \lambda\sigma^{-1}(y - \mu)$, $\phi(\cdot\,; \mu, \sigma^2)$ stands for the pdf of the univariate normal distribution with mean μ and variance σ^2, and $\Phi(\cdot)$ represents the cumulative distribution function of the standard univariate normal distribution.

Definition 7.4.2. The distribution of the random variable Y belongs to the family of SMSN distributions when

$$Y = \mu + K(U)^{1/2}X,$$

where μ is a location parameter, $X \sim SN(0, \sigma^2, \lambda)$, $K(\cdot)$ is a positive weight function and U is a positive random variable with a cdf $H(u; \boldsymbol{\nu})$, where $\boldsymbol{\nu}$ is a parameter indexing the distribution of U, known as the scale factor parameter, which is independent of X.

The SMSN distribution is written as $Y \sim SMSN(\mu, \sigma^2, \lambda; H)$. The name of the class becomes clear when we note that the conditional distribution of Y given $U = u$ is skew-normal. Specifically, we have that

$$Y|U = u \sim SN(\mu, K(u)\sigma^2, \lambda), \; U \sim H(.; \boldsymbol{\nu}).$$

As a result, the density of an SMSN distribution is

$$g(y) = 2\int_0^\infty \phi(y; \mu, K(u)\sigma^2)\Phi(K(u)^{-1/2}a)dH(u; \boldsymbol{\nu}).$$

Due to the additional parameters, the SMSN distribution can be more flexible in accommodating different error densities than the t-distribution and the Laplace distribution, at the expense of more complicated computation.

Zeller et al. (2016) proposed a robust mixture regression modeling based on SMSN distributions as follows:

$$f(y; \mathbf{x}, \boldsymbol{\theta}) = \sum_{c=1}^{C} \pi_c g(y; \mathbf{x}, \boldsymbol{\theta}_c), \tag{7.17}$$

where $g(\cdot; \mathbf{x}, \boldsymbol{\theta}_c)$ is the density function of $SMSN(\mathbf{x}^\top \boldsymbol{\beta}_c + b\Delta_c, \sigma_c^2, \lambda_c, \boldsymbol{\nu}_c)$, $\boldsymbol{\theta}_c = (\pi_c, \boldsymbol{\beta}_c, \sigma_c^2, \lambda_c, \boldsymbol{\nu}_c)^\top$, $\Delta_c = \sigma_c \delta_c$, $\delta_c = \frac{\lambda_c}{\sqrt{1+\lambda_c^2}}$, $b = -\sqrt{\frac{2}{\pi}} K_1$, with $K_r = E[K^{r/2}(U)], r = 1, 2, \ldots$. To simplify the computation process, the assumption is made that $\boldsymbol{\nu}_1 = \boldsymbol{\nu}_2 = \ldots = \boldsymbol{\nu}_c = \boldsymbol{\nu}$. This allows for a more efficient estimation of the model parameters using the ECME algorithm proposed by Zeller et al. (2016). The ECME algorithm, an extension of the EM and ECM algorithms (Liu and Rubin, 1994; Meng and Rubin, 1993), consists of three steps: E-step, CM-step, and CML-step. In the E-step, the algorithm computes the conditional expectation of the complete log-likelihood function. The CM-step updates the parameter vector $\boldsymbol{\theta}$ by maximizing the constrained conditional expectation of the complete log-likelihood. Finally, the CML-step updates $\hat{\boldsymbol{\nu}}$ by maximizing the actual marginal likelihood function. Together, these steps provide an effective and robust approach for estimating the parameters of the SMSN-based mixture regression model. We summarize the algorithm in Algorithm 7.4.1.

7.5   Robust EM-type algorithm for log-concave mixture regression models

By extending the idea of Hu et al. (2016), Hu et al. (2017) proposed a robust EM-type algorithm for mixture regression models by assuming the component error densities to be log-concave. A density function $g(x)$ is said to be log-concave if its logarithm, denoted by $\phi(x) = \log g(x)$, is concave. Several well-known distributions, such as normal, Laplace, chi-square, logistic, gamma (with shape parameter greater than 1), and beta (with both parameters greater than 1) distributions, exhibit log-concavity. Log-concave densities possess numerous desirable properties, as discussed by Balabdaoui et al. (2009) and other authors in the literature.

Given iid observations $\{x_1, \ldots, x_n\}$ from a log-concave density $g(x)$, i.e., $\phi(x) = \log g(x)$ is concave, Dümbgen et al. (2011) proposed estimating $\phi(x)$ nonparametrically by maximizing a log-likelihood-type functional:

$$L(\phi) = \frac{1}{n} \sum_{i=1}^{n} \phi(x_i) - \int \exp\{\phi(x)\} dx + 1. \tag{7.18}$$

Algorithm 7.4.1 ECME algorithm

Given an initial value $\widehat{\boldsymbol{\theta}}^{(0)}$, at the $(k+1)th$ step, iterate the following three steps until convergence.

E-step: Compute $p_{ic}^{(k+1)}$, $\widehat{zu}_{ic}^{(k+1)}$, $\widehat{zut}_{ic}^{(k+1)}$ and $\widehat{zut^2}_{ic}^{(k+1)}$ for $i = 1, \ldots, n$, using the following formulas:

$$p_{ic}^{(k+1)} = \frac{\pi_c^{(k)} g(y_i; \mathbf{x}_i, \widehat{\boldsymbol{\theta}}_c^{(k)})}{\sum_{c=1}^{C} \pi_c^{(k)} g(y_i; \mathbf{x}_i, \widehat{\boldsymbol{\theta}}_c^{(k)})}, \qquad \widehat{zu}_{ic}^{(k+1)} = p_{ic}^{(k+1)} \widehat{u}_{ic}^{(k)},$$

$$\widehat{zut}_{ic}^{(k+1)} = p_{ic}^{(k+1)} \widehat{ut}_{ic}^{(k)}, \qquad \widehat{zut^2}_{ic}^{(k+1)} = p_{ic}^{(k+1)} \widehat{ut^2}_{ic}^{(k)},$$

where $\widehat{ut}_{ic}^{(k)} = \widehat{u}_{ic}^{(k)}(\widehat{m}_{ic}^{(k)} + b) + \widehat{M}_c^{(k)}\widehat{\eta}_{ic}^{(k)}$, and $\widehat{ut^2}_{ic}^{(k)} = \widehat{u}_{ic}^{(k)}(\widehat{m}_{ic}^{(k)} + b)^2 + \widehat{M}_c^{2(k)} + \widehat{M}_c^{(k)}(\widehat{m}_{ic}^{(k)} + 2b)\widehat{\eta}_{ic}^{(k)}$, with $\widehat{M}_c^2 = \widehat{\gamma}_c^2/\{\widehat{\gamma}_c^2 + \widehat{\Delta}_c^2\}$ and $\widehat{m}_{ic} = \widehat{M}_c^2 \widehat{\Delta}_c \widehat{\gamma}_c^{-2}(y_i - \mathbf{x}_i^{\top} \widehat{\boldsymbol{\beta}}_c - b\widehat{\Delta}_c)$, evaluated at $\boldsymbol{\theta} = \widehat{\boldsymbol{\theta}}^{(k)}$.

CM-step: For $c = 1, \ldots, C$, update component parameters:

$$\boldsymbol{\beta}_c^{(k+1)} = \left(\sum_{i=1}^{n} p_{ic}^{(k+1)} \mathbf{x}_i \mathbf{x}_i^{\top}\right)^{-1} \sum_{i=1}^{n} \left(\widehat{zu}_i^{(k+1)} y_i - \widehat{zut}_{ic}^{(k+1)} \widehat{\Delta}_c^{(k)}\right) \mathbf{x}_i,$$

$$\widehat{\gamma}_c^{2(k+1)} = \frac{\sum_{i=1}^{n} \left[\widehat{zu}_{ic}^{(k+1)}(r_{ic}^{(k+1)})^2 - 2\widehat{zut}_{ic}^{(k+1)} \widehat{\Delta}_c^{(k)} r_{ic}^{(k+1)} + \widehat{zut^2}_{ic}^{(k+1)} \widehat{\Delta}_c^{2(k)}\right]}{\sum_{i=1}^{n} p_{ic}^{(k+1)}},$$

$$\widehat{\Delta}_c^{(k+1)} = \frac{\widehat{zut}_{ic}^{(k+1)}(y_i - \mathbf{x}_i^{\top}\boldsymbol{\beta}_c^{(k+1)})}{\sum_{i=1}^{n} \widehat{zut^2}_{ic}^{(k+1)}}, \quad \pi_c^{(k+1)} = \frac{\sum_{i=1}^{n} p_{ic}^{(k+1)}}{n},$$

$$\sigma_c^{2(k+1)} = \widehat{\gamma}_c^{2(k+1)} + \widehat{\Delta}_c^{2(k+1)}, \quad \widehat{\lambda}_c^{(k+1)} = \frac{\widehat{\Delta}_c^{(k+1)}}{\sqrt{\widehat{\gamma}_c^{2(k+1)}}}.$$

where $r_{ic}^{(k+1)} = y_i - \mathbf{x}_i^{\top}\boldsymbol{\beta}_c^{(k+1)}$.

CML-step: Update $\widehat{\boldsymbol{\nu}}$ using the following

$$\widehat{\boldsymbol{\nu}}^{(k+1)} = \arg\max_{\boldsymbol{\nu}} \sum_{i=1}^{n} \log\left(\sum_{c=1}^{C} \pi_c^{(k+1)} g(y_i; \mathbf{x}_i, \boldsymbol{\beta}_c^{(k+1)}, \widehat{\sigma}_c^{2(k+1)}, \widehat{\lambda}_c^{(k+1)}, \boldsymbol{\nu})\right).$$

Dümbgen et al. (2011) also extended the method of (7.18) to the linear regression setting $y_i = \mathbf{x}_i^\top \boldsymbol{\beta} + \epsilon_i$, where ϵ_i is assumed to have a log-concave density, and proposed estimating the log-density $\phi(x)$ and the regression parameter $\boldsymbol{\beta}$ by maximizing:

$$L(\phi, \boldsymbol{\beta}) = \frac{1}{n} \sum_{i=1}^{n} \phi(y_i - \mathbf{x}_i^\top \boldsymbol{\beta}) - \int \exp\{\phi(x)\} dx + 1. \qquad (7.19)$$

Such estimators, like the maximizers of (7.18) and (7.19), are called log-concave maximum likelihood estimators (LCMLEs). Dümbgen et al. (2011) proved the existence, uniqueness, and consistency of LCMLEs for (7.18) and (7.19) under fairly general conditions. These estimators provide more generality and flexibility without any tuning parameter and were extensively studied by many researchers recently, such as Dümbgen et al. (2009), Cule et al. (2010b), Cule et al. (2010a), Chen and Samworth (2013), and Dümbgen et al. (2011).

Based on Hu et al. (2017), if component error densities of mixture regressions are assumed to be log-concave, the corresponding log-likelihood function can be written as

$$\ell(\boldsymbol{\theta}, \mathbf{g} | \mathbf{X}, \mathbf{y}) = \sum_{i=1}^{n} \log \sum_{c=1}^{C} \pi_c g_c(y_i - \mathbf{x}_i^\top \boldsymbol{\beta}_c) \qquad (7.20)$$

where $\mathbf{y} = (y_1, \ldots, y_n)^\top, \mathbf{X} = (\mathbf{x}_1, \ldots, \mathbf{x}_n)^\top, \boldsymbol{\theta} = (\pi_1, \boldsymbol{\beta}_1, \ldots, \pi_C, \boldsymbol{\beta}_C)^\top$, and $g_c(x) = \exp\{\phi_c(x)\}$ for some unknown concave function $\phi_c(x)$. The identifiability of the above model has been established by Ma et al. (2021).

Unlike other robust mixture regression methods, the mixture regression model (7.20) is semiparametric because it does not require a specific parametric assumption for error densities and can be adaptive to different error densities with a log-concave assumption.

Hu et al. (2017) proposed maximizing the objective function (7.20) by an EM-type algorithm. Specifically, in the $(k+1)th$ step, E-step computes the posterior probabilities:

$$p_{ic}^{(k+1)} = \frac{\pi_c^{(t)} g_c^{(k)}(y_i - \mathbf{x}_i^\top \boldsymbol{\beta}_c^{(k)})}{\sum_{h=1}^{C} \pi_h^{(k)} g_h^{(k)}(y_i - \mathbf{x}_i^\top \boldsymbol{\beta}_h^{(k)})}, \quad i = 1, \ldots, n, \ c = 1, \ldots, C.$$

The M-step updates the parameter estimates and the error density by

$$\pi_c^{(k+1)} = \frac{1}{n} \sum_{i=1}^{n} p_{ic}^{(k+1)},$$

$$\boldsymbol{\beta}_c^{(k+1)} = \arg\max_{\boldsymbol{\beta}_c} \sum_{i=1}^{n} p_{ic}^{(k+1)} \log g_c^{(k)}(y_i - \mathbf{x}_i^\top \boldsymbol{\beta}_c),$$

$$g_c^{(k+1)} \leftarrow \arg\max_{g_c \in \mathbb{G}} \sum_{i=1}^{n} p_{ic}^{(k+1)} \log g_c(y_i - \mathbf{x}_i^\top \boldsymbol{\beta}_c^{(k+1)}), \qquad c = 1, \ldots, C,$$

where $\mathbb{G}$ is the family of all log-concave densities. If $g_c \equiv g$, then the g_c in the algorithm should be replaced by g, and the update of g in M-step becomes

$$g^{(k+1)} \leftarrow \arg\max_{g \in \mathbb{G}} \sum_{i=1}^{n} \sum_{c=1}^{C} p_{ic}^{(k+1)} \log g(y_i - \mathbf{x}_i^\top \widehat{\boldsymbol{\beta}}_c^{(k+1)}).$$

The evaluation of $g_c^{(k)}(y_i - \mathbf{x}_i^\top \boldsymbol{\beta}_c^{(k)})$ can be done based on the function `dlcd` in R package `LogConcDEAD`. The computation of $g_c^{(k+1)}$ in M step can be implemented through the function called `mlelcd` in the R package `LogConcDEAD` (Cule et al., 2009) based on fitted residuals $y_i - \mathbf{x}_i^\top \boldsymbol{\beta}_c^{(k+1)}$ with posterior probabilities $p_{ic}^{(k+1)}$s as weights.

By estimating the log-concave error density data adaptively, the log-concave maximum likelihood estimator corrects the possible model misspecification of traditional, fully parametric models, e.g. adjusting the skewness and heavy tails when the error distribution is not normal. Please see Hu et al. (2017) for more detailed discussions and numerical comparisons.

7.6   Robust estimators based on trimming

Neykov et al. (2007) proposed a robust mixture modeling based on the trimmed likelihood method (TLE). The TLE is defined as

$$\widehat{\boldsymbol{\theta}}_{TLE} = \arg\max_{\boldsymbol{\theta}, S_h} \sum_{i \in S_h} \log \left\{ \sum_{c=1}^{C} \pi_c \phi(y_i; \mathbf{x}_i^\top \boldsymbol{\beta}_c, \sigma_c^2) \right\},$$

where S_h is a subset of $(1, 2, \cdots, n)$ with cardinality h, h is the round number of $n(1 - \alpha)$ and α the is percentage of the data to be trimmed.

Neykov et al. (2007) used the FAST-TLE algorithm to compute the $\widehat{\boldsymbol{\theta}}_{TLE}$ based on a two-step procedure: a trial step and a refinement step. In the trial step, we randomly select several subsamples of size h^* from the data followed by fitting the model to each subsample to get multiple trial/initial MLEs. The trial subsample size h^* should be greater than or equal to $C(p + 1)$, where p is the number of covariates. In the refinement step, each trial MLE is used as an initial estimator, and the following two steps are repeated until convergence: (1) Identify the h observations with the smallest negative log-likelihoods based on the current estimate. (2) Fit the mixture regression model to these h observations to obtain an improved fit. The final estimate is selected as the converged estimate with the smallest negative log-likelihood when multiple initial estimates are used.

To apply the TLE, we first need to choose the proportion of trimming α, which plays an important role for the TLE. If α is too large, the TLE will lose efficiency. If α is too small and the percentage of outliers is more than α, then the TLE will fail. In practice, a conservative α is usually used so that it is expected that the proportion of outliers is no more than α.

To enhance the performance of the TLE method, García-Escudero et al. (2010) proposed an adaptive approach to "second" trimming and protection against bad leverage points. The first trimming parameter α_1 in TLE is effective in reducing the impact of outliers in the response variable y, but it has minimal effect on high-leverage outliers. To address this, a "second" trimming of size α_2 is proposed to remove the high-leverage observations that survived the first trimming. To ensure effective trimming, it is recommended to choose the trimming parameters α_1 and α_2 in a preventive fashion, i.e., slightly larger than needed, to ensure that all outlying observations are removed. This adaptive approach can significantly improve the robustness and efficiency of TLE in the presence of outliers and high-leverage points.

By extending the least trimmed squares (LTS) method (Rousseeuw, 1984), Doğru and Arslan (2018) proposed the trimmed complete data log-likelihood function:

$$\ell_n^c(\boldsymbol{\theta}) = \sum_{i=1}^{n}\sum_{c=1}^{C} z_{ic}\left\{\log(\pi_c) - \frac{1}{2}\log(2\pi) - \frac{1}{2}\log(\sigma_c^2)\right\} - \sum_{c=1}^{C}\sum_{i\in S_{ch}} z_{ic}\frac{r_{ic}^2}{2\sigma_c^2},$$

$$(7.21)$$

where $r_{ic} = y_i - \mathbf{x}_i^\top \boldsymbol{\beta}_c$, and S_{ch} is a subset of $(1, 2, \cdots, n)$ with cardinality h, such that $\{r_{ic}^2, i \in S_{ch}\}$ are the h smallest squared residuals among $\{r_{ic}^2, i = 1, \ldots, n\}$ for $c = 1, \ldots, C$. Note that the set of h observations varies across the components. To implement the EM algorithm, z_{ic} in (7.21) is replaced by its expectation, and the mixture regression parameters are estimated by the LTS criterion (7.21).

7.7 Robust mixture regression modeling by cluster-weighted modeling

Let $f(\mathbf{x}, y)$ be the joint density of $(\mathbf{x}, y)$. The general cluster-weighted models (García-Escudero et al., 2017; Punzo and McNicholas, 2017, CWM) are mixture models whose density is

$$f(\mathbf{x}, y; \boldsymbol{\theta}) = \sum_{c=1}^{C} \pi_c g(y|\mathbf{x}; \xi_c) g(\mathbf{x}; \psi_c),$$

where $\boldsymbol{\theta}$ collects all unknown parameters, $g(y|\mathbf{x}; \xi_c)$ is the conditional density of Y given $\mathbf{x}$ in the cth component, $g(\mathbf{x}; \psi_c)$ is the marginal density of $\mathbf{x}$ in the cth component and π_c is the cth component proportion. Compared to traditional mixture regression models, the covariates in CWM are jointly modeled with the response variable, which makes it easy to detect outliers in the $\mathbf{x}$ direction and to handle high-leverage outliers.

More specifically, the linear Gaussian cluster-weighted model assumes that the conditional relationship between Y and $\mathbf{x}$ in the cth group is $Y = \alpha_c + \mathbf{x}^\top \boldsymbol{\beta}_c + \epsilon_c$, where $\epsilon_c \sim N(0, \sigma_c^2)$, and the joint density of $(\mathbf{x}, y)$ is

$$f(\mathbf{x}, y; \boldsymbol{\theta}) = \sum_{c=1}^{C} \pi_c \phi \left(y; \alpha_c + \mathbf{x}^\top \boldsymbol{\beta}_c, \sigma_c^2 \right) \phi_p(\mathbf{x}; \boldsymbol{\mu}_c, \boldsymbol{\Sigma}_c),$$

where $\phi_p(\cdot; \boldsymbol{\mu}_c, \boldsymbol{\Sigma}_c)$ denotes the density of the p-variate Gaussian distribution with mean vector $\boldsymbol{\mu}_c$ and covariance matrix $\boldsymbol{\Sigma}_c$.

García-Escudero et al. (2017) proposed a new robust estimator for mixtures of regression based on the trimmed cluster-weighted restricted model (CWRM). The trimmed CWRM methodology maximizes

$$\sum_{i=1}^{n} z(\mathbf{x}_i, y_i) \log \left[\sum_{c=1}^{C} \pi_c \phi \left(y_i; \alpha_c + \mathbf{x}^\top \boldsymbol{\beta}_c, \sigma_c^2 \right) \phi_p(\mathbf{x}_i; \boldsymbol{\mu}_c, \boldsymbol{\Sigma}_c) \right], \qquad (7.22)$$

where $z(\cdot, \cdot)$ is a $0-1$ trimming indicator function which equals to 0 if the observation $(\mathbf{x}_i, y_i)$ is trimmed off and equals to 1, otherwise.

The fraction of trimmed observations is denoted by α, where $\sum_{i=1}^{n} z(\mathbf{x}_i, y_i) = n(1-\alpha)$. To maximize (7.22) for any fixed α, García-Escudero et al. (2017) proposed a trimmed EM algorithm and an adaptive way of trimming to gain protection against bad leverage points. However, one difficulty of using this method in practice is the interrelatedness of tuning parameters with the number of components C. For instance, a large trimming level α could result in components with fewer observations being trimmed off, leading to smaller C values. Similarly, larger values of c_1 and c_2 may increase C by including components with very few observations, approaching local collinearity.

Punzo and McNicholas (2017) proposed a robust estimation of mixture regression via the contaminated Gaussian cluster-weighted model. A contaminated Gaussian distribution for a random variable $\mathbf{z}$ is given by

$$f(\mathbf{z}; \boldsymbol{\mu}_\mathbf{z}, \boldsymbol{\Sigma}_\mathbf{z}, \alpha_\mathbf{z}, \eta_\mathbf{z}) = (1 - \alpha_\mathbf{z})\phi(\mathbf{z}; \boldsymbol{\mu}_\mathbf{z}, \boldsymbol{\Sigma}_\mathbf{z}) + \alpha_\mathbf{z}\phi(\mathbf{z}; \boldsymbol{\mu}_\mathbf{z}, \eta_\mathbf{z}\boldsymbol{\Sigma}_\mathbf{z}),$$

where $\alpha_\mathbf{z} \in (0, 1)$ is the proportion of contamination and $\eta_\mathbf{z} > 1$ denotes the degree of contamination. The conditional distribution of $\mathbf{y}|\mathbf{x}$ from a contaminated Gaussian CWM is:

$$f(\mathbf{y}|\mathbf{x}; \boldsymbol{\theta}) = \frac{\sum_{c=1}^{C} \pi_c f(\mathbf{x}; \boldsymbol{\mu}_{\mathbf{x}|c}, \boldsymbol{\Sigma}_{\mathbf{x}|c}, \alpha_{\mathbf{x}|c}, \eta_{\mathbf{x}|c}) f(\mathbf{y}; \boldsymbol{\mu}_\mathbf{y}(\mathbf{x}; \boldsymbol{\beta}_c), \boldsymbol{\Sigma}_{\mathbf{y}|c}, \alpha_{\mathbf{y}|c}, \eta_{\mathbf{y}|c})}{\sum_{l=1}^{C} \pi_l f(\mathbf{x}; \boldsymbol{\mu}_{\mathbf{x}|l}, \boldsymbol{\Sigma}_{\mathbf{x}|l}, \alpha_{\mathbf{x}|l}, \eta_{\mathbf{x}|l})},$$

$$(7.23)$$

where $\boldsymbol{\theta}$ collects all the unknown parameters and both $\mathbf{x}$ and $\mathbf{y}$ can be a vector.

An EM algorithm can be used to compute the MLE of the model (7.23). There are three sources of incompleteness. The first source is a component indicator z_{ic}, where $z_{ic} = 1$ if $(\mathbf{x}_i, \mathbf{y}_i)$ comes from component c and $z_{ic} = 0$ otherwise. The second source can be defined as an outlier indicator: $u_{ic} = 1$ if $(\mathbf{x}_i, \mathbf{y}_i)$ is not an outlier in component c and $u_{ic} = 0$ otherwise. The third source is a leverage point indicator: $v_{ic} = 1$ if $(\mathbf{x}_i, \mathbf{y}_i)$ is not a leverage point in component c and $v_{ic} = 0$ otherwise. The complete-data likelihood can be written as

$$\ell_c(\boldsymbol{\theta}) = \ell_{1c}(\boldsymbol{\pi}) + \ell_{2c}(\boldsymbol{\alpha}_{\mathbf{x}}) + \ell_{3c}(\boldsymbol{\mu}_{\mathbf{x}}, \boldsymbol{\Sigma}_{\mathbf{x}}, \boldsymbol{\eta}_{\mathbf{x}}) + \ell_{4c}(\boldsymbol{\alpha}_{\mathbf{y}}) + \ell_{5c}(\boldsymbol{\beta}, \boldsymbol{\Sigma}_{\mathbf{y}}, \boldsymbol{\eta}_{\mathbf{y}}),$$

where $\boldsymbol{\pi} = (\pi_1, \ldots, \pi_C)$, $\boldsymbol{\mu}_{\mathbf{x}} = (\boldsymbol{\mu}_{\mathbf{x}|1}, \ldots, \boldsymbol{\mu}_{\mathbf{x}|C})$, $\boldsymbol{\Sigma}_{\mathbf{x}} = (\boldsymbol{\Sigma}_{\mathbf{x}|1}, \ldots, \boldsymbol{\Sigma}_{\mathbf{x}|C})$, $\boldsymbol{\alpha}_{\mathbf{x}} = (\alpha_{\mathbf{x}|1}, \ldots, \alpha_{\mathbf{x}|C})$, $\boldsymbol{\eta}_{\mathbf{x}} = (\eta_{\mathbf{x}|1}, \ldots, \eta_{\mathbf{x}|C})$, $\boldsymbol{\beta} = (\boldsymbol{\beta}_1, \ldots, \boldsymbol{\beta}_C)$, $\boldsymbol{\Sigma}_{\mathbf{y}} = (\boldsymbol{\Sigma}_{\mathbf{y}|1}, \ldots, \boldsymbol{\Sigma}_{\mathbf{y}|C})$, $\boldsymbol{\alpha}_{\mathbf{y}} = (\alpha_{\mathbf{y}|1}, \ldots, \alpha_{\mathbf{y}|C})$, $\boldsymbol{\eta}_{\mathbf{y}} = (\eta_{\mathbf{y}|1}, \ldots, \eta_{\mathbf{y}|C})$, $\ell_{1c}(\boldsymbol{\pi}) = \sum_{i=1}^{n} \sum_{c=1}^{C} z_{ic} \log(\pi_c)$,

$$\ell_{2c}(\boldsymbol{\alpha}_{\mathbf{x}}) = \sum_{i=1}^{n} \sum_{c=1}^{C} z_{ic} \left[v_{ic} \log(\alpha_{\mathbf{x}|c}) + (1 - v_{ic}) \log(1 - \alpha_{\mathbf{x}|c}) \right],$$

$$\ell_{3c}(\boldsymbol{\mu}_{\mathbf{x}}, \boldsymbol{\Sigma}_{\mathbf{x}}, \boldsymbol{\eta}_{\mathbf{x}}) = -\frac{1}{2} \sum_{i=1}^{n} \sum_{c=1}^{C} \left\{ z_{ic} \log |\boldsymbol{\Sigma}_{\mathbf{x}|c}| + d_{\mathbf{x}} z_{ic} (1 - v_{ic}) \log(\eta_{\mathbf{x}|c}) \right. $$
$$\left. + z_{ic} \left(v_{ic} + \frac{1 - v_{ic}}{\eta_{\mathbf{x}|c}} \right) \delta(\mathbf{x}_i; \boldsymbol{\mu}_{\mathbf{x}|c}, \boldsymbol{\Sigma}_{\mathbf{x}|c}) \right\},$$

$$\ell_{4c}(\boldsymbol{\alpha}_{\mathbf{y}}) = \sum_{i=1}^{n} \sum_{c=1}^{C} z_{ic} [u_{ic} \log(\alpha_{\mathbf{y}|c}) + (1 - u_{ic}) \log(1 - \alpha_{\mathbf{y}|c})],$$

$$\ell_{5c}(\boldsymbol{\beta}, \boldsymbol{\Sigma}_{\mathbf{y}}, \boldsymbol{\eta}_{\mathbf{y}}) = -\frac{1}{2} \sum_{i=1}^{n} \sum_{c=1}^{C} \left\{ z_{ic} \log |\boldsymbol{\Sigma}_{\mathbf{y}|c}| + d_{\mathbf{y}} z_{ic} (1 - u_{ic}) \log(\eta_{\mathbf{y}|c}), \right. $$
$$\left. + z_{ic} \left(u_{ic} + \frac{1 - u_{ic}}{\eta_{\mathbf{y}|c}} \right) \delta(\mathbf{y}_i; \boldsymbol{\mu}_{\mathbf{y}}(\mathbf{x}_i; \boldsymbol{\beta}_c), \boldsymbol{\Sigma}_{\mathbf{y}|c}) \right\},$$

and where $\delta(\mathbf{w}; \boldsymbol{\mu}, \boldsymbol{\Sigma}) = (\mathbf{w} - \boldsymbol{\mu})^{\top} \boldsymbol{\Sigma}^{-1} (\mathbf{w} - \boldsymbol{\mu})$ denotes the squared Mahalanobis distance between $\mathbf{w}$ and $\boldsymbol{\mu}$, with covariance matrix $\boldsymbol{\Sigma}$.

The usual parameter $\boldsymbol{\theta}$ is partitioned into $\boldsymbol{\theta}_1$ and $\boldsymbol{\theta}_2$, where $\boldsymbol{\theta}_1 = (\boldsymbol{\pi}, \boldsymbol{\mu}_{\mathbf{x}}, \boldsymbol{\Sigma}_{\mathbf{x}}, \boldsymbol{\alpha}_{\mathbf{x}}, \boldsymbol{\beta}, \boldsymbol{\Sigma}_{\mathbf{y}}, \boldsymbol{\alpha}_{\mathbf{y}})$ and $\boldsymbol{\theta}_2 = (\boldsymbol{\eta}_{\mathbf{x}}, \boldsymbol{\eta}_{\mathbf{y}})$. Punzo and McNicholas (2017) proposed an ECM algorithm to estimate the model (7.23) by updating $\boldsymbol{\theta}_1$ and $\boldsymbol{\theta}_2$ alternately in M step.

When the contaminated Gaussian CWM is used for the detection of atypical points in each component, $\alpha_{\mathbf{x}|c}$ and $\alpha_{\mathbf{y}|c}$ represent the proportion of leverage points and outliers, respectively, in the cth component. Similar to TLE, one could require that in the cth group, $c = 1, \ldots, C$, the proportion of atypical observations, with respect to $\mathbf{x}$ and $\mathbf{y}$, respectively, to be no more than a predetermined value. However, pre-specifying points as outliers and/or leverage a priori may be difficult or unrealistic in many practical scenarios.

7.8   Robust mixture regression via mean-shift penalization

Note that the traditional finite mixture regression (FMR) model has the following conditional density:

$$f(y \mid \mathbf{x}, \boldsymbol{\theta}, \boldsymbol{\gamma}_i) = \sum_{c=1}^{C} \pi_c \phi(y; \mathbf{x}^\top \boldsymbol{\beta}_c, \sigma_c^2).$$

Yu et al. (2017) proposed the following robust mixture regression via a mean-shift penalization approach (RM^2) to conduct simultaneous outlier detection and robust mixture model estimation:

$$f(y_i \mid \mathbf{x}_i, \boldsymbol{\theta}, \boldsymbol{\gamma}_i) = \sum_{c=1}^{C} \pi_c \phi(y_i; \mathbf{x}_i^\top \boldsymbol{\beta}_c + \gamma_{ic}\sigma_c, \sigma_c^2), \qquad i = 1, \ldots, n, \quad (7.24)$$

where $\boldsymbol{\theta} = (\pi_1, \boldsymbol{\beta}_1, \sigma_1, \ldots, \pi_C, \boldsymbol{\beta}_C, \sigma_C)^\top$, and $\boldsymbol{\gamma}_i = (\gamma_{i1}, \ldots, \gamma_{iC})^\top$ is the defined mean-shift vector for the ith observation for $i = 1, \ldots, n$. Here, an additional mean-shift term, $\gamma_{ic}\sigma_c$, is added to each component regression to incorporate the possible outlying effect of the ith observation. The dependence of $\gamma_{ic}\sigma_c$ on the component scale σ_c is to ensure each γ_{ic} parameter is scale-free and can be interpreted as the number of standard deviations shifted from the mixture regression model. The mean-shift γ_{ic} is nonzero when the ith observation from the cth component is an outlier and zero, otherwise.

The model (7.24) is considered over-parameterized. However, because most of the observations are expected to be typical, only a small proportion of γ_{ic}'s will be nonzero, corresponding to the outliers. Therefore, promoting sparsity of γ_{ic} in estimation provides a direct way for identifying and accommodating outliers in the mixture regression model.

The model framework developed in (7.24) inherits the simplicity of the traditional normal mixture regression model and allows us to use the celebrated penalized estimation approaches (Tibshirani, 1996; Fan et al., 2001; Zou, 2006; Huang et al., 2008) for robust estimation. For a comprehensive account of the penalized regression and variable selection techniques, see, e.g., Bühlmann and Van De Geer (2011).

Let $\boldsymbol{\Gamma} = (\boldsymbol{\gamma}_1^\top, \ldots, \boldsymbol{\gamma}_n^\top)^\top$ collect all the mean-shift parameters. Yu et al. (2017) proposed estimating the unknown parameters $\boldsymbol{\theta}$ and $\boldsymbol{\Gamma}$ by maximizing the following penalized log-likelihood

$$\ell(\boldsymbol{\theta}, \boldsymbol{\Gamma}) = \sum_{i=1}^{n} \log\{\sum_{c=1}^{C} \pi_c \phi(y_i; \mathbf{x}_i^\top \boldsymbol{\beta}_c + \gamma_{ic}\sigma_c, \sigma_c^2)\} - \sum_{i=1}^{n} P_\lambda(\boldsymbol{\gamma}_i), \qquad (7.25)$$

where $P_\lambda(\cdot)$ is a penalty function chosen to induce either element-wise or vector-wise sparsity of the enclosed vector with λ being a tuning parameter that controls the degrees of penalization. There are many choices of the penalty function in (7.25). To impose element-wise sparsity, one may take $P_\lambda(\boldsymbol{\gamma}_i) = \sum_{c=1}^{C} P_\lambda(|\gamma_{ic}|)$, where $P_\lambda(|\gamma_{ic}|)$ is a penalty function to induce element-wise sparsity. Alternatively, to impose vector-wise sparsity, we can use the group ℓ_0 penalty $P_\lambda(\boldsymbol{\gamma}_i) = \lambda^2 I(\|\boldsymbol{\gamma}_i\|_2 \neq 0)/2$ or the group lasso penalty of the form $P_\lambda(\boldsymbol{\gamma}_i) = \lambda\|\boldsymbol{\gamma}_i\|_2$, where $\|\cdot\|_q$ denotes the ℓ_q norm for $q \geq 0$, and $I(\cdot)$ is the indicator function. These penalty functions penalize the ℓ_2-norm of each $\boldsymbol{\gamma}_i$ vector to promote the entire vector to be a zero vector.

For simplicity of explanation, we will describe the EM algorithm for the element-wise penalty. Some commonly used sparsity-inducing penalty functions include the ℓ_1 penalty (Donoho and Johnstone, 1994; Tibshirani, 1996),

$$P_\lambda(\gamma) = \lambda|\gamma|,$$

the ℓ_0 penalty (Antoniadis, 1997),

$$P_\lambda(\gamma) = \frac{\lambda^2}{2} I(\gamma \neq 0),$$

the Smooth Clipped Absolute Deviation penalty (Fan et al., 2001, SCAD),

$$P_\lambda(\gamma) = \begin{cases} \lambda|\gamma|, & \text{if } |\gamma| \leq \lambda; \\ -\left(\frac{\gamma^2 - 2a\lambda|\gamma| + \lambda^2}{2(a-1)}\right), & \text{if } \lambda < |\gamma| \leq a\lambda; \\ \frac{(a+1)\lambda^2}{2}, & \text{if } |\gamma| > a\lambda, \end{cases}$$

with

$$p_\lambda'(\theta) = \lambda\{I(\theta \leq \lambda) + \frac{(a\lambda - \theta)_+}{(a-1)\lambda} I(\theta > \lambda)\},$$

where a is a constant usually set to be 3.7, and the minimax concave penalty (Zhang et al., 2010, MCP)

$$p_\lambda'(\theta) = \lambda(1 - \theta/(\gamma\lambda))_+,$$

where $\gamma > 0$. Each of the above penalties corresponds to a thresholding rule, e.g., ℓ_1 penalization corresponds to a soft-thresholding rule, and ℓ_0 penalization corresponds to a hard-thresholding rule.

Yu et al. (2017) proposed an EM algorithm to maximize (7.25). Specifically, let

$$z_{ic} = \begin{cases} 1, & \text{if the } i\text{th observation is from the } c\text{th component;} \\ 0, & \text{otherwise,} \end{cases}$$

and $\mathbf{z}_i = (z_{i1}, z_{i2}, \ldots, z_{iC})$. The penalized complete log-likelihood function for the complete data $\{(\mathbf{x}_i, \mathbf{z}_i, y_i) : i = 1, 2, \ldots, n\}$ is

$$\ell_n^c(\boldsymbol{\theta}, \boldsymbol{\Gamma}) = \sum_{i=1}^{n} \sum_{c=1}^{C} z_{ic} \log \left\{ \pi_c \phi(y_i; \mathbf{x}_i^\top \boldsymbol{\beta}_c + \gamma_{ic}\sigma_c, \sigma_c^2) \right\} - \sum_{i=1}^{n} \sum_{c=1}^{C} P_\lambda(|\gamma_{ic}|).$$

(7.26)

Given the current estimates $\boldsymbol{\theta}^{(k)}$ and $\boldsymbol{\Gamma}^{(k)}$, E-step computes the conditional expectation of the penalized complete log-likelihood (7.26):

$$Q(\boldsymbol{\theta}, \boldsymbol{\Gamma} \mid \boldsymbol{\theta}^{(k)}, \boldsymbol{\Gamma}^{(k)})$$
$$= \sum_{i=1}^{n} \sum_{c=1}^{C} p_{ic}^{(k+1)} \left\{ \log \pi_c + \log \phi(y_i; \mathbf{x}_i^\top \boldsymbol{\beta}_c + \gamma_{ic}\sigma_c, \sigma_c^2) \right\} - \sum_{i=1}^{n} \sum_{c=1}^{C} P_\lambda(|\gamma_{ic}|)$$

(7.27)

where

$$p_{ic}^{(k+1)} = \mathbb{E}(z_{ic} | y_i; \boldsymbol{\theta}^{(k)}, \boldsymbol{\Gamma}^{(k)}) = \frac{\pi_c^{(k)} \phi(y_i; \mathbf{x}_i^\top \boldsymbol{\beta}_c^{(k)} + \gamma_{ic}^{(k)} \sigma_c^{(k)}, \sigma_c^{2^{(k)}})}{\sum_{c=1}^{C} \pi_c^{(k)} \phi(y_i; \mathbf{x}_i^\top \boldsymbol{\beta}_c^{(k)} + \gamma_{ic}^{(k)} \sigma_c^{(k)}, \sigma_c^{2^{(k)}})}.$$

(7.28)

In the M-step, we then maximize (7.27) with respect to $(\boldsymbol{\theta}, \boldsymbol{\Gamma})$. Specifically, in the M-step, $\boldsymbol{\theta}$ and $\boldsymbol{\Gamma}$ are alternatingly updated until convergence. For fixed $\boldsymbol{\theta}$, $\boldsymbol{\Gamma}$ is updated by maximizing

$$\sum_{i=1}^{n} \sum_{c=1}^{C} p_{ic}^{(k+1)} \log \phi(y_i; \mathbf{x}_i^\top \boldsymbol{\beta}_c + \gamma_{ic}\sigma_c, \sigma_c^2) - \sum_{i=1}^{n} \sum_{c=1}^{C} P_\lambda(|\gamma_{ic}|),$$

which is separable in γ_{ic} and is equivalent to minimizing

$$\frac{1}{2} \left(\gamma_{ic} - \frac{y_i - \mathbf{x}_i^\top \boldsymbol{\beta}_c}{\sigma_c} \right)^2 + \frac{1}{p_{ic}^{(k+1)}} P_\lambda(|\gamma_{ic}|),$$

(7.29)

for each γ_{ic}. The solution of (7.29) for using the ℓ_1 penalty or the ℓ_0 penalty is given by the following corresponding thresholding rule Θ_{soft} or Θ_{hard}, respectively:

$$\widehat{\gamma}_{ic} = \Theta_{soft}(\xi_{ic}; \lambda_{ic}^*) = \text{sgn}(\xi_{ic})(|\xi_{ic}| - \lambda_{ic}^*)_+, \qquad (7.30)$$
$$\widehat{\gamma}_{ic} = \Theta_{hard}(\xi_{ic}; \lambda_{ic}^*) = \xi_{ic} I(|\xi_{ic}| > \lambda_{ic}^*), \qquad (7.31)$$

where $\xi_{ic} = (y_i - \mathbf{x}_i^\top \boldsymbol{\beta}_c)/\sigma_c$, $a_+ = \max(a, 0)$, λ_{ic}^* is taken as $\lambda/p_{ic}^{(k+1)}$ in Θ_{soft}, and λ_{ic}^* is set as $\lambda/\sqrt{p_{ic}^{(k+1)}}$ in Θ_{hard}. See Yu et al. (2017) for more details about how to handle group penalties on $\boldsymbol{\gamma}_i$'s, such as the group ℓ_1 penalty and the group ℓ_0 penalty.

Algorithm 7.8.1 Thresholding-embedded EM algorithm (Yu et al., 2017)

Initialize $\boldsymbol{\theta}^{(0)}$ and $\boldsymbol{\Gamma}^{(0)}$. Set $k = 0$.

repeat

  E-Step: Compute $Q(\boldsymbol{\theta}, \boldsymbol{\Gamma} \mid \boldsymbol{\theta}^{(k)}, \boldsymbol{\Gamma}^{(k)})$ based on (7.27) and (7.28).

  M-Step: Compute

$$\pi_c^{(k+1)} = \frac{\sum_{i=1}^n p_{ic}^{(k+1)}}{n},$$

and obtain $(\boldsymbol{\beta}^{(k+1)}, \sigma_c^{2^{(k+1)}}, \boldsymbol{\Gamma}^{(k+1)})$ by maximizing $Q(\boldsymbol{\theta}, \boldsymbol{\Gamma}|\boldsymbol{\theta}^{(k)}, \boldsymbol{\Gamma}^{(k)})$, i.e., iterate the following steps until convergence:

$$\boldsymbol{\beta}_c = \left(\sum_{i=1}^n \mathbf{x}_i \mathbf{x}_i^\top p_{ic}^{(k+1)}\right)^{-1} \left(\sum_{i=1}^n \mathbf{x}_i p_{ic}^{(k+1)} (y_i - \gamma_{ic}\sigma_c)\right), c = 1, \ldots, C,$$

$$(\sigma_1, \ldots, \sigma_C) = \arg \max_{(\sigma_1, \ldots, \sigma_C)} \sum_{i=1}^n \sum_{c=1}^C p_{ic}^{(k+1)} \log \phi(y_i; \mathbf{x}_i^\top \boldsymbol{\beta}_c + \gamma_{ic}\sigma_c, \sigma_c^2),$$

$$\gamma_{ic} = \Theta(\xi_{ic}; \lambda_{ic}^*), i = 1, \ldots, n, c = 1, \ldots, C,$$

where Θ denotes one of the thresholding rules in (7.30–7.31) depending on the penalty form adopted.

  $k = k + 1$.

until convergence

The detailed thresholding-embedded EM algorithm (Yu et al., 2017) for any fixed tuning parameter λ is summarized as in Algorithm 7.8.1. If $\sigma_1 = \sigma_2 = \ldots = \sigma_c = \sigma$, in the algorithm, σ_cs shall be replaced by σ and can be updated in M-step by

$$\sigma = \arg \max_{\sigma > 0} \sum_{i=1}^n \sum_{c=1}^C p_{ic}^{(k+1)} \log \phi(y_i; \mathbf{x}_i^\top \boldsymbol{\beta}_c + \gamma_{ic}\sigma, \sigma^2).$$

Note that the parameter λ in (7.25) controls the number of outliers, i.e., the number of nonzero γ_{ic} values. To choose a tuning parameter λ, we can fit the model with a grid of λ values that are equally spaced on the log scale within an interval $(\lambda_{\min}, \lambda_{\max})$, where $\lambda_{\min}$ yields roughly 50% of nonzero γ_{ic} values, and $\lambda_{\max}$ corresponds to estimates where all γ_{ic} values are estimated as zero. We can then use a cross-validation method or an information criterion, such as the Bayesian information criterion (BIC), to select the tuning parameter:

$$\mathrm{BIC}(\lambda) = -\ell(\hat{\boldsymbol{\theta}}_\lambda) + \log(n)\mathrm{df}(\lambda),$$

where $\ell(\hat{\boldsymbol{\theta}}_\lambda)$ is the mixture log-likelihood function evaluated at the maximum likelihood estimate $\hat{\boldsymbol{\theta}}_\lambda$ for a given tuning parameter λ, and $df(\lambda)$ is the sum of the number of nonzero elements in $\widehat{\boldsymbol{\Gamma}}$ and the number of component parameters in the mixture model.

According to Yu et al. (2017), the trimmed likelihood estimator (TLE) can be seen as a specific instance of RM^2 when using a group ℓ_0 penalty. However, compared to TLE, RM^2 is computationally simpler and does not require subsampling, and it can employ the Bayesian information criterion (BIC) for data-adaptive selection of the trimming proportion α. Additionally, Yu et al. (2017) showed that there is a general correspondence between thresholding rules and M-estimators. For instance, the soft-thresholding rule Θ_{soft} corresponds to Huber's ψ function, while the hard-thresholding rule Θ_{hard} corresponds to the skipped mean loss, and the SCAD thresholding corresponds to a particular case of the Hampel loss. To achieve robust estimation, a redescending ψ function is generally preferred, and the use of a nonconvex penalty in RM^2 corresponds to such a function. This is due to the fact that a redescending ψ function offers a higher achievable breakdown point if the initial value has a high breakdown point. These findings provide strong justifications for the RM^2 method and shed light on its robustness properties.

7.9   Some numerical comparisons

In this section, we summarize the numerical comparisons conducted by Yu et al. (2020). Due to the limited space, not all methods are included, and we will mainly focus on those that are popular or have R packages. More specifically, the following seven methods are compared:

1. traditional MLE assuming the normal error density (MLE);

2. robust mixture regression based on t-distribution (Yao et al., 2014, Mixregt);

3. robust mixture regression based on Laplace distribution (Song et al., 2014, MixregL);

4. trimmed likelihood estimator with the percentage of trimmed data α set to be 0.1 (Neykov et al., 2007, TLE) ;

5. robust modified EM algorithm based on bisquare (Bai et al., 2012, MEM-bisquare);

6. robust mixture regression model via mean-shift penalization using the ℓ_0 penalty (Yu et al. (2017), $RM^2(\ell_0)$);

7. trimmed cluster-weighted restricted model (García-Escudero et al., 2017, CWRM).

To compare different methods, we use the mean squared errors (MSE) and absolute robust bias (RB) of the parameter estimates for each estimation method, where robust bias is defined as:

$$RB_c = \text{median}(\hat{\beta}_c) - \beta_c.$$

Example 7.9.1. We generate iid observations $\{(\mathbf{x}_i, y_i), i = 1, \ldots, n\}$ from the following model:

$$y_i = \begin{cases} 1 - x_{i1} + x_{i2} + \gamma_{i1}\sigma + \epsilon_{i1}, & \text{if } z_{i1} = 1; \\ 1 + 3x_{i1} + x_{i2} + \gamma_{i2}\sigma + \epsilon_{i2}, & \text{if } z_{i1} = 0. \end{cases}$$

where $\sigma = 1$. The first component indicator z_{i1} is generated from a Bernoulli distribution with a probability of success 0.3, the predictors x_{i1} and x_{i2} are independently generated from $N(0,1)$, and the error terms ϵ_{i1} and ϵ_{i2} have the same distribution as ϵ. We investigate seven different cases for the density of the error term ϵ, which are as follows:

Case I: $\epsilon \sim N(0,1)$ – standard normal distribution;

Case II: $\epsilon \sim t_3$ – t-distribution with degrees of freedom 3;

Case III: $\epsilon \sim t_1$ – t-distribution with degrees of freedom 1 (Cauchy distribution);

Case IV: $\epsilon \sim skew - t_3$ – skew-t distribution with degrees of freedom 3 and slant parameter 1;

Case V: $\epsilon \sim N(0,1)$ with 5% of low leverage outliers;

Case VI: $\epsilon \sim N(0,1)$ with 5% of high leverage outliers;

Case VII: $\epsilon \sim N(0,1)$ with 20% of gross outliers concentrated in a very remote position.

In Case I to IV, we set all mean-shift parameters γ_{ic} to zero. In Case V, we randomly generate 5% of nonzero mean-shift parameters from a uniform distribution between 11 and 13 for the second component and set the remaining 95% of γ_{ic} to zero. In Case VI, we introduce high-leverage outliers by replacing 5% of the observations with $x_{i1} = 20, x_{i2} = 20$ and $y_i = 100$. In Case VII, we randomly generate 20% of nonzero mean-shift parameters from a uniform distribution between 101 and 102 for the second component and set the remaining 80% of γ_{ic} to zero. For CWRM, we use 10% trimming proportion for Case I to VI and use 20% trimming proportion for Case VII.

Tables 7.2 and 7.3 display MSE(RB) of the parameter estimators for each estimation method with sample sizes of $n = 200$ and $n = 400$, respectively,

TABLE 7.2

MSE (RB) of point estimates for $n = 200$ in Example 7.9.1

TRUE	MLE	Mixregt	MixregL	TLE	MEM-bisquare	$RM^2(\ell_0)$	CWRM
\multicolumn{8}{c}{Case I: $\epsilon \sim N(0,1)$}							
$\beta_{10}:1$	0.022 (0.010)	0.055 (0.009)	0.041 (0.004)	0.102 (0.009)	0.031 (0.009)	0.022 (0.008)	0.025 (-0.007)
$\beta_{20}:1$	0.010 (0.002)	0.017 (0.017)	0.012 (0.004)	0.025 (0.011)	0.011 (0.008)	0.010 (0.003)	0.010 (0.011)
$\beta_{11}:-1$	0.023 (0.016)	0.116 (0.045)	0.039 (0.032)	0.323 (0.123)	0.081 (0.018)	0.023 (0.014)	0.116 (0.186)
$\beta_{21}:3$	0.008 (0.009)	0.015 (0.018)	0.011 (0.012)	0.016 (0.033)	0.009 (0.009)	0.008 (0.009)	0.021 (-0.050)
$\beta_{12}:1$	0.027 (0.025)	0.060 (0.013)	0.049 (0.004)	0.082 (0.005)	0.038 (0.025)	0.027 (0.024)	0.061 (-0.013)
$\beta_{22}:1$	0.009 (0.008)	0.015 (0.011)	0.012 (0.002)	0.014 (0.006)	0.009 (0.007)	0.009 (0.008)	0.025 (-0.004)
$\pi_1:0.3$	0.002 (0.000)	0.002 (0.008)	0.002 (0.003)	0.003 (0.003)	0.002 (0.000)	0.002 (0.001)	0.003 (0.003)
\multicolumn{8}{c}{Case II: $\epsilon \sim t_3$}							
$\beta_{10}:1$	5.095 (0.007)	0.079 (0.031)	0.070 (0.033)	0.109 (0.028)	0.069 (0.006)	0.035 (0.011)	0.027 (-0.007)
$\beta_{20}:1$	2.488 (0.003)	0.016 (0.004)	0.016 (0.005)	0.016 (0.002)	0.014 (0.002)	0.010 (0.005)	0.010 (0.005)
$\beta_{11}:-1$	2.216 (0.012)	0.049 (0.015)	0.058 (0.005)	0.160 (0.084)	0.120 (0.002)	0.023 (0.035)	0.055 (0.026)
$\beta_{21}:3$	1.074 (0.001)	0.018 (0.009)	0.013 (0.006)	0.017 (0.020)	0.014 (0.018)	0.013 (0.026)	0.021 (-0.007)
$\beta_{12}:1$	5.528 (0.003)	0.069 (0.007)	0.068 (0.007)	0.089 (0.034)	0.072 (0.012)	0.028 (0.013)	0.048 (-0.005)
$\beta_{22}:1$	5.678 (0.020)	0.017 (0.003)	0.019 (0.017)	0.015 (0.005)	0.014 (0.011)	0.012 (0.012)	0.090 (0.018)
$\pi_1:0.3$	0.015 (0.002)	0.003 (0.005)	0.002 (0.007)	0.003 (0.018)	0.004 (0.021)	0.002 (0.015)	0.002 (0.010)
\multicolumn{8}{c}{Case III: $\epsilon \sim t_1$}							
$\beta_{10}:1$	7.2e+4 (0.081)	0.175 (0.018)	0.115 (0.057)	1.733 (0.027)	0.288 (0.027)	0.037 (0.010)	0.138 (0.014)
$\beta_{20}:1$	2.7e+5 (0.026)	0.020 (0.019)	0.045 (0.067)	1.711 (0.013)	0.038 (0.028)	0.018 (0.011)	0.039 (0.020)
$\beta_{11}:-1$	1.9e+5 (1.682)	0.251 (0.033)	4.504 (0.486)	1.141 (0.055)	0.657 (0.034)	0.032 (0.019)	0.163 (0.208)
$\beta_{21}:3$	1.1e+5 (1.241)	0.020 (0.003)	0.166 (0.271)	0.185 (0.006)	0.078 (0.031)	0.019 (0.006)	0.095 (0.229)
$\beta_{12}:1$	4.0e+4 (0.044)	0.142 (0.016)	0.080 (0.035)	1.178 (0.013)	0.279 (0.052)	0.033 (0.021)	0.170 (-0.017)
$\beta_{22}:1$	2.5e+5 (0.047)	0.023 (0.010)	0.030 (0.002)	0.220 (0.049)	0.066 (0.023)	0.020 (0.003)	0.154 (0.083)
$\pi_1:0.3$	0.253 (0.199)	0.003 (0.002)	0.019 (0.045)	0.014 (0.011)	0.008 (0.042)	0.003 (0.020)	0.004 (0.043)
\multicolumn{8}{c}{Case IV: $\epsilon \sim skew-t_3$}							
$\beta_{10}:1$	3.158 (0.839)	0.372 (0.556)	0.435 (0.614)	0.389 (0.581)	0.482 (0.670)	0.605 (0.740)	0.250 (0.452)
$\beta_{20}:1$	4.668 (0.722)	0.315 (0.559)	0.361 (0.599)	0.333 (0.578)	0.381 (0.605)	0.483 (0.679)	0.312 (0.540)
$\beta_{11}:-1$	0.904 (-0.306)	0.069 (-0.026)	0.303 (-0.021)	0.076 (0.049)	0.091 (-0.015)	0.030 (-0.031)	0.078 (-0.103)
$\beta_{21}:3$	2.282 (0.006)	0.014 (0.021)	0.051 (0.012)	0.034 (-0.047)	0.103 (0.050)	0.014 (0.013)	0.016 (0.015)
$\beta_{12}:1$	4.127 (0.018)	0.061 (-0.034)	0.152 (-0.083)	0.051 (-0.021)	0.124 (-0.090)	0.053 (0.021)	0.052 (-0.033)
$\beta_{22}:1$	8.475 (0.005)	0.073 (0.047)	0.114 (-0.101)	0.073 (-0.082)	0.011 (0.012)	0.023 (0.013)	0.089 (0.054)
$\pi_1:0.3$	0.011 (-0.010)	0.007 (-0.010)	0.002 (-0.002)	0.006 (-0.010)	0.005 (0.010)	0.002 (-0.005)	0.002 (-0.002)
\multicolumn{8}{c}{Case V: $\epsilon \sim N(0,1)$ with 5% of low leverage outliers}							
$\beta_{10}:1$	2.866 (0.058)	0.052 (0.058)	1.401 (0.940)	0.049 (0.018)	0.027 (0.022)	0.037 (0.038)	0.023 (0.019)
$\beta_{20}:1$	119.7 (11.54)	0.015 (0.010)	0.936 (0.053)	0.011 (0.007)	0.010 (0.005)	0.009 (0.003)	0.009 (0.010)
$\beta_{11}:-1$	7.891 (0.890)	0.043 (0.025)	2.771 (2.033)	0.046 (0.074)	0.025 (0.015)	0.039 (0.031)	0.057 (0.074)
$\beta_{21}:3$	0.850 (0.585)	0.014 (0.011)	0.273 (0.293)	0.011 (0.020)	0.009 (0.011)	0.009 (0.003)	0.021 (0.007)
$\beta_{12}:1$	0.510 (0.006)	0.060 (0.027)	2.334 (1.985)	0.057 (0.002)	0.034 (0.007)	0.043 (0.010)	0.046 (0.001)
$\beta_{22}:1$	0.368 (0.005)	0.017 (0.006)	0.572 (0.017)	0.015 (0.014)	0.011 (0.015)	0.011 (0.011)	0.097 (0.055)
$\pi_1:0.3$	0.362 (0.635)	0.003 (0.015)	0.012 (0.016)	0.003 (0.020)	0.003 (0.033)	0.002 (0.026)	0.003 (0.013)
\multicolumn{8}{c}{Case VI: $\epsilon \sim N(0,1)$ with 5% of high leverage outliers}							
$\beta_{10}:1$	0.090 (0.002)	0.064 (-0.017)	0.037 (0.015)	0.079 (-0.012)	0.044 (0.003)	0.027 (-0.006)	0.020 (-0.012)
$\beta_{20}:1$	0.015 (0.018)	0.040 (0.025)	0.020 (0.033)	0.037 (0.021)	0.013 (0.015)	0.010 (0.003)	0.009 (-0.007)
$\beta_{11}:-1$	3.037 (1.830)	0.620 (0.069)	0.031 (0.056)	1.421 (0.130)	0.146 (0.023)	0.022 (-0.010)	0.071 (0.070)
$\beta_{21}:3$	0.181 (0.416)	0.187 (0.422)	0.190 (0.422)	0.123 (0.325)	0.087 (0.046)	0.006 (-0.003)	0.018 (-0.038)
$\beta_{12}:1$	1.037 (0.550)	0.111 (-0.026)	0.050 (-0.023)	0.130 (-0.036)	0.111 (-0.009)	0.034 (-0.015)	0.053 (0.019)
$\beta_{22}:1$	0.276 (0.517)	0.293 (0.522)	0.282 (0.531)	0.201 (0.445)	0.106 (0.107)	0.006 (0.009)	0.110 (-0.015)
$\pi_1:0.3$	0.008 (0.023)	0.006 (0.025)	0.003 (0.022)	0.008 (0.033)	0.003 (-0.002)	0.002 (-0.005)	0.002 (0.002)
\multicolumn{8}{c}{Case VII: 20% gross outliers}							
$\beta_{10}:1$	2.125 (-0.010)	117.4 (0.039)	147.3 (0.030)	154.2 (0.008)	13.26 (-0.016)	0.028 (0.004)	0.019 (-0.005)
$\beta_{20}:1$	1.0e+4 (101.5)	305.4 (0.005)	336.3 (6.290)	154.5 (-0.008)	259.0 (0.013)	0.010 (-0.007)	0.010 (-0.003)
$\beta_{11}:-1$	6.389 (2.541)	17.72 (0.075)	38.65 (3.278)	0.483 (0.005)	14.81 (2.571)	0.023 (-0.006)	0.040 (0.015)
$\beta_{21}:3$	0.050 (-0.006)	12.70 (-0.070)	187.5 (9.991)	0.013 (-0.004)	13.57 (-1.277)	0.014 (-0.036)	0.024 (-0.017)
$\beta_{12}:1$	0.037 (-0.007)	20.69 (0.001)	506.2 (0.009)	0.025 (0.006)	0.045 (0.002)	0.025 (-0.002)	0.040 (-0.010)
$\beta_{22}:1$	0.031 (0.034)	46.75 (-0.019)	994.8 (30.15)	0.016 (-0.018)	9.045 (0.002)	0.015 (-0.015)	0.108 (-0.010)
$\pi_1:0.3$	0.249 (0.500)	0.023 (0.083)	0.184 (0.470)	0.012 (0.076)	0.044 (0.200)	0.002 (-0.008)	0.005 (0.050)

based on 200 replicates. We can see that the MLE works the best for Case I ($\epsilon \sim N(0,1)$), but fails to provide reasonable estimators for Case II to VI. In addition, $RM^2(\ell_0)$ has the best performance overall. CWRM and TLE also work well if the trimming proportion is properly chosen. For the rest

TABLE 7.3
MSE (RB) of point estimates for $n = 400$ in Example 7.9.1

TRUE	MLE	Mixregt	MixregL	TLE	MEM-bisquare	$\mathrm{RM}^2(\ell_0)$	CWRM
			Case I: $\epsilon \sim N(0,1)$				
$\beta_{10}:1$	0.012 (0.008)	0.026 (0.001)	0.025 (0.038)	0.032 (0.005)	0.013 (0.002)	0.012 (0.007)	0.014 (0.017)
$\beta_{20}:1$	0.004 (0.006)	0.007 (0.001)	0.006 (0.023)	0.006 (0.007)	0.005 (0.009)	0.004 (0.006)	0.005 (0.002)
$\beta_{11}:-1$	0.010 (0.003)	0.020 (0.017)	0.013 (0.023)	0.040 (0.095)	0.011 (0.006)	0.010 (0.003)	0.098 (0.142)
$\beta_{21}:3$	0.004 (0.004)	0.007 (0.009)	0.007 (0.004)	0.008 (0.009)	0.005 (0.003)	0.004 (0.002)	0.012 (0.057)
$\beta_{12}:1$	0.015 (0.010)	0.029 (0.019)	0.024 (0.031)	0.032 (0.028)	0.016 (0.011)	0.015 (0.010)	0.042 (0.017)
$\beta_{22}:1$	0.005 (0.013)	0.007 (0.018)	0.007 (0.002)	0.007 (0.017)	0.005 (0.012)	0.005 (0.013)	0.009 (-0.002)
$\pi_1:0.3$	0.001 (0.001)	0.001 (0.012)	0.001 (0.002)	0.001 (0.010)	0.001 (0.001)	0.001 (0.001)	0.003 (-0.035)
			Case II: $\epsilon \sim t_3$				
$\beta_{10}:1$	8.523 (0.009)	0.026 (0.001)	0.023 (0.009)	0.031 (0.002)	0.024 (0.003)	0.015 (0.003)	0.015 (0.019)
$\beta_{20}:1$	0.097 (0.014)	0.008 (0.008)	0.009 (0.002)	0.008 (0.009)	0.007 (0.003)	0.010 (0.001)	0.005 (0.002)
$\beta_{11}:-1$	0.603 (0.006)	0.029 (0.008)	0.021 (0.008)	0.036 (0.078)	0.020 (0.005)	0.016 (0.038)	0.041 (0.092)
$\beta_{21}:3$	0.832 (0.002)	0.008 (0.009)	0.008 (0.006)	0.008 (0.022)	0.008 (0.016)	0.007 (0.013)	0.012 (-0.022)
$\beta_{12}:1$	4.268 (0.030)	0.032 (0.012)	0.031 (0.020)	0.034 (0.003)	0.027 (0.004)	0.016 (0.007)	0.028 (-0.014)
$\beta_{22}:1$	1.959 (0.004)	0.007 (0.012)	0.009 (0.010)	0.008 (0.003)	0.007 (0.001)	0.006 (0.009)	0.045 (0.004)
$\pi_1:0.3$	0.005 (0.013)	0.001 (0.006)	0.001 (0.003)	0.001 (0.004)	0.001 (0.013)	0.001 (0.010)	0.001 (0.003)
			Case III: $\epsilon \sim t_1$				
$\beta_{10}:1$	5.1e+4 (0.032)	0.040 (0.006)	0.037 (0.024)	0.088 (0.053)	0.055 (0.027)	0.017 (0.004)	0.073 (0.015)
$\beta_{20}:1$	8.5e+5 (0.141)	0.007 (0.014)	0.010 (0.029)	0.017 (0.010)	0.012 (0.001)	0.008 (0.011)	0.016 (0.005)
$\beta_{11}:-1$	2.7e+4 (1.171)	0.079 (0.056)	4.413 (0.421)	0.066 (0.024)	0.421 (0.024)	0.017 (0.049)	0.127 (0.225)
$\beta_{21}:3$	7.7e+5 (0.651)	0.008 (0.015)	0.207 (0.270)	0.020 (0.038)	0.071 (0.045)	0.010 (0.015)	0.059 (0.178)
$\beta_{12}:1$	1.7e+5 (0.112)	0.068 (0.020)	0.075 (0.034)	0.130 (0.032)	0.144 (0.020)	0.021 (0.021)	0.087 (0.007)
$\beta_{22}:1$	6.3e+5 (0.015)	0.010 (0.023)	0.013 (0.039)	0.021 (0.006)	0.014 (0.007)	0.008 (0.009)	0.029 (0.010)
$\pi_1:0.3$	0.231 (0.054)	0.001 (0.002)	0.022 (0.069)	0.002 (0.015)	0.005 (0.047)	0.001 (0.023)	0.003 (0.042)
			Case IV: $\epsilon \sim skew - t_3$				
$\beta_{10}:1$	10.17 (0.815)	0.331 (0.554)	0.377 (0.574)	0.338 (0.564)	0.437 (0.644)	0.582 (0.736)	0.265 (0.494)
$\beta_{20}:1$	4.947 (0.728)	0.302 (0.537)	0.348 (0.580)	0.322 (0.565)	0.377 (0.611)	0.490 (0.689)	0.317 (0.561)
$\beta_{11}:-1$	1.332 (-0.007)	0.053 (0.026)	0.056 (0.013)	0.017 (0.053)	0.072 (-0.028)	0.017 (-0.021)	0.059 (0.098)
$\beta_{21}:3$	1.245 (0.016)	0.005 (0.001)	0.036 (0.019)	0.106 (0.083)	0.015 (0.011)	0.007 (0.023)	0.011 (-0.034)
$\beta_{12}:1$	15.11 (-0.006)	0.021 (-0.003)	0.024 (-0.017)	0.020 (-0.017)	0.017 (-0.012)	0.025 (0.007)	0.045 (-0.001)
$\beta_{22}:1$	4.071 (0.021)	0.005 (0.019)	0.077 (-0.014)	0.065 (0.036)	0.024 (0.009)	0.006 (0.014)	0.061 (-0.029)
$\pi_1:0.3$	0.007 (-0.005)	0.001 (-0.002)	0.001 (0.004)	0.001 (0.007)	0.001 (0.016)	000.1 (0.006)	0.001 (-0.011)
			Case V: $\epsilon \sim N(0,1)$ with 5% of low leverage outliers				
$\beta_{10}:1$	0.450 (0.076)	0.024 (0.027)	1.709 (1.042)	0.023 (0.019)	0.015 (0.001)	0.017 (0.019)	0.011 (0.003)
$\beta_{20}:1$	131.5 (11.59)	0.008 (0.006)	0.489 (0.003)	0.007 (0.002)	0.005 (0.004)	0.007 (0.006)	0.004 (0.001)
$\beta_{11}:-1$	8.385 (2.916)	0.016 (0.027)	2.434 (2.195)	0.021 (0.073)	0.011 (0.021)	0.052 (0.030)	0.032 (0.048)
$\beta_{21}:3$	0.635 (0.644)	0.008 (0.011)	0.374 (0.301)	0.006 (0.018)	0.005 (0.003)	0.004 (0.001)	0.010 (0.021)
$\beta_{12}:1$	0.024 (0.014)	0.021 (0.007)	2.575 (1.551)	0.021 (0.023)	0.014 (0.004)	0.017 (0.010)	0.025 (0.005)
$\beta_{22}:1$	0.190 (0.001)	0.009 (0.006)	0.779 (0.060)	0.008 (0.006)	0.006 (0.007)	0.005 (0.005)	0.047 (0.014)
$\pi_1:0.3$	0.397 (0.638)	0.001 (0.013)	0.012 (0.017)	0.001 (0.017)	0.001 (0.028)	0.003 (0.022)	0.001 (0.010)
			Case VI: $\epsilon \sim N(0,1)$ with 5% of high leverage outliers				
$\beta_{10}:1$	0.081 (-0.041)	0.022 (0.032)	0.015 (0.006)	0.023 (0.022)	0.011 (0.016)	0.010 (0.005)	0.011 (0.002)
$\beta_{20}:1$	0.007 (0.023)	0.014 (0.032)	0.010 (0.035)	0.011 (0.012)	0.006 (0.009)	0.004 (0.002)	0.005 (0.005)
$\beta_{11}:-1$	2.016 (1.730)	0.020 (0.046)	0.014 (0.069)	0.297 (0.082)	0.073 (0.012)	0.009 (-0.003)	0.029 (0.074)
$\beta_{21}:3$	0.178 (0.419)	0.195 (0.432)	0.132 (0.410)	0.124 (0.365)	0.074 (0.036)	0.003 (0.002)	0.008 (0.029)
$\beta_{12}:1$	0.450 (0.480)	0.024 (0.006)	0.013 (0.005)	0.028 (0.003)	0.013 (-0.001)	0.012 (0.004)	0.029 (0.001)
$\beta_{22}:1$	0.971 (0.517)	0.268 (0.515)	0.216 (0.537)	0.168 (0.433)	0.107 (0.061)	0.003 (-0.001)	0.044 (0.014)
$\pi_1:0.3$	0.001 (0.023)	0.001 (0.025)	0.001 (0.022)	0.003 (0.025)	0.002 (0.002)	0.001 (-0.006)	0.001 (0.002)
			Case VII: 20% gross outliers				
$\beta_{10}:1$	0.015 (-0.011)	34.70 (0.017)	18.39 (0.038)	0.013 (0.009)	35.96 (-0.011)	0.014 (0.006)	0.019 (0.004)
$\beta_{20}:1$	1.0e+4 (101.5)	775.6 (0.018)	41.10 (12.81)	257.1 (-0.003)	123.1 (-0.011)	0.007 (-0.001)	0.007 (-0.004)
$\beta_{11}:-1$	6.395 (2.532)	16.17 (0.069)	2.934 (3.348)	0.374 (-0.015)	10.38 (2.623)	0.010 (-0.020)	0.018 (-0.008)
$\beta_{21}:3$	0.014 (0.015)	5.582 (-0.049)	16.28 (10.02)	0.005 (0.002)	2.171 (-1.274)	0.006 (-0.032)	0.009 (0.004)
$\beta_{12}:1$	0.018 (0.014)	1.929 (0.015)	45.93 (0.043)	0.015 (0.012)	6.839 (0.000)	0.016 (0.006)	0.021 (0.012)
$\beta_{22}:1$	0.013 (-0.005)	8.953 (-0.001)	118.8 (30.27)	0.007 (-0.006)	6.701 (0.017)	0.007 (-0.009)	0.046 (-0.014)
$\pi_1:0.3$	0.250 (0.500)	0.031 (0.067)	0.026 (0.472)	0.012 (0.071)	0.038 (0.200)	0.001 (-0.015)	0.005 (0.063)

of the robust methods, none of them has a clear advantage over the others. Furthermore, all methods broke down except for $\mathrm{RM}^2(\ell_0)$ and CWRM in Case VII where there are 20% gross outliers in the simulated data. □

Example 7.9.2. We generate independent and identically distributed (i.i.d.) data $\{(x_i, y_i), i = 1, \ldots, n\}$ from the following model:

$$
y_i = \begin{cases}
1 + x_{i1} + \gamma_{i1}\sigma + \epsilon_{i1}, & \text{if } z_{i1} = 1; \\
1 + 2x_{i1} + \gamma_{i2}\sigma + \epsilon_{i2}, & \text{if } z_{i2} = 2; \\
1 + 3x_{i1} + \gamma_{i3}\sigma + \epsilon_{i3}, & \text{if } z_{i3} = 3;
\end{cases}
$$

where $P(z_{i1} = 1) = P(z_{i2} = 2) = 0.3, P(z_{i3} = 3) = 0.4$, and $\sigma = 1$. The predictor x_{i1} is independently generated from $N(0, 1)$. It is worth noting that in this particular example, all three components exhibit the same sign of slopes and are located close to one another. To further explore the robustness of our approach, we consider the following seven cases for the component error densities:

Case I: ϵ_1, ϵ_2, and ϵ_3 have the same distribution from $N(0, 1)$;

Case II: $\epsilon_1 \sim t_9$, $\epsilon_2 \sim t_6$ and $\epsilon_3 \sim t_3$;

Case III: $\epsilon_1 \sim N(0, 1)$, $\epsilon_2 \sim N(0, 1)$, and $\epsilon_3 \sim t_3$;

Case IV: $\epsilon_1 \sim N(0, 1)$, $\epsilon_2 \sim N(0, 1)$, and $\epsilon_3 \sim skew - t_3$;

Case V: ϵ_1, ϵ_2, and ϵ_3 have the same distribution from $N(0, 1)$ with 5% of low leverage outliers;

Case VI: ϵ_1, ϵ_2, and ϵ_3 have the same distribution from $N(0, 1)$ with 5% of high leverage outliers;

Case VII: ϵ_1, ϵ_2, and ϵ_3 have the same distribution from $N(0, 1)$ with 20% of gross outliers where outliers are concentrated in a very remote position.

In Case V, we randomly generate 5% of nonzero mean-shift parameters for the third component from a uniform distribution between 11 and 13, while setting all other γ_{ic}'s to zero. In Case VI, we introduce high-leverage outliers by replacing 5% of the observations with $x_{i1} = 20$ and $y_i = 200$. In Case VII, we randomly generate 20% of nonzero mean-shift parameters for the third component from a uniform distribution between 101 and 102, while setting the remaining 80% of γ_{ic}'s to zero. The trimming proportion for CWRM is the same as those used in Example 7.9.1.

Tables 7.4 and 7.5 report MSE(RB) of the parameter estimators with sample sizes $n = 200$ and $n = 400$, respectively. The findings are similar to Example 7.9.1.

□

TABLE 7.4
MSE (RB) of point estimates for $n = 200$ in Example 7.9.2

TRUE	MLE	Mixregt	MixregL	TLE	MEM-bisquare	$RM^2(\ell_0)$	CWRM
	Case I: $\epsilon_1 \sim N(0,1), \epsilon_2 \sim N(0,1)$, and $\epsilon_3 \sim N(0,1)$						
$\beta_{10} : 1$	0.092 (0.092)	0.599 (0.071)	0.395 (0.021)	0.663 (0.167)	0.537 (0.129)	0.137 (0.066)	0.129 (0.017)
$\beta_{20} : 1$	0.092 (0.003)	0.640 (0.091)	0.601 (0.014)	0.598 (0.155)	0.813 (0.170)	0.100 (0.003)	0.114 (0.004)
$\beta_{30} : 1$	0.041 (0.017)	0.390 (0.079)	0.324 (0.003)	0.493 (0.115)	0.415 (0.064)	0.140 (0.016)	0.421 (-0.012)
$\beta_{11} : 1$	0.023 (0.005)	0.225 (0.357)	0.171 (0.148)	0.286 (0.489)	0.153 (0.048)	0.095 (0.040)	0.263 (0.448)
$\beta_{21} : 2$	0.017 (0.016)	0.120 (0.279)	0.134 (0.253)	0.106 (0.261)	0.154 (0.304)	0.060 (0.167)	0.054 (-0.019)
$\beta_{31} : 3$	0.183 (0.089)	1.044 (1.028)	0.799 (0.897)	1.087 (1.055)	0.753 (0.857)	0.660 (0.820)	0.980 (0.985)
$\pi_1 : 0.3$	0.008 (0.009)	0.009 (0.015)	0.013 (0.021)	0.007 (0.002)	0.014 (0.052)	0.007 (0.028)	0.010 (0.035)
	Case II: $\epsilon_1 \sim t_9, \epsilon_2 \sim t_6$, and $\epsilon_3 \sim t_3$						
$\beta_{10} : 1$	14.66 (0.006)	0.808 (0.037)	0.390 (0.038)	0.855 (0.012)	1.028 (0.010)	0.398 (0.005)	0.528 (0.021)
$\beta_{20} : 1$	2.195 (0.021)	0.168 (0.014)	0.164 (0.046)	0.092 (0.017)	0.124 (0.020)	0.097 (0.001)	0.115 (0.013)
$\beta_{30} : 1$	10.63 (0.076)	0.597 (0.063)	0.352 (0.076)	0.679 (0.023)	0.867 (0.038)	0.527 (0.012)	0.823 (0.016)
$\beta_{11} : 1$	13.46 (0.205)	0.266 (0.302)	0.197 (0.064)	0.278 (0.405)	0.268 (0.008)	0.241 (0.045)	0.121 (0.231)
$\beta_{21} : 2$	0.365 (0.245)	0.158 (0.261)	0.183 (0.291)	0.134 (0.267)	0.216 (0.274)	0.065 (0.170)	0.053 (0.155)
$\beta_{31} : 3$	8.169 (0.529)	0.956 (0.990)	0.670 (0.771)	0.989 (1.024)	0.692 (0.729)	0.805 (0.776)	0.800 (0.951)
$\pi_1 : 0.3$	0.076 (0.166)	0.012 (0.027)	0.017 (0.018)	0.009 (0.001)	0.016 (0.060)	0.013 (0.031)	0.010 (0.048)
	Case III: $\epsilon_1 \sim N(0,1)$, $\epsilon_2 \sim N(0,1)$, and $\epsilon_3 \sim t_3$						
$\beta_{10} : 1$	18.13 (0.015)	0.652 (0.010)	0.295 (0.063)	0.669 (0.034)	0.631 (0.015)	0.195 (0.005)	0.425 (0.034)
$\beta_{20} : 1$	2.889 (0.027)	0.712 (0.184)	0.410 (0.044)	0.708 (0.218)	1.021 (0.035)	0.086 (0.008)	0.015 (0.004)
$\beta_{30} : 1$	16.56 (0.003)	0.470 (0.017)	0.375 (0.027)	0.507 (0.021)	0.578 (0.035)	0.852 (0.001)	0.122 (-0.009)
$\beta_{11} : 1$	78.58 (0.159)	0.189 (0.299)	0.142 (0.049)	0.207 (0.325)	0.173 (0.017)	0.126 (0.041)	0.161 (0.323)
$\beta_{21} : 2$	0.553 (0.245)	0.162 (0.293)	0.153 (0.264)	0.169 (0.292)	0.181 (0.291)	0.064 (0.167)	0.065 (-0.191)
$\beta_{31} : 3$	19.42 (0.479)	0.929 (0.981)	0.698 (0.792)	0.935 (0.965)	0.701 (0.761)	0.765 (0.694)	0.885 (-0.959)
$\pi_1 : 0.3$	0.068 (0.115)	0.011 (0.009)	0.015 (0.038)	0.009 (0.001)	0.022 (0.031)	0.011 (0.042)	0.005 (0.052)
	Case IV: $\epsilon_1 \sim N(0,1)$, $\epsilon_2 \sim N(0,1)$ and $\epsilon_3 \sim skew - t_3$						
$\beta_{10} : 1$	46.23 (0.211)	0.627 (-0.051)	0.408 (0.075)	0.571 (-0.025)	0.745 (-0.018)	0.477 (0.157)	0.659 (0.146)
$\beta_{20} : 1$	1.801 (0.285)	0.725 (0.348)	0.488 (0.337)	0.771 (0.235)	1.058 (0.285)	0.384 (0.232)	0.587 (0.273)
$\beta_{30} : 1$	24.01 (0.495)	0.524 (0.341)	0.489 (0.402)	0.499 (0.344)	0.838 (0.391)	1.161 (0.488)	0.101 (0.298)
$\beta_{11} : 1$	33.80 (-0.135)	0.186 (0.204)	0.248 (0.025)	0.185 (0.208)	0.163 (-0.050)	0.198 (-0.018)	0.185 (0.349)
$\beta_{21} : 2$	0.525 (-0.216)	0.161 (-0.273)	0.144 (-0.285)	0.153 (-0.267)	0.239 (-0.322)	0.073 (-0.154)	0.050 (-0.160)
$\beta_{31} : 3$	22.45 (-0.537)	0.940 (-0.964)	0.710 (-0.798)	0.967 (-0.974)	0.725 (-0.798)	0.625 (-0.747)	0.963 (-0.978)
$\pi_1 : 0.3$	0.059 (-0.071)	0.010 (-0.023)	0.015 (-0.021)	0.008 (-0.030)	0.018 (-0.042)	0.012 (0.026)	0.009 (0.043)
	Case V: $\epsilon_1, \epsilon_2, \epsilon_3, \sim N(0,1)$ with 5% of low leverage outliers						
$\beta_{10} : 1$	17.02 (0.103)	1.208 (0.150)	0.352 (0.502)	0.374 (0.231)	11.17 (0.209)	0.348 (0.231)	0.272 (0.482)
$\beta_{20} : 1$	78.27 (1.234)	1.305 (0.951)	13.21 (0.974)	1.354 (0.903)	6.120 (0.486)	1.003 (0.959)	0.683 (0.822)
$\beta_{30} : 1$	74.59 (1.205)	2.443 (1.203)	0.945 (0.946)	1.672 (1.167)	9.435 (0.960)	1.490 (1.031)	0.804 (0.903)
$\beta_{11} : 1$	0.161 (0.127)	0.194 (0.045)	0.197 (0.353)	0.138 (0.105)	6.923 (0.211)	0.535 (0.023)	0.189 (0.353)
$\beta_{21} : 2$	0.100 (0.222)	0.111 (0.183)	0.069 (0.216)	0.125 (0.231)	0.505 (0.563)	0.084 (0.153)	0.033 (-0.123)
$\beta_{31} : 3$	0.847 (0.924)	0.897 (0.948)	1.036 (1.069)	0.941 (0.970)	3.014 (0.954)	1.015 (0.855)	0.983 (-0.979)
$\pi_1 : 0.3$	0.022 (0.036)	0.008 (0.037)	0.012 (0.003)	0.008 (0.051)	0.032 (0.124)	0.009 (0.008)	0.003 (0.047)
	Case VI: $\epsilon_1, \epsilon_2, \epsilon_3, \sim N(0,1)$ with 5% of high leverage outliers						
$\beta_{10} : 1$	0.098 (-0.037)	0.262 (0.053)	0.125 (0.033)	0.381 (-0.211)	0.149 (-0.073)	0.784 (0.856)	0.058 (0.098)
$\beta_{20} : 1$	1.222 (1.063)	1.434 (1.203)	1.422 (1.122)	1.282 (0.710)	1.244 (1.020)	0.617 (0.778)	1.253 (1.064)
$\beta_{30} : 1$	15.29 (0.850)	10.45 (0.177)	10.85 (0.659)	1.720 (1.300)	7.065 (1.109)	0.337 (0.476)	1.374 (1.161)
$\beta_{11} : 1$	0.091 (-0.001)	0.261 (0.455)	0.165 (0.226)	0.164 (0.155)	0.140 (-0.009)	0.492 (0.694)	0.073 (0.136)
$\beta_{21} : 2$	0.026 (0.016)	0.053 (-0.095)	0.040 (-0.051)	0.136 (-0.257)	0.078 (-0.112)	0.102 (-0.301)	0.026 (-0.055)
$\beta_{31} : 3$	67.87 (6.907)	48.07 (6.941)	48.03 (6.917)	0.981 (-1.000)	32.53 (6.793)	1.676 (-1.292)	0.257 (-0.454)
$\pi_1 : 0.3$	0.009 (0.037)	0.022 (0.116)	0.022 (0.086)	0.008 (-0.390)	0.014 (0.028)	0.013 (-0.113)	0.012 (0.095)
	Case VII: 20% gross outliers						
$\beta_{10} : 1$	8.937 (0.175)	317.9 (0.227)	0.026 (0.078)	373.7 (-0.011)	0.397 (0.629)	0.080 (0.072)	0.111 (0.287)
$\beta_{20} : 1$	4525 (1.195)	892.7 (0.765)	472.5 (1.429)	420.5 (0.774)	52.96 (0.630)	0.629 (0.796)	0.602 (0.768)
$\beta_{30} : 1$	5787 (102.3)	1069 (0.945)	577.7 (102.3)	735.8 (1.186)	0.404 (0.632)	0.855 (0.888)	0.797 (0.883)
$\beta_{11} : 1$	0.171 (0.172)	68.57 (0.077)	0.011 (0.041)	18.76 (-0.012)	0.397 (0.618)	0.076 (-0.064)	0.049 (0.124)
$\beta_{21} : 2$	0.076 (-0.182)	0.207 (-0.308)	0.005 (-0.087)	0.253 (-0.363)	0.158 (-0.379)	0.086 (-0.144)	0.063 (-0.167)
$\beta_{31} : 3$	0.914 (-0.937)	26.23 (-0.934)	0.082 (-0.913)	0.841 (-0.907)	1.906 (-1.378)	1.060 (-1.041)	0.975 (-0.987)
$\pi_1 : 0.3$	0.011 (0.062)	0.011 (0.021)	0.001 (0.052)	0.020 (-0.017)	0.001 (0.033)	0.009 (-0.033)	0.007 (0.064)

TABLE 7.5
MSE (RB) of point estimates for $n = 400$ in Example 7.9.2

TRUE	MLE	Mixregt	MixregL	TLE	MEM-bisquare	$RM^2(\ell_0)$	CWRM
		Case I: $\epsilon_1 \sim N(0,1), \epsilon_2 \sim N(0,1)$, and $\epsilon_3 \sim N(0,1)$					
$\beta_{10} : 1$	0.038 (0.005)	0.064 (0.021)	0.056 (0.020)	0.070 (0.031)	0.043 (0.080)	0.059 (0.025)	0.085 (0.010)
$\beta_{20} : 1$	0.064 (0.023)	0.056 (0.064)	0.041 (0.038)	0.054 (0.004)	0.072 (0.021)	0.053 (0.012)	0.068 (-0.014)
$\beta_{30} : 1$	0.026 (0.007)	0.047 (0.113)	0.045 (0.081)	0.049 (0.011)	0.275 (0.045)	0.046 (0.017)	0.051 (-0.014)
$\beta_{11} : 1$	0.015 (0.007)	0.275 (0.052)	0.163 (0.155)	0.033 (0.058)	0.116 (0.017)	0.039 (0.004)	0.237 (0.448)
$\beta_{21} : 2$	0.018 (0.030)	0.095 (0.247)	0.107 (0.100)	0.094 (0.236)	0.159 (0.298)	0.034 (0.143)	0.041 (0.163)
$\beta_{31} : 3$	0.827 (0.094)	1.140 (1.082)	0.985 (0.891)	1.152 (1.093)	0.827 (0.919)	0.731 (0.851)	1.015 (0.998)
$\pi_1 : 0.3$	0.007 (0.008)	0.008 (0.013)	0.010 (0.012)	0.006 (0.009)	0.014 (0.068)	0.005 (0.017)	0.005 (0.035)
		Case II: $\epsilon_1 \sim t_9, \epsilon_2 \sim t_6$, and $\epsilon_3 \sim t_3$					
$\beta_{10} : 1$	58.11 (0.001)	0.685 (0.052)	0.208 (0.017)	0.676 (0.040)	0.527 (0.032)	0.116 (0.003)	0.617 (0.011)
$\beta_{20} : 1$	30867 (0.001)	0.680 (0.132)	0.362 (0.013)	0.826 (0.036)	1.511 (0.044)	0.028 (0.002)	0.108 (0.009)
$\beta_{30} : 1$	26.91 (0.024)	0.421 (0.061)	0.184 (0.030)	0.413 (0.033)	0.574 (0.027)	0.247 (0.081)	0.012 ()0.011
$\beta_{11} : 1$	136.9 (0.131)	0.225 (0.341)	0.103 (0.002)	0.201 (0.297)	0.125 (0.022)	0.165 (0.132)	0.104 (0.249)
$\beta_{21} : 2$	0.426 (0.240)	0.151 (0.285)	0.174 (0.285)	0.160 (0.302)	0.195 (0.301)	0.058 (0.188)	0.039 (0.126)
$\beta_{31} : 3$	20.12 (0.177)	1.047 (1.025)	0.735 (0.868)	1.050 (1.017)	0.758 (0.893)	0.706 (0.807)	0.863 (0.987)
$\pi_1 : 0.3$	0.094 (0.148)	0.011 (0.019)	0.014 (0.031)	0.008 (0.023)	0.016 (0.036)	0.008 (0.044)	0.003 (0.044)
		Case III: $\epsilon_1 \sim N(0,1), \epsilon_2 \sim N(0,1)$, and $\epsilon_3 \sim t_3$					
$\beta_{10} : 1$	19.10 (0.008)	0.679 (0.039)	0.325 (0.019)	0.627 (0.060)	0.331 (0.023)	0.702 (0.044)	0.725 (0.016)
$\beta_{20} : 1$	5.947 (0.005)	0.640 (0.172)	0.458 (0.140)	0.720 (0.169)	0.760 (0.026)	0.024 (0.008)	0.012 (0.005)
$\beta_{30} : 1$	13.48 (0.016)	0.435 (0.077)	0.236 (0.165)	0.407 (0.085)	0.686 (0.012)	0.418 (0.047)	0.913 (0.012)
$\beta_{11} : 1$	14.19 (0.170)	0.226 (0.426)	0.125 (0.146)	0.212 (0.366)	0.129 (0.078)	0.161 (0.090)	0.145 (0.341)
$\beta_{21} : 2$	10233 (0.229)	0.118 (0.268)	0.121 (0.290)	0.131 (0.281)	0.159 (0.237)	0.057 (0.172)	0.042 (0.146)
$\beta_{31} : 3$	15.59 (0.374)	1.130 (1.093)	0.729 (0.860)	1.099 (1.027)	0.757 (0.866)	0.744 (0.801)	0.967 (0.989)
$\pi_1 : 0.3$	0.076 (0.176)	0.008 (0.017)	0.006 (0.003)	0.006 (0.024)	0.019 (0.032)	0.010 (0.045)	0.005 (0.048)
		Case IV: $\epsilon_1 \sim N(0,1), \epsilon_2 \sim N(0,1)$ and $\epsilon_3 \sim skew - t_3$					
$\beta_{10} : 1$	18.14 (0.222)	0.567 (-0.382)	0.091 (0.100)	0.550 (-0.275)	0.611 (0.024)	0.926 (0.121)	0.381 (0.146)
$\beta_{20} : 1$	7.983 (0.292)	0.782 (0.543)	0.325 (0.282)	0.795 (0.451)	0.915 (0.146)	0.246 (0.220)	0.820 (0.272)
$\beta_{30} : 1$	12.94 (0.936)	0.491 (0.346)	0.356 (0.539)	0.512 (0.309)	0.541 (0.348)	0.849 (0.504)	0.094 (0.289)
$\beta_{11} : 1$	3.274 (-0.248)	0.206 (0.413)	0.084 (-0.108)	0.198 (0.348)	0.220 (-0.041)	0.111 (0.087)	0.156 (0.347)
$\beta_{21} : 2$	1.185 (-0.224)	0.162 (-0.303)	0.246 (-0.182)	0.132 (-0.274)	0.248 (-0.341)	0.050 (-0.017)	0.028 (-0.115)
$\beta_{31} : 3$	4.579 (-0.489)	1.122 (-1.051)	0.746 (-0.900)	1.089 (-1.022)	0.797 (-0.916)	0.756 (-0.818)	0.989 (-0.996)
$\pi_1 : 0.3$	0.062 (-0.116)	0.007 (-0.033)	0.012 (-0.075)	0.005 (-0.035)	0.018 (-0.070)	0.010 (0.040)	0.006 (0.038)
		Case V: $\epsilon_1, \epsilon_2, \epsilon_3, \sim N(0,1)$ with 5% of low leverage outliers					
$\beta_{10} : 1$	5.250 (0.098)	0.286 (0.260)	6.355 (0.229)	0.264 (0.324)	7.402 (0.112)	0.135 (0.035)	0.242 (0.467)
$\beta_{20} : 1$	69.39 (1.120)	1.118 (0.709)	16.87 (0.850)	1.201 (0.662)	10.36 (0.302)	1.405 (0.916)	0.678 (0.818)
$\beta_{30} : 1$	95.49 (12.63)	1.689 (1.318)	7.742 (1.262)	1.786 (1.330)	11.04 (0.913)	1.673 (1.027)	0.778 (0.875)
$\beta_{11} : 1$	0.125 (0.080)	0.138 (0.228)	0.083 (0.017)	0.150 (0.226)	5.432 (0.100)	0.377 (0.003)	0.169 (0.368)
$\beta_{21} : 2$	0.068 (0.129)	0.111 (0.225)	0.059 (0.157)	0.125 (0.244)	0.584 (0.750)	0.063 (0.118)	0.019 (0.093)
$\beta_{31} : 3$	0.897 (0.930)	1.057 (1.043)	0.810 (0.846)	1.070 (1.048)	2.181 (1.038)	1.024 (0.932)	1.004 (1.001)
$\pi_1 : 0.3$	0.013 (0.300)	0.007 (0.046)	0.015 (0.074)	0.007 (0.062)	0.025 (0.112)	0.006 (0.007)	0.003 (0.041)
		Case VI: $\epsilon_1, \epsilon_2, \epsilon_3, \sim N(0,1)$ with 5% of high leverage outliers					
$\beta_{10} : 1$	0.084 (0.001)	0.343 (0.027)	0.295 (-0.035)	0.531 (0.047)	0.093 (0.025)	0.110 (0.227)	0.018 (-0.009)
$\beta_{20} : 1$	0.044 (-0.020)	0.300 (-0.055)	0.240 (-0.045)	0.695 (0.046)	0.070 (-0.032)	0.019 (-0.036)	0.009 (-0.005)
$\beta_{30} : 1$	9.697 (0.284)	8.753 (0.511)	10.36 (0.544)	0.400 (-0.045)	4.253 (0.094)	0.087 (-0.231)	0.013 (0.033)
$\beta_{11} : 1$	0.076 (-0.036)	0.323 (0.584)	0.246 (0.524)	0.191 (0.278)	0.080 (0.004)	0.468 (0.683)	0.142 (0.326)
$\beta_{21} : 2$	0.029 (0.068)	0.069 (-0.195)	0.060 (-0.137)	0.113 (-0.262)	0.065 (0.043)	0.096 (-0.297)	0.037 (-0.127)
$\beta_{31} : 3$	78.08 (6.936)	48.20 (6.924)	48.07 (6.923)	1.030 (-1.025)	38.11 (6.891)	1.679 (-1.292)	0.955 (-0.977)
$\pi_1 : 0.3$	0.017 (0.047)	0.031 (0.144)	0.027 (0.124)	0.010 (-0.045)	0.018 (0.051)	0.002 (0.041)	0.004 (0.046)
		Case VII: 20% gross outliers					
$\beta_{10} : 1$	0.112 (0.089)	111.4 (0.177)	0.009 (-0.006)	126.9 (-0.129)	0.402 (0.632)	0.049 (0.038)	0.085 (0.281)
$\beta_{20} : 1$	4991 (1.244)	630.5 (0.751)	368.1 (102.44)	524.4 (0.895)	0.402 (0.632)	0.655 (0.819)	0.639 (0.791)
$\beta_{30} : 1$	5519 (102.3)	1320 (0.985)	314.8 (1.175)	1001 (1.186)	0.403 (0.632)	0.792 (0.867)	0.816 (0.899)
$\beta_{11} : 1$	0.113 (0.064)	23.89 (0.114)	0.017 (0.085)	21.35 (0.010)	0.415 (0.634)	0.043 (-0.085)	0.030 (0.120)
$\beta_{21} : 2$	0.040 (-0.092)	0.121 (-0.235)	0.003 (-0.126)	0.201 (-0.267)	0.136 (-0.366)	0.047 (-0.132)	0.047 (-0.147)
$\beta_{31} : 3$	0.915 (-0.954)	12.97 (-0.954)	0.058 (-0.867)	13.64 (-0.904)	1.856 (-1.365)	1.120 (-1.071)	1.006 (-1.014)
$\pi_1 : 0.3$	0.007 (0.036)	0.008 (0.021)	0.001 (0.018)	0.017 (-0.003)	0.002 (0.033)	0.009 (-0.070)	0.007 (0.071)

7.10   Fitting robust mixture regression models using R

To apply the aforementioned robust mixture regression techniques, let us consider the **tone** data, which compares perceived tone and actual tone for a

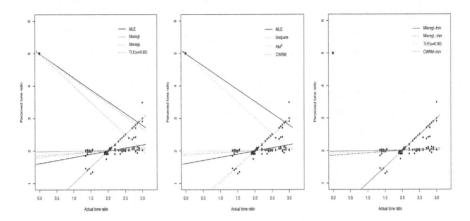

FIGURE 7.1
Robust FMR on tone data.

trained musician. The tone perception data stems from an experiment. A pure fundamental tone was played to a trained musician. Electronically generated overtones were added, determined by a stretching ratio of 2.0 corresponding to the harmonic pattern usually heard in traditional definite-pitched instruments. The musician was asked to tune an adjustable tone to the octave above the fundamental tone. The variable "tuned" gives the ratio of the adjusted tone to the fundamental, i.e. tuned=2.0 would be the correct tuning for all stretch ratio values. The data analyzed here belongs to 150 trials with the same musician.

The type of outliers included in this data set is not very harmful and, thus, no dramatic differences can be expected in terms of the estimated parameters, when using any (robust) mixture of regressions approach. As a result, we proceed to artificially contaminate the data, by adding 10 points at $(0, 5)$. The results are plotted in Figure 7.1.

```
library(MixSemiRob)
data(tone, package = "MixSemiRob")
y = tone$tuned
x = tone$stretchratio
k = nrow(tone) # 160
x[151:k] = 0
y[151:k] = 5

mle = mixreg(x, y, C = 2, nstart = 20)
# Robust mixture regression based on t-distribution
est_t = mixregT(x, y, C = 2, nstart = 20)
# Robust mixture regression based on Laplace distribution
```

```
est_lap = mixregLap(x, y, C = 2, nstart = 20)
# Trimmed likelihood estimator
est_TLE = mixregTrim(x, y, C = 2, keep = 0.95, nstart = 20)
# Robust modified EM algorithm based on bisquare function
est_bi = mixregBisq(x, y, C = 2, nstart = 20)
# Trimmed cluster weighted restricted model
est_CWRM = mixregCWRM(x, y, C = 2, nstart = 20)
# Robust mixture regression via mean-shift
est_RM2 = mixregRM2(x, y, method = "HARD")
```

8

Mixture models for high-dimensional data

8.1 Challenges of high-dimensional mixture models

Clustering in high-dimensional spaces is a recurrent problem in many domains, for example, in genotype analysis, object recognition, and so on, especially for model-based methods. Due to the famous "curse of dimensionality," model-based clustering methods are often over-parameterized with unsatisfactory performance for high-dimensional data.

Since high-dimensional data usually lives in low-dimensional subspaces hidden in the original space, many efforts have been made to allow model-based methods to efficiently cluster high-dimensional data.

8.2 Mixtures of factor analyzers

To reduce the dimension of multivariate data, it is common practice to perform a principal component analysis (PCA) to find an orthogonal projection of the data set into a low-dimensional linear subspace such that the variance of the projected data is maximized. However, as McLachlan and Peel (2000) pointed out, when the data is heterogeneous, the first few principal axes of the projection of **x** may not be able to portray the group structure.

Another popularly used dimension reduction technique is factor analysis (FA), which aims to reduce the dimension of the space while keeping the observed covariance structure of the data. However, one main drawback of this methodology is that its maximum likelihood estimator is not conducive to closed-form estimators. Mixtures of factor analyzers (MFA) assume that, within each component, the data is generated according to a factor model, thus reducing the number of parameters on which the covariance matrices depend.

8.2.1 Factor analysis (FA)

Factor analysis (FA) is a statistical dimension-reduction technique for modeling the covariance structure of high-dimensional data using a small number

DOI: 10.1201/9781003038511-8

of latent variables (Hinton et al., 1997). Let $(\mathbf{x}_1, \ldots, \mathbf{x}_n) \in \mathbb{R}^p$ be a given dataset. A typical FA model is defined as:

$$\mathbf{x}_i = \boldsymbol{\mu} + \boldsymbol{\Lambda}\mathbf{z}_i + \mathbf{e}_i, i = 1, \ldots, n, \tag{8.1}$$

where $\boldsymbol{\mu}$ is the mean vector, $\mathbf{e}_i$ is independent and identically distributed with $N_p(\mathbf{0}, \boldsymbol{\Psi})$, $\boldsymbol{\Psi} = \text{diag}(\sigma_1^2, \ldots, \sigma_p^2)$, and $\mathbf{z}_i$ is a d-dimensional vector $(d < p)$ of latent variables (called factors), independently and identically distributed with $N_d(\mathbf{0}, \mathbf{I}_d)$ and is independent of $\mathbf{e}_i$. $\boldsymbol{\Lambda}$ is a factor loading matrix of size $p \times d$. Then, the marginal density of $\mathbf{x}_i$ is a p-dimensional normal with mean $\boldsymbol{\mu}$ and covariance matrix

$$\boldsymbol{\Sigma} = \boldsymbol{\Lambda}\boldsymbol{\Lambda}^\top + \boldsymbol{\Psi}. \tag{8.2}$$

The traditional FA is a popular method used for explaining data, in particular, correlations between variables in multivariate observations. Note that the probabilistic principal component analysis (PPCA) model (Tipping and Bishop, 1999) is a special case of the FA model with isotropic covariance structure $\boldsymbol{\Psi} = \psi\mathbf{I}_p$.

The log-likelihood of the FA model (8.1) is:

$$\ell(\boldsymbol{\theta}) = \sum_{i=1}^{n} \log\{(2\pi)^{p/2}|\boldsymbol{\Lambda}\boldsymbol{\Lambda}^\top + \boldsymbol{\Psi}|^{-1/2}\exp[-\frac{1}{2}(\mathbf{x}_i - \boldsymbol{\mu})^\top(\boldsymbol{\Lambda}\boldsymbol{\Lambda}^\top + \boldsymbol{\Psi})^{-1}(\mathbf{x}_i - \boldsymbol{\mu})]\},$$

with $\boldsymbol{\theta} = (\boldsymbol{\mu}^\top, \boldsymbol{\Lambda}^\top, \boldsymbol{\Psi}^\top)^\top$ which can be estimated by the EM algorithm of Algorithm 8.2.1 if $\mathbf{z}_i$ is considered as the missing data.

8.2.2 Mixtures of factor analyzers (MFA)

As an extension and generalization of the factor analysis and the finite mixture models, the MFA model is attracting much interest from various scientific communities for a variety of applications. Hinton et al. (1997); McLachlan and Peel (2000); McLachlan et al. (2003), among others, studied the mixture of factor analyzers (MFA), which both clusters the data and reduce the dimension of each cluster. Since MFA allows for different local factor models in different regions of the input space, it is different from the factor analysis (FA) model.

Assume we have a mixture of C factor analyzers with mixing proportion $\pi_c, c = 1, \ldots, C$. Then, Hinton et al. (1997) proposed to model the marginal density of $\mathbf{x}$ as

$$f(\mathbf{x}; \boldsymbol{\theta}) = \sum_{c=1}^{C} \pi_c N_p(\mathbf{x}; \boldsymbol{\mu}_c, \boldsymbol{\Lambda}_c\boldsymbol{\Lambda}_c^\top + \boldsymbol{\Psi}_c), \tag{8.3}$$

where $\boldsymbol{\theta} = (\boldsymbol{\pi}^\top, \boldsymbol{\mu}^\top, \boldsymbol{\Lambda}^\top, \boldsymbol{\Psi}_c^\top)^\top, \boldsymbol{\pi} = (\pi_1, \ldots, \pi_{C-1})^\top, \boldsymbol{\mu} = (\boldsymbol{\mu}_1^\top, \ldots, \boldsymbol{\mu}_C^\top)^\top, \boldsymbol{\Lambda} = (\boldsymbol{\Lambda}_1^\top, \ldots, \boldsymbol{\Lambda}_C^\top)^\top$, and $\boldsymbol{\Psi} = (\boldsymbol{\Psi}_1^\top, \ldots, \boldsymbol{\Psi}_C^\top)^\top$. To be more specific, $\boldsymbol{\mu}_c$ is the mean of the c-th component, $\boldsymbol{\Lambda}_c$ is a $p \times d$ dimensional factor loading matrix and $\boldsymbol{\Psi}_c$

Algorithm 8.2.1 EM algorithm for FA

For $k = 0, 1, \ldots,$

E-step: Given the current estimator $\boldsymbol{\theta}^{(k)}$, compute the conditional expectations:

$$
\begin{aligned}
\mathbf{a}_i^{(k)} &= \mathrm{E}(\mathbf{z}_i | \mathbf{x}_i, \boldsymbol{\theta}^{(k)}) = \boldsymbol{\Lambda}^{(k)\top}(\boldsymbol{\Psi}^{(k)} + \boldsymbol{\Lambda}^{(k)}\boldsymbol{\Lambda}^{(k)\top})^{-1}(\mathbf{x}_i - \boldsymbol{\mu}^{(k)}), \\
\mathbf{b}_i^{(k)} &= \mathrm{E}(\mathbf{z}_i \mathbf{z}_i^\top | \mathbf{x}_i, \boldsymbol{\theta}^{(k)}) = \mathbf{I} - \boldsymbol{\Lambda}^{(k)\top}(\boldsymbol{\Psi}^{(k)} + \boldsymbol{\Lambda}^{(k)}\boldsymbol{\Lambda}^{(k)\top})^{-1}\boldsymbol{\Lambda}^{(k)} \\
&\quad + \{\boldsymbol{\Lambda}^{(k)\top}(\boldsymbol{\Psi}^{(k)} + \boldsymbol{\Lambda}^{(k)}\boldsymbol{\Lambda}^{(k)\top})^{-1}(\mathbf{x}_i - \boldsymbol{\mu}^{(k)})\} \\
&\quad \times \{\boldsymbol{\Lambda}^{(k)\top}(\boldsymbol{\Psi}^{(k)} + \boldsymbol{\Lambda}^{(k)}\boldsymbol{\Lambda}^{(k)\top})^{-1}(\mathbf{x}_i - \boldsymbol{\mu}^{(k)})\}^\top.
\end{aligned}
$$

M-step: Calculate

$$
\boldsymbol{\mu}^{(k+1)} = \sum_{i=1}^n (\mathbf{x}_i - \boldsymbol{\Lambda}^{(k)}\mathbf{a}_i^{(k)}),
$$

$$
\boldsymbol{\Lambda}^{(k+1)} = \left\{\sum_{i=1}^n (\mathbf{x}_i - \boldsymbol{\mu}^{(k+1)})\mathbf{a}_i^{(k)\top}\right\}\left\{\sum_{i=1}^n \mathbf{b}_i^{(k)}\right\}^{-1},
$$

$$
\boldsymbol{\Psi}^{(k+1)} = \frac{1}{n}\mathrm{diag}\left[\sum_{i=1}^n \left\{(\mathbf{x}_i - \boldsymbol{\mu}^{(k+1)})(\mathbf{x}_i - \boldsymbol{\mu}^{(k+1)})^\top \right.\right. \\
\left.\left. - \boldsymbol{\Lambda}^{(k+1)}\mathbf{a}_i^{(k)}(\mathbf{x}_i - \boldsymbol{\mu}^{(k+1)})^\top\right\}\right].
$$

---

is a diagonal matrix. That is to say, within the c-th component, the conditional density of $\mathbf{x}$ given $\mathbf{z}$ is

$$
f_c(\mathbf{x}|\mathbf{z}) = N_p(\mathbf{x}; \boldsymbol{\mu}_c + \boldsymbol{\Lambda}_c \mathbf{z}, \boldsymbol{\Psi}_c),
$$

and the joint density of $\mathbf{x}$ and $\mathbf{z}$ is:

$$
\begin{bmatrix} \mathbf{x} \\ \mathbf{z} \end{bmatrix} \sim N_{p+d}\left(\begin{bmatrix} \boldsymbol{\mu}_c \\ \mathbf{0} \end{bmatrix}, \begin{bmatrix} \boldsymbol{\Lambda}_c \boldsymbol{\Lambda}_c^\top + \boldsymbol{\Psi}_c & \boldsymbol{\Lambda}_c \\ \boldsymbol{\Lambda}_c^\top & \mathbf{I}_d \end{bmatrix}\right).
$$

Through this approach, the number of free parameters is controlled by the dimension of the latent factor space. Since there are pd and p quantities in $\boldsymbol{\Lambda}_c$ and $\boldsymbol{\Psi}_c$, respectively, and $d(d-1)/2$ constraints are needed for $\boldsymbol{\Lambda}_c$ to be unique, the total number of free parameters in (8.2) is

$$
pd + p - d(d-1)/2.
$$

Therefore, the total number of parameters in the model (8.3) is

$$
(C-1) + Cp + Cd\{p - (d-1)/2 + Cp\}.
$$

MFA has been traditionally fit using the maximum likelihood estimator (MLE) based on the normality assumptions of the random terms. Montanari and Viroli (2011) derived the information matrix for the MFA, and developed a direct maximum likelihood procedure for estimation. Ghahramani and Hinton (1997) introduced an exact EM algorithm to compute the MLE of MFA. The log-likelihood of the above model is

$$
\ell(\boldsymbol{\theta}) = \sum_{i=1}^{n} \log \sum_{c=1}^{C} \pi_c \Big[(2\pi)^{p/2} |\boldsymbol{\Lambda}_c \boldsymbol{\Lambda}_c^\top + \boldsymbol{\Psi}_c|^{-1/2}
$$
$$
\times \exp\{-\frac{1}{2}(\mathbf{x}_i - \boldsymbol{\mu}_c)^\top (\boldsymbol{\Lambda}_c \boldsymbol{\Lambda}_c^\top + \boldsymbol{\Psi}_c)^{-1}(\mathbf{x}_i - \boldsymbol{\mu}_c)\}\Big], \qquad (8.4)
$$

the maximizer of which does not have an explicit solution. Hinton et al. (1997) suggested to maximize (8.4) by an EM algorithm. Let ω_{ic} be an indicator variable such that

$$
\omega_{ic} = \begin{cases} 1, & \text{if } \mathbf{x}_i \text{ is from the } c\text{-th component;} \\ 0, & \text{otherwise.} \end{cases}
$$

Then, the complete log-likelihood of $\{(\mathbf{x}_i, \mathbf{z}_i, \omega_{ic}), i = 1, \ldots, n, c = 1, \ldots, C\}$ is

$$
\ell_c(\boldsymbol{\theta}) = \sum_{i=1}^{n} \log \prod_{c=1}^{C} \pi_c^{\omega_{ic}} \Big[(2\pi)^{p/2} |\boldsymbol{\Psi}_c|^{-1/2}
$$
$$
\times \exp\{-\frac{1}{2}(\mathbf{x}_i - \boldsymbol{\mu}_c - \boldsymbol{\Lambda}_c \mathbf{z}_i)^\top \boldsymbol{\Psi}_c^{-1}(\mathbf{x}_i - \boldsymbol{\mu}_c - \boldsymbol{\Lambda}_c \mathbf{z}_i)\}\Big]^{\omega_{ic}},
$$

and the corresponding EM algorithm is summarized in Algorithm 8.2.2.

Different from the above algorithm where the missing data contains component-indicator vectors as well as latent factors, Zhao and Yu (2008) developed another fast ECM algorithm where the missing data now consists of component-indicator vectors only. In addition, instead of resorting to numerical optimization methods, closed-form expressions in all conditional maximization steps are explicitly obtained in the fast ECM algorithm. Salah and Alpaydm (2004) proposed an incremental algorithm for MFA, in which factors and components were added to the mixture one by one. Kaya and Salah (2015) presented an adaptive MFA algorithm, which is a robust and parsimonious model selection method for training an MFA. The algorithm starts with a 1-component, 1-factor mixture model, where the factor is nothing but the first principal component of the data. Then, at each step, the algorithm adds more components and factors to the mixture model and runs EM algorithms to fit the model, during which a minimum message length (MML) criterion is applied to determine whether any components should be deleted. If the MML does not increase when the number of factors and/or components increases, a downsizing component annihilation process begins with components eliminated one by one.

Algorithm 8.2.2 EM algorithm for MFA

For $k = 0, 1, \ldots,$

E-step: Given the current estimator $\boldsymbol{\theta}^{(k)}$, compute the conditional expectations:

$$p_{ic}^{(k)} = \mathrm{E}(\omega_{ic}|\mathbf{x}_i, \boldsymbol{\theta}^{(k)}) = \frac{\pi_c^{(k)} N_p(\mathbf{x}_i; \boldsymbol{\mu}_c^{(k)}, \boldsymbol{\Lambda}_c^{(k)} \boldsymbol{\Lambda}_c^{(k)\top} + \boldsymbol{\Psi}_c^{(k)})}{\sum_{c=1}^{C} \pi_c^{(k)} N_p(\mathbf{x}_i; \boldsymbol{\mu}_c^{(k)}, \boldsymbol{\Lambda}_c^{(k)} \boldsymbol{\Lambda}_c^{(k)\top} + \boldsymbol{\Psi}_c^{(k)})},$$

$$\mathbf{a}_{ic}^{(k)} = \mathrm{E}(\mathbf{z}_i|\mathbf{x}_i, \omega_{ic} = 1, \boldsymbol{\theta}^{(k)}) = \boldsymbol{\Gamma}_c^{(k)}(\mathbf{x}_i - \boldsymbol{\mu}_c^{(k)}),$$

$$\mathbf{b}_{ic}^{(k)} = \mathrm{E}(\mathbf{z}_i \mathbf{z}_i^\top|\mathbf{x}_i, \omega_{ic} = 1, \boldsymbol{\theta}^{(k)})$$

$$= \mathbf{I}_d - \boldsymbol{\Gamma}_c^{(k)} \boldsymbol{\Lambda}_c^{(k)} + \boldsymbol{\Gamma}_c^{(k)}(\mathbf{x}_i - \boldsymbol{\mu}_c^{(k)})\{\boldsymbol{\Gamma}_c^{(k)}(\mathbf{x}_i - \boldsymbol{\mu}_c^{(k)})\}^\top,$$

where $\boldsymbol{\Gamma}_c = \boldsymbol{\Lambda}_c^\top (\boldsymbol{\Psi}_c + \boldsymbol{\Lambda}_c \boldsymbol{\Lambda}_c^\top)^{-1}$.

M-step: Calculate

$$\pi_c^{(k+1)} = \frac{1}{n} \sum_{i=1}^{n} p_{ic}^{(k)},$$

$$\boldsymbol{\mu}_c^{(k+1)} = \left\{ \sum_{i=1}^{n} p_{ic}^{(k)}(\mathbf{x}_i - \boldsymbol{\Lambda}_c^{(k)} \mathbf{a}_{ic}^{(k)}) \right\} \left\{ \sum_{i=1}^{n} p_{ic}^{(k)} \right\}^{-1},$$

$$\boldsymbol{\Lambda}_c^{(k+1)} = \left\{ \sum_{i=1}^{n} p_{ic}^{(k)}(\mathbf{x}_i - \boldsymbol{\mu}_c^{(k+1)}) \mathbf{a}_{ic}^{(k)\top} \right\} \left\{ \sum_{i=1}^{n} p_{ic}^{(k)} \mathbf{b}_{ic}^{(k)} \right\}^{-1},$$

$$\boldsymbol{\Psi}_c^{(k+1)} = \mathrm{diag}\left\{ \sum_{i=1}^{n} p_{ic}^{(k)}(\mathbf{x}_i - \boldsymbol{\mu}_c^{(k+1)} - \boldsymbol{\Lambda}_c^{(k+1)} \mathbf{a}_{ic}^{(k)})(\mathbf{x}_i - \boldsymbol{\mu}_c^{(k+1)})^\top \right\}$$

$$\times \left\{ \sum_{i=1}^{n} p_{ic}^{(k)} \right\}^{-1}.$$

Given the number of components C, the number of factors d can be decided through information criteria, such as BIC, or a likelihood ratio test. Given the hypotheses $H_0 : d = d_0$ v.s. $H_1 : d = d_0 + 1$, -2 times the logarithm of the likelihood ratio is shown to be asymptotically chi-squared distributed with degrees of freedom $C(p - d_0)$. However, as McLachlan and Baek (2010) pointed out when the sample size n is not large enough relative to the number of parameters, BIC is more likely to underestimate d. If the dimensionality p is large and the number of components C is not small, MFA may not be able to manage the scale of the analysis.

McNicholas et al. (2017) extended the MFA model to variance-gamma mixtures, where the formation of the variance-gamma distribution utilizes the limiting case of the generalized hyperbolic distribution, to make the model more robust, and the parameters are estimated through an alternating expectation-conditional maximization algorithm. Yang et al. (2017) applied the trimmed likelihood estimator (TLE) to MFA, so that only the majority of the data are used to fit the model, whereas the remaining data are considered outliers and are not used for model fitting.

Viroli (2010) proposed a mixture of factor mixture analyzers (MFMA), which includes the MFA as special cases, and is powerful for nonGaussian latent variables scenarios. With a certain prior probability, the data is assumed to be generated from several factor models, and within each factor model, the factors are assumed to follow a multivariate mixture of Gaussians. To be more specific, with a probability π_c, it is assumed that

$$\mathbf{x} = \boldsymbol{\mu}_c + \boldsymbol{\Lambda}_c \mathbf{z} + \mathbf{e}_c,$$

where $\mathbf{e}_c \sim N_p(\mathbf{0}, \boldsymbol{\Psi}_c)$ and $\boldsymbol{\Psi}_c$ is a diagonal covariance matrix that is different for each factor mixture model. $\boldsymbol{\Lambda}_c$ is a $p \times q$ factor loading matrix and $\mathbf{z}$ is a random vector of factors, mutually independent from $\mathbf{e}_c$. Within each factor model, the factors are assumed to follow a mixture of the Gaussian model:

$$f(\mathbf{z}; \boldsymbol{\gamma}, \boldsymbol{\eta}, \boldsymbol{\Sigma}) = \sum_{j=1}^{m} \gamma_j N_q(\boldsymbol{\eta}_j, \boldsymbol{\Sigma}_j).$$

By variational approximation, Ghahramani and Beal (2000) proposed a Bayesian treatment for MFA. Ueda et al. (2000) applied the MFA to image compression and handwritten digit recognition, and offered a split-and-merge-EM (SMEM) algorithm. Fokoué (2005) extended the MFA model, where both the continuous latent variable and the categorical latent variable are assumed to be influenced by the effects of fixed, observed covariates.

Many efforts have been made to apply the MFA to clustering tissue samples with microarray gene expression data. McLachlan et al. (2002, 2003) considered a method, which starts with a univariate screening. The selected subset of genes is then modeled by MFA to effectively reduce the dimension of the feature space; see McLachlan et al. (2007) and Baek et al. (2010a) for more of such recent applications and extensions. Xie et al. (2010) generalized the MFA to incorporate penalization so that the model can effectively perform variable selection. For centered and standardized data, with a common diagonal covariance matrix, the variable j is irrelevant to clustering if and only if all the cluster centers are the same across the C clusters, i.e., $\mu_{1j} = \mu_{2j} = \ldots = \mu_{Cj} = 0$. Let $p_1(\boldsymbol{\mu}) = \sum_c \sum_j |\mu_{cj}|$ denote the L_1 penalty function for mean parameters and $p_2(\boldsymbol{\Lambda}) = \sum_c \sum_j \|\boldsymbol{\lambda}_{cj.}\|_2$ be the penalty function for factor loading $\boldsymbol{\Lambda}$'s, where $\|\boldsymbol{\lambda}_{cj.}\|_2 = \sqrt{\sum_l \lambda_{cjl}^2}$. Xie et al. (2010) proposed the maximum penalized likelihood estimation (MPLE), which maximizes the following penalized log-likelihood:

$$\ell(\boldsymbol{\theta}) - \gamma_1 p_1(\boldsymbol{\mu}) - \gamma_2 p_2(\boldsymbol{\Lambda}),$$

TABLE 8.1
Parsimonious covariance structures derived from the mixture of factor analyzers model

Model ID	Loading Matrix	Cov. structure	Isotropic	No. of parameters
CCC	Constrained	Constrained	Constrained	$\{pd - d(d-1)/2\} + 1$
CCU	Constrained	Constrained	Unconstrained	$\{pd - d(d-1)/2\} + p$
CUC	Constrained	Unconstrained	Constrained	$\{pd - d(d-1)/2\} + C$
CUU	Constrained	Unconstrained	Unconstrained	$\{pd - d(d-1)/2\} + Cp$
UCC	Unconstrained	Constrained	Constrained	$C\{pd - d(d-1)/2\} + 1$
UCU	Unconstrained	Constrained	Unconstrained	$C\{pd - d(d-1)/2\} + p$
UUC	Unconstrained	Unconstrained	Constrained	$C\{pd - d(d-1)/2\} + C$
UUU	Unconstrained	Unconstrained	Unconstrained	$C\{pd - d(d-1)/2\} + Cp$

where $\ell(\boldsymbol{\theta})$ is the log-likelihood. If a variable j is independent of all other variables with $\boldsymbol{\lambda}_{cj.} = \mathbf{0}$ for all c values, this variable is effectively treated as irrelevant.

8.2.3 Parsimonious mixtures of factor analyzers

Depending on whether to put constraints (such as homogeneous constraints) across groups on $\boldsymbol{\Lambda}_c$ and $\boldsymbol{\Psi}_c$ and on whether or not $\boldsymbol{\Psi}_c = \psi_c \mathbf{I}_p$, McNicholas and Murphy (2008) proposed a family of eight parsimonious Gaussian mixture models (PGMMs), summarized in Table 8.1. McNicholas and Murphy (2010) further extended the factor analysis covariance structure by letting $\boldsymbol{\Psi}_c = \omega_c \boldsymbol{\Delta}_c$, where $\omega_c \in \mathbb{R}^+$ and $\boldsymbol{\Delta}_c = \text{diag}\{\delta_1, \delta_2, \ldots, \delta_p\}$ such that $|\boldsymbol{\Delta}_c| = 1$ for $c = 1, \ldots, C$. By imposing constraints on the parameters $\boldsymbol{\Lambda}_c, \omega_c$ and $\boldsymbol{\Delta}_c$, McNicholas and Murphy (2010) expanded the GMM to a family of 12 expanded PGMM (EPGMM), listed in Table 8.2. In particular, in the PGMM family, the first letter indicates whether the loading matrix is constrained to be common (C..) or not (U..) between groups, the second term refers to whether the noise variance is common (.C.) between factors or not (.U.), and the last one indicates whether the covariance structure is isotropic (..C) or not (..U). On the other hand, in the EPGMM family, the terminology has the following means: the loading matrix is common (C...) or not (U...), the noise covariance matrix $\boldsymbol{\Delta}_c$ is common (.C..) or not (.U..), ω_c is common (..C.) or not (..U.), and finally $\boldsymbol{\Delta}_c = \mathbf{I}_p$ (...C) or not (...U).

Some previously developed models are special cases of the EPGMM family. For example, the famous mixture of probabilistic PCA of Tipping and Bishop (1999) is the UUUC-UUC model. The MFA developed by Ghahramani and Hinton (1997) is equivalent to the UCCU-UCU, while the MFA model proposed by McLachlan et al. (2003) is equivalent to the UUUU-UUU model.

Since the multivariate normal distribution is assumed for the component error and factor distributions, the traditional MFA model is sensitive to

Mixture models for high-dimensional data

TABLE 8.2
The covariance structure and number of parameters of PGMM and EPGMM
family with $K = (C-1) + Cp$ and $M = d\{p - (d-1)/2\}$

EPGMM	PGMM equiv.	Cov. structure	No. of parameters
UUUU	UUU	$\Sigma_c = \Lambda_c\Lambda_c^\top + \omega_c\Delta_c$	$K + CM + Cp$
UUCU	-	$\Sigma_c = \Lambda_c\Lambda_c^\top + \omega\Delta_c$	$K + CM + \{1 + C(p-1)\}$
UCUU	-	$\Sigma_c = \Lambda_c\Lambda_c^\top + \omega_c\Delta$	$K + CM + \{C + (p-1)\}$
UCCU	UCU	$\Sigma_c = \Lambda_c\Lambda_c^\top + \Psi$	$K + CM + p$
UCUC	UUC	$\Sigma_c = \Lambda_c\Lambda_c^\top + \omega_c I_p$	$K + CM + C$
UCCC	UCC	$\Sigma_c = \Lambda_c\Lambda_c^\top + \omega I_p$	$K + CM + 1$
CUUU	CUU	$\Sigma_c = \Lambda\Lambda^\top + \omega_c\Delta_c$	$K + M + Cp$
CUCU	-	$\Sigma_c = \Lambda\Lambda^\top + \omega\Delta_c$	$K + M + \{1 + C(p-1)\}$
CCUU	-	$\Sigma_c = \Lambda\Lambda^\top + \omega_c\Delta$	$K + M + \{C + (p-1)\}$
CCCU	CCU	$\Sigma_c = \Lambda\Lambda^\top + \omega\Delta$	$K + M + p$
CCUC	CUC	$\Sigma_c = \Lambda\Lambda^\top + \omega_c I_p$	$K + M + C$
CCCC	CCC	$\Sigma_c = \Lambda\Lambda^\top + \omega I_p$	$K + M + 1$

outliers. As a result, McLachlan et al. (2007) extended the MFA model to
incorporate the multivariate t-distribution, and proposed a mixture of multivariate t-factor analyzers (MMtFA). The MMtFA assumes that $\{x_1, \ldots, x_n\}$
is an observed random sample from the t-mixture density:

$$f(\mathbf{x}; \boldsymbol{\theta}) = \sum_{c=1}^{C} \pi_c g_t(\mathbf{x}; \boldsymbol{\mu}_c, \boldsymbol{\Sigma}_c, \nu_c),$$

where $\boldsymbol{\Sigma}_c = \boldsymbol{\Lambda}_c\boldsymbol{\Lambda}_c^\top + \boldsymbol{\Psi}_c$, and $g_t(\mathbf{x}; \boldsymbol{\mu}, \boldsymbol{\Sigma}, \nu)$ denotes the density of a t-
distribution:

$$g_t(\mathbf{x}; \boldsymbol{\mu}, \boldsymbol{\Sigma}, \nu) = \frac{\Gamma(\frac{\nu+p}{2})|\boldsymbol{\Sigma}|^{-1/2}}{(\pi\nu)^{p/2}\Gamma(\frac{\nu}{2})\{1 + \delta(\mathbf{x}, \boldsymbol{\mu}; \boldsymbol{\Sigma})/\nu\}^{(\nu+p)/2}},$$

where $\delta(\mathbf{x}, \boldsymbol{\mu}; \boldsymbol{\Sigma}) = (\mathbf{x} - \boldsymbol{\mu})^\top\boldsymbol{\Sigma}^{-1}(\mathbf{x} - \boldsymbol{\mu})$ denotes the Mahalanobis squared
distance between $\mathbf{x}$ and $\boldsymbol{\mu}$. In this case, in addition to $\pi_c, \boldsymbol{\mu}_c, \boldsymbol{\Sigma}_c$, and $\boldsymbol{\Psi}_c$, the
vector of unknown parameters $\boldsymbol{\theta}$ consists of the degrees of freedom ν_c. The
readers are referred to McLachlan et al. (2007) for an AECM algorithm to
estimate mixtures of t-factor analyzers. By putting constraints on the degrees
of freedom ν_c's, the factor loadings $\boldsymbol{\Lambda}_c$'s, and error variance matrices $\boldsymbol{\Psi}_c$'s,
Andrews and McNicholas (2011) extended the MMtFA to a family of six
mixture models, including parsimonious models.

Baek et al. (2010b) proposed a mixture of common factor analyzer approach, which assumes

$$\mathbf{X}_i = \mathbf{A}\mathbf{Z}_{ci} + \boldsymbol{\eta}_{ci}$$

with probability π_c, where $\boldsymbol{\eta}_{ci}$ is independently distributed with $N(\mathbf{0}, \boldsymbol{\Psi})$ with $\boldsymbol{\Psi}$ being a diagonal matrix, $\mathbf{Z}_{ci}$ is an unobservable factor, independently distributed with $N(\boldsymbol{\xi}_c, \mathbf{D}_c)$, and is independent of $\boldsymbol{\eta}_{ci}$, and $\mathbf{A}$ is a factor loading matrix of dimension $p \times m$. This method lowers the complexity of the MFA model by introducing common factor loadings. Application of it can also be found in Yoshida et al. (2004, 2006).

8.3 Model-based clustering based on reduced projections

8.3.1 Clustering with envelope mixture models

The envelope technique, first introduced in the regression settings, is aimed at dimension reduction and efficient parameter estimation, simultaneously, and has been widely used in discriminant subspace for classification and discriminant analysis. As a result, a set of clustering with envelope mixture models (CLEMM) was proposed by Wang et al. (2020), which combines the ideas of the Gaussian mixture model and envelope methodology.

Similar to the envelope modeling techniques in regression, in a CLEMM, it is assumed that there exists a low-dimensional subspace that fully captures the variation of data across all clusters. Without loss of generality, assume $\mathbb{E}(\mathbf{X}) = \mathbf{0}$. Let $(\boldsymbol{\Gamma}, \boldsymbol{\Gamma}_0) \in \mathbb{R}^{p \times p}$ be an orthogonal matrix where $\boldsymbol{\Gamma} \in \mathbb{R}^{p \times q}$ $(q \leq p)$ is the semi-orthogonal basis for the subspace of interest. Define $\mathbf{X}_M = \boldsymbol{\Gamma}^\top \mathbf{X} \in \mathbb{R}^q$ as the material part of $\mathbf{X}$, that is the part that contains all the information about clustering, and $\mathbf{X}_{IM} = \boldsymbol{\Gamma}_0^\top \mathbf{X} \in \mathbb{R}^{p-q}$, independent of $\mathbf{X}_M$, as the immaterial part of $\mathbf{X}$, which is homogeneous through all components. Then, a CLEMM is assumed as

$$\mathbf{X}_M \sim \sum_{c=1}^{C} \pi_c N(\boldsymbol{\alpha}_c, \boldsymbol{\Omega}_c), \quad \mathbf{X}_{IM} \sim N(\mathbf{0}, \boldsymbol{\Omega}_0), \quad \mathbf{X}_M \perp \mathbf{X}_{IM}, \quad (8.5)$$

where $\boldsymbol{\alpha}_c \in \mathbb{R}^q, \boldsymbol{\Omega}_c \in \mathbb{R}^{q \times q}$ and $\boldsymbol{\Omega}_0 \in \mathbb{R}^{(p-q) \times (p-q)}$ are symmetric positive definite matrices, and $\boldsymbol{\Psi} = \{\pi_1, \ldots, \pi_C, \boldsymbol{\Gamma}, \boldsymbol{\alpha}_1, \ldots, \boldsymbol{\alpha}_C, \boldsymbol{\Omega}_1, \ldots, \boldsymbol{\Omega}_C\}$ contains all the unknown parameters.

In fact, the model (8.5) can be considered as a parsimonious parameterization of the original GMM model (1.24):

$$\mathbf{X} \sim \sum_{c=1}^{C} \pi_c N(\boldsymbol{\mu}_c, \boldsymbol{\Sigma}_c), \quad \boldsymbol{\mu}_c = \boldsymbol{\Gamma}\boldsymbol{\alpha}_c, \quad \boldsymbol{\Sigma}_c = \boldsymbol{\Gamma}\boldsymbol{\Omega}_c\boldsymbol{\Gamma}^\top + \boldsymbol{\Gamma}_0\boldsymbol{\Omega}_0\boldsymbol{\Gamma}_0^\top, \quad c = 1, \ldots, C,$$

indicating that the centers of each cluster lie with the lower dimensional subspace span($\boldsymbol{\Gamma}$). Note that the marginal covariance of $\mathbf{X}$ can be written as $\Sigma_X = \Gamma\Omega_X\Gamma^\top + \Gamma_0\Omega_0\Gamma_0^\top$, indicating that span($\Gamma$) is a reducing subspace of both Σ_c and Σ_X, where $\Omega_X = \mathrm{Cov}(\mathbf{X}_M)$. Compared to the original GMM (1.24) to CLEMM (8.5), the number of free parameters reduced from $(C-1)+C(p-1)+Cp(p+1)/2$ (note that the number of free parameters in $\boldsymbol{\mu}_c$ is $(p-1)$ due to $\mathrm{E}(\mathbf{X}) = \mathbf{0}$) to $(p-q)q+(C-1)q+Cq(q+1)/2+(p-q)(p-q+1)/2$, which results in a reduction of $(C-1)[(p-q) + \{p(p+1) - q(q+1)\}/2]$.

Note that the MFA with common factors is a special case of CLEMM (8.5), with CLEMM being more flexible. However, this flexibility is compensated by the difficulty in the EM algorithm.

8.3.2　Envelope EM algorithm for CLEMM

Assume Z to be an unknown component indicator variable, defined as

$$P(Z = c) = \pi_c, \; \Gamma^\top\mathbf{X}|Z = c \sim N(\boldsymbol{\alpha}_c, \boldsymbol{\Omega}_c),$$
$$\Gamma_0^\top\mathbf{X}|Z = c \sim N(\mathbf{0}, \boldsymbol{\Omega}_0), \; \Gamma^\top\mathbf{X} \perp \Gamma_0^\top\mathbf{X}.$$

The difficulty of the estimation procedure lies in the estimation of Γ, which involves optimizing a nonconvex objective function on a Grassmann manifold. The detailed computation of CLEMM is described in Algorithm 8.3.1.

8.4　Regularized mixture modeling

Since estimation and classification of a GMM involve the inversion of Σ_c's, many researchers proposed to numerically regularize the estimators of Σ_c's before inversion, to solve the problem of "curse of dimensionality." Most commonly used regularized mixture models focus on modifying the objective function or constraining covariance matrix structure. Readers are also referred to Mkhadri et al. (1997) for an overview of regularized mixture modeling.

The simplest way to regularize the $\hat{\Sigma}_c$'s is to consider a ridge regularization by

$$\tilde{\boldsymbol{\Sigma}}_c = \hat{\boldsymbol{\Sigma}}_c + \sigma_c\mathbf{I}_p,$$

where $\mathbf{I}_p$ is a p-dimensional diagonal matrix and σ_c is a positive quantity to be determined. This methodology has been popularly used in R (Venables and Ripley, 2002), among many other statistical software.

A more commonly used method was proposed by Hastie et al. (1995), which assumes

$$\tilde{\boldsymbol{\Sigma}}_c = \hat{\boldsymbol{\Sigma}}_c + \sigma_c\boldsymbol{\Omega}, \qquad (8.6)$$

where $\boldsymbol{\Omega}$ is a $p \times p$ regularization matrix. Compared to the previous one, (8.6) also penalizes the correlations between covariates.

Algorithm 8.3.1 Envelope EM algorithm (Wang et al., 2020)

Starting with an initial value $\boldsymbol{\theta}^{(0)}$, in the $(k+1)$th step,

E-step: Finding the conditional probabilities:

$$p_{ic}^{(k)} = \frac{\pi_c^{(k)}\phi(\mathbf{x}_i; \boldsymbol{\mu}_c^{(k)}, \boldsymbol{\Sigma}_c^{(k)})}{\sum_{l=1}^{C}\pi_l^{(k)}\phi(\mathbf{x}_i; \boldsymbol{\mu}_l^{(k)}, \boldsymbol{\Sigma}_l^{(k)})}, \quad i = 1,\ldots,n, \quad c = 1,\ldots,C.$$

M-step: Update the parameters:

$$\pi_c^{(k+1)} = \frac{\sum_{i=1}^{n} p_{ic}^{(k)}}{n}, \quad \tilde{\boldsymbol{\mu}}_c^{(k+1)} = \frac{\sum_{i=1}^{n} p_{ic}^{(k)}\mathbf{x}_i}{\sum_{i=1}^{n} p_{ic}^{(k)}},$$

$$\mathbf{S}_c^{(k+1)} = \frac{\sum_{i=1}^{n} p_{ic}^{(k)}(\mathbf{x}_i - \tilde{\boldsymbol{\mu}}_c^{(k+1)})(\mathbf{x}_i - \tilde{\boldsymbol{\mu}}_c^{(k+1)})^\top}{\sum_{i=1}^{n} p_{ic}^{(k)}},$$

$$\boldsymbol{\Gamma}^{(k+1)} = \arg\min_{\boldsymbol{\Gamma}\in\mathcal{G}(p,u)} \left\{ \sum_{c=1}^{C} \pi_c^{(k+1)}\log|\boldsymbol{\Gamma}^\top\mathbf{S}_c^{(k+1)}\boldsymbol{\Gamma}| + \log|\boldsymbol{\Gamma}^\top\mathbf{S}_X^{-1}\boldsymbol{\Gamma}| \right\},$$

$$\boldsymbol{\mu}_c^{(k+1)} = \boldsymbol{\Gamma}^{(k+1)}\boldsymbol{\Gamma}^{(k+1)\top}\tilde{\boldsymbol{\mu}}_c^{(k+1)}$$

$$\boldsymbol{\Sigma}_c^{(k+1)} = \boldsymbol{\Gamma}^{(k+1)}\boldsymbol{\Gamma}^{(k+1)\top}S_c^{(k+1)}\boldsymbol{\Gamma}^{(k+1)}\boldsymbol{\Gamma}^{(k+1)\top}$$
$$+ \boldsymbol{\Gamma}_0^{(k+1)}\boldsymbol{\Gamma}_0^{(k+1)\top}S_X\boldsymbol{\Gamma}_0^{(k+1)}\boldsymbol{\Gamma}_0^{(k+1)\top},$$

where $c = 1,\ldots,C$, and S_X is the sample covariance matrix of $\{\mathbf{x}_i, i = 1,\ldots,n\}$, $\mathcal{G}(p,u)$ is the Grassmann manifold. $\boldsymbol{\Gamma}^{(k+1)}$ can be updated by many manifold optimization packages, and $\boldsymbol{\Gamma}_0^{(k+1)}$ is updated by the fact that $(\boldsymbol{\Gamma}^{(k+1)}, \boldsymbol{\Gamma}_0^{(k+1)})$ is an orthogonal matrix.

Define

$$\hat{\boldsymbol{\Sigma}}_c(\alpha) = \frac{(1-\alpha)(n_c-1)\hat{\boldsymbol{\Sigma}}_c + \alpha(n-C)\hat{\boldsymbol{\Sigma}}}{(1-\alpha)(n_c-1) + \alpha(n-C)},$$

where $\hat{\boldsymbol{\Sigma}}$ estimates the within covariance matrix and $0 \leq \alpha \leq 1$. Friedman (1989) proposed the regularized discriminant analysis in which the component-wise covariance matrix is shrunk towards a diagonal or a common covariance matrix across components:

$$\tilde{\boldsymbol{\Sigma}}_c(\alpha,\gamma) = (1-\gamma)\hat{\boldsymbol{\Sigma}}_c(\alpha) + \gamma\left(\frac{\text{tr}(\hat{\boldsymbol{\Sigma}}_c(\alpha))}{p}\right)\mathbf{I}_p,$$

where γ represents the ridge regularization, and α controls the relative contribution of $\hat{\boldsymbol{\Sigma}}_c$ and $\hat{\boldsymbol{\Sigma}}$.

By employing local consistency or Laplacian regularizer to the objective function, manifold regularized GMMs become increasingly popular regularization methods. Generally speaking, a smoothness term is incorporated into the original objective of GMM, leading to a regularized objective function as

$$\ell_{gmm}(\boldsymbol{\theta}) - \lambda \mathcal{R},$$

where $\ell_{gmm}(\boldsymbol{\theta})$ is the log-likelihood of a GMM and $\mathcal{R}$ is the regularizer. Liu et al. (2010) constructed the nearest neighbor graph and adopted Kullback-Leibler divergence as the distance measure when defining $\mathcal{R}$, that is,

$$\mathcal{R} = \frac{1}{2} \sum_{i,j=1}^{n} \left(D(P_i \| P_j) + D(P_j \| P_i) \right) W_{ij},$$

where $P_i = (p_{i1}, \ldots, p_{iC})^T$ and $p_{ic} = P(c|\mathbf{x}_i)$ is the possibility of observation $\mathbf{x}_i$ belonging to the c-th component,

$$D(P_i \| P_j) = \sum_{c=1}^{C} p_{ic} \log \frac{p_{ic}}{p_{jc}}$$

is the KL-divergence between P_i and P_j, and

$$W_{ij} = \begin{cases} 1, & \text{if } \mathbf{x}_i \in N_q(\mathbf{x}_j) \text{ or } \mathbf{x}_j \in N_q(\mathbf{x}_i); \\ 0, & \text{otherwise}, \end{cases}$$

is the edge weight matrix with $N_q(\mathbf{x}_i)$ being the data sets of q nearest neighbors of $\mathbf{x}_i$. A similar idea has also been used by He et al. (2011) where Euclidean distance is used instead of the divergence measure, which leads to a Laplacian regularized GMM. In this case, $\mathcal{R}$ is now defined as

$$\mathcal{R} = \sum_{c=1}^{C} \mathcal{R}_c,$$

where

$$\mathcal{R}_c = \frac{1}{2} \sum_{i,j=1}^{n} (p_{ic} - p_{jc})^2 \, \mathbf{x}_i^\top \mathbf{x}_j.$$

Ruan and Yuan (2011) employed the L_1 penalty on the inverse covariance matrices and therefore reduced the effective dimensionality of the estimation of the GMM. To be more specific, the following penalized likelihood estimation is suggested for GMM:

$$\hat{\boldsymbol{\theta}} = \arg\min_{\boldsymbol{\mu}_c, \boldsymbol{\Sigma}_c} \left\{ -\sum_{i=1}^{n} \log \left(\sum_{c=1}^{C} \pi_c \phi(\mathbf{x}_i; \boldsymbol{\mu}_c, \boldsymbol{\Sigma}_c) \right) + \lambda \sum_{c=1}^{C} \| \boldsymbol{\Sigma}_c^{-1} \|_{L_1} \right\},$$

where $\lambda \geq 0$ is a tuning parameter, and $\|\mathbf{A}\|_{L_1} = \sum_{i \neq j} |a_{ij}|$. In order to choose C and λ, the authors suggested a BIC-type criterion, where the degree of freedom is approximated by

$$\mathrm{df}(C, \lambda) = \sum_{c=1}^{C} \left(p + \sum_{i \leq j} I\{(\hat{\boldsymbol{\Sigma}}_c^{-1})_{ij} \neq 0\} \right).$$

This method works by driving most off-diagonal entries in the inverse covariance matrices to zero. The result of such a strategy is that most correlations are ignored. This is a strong assumption as correlations are nonnegligible in many applications. For example, in microarray data analysis, correlations between gene expression levels are of interest.

Recently, research efforts have been devoted to constraining the mean vectors as well. It is found that when the dimension is extremely high, for instance, larger than the sample size, regularizing the mean vector results in better classification even when the covariance structure is maintained highly parsimoniously or when covariance is not part of the estimation. Tibshirani et al. (2003) proposed the nearest shrunken centroid method based on denoised versions of the centroids as prototypes for each class and further applied it to the DNA microarray studies. Let x_{ij} be the observed value, and then, the j-th component of the centroid for class c is $\bar{x}_{jc} = \sum_{i \in I_c} x_{ij}/n_c$, where I_c is the index set of component c, n_c is the number of samples within the component c, and the j-th variable of the overall centroid is $\bar{x}_j = \sum_{i=1}^{n} x_{ij}/n$. Let

$$d_{jc} = \frac{\bar{x}_{jc} - \bar{x}_j}{m_c \cdot s_j}$$

be the normalized value of each gene, where s_j is the pooled within-class standard deviation for gene j

$$s_j^2 = \frac{1}{n - C} \sum_{c=1}^{C} \sum_{i \in I_c} (x_{ij} - \bar{x}_{jc})^2,$$

and $m_c = \sqrt{1/n_c - 1/n}$. Then, the authors proposed shrinking each d_{jc} towards zero by soft-thresholding as

$$d'_{jc} = sgn(d_{jc})(|d_{jc} - \Delta|)_+, \tag{8.7}$$

where Δ is a reduced value in absolute. Then, the new shrunken centroids are given by

$$\bar{x}'_{jc} = \bar{x}_j + m_c s_j d'_{jc}.$$

Guo et al. (2007) extended the centroid shrinkage idea of Tibshirani et al. (2003), and proposed regularizing the class means under the linear discriminant analysis (LDA) model. First, instead of using $\boldsymbol{\Sigma}$ directly, they proposed to use

$$\tilde{\boldsymbol{\Sigma}} = \alpha \boldsymbol{\Sigma} + (1 - \alpha) \mathbf{I}_p$$

for some $\alpha, 0 \leq \alpha \leq 1$, or the modified correlation matrix

$$\tilde{\mathbf{R}} = \alpha \mathbf{R} + (1 - \alpha)\mathbf{I}_p.$$

Then, the soft-thresholding method (8.7) is applied to shrink the centroids. Some dimensions of the mean vectors are shrunk to common values so that they become irrelevant to class labels, achieving variable selection.

Assuming the covariance matrices to be a common diagonal matrix, Pan and Shen (2007) employed the L_1 penalty to shrink mixture component means towards the global mean, automatically realizing variable selection and delivering a sparse solution. Specifically, assume that the data has been standardized so that each attribute has a mean of 0 and a variance of 1. That is, consider the GMM in (1.24) where the variance of each component is assumed to be the same as $\boldsymbol{\Sigma} = \mathrm{diag}(\sigma_1, \ldots, \sigma_p)$. Then, consider the penalized log-likelihood

$$\ell_{gmm}(\boldsymbol{\theta}) - p_\lambda(\boldsymbol{\theta}),$$

where $\ell_{gmm}(\boldsymbol{\theta})$ is the Gaussian mixture log-likelihood and $p_\lambda(\boldsymbol{\theta})$ is the L_1 penalty:

$$p_\lambda(\boldsymbol{\theta}) = \lambda \sum_{c=1}^{C} \sum_{j=1}^{p} |\mu_{cj}|.$$

In this way, the L_1 norm penalty shrinks some of the fitted means μ_{cj} to be exactly zero when λ is sufficiently large. If for the j-th variable, all the cluster-specific means $\mu_{cj}, c = 1, \ldots, C$ are shrunken to zero, then, the j-th variable does not contribute to the clustering and can be removed.

However, the method proposed by Pan and Shen (2007) failed to consider the fact that μ_{cj} and $\mu_{c'j}$ are associated with the same variable $\mathbf{x}_j$, hence intuitively, should be in one "group." By ignoring the "group" information, the L_1-norm penalty tends to shrink only some but not all μ_{cj}'s to zero, hence, failing to exclude the j-th variable.

As a result, along this line of research, Wang and Zhu (2008) proposed a L_∞-norm penalty and a hierarchical penalty, that incorporate the "group" information into the modeling procedure. First, an adaptive L_∞-norm penalized GMM is proposed by considering the following penalized log-likelihood function:

$$\ell_{gmm}(\boldsymbol{\theta}) - \lambda \sum_{j=1}^{p} \max_c(|\mu_{1j}|, \ldots, |\mu_{Cj}|),$$

where $\max_c(|\mu_{1j}|, \ldots, |\mu_{Cj}|) = \|(\mu_{1j}, \ldots, \mu_{Cj})\|_\infty$. If the maximum of $|\mu_{cj}|$ is shrunken to zero, then all μ_{cj}'s automatically shrink to zero. To further improve the model, Wang and Zhu (2008) also considered penalizing variables differently. That is,

$$\ell_{gmm}(\boldsymbol{\theta}) - \lambda \sum_{j=1}^{p} \omega_j \max_c(|\mu_{1j}|, \ldots, |\mu_{Cj}|),$$

where ω_j are pre-specified weights. The authors also considered an adaptive hierarchically penalized GMM, where μ_{cj} is modeled more generally by

$$\mu_{cj} = \gamma_j \theta_{cj}, \quad c = 1, \ldots, C, j = 1, \ldots, p,$$

where $\gamma_j \geq 0$ for identifiability reason. Then, consider the penalized log-likelihood:

$$\ell_{gmm}(\boldsymbol{\theta}) - \lambda_\gamma \sum_{j=1}^{p} \gamma_j - \lambda_\theta \sum_{c=1}^{C} \sum_{j=1}^{p} |\theta_{cj}|,$$

subject to $\gamma_j \geq 0$. Note that λ_γ controls the estimates at the variable-specific level. If γ_j is shrunken to zero, then all μ_{cj}'s will be automatically removed. Alternatively, λ_θ controls the estimators at the cluster-specific. Even if the γ_j is not zero, some of the μ_{cj}'s might still be able to be shrunken to zero, and thus, keep the flexibility of the L_1-norm.

Another commonly used regularized GMM method is based on constraining covariance matrix structure. Early efforts focused on controlling the complexity of the covariance matrices, partly driven by the frequent occurrences of singular matrices in estimation. Commonly used methods include $\boldsymbol{\Sigma}_c = \lambda \mathbf{I}$, where all clusters are spherical and of the same size; $\boldsymbol{\Sigma}_c = \boldsymbol{\Sigma}$ constant across components, where all clusters have the same geometry but do not need to be spherical (Fraley and Raftery, 2002). Please also see Section 1.7.2 for a more detailed introduction in this regard.

8.5 Subspace methods for mixture models

8.5.1 Introduction

When clusters are found in different subspaces, techniques like feature selection algorithms often fail. It is this type of data that motivated the evolution of subspace clustering algorithms. Figure 8.1 illustrates the idea of subspace clustering.

Many subspace clustering methods use heuristic search techniques to find the subspaces. These algorithms use the concepts of feature selection, but find clusters in different subspaces of the same dataset. There are two major branches of subspace clustering. The first one can be called the top-down algorithm, which finds an initial clustering in the full set of dimensions and evaluates the subspaces of each cluster. The other one is the bottom-up methodology, which finds dense regions in low-dimensional spaces, followed by combining them to form clusters. Both algorithms need to be iterated to improve results.

The bottom-up algorithms first create a histogram for each dimension followed by selecting those bins whose densities are above a given threshold.

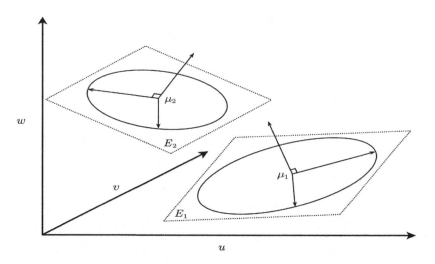

FIGURE 8.1
Subspace clustering (Bouveyron et al., 2007).

Candidate subspaces are made up of dimensions with dense units. The algorithm continues until dense units cannot be found. Adjacent dense units are then combined into clusters. Overlapping clusters can be found by bottom-up approaches. There are generally two ways to form multidimensional grids: static grid, which includes CLIQUE (Agrawal et al., 1998), ENCLUS (Cheng et al., 1999) and adaptive grid, such as MAFIA (Goil et al., 1999), cell-based clustering (Chang and Jin, 2002, CBF), CLTree (Liu et al., 2000), and density-based optimal projective clustering (Procopiuc et al., 2002, DOC).

The iterative top-down subspace search methods start with an initial approximation of the clusters in the full dimension with equal weights. Then, each dimension is assigned a weight for each cluster, and the updated weights are then used in the next iteration to generate new clusters. The number of clusters and the size of the subspaces are difficult to determine, but are critical parameters for these types of methods. Depending on the ways to determine weights, popularly used top-down methods can be categorized into two categories. PROCLUS (Aggarwal et al., 1999), ORCLUS (Aggarwal and Yu, 2000), FINDIT (Woo and Lee, 2002), and δ-Clusters (Yang et al., 2002) determine the weights of instances for each cluster, while COSA (Friedman, 2004) determines the weights for each dimension for that particular instance. Readers are referred to Parsons et al. (2004) for a comprehensive overview of the heuristic methods.

On the other hand, there is also a huge family of model-based subspace clustering methods, which are closely connected to one another. Fisher (1936) applied the idea of restricting component means to a linear subspace and proposed the linear discriminant analysis (LDA). Since the marginal distribution of the observation without the class label is a mixture distribution, LDA involves the idea of the mixture model. Hastie and Tibshirani (1996) applied the idea of reduced rank LDA to GMM, and proposed a reduced rank version of the mixture discriminant analysis (MDA), which performs a reduced rank weighted LDA in each iteration of the EM algorithm. MFA may be considered one of the most famous subspace clustering methods, clustering the data and locally reducing the dimensionality of each cluster. See Section 8.2.2 for detailed discussions on MFA.

8.5.2 High-dimensional GMM

Based on the singular value decomposition (SVD), by constraining the covariance structure of the GMM as $\boldsymbol{\Sigma}_c = \mathbf{Q}_c \boldsymbol{\Lambda}_c \mathbf{Q}_c^\top$, where $\mathbf{Q}_c$ is a p-dimensional orthogonal matrix containing the eigenvectors, $\boldsymbol{\Lambda}_c$ is a p-dimensional diagonal matrix containing the corresponding eigenvalues (in descending order), Bouveyron et al. (2007) proposed a series of 28 high-dimensional GAMM (HD-GMM), which are both flexible and parsimonious. The key idea of this methodology is to assume the diagonal matrix as $\boldsymbol{\Lambda}_c = \text{diag}\{a_{c1}, a_{cd_c}, b_c, \ldots, b_c\}$ $(d_c < p)$, where $a_{c1}, \ldots, a_{cd_c}$ parameterize the variance of the c-th component of the j-th subspace, $j = 1, \ldots, d_c$, and b_c's model the variance of the noise term. That is to say, the component-specific subspace $\mathbb{E}_c$ is spanned by the d_c eigenvectors associated with a_{cj} such that $\boldsymbol{\mu}_c \in \mathbb{E}_c$, and $\mathbb{E}_c$ is called the specific subspace of the c-th component since most of the data live on or near this subspace.

By adjusting certain parameters to be common within or between components, Bouveyron et al. (2007) studied a family of HD-GMM models, denoted by $[a_{cj} b_c Q_c d_c]$, whose details are listed in Table 8.3.

If we assume that $d_c = p - 1$, then, $[a_{cj} b_c Q_c d_c]$ is nothing but the classical GMM with full covariance matrices for each component. The $[a_c b_c Q_c d_c]$ model, which assumes that each $\boldsymbol{\Lambda}_c$ only has two different eigenvalues, is an efficient way to regularize $\boldsymbol{\Lambda}_c$, and often gives satisfying results. In addition, the $[a_c b Q_c d_c]$ model, which models the noise in $\mathbb{E}_c$ in a single parameter b, is a natural model to use when data is obtained in a common acquisition process. On the other hand, the $[a_{cj} b_c Q(p-1)]$ model is equivalent to the GMM with diagonal matrices and $[a_{cj} b Q(p-1)]$ is equivalent to the GMM with a common diagonal matrix. The $[a_{cj} b_c Q_c d]$ method is equivalent to the mixture of principal component analyzer (Tipping and Bishop, 1999).

8.6 Variable selection for mixture models

In the "high dimension, low sample size" scenario, since there is no suitable hypothesis test to assess the statistical significance of an attribute (to the

TABLE 8.3

Members of the HD-GMM family (Bouveyron et al., 2007)

Model	No. of parameters
$[a_{cj}b_cQ_cd_c]$	$(C-1) + Cp + \sum_c d_c\{p - (d_c+1)/2\} + \sum_c d_c + 2C$
$[a_{cj}bQ_cd_c]$	$(C-1) + Cp + \sum_c d_c\{p - (d_c+1)/2\} + \sum_c d_c + 1 + C$
$[a_cb_cQ_cd_c]$	$(C-1) + Cp + \sum_c d_c\{p - (d_c+1)/2\} + 3C$
$[ab_cQ_cd_c]$	$(C-1) + Cp + \sum_c d_c\{p - (d_c+1)/2\} + 1 + 2C$
$[a_cbQ_cd_c]$	$(C-1) + Cp + \sum_c d_c\{p - (d_c+1)/2\} + 1 + 2C$
$[abQ_cd_c]$	$(C-1) + Cp + \sum_c d_c\{p - (d_c+1)/2\} + 2 + C$
$[a_{cj}b_cQ_cd]$	$(C-1) + Cp + Cd\{p - (d+1)/2\} + Cd + C + 1$
$[a_jb_cQ_cd]$	$(C-1) + Cp + Cd\{p - (d+1)/2\} + d + C + 1$
$[a_{cj}bQ_cd]$	$(C-1) + Cp + Cd\{p - (d+1)/2\} + Cd + 2$
$[a_jbQ_cd]$	$(C-1) + Cp + Cd\{p - (d+1)/2\} + d + 2$
$[a_cb_cQ_cd]$	$(C-1) + Cp + Cd\{p - (d+1)/2\} + 2C + 1$
$[ab_cQ_cd]$	$(C-1) + Cp + Cd\{p - (d+1)/2\} + C + 2$
$[a_cbQ_cd]$	$(C-1) + Cp + Cd\{p - (d+1)/2\} + C + 2$
$[abQ_cd]$	$(C-1) + Cp + Cd\{p - (d+1)/2\} + 3$
$[a_{cj}b_cQd_c]$	$(C-1) + Cp + d\{p - (d+1)/2\} + \sum_c d_c + 2C$
$[a_{cj}bQd_c]$	$(C-1) + Cp + d\{p - (d+1)/2\} + \sum_c d_c + 1 + C$
$[a_cb_cQd_c]$	$(C-1) + Cp + d\{p - (d+1)/2\} + 3C$
$[ab_cQd_c]$	$(C-1) + Cp + d\{p - (d+1)/2\} + 1 + 2C$
$[a_cbQd_c]$	$(C-1) + Cp + d\{p - (d+1)/2\} + 1 + 2C$
$[abQd_c]$	$(C-1) + Cp + d\{p - (d+1)/2\} + 2 + C$
$[a_{cj}b_cQd]$	$(C-1) + Cp + d\{p - (d+1)/2\} + Cd + C + 1$
$[a_jb_cQd]$	$(C-1) + Cp + d\{p - (d+1)/2\} + d + C + 1$
$[a_{cj}bQd]$	$(C-1) + Cp + d\{p - (d+1)/2\} + Cd + 2$
$[a_jbQd]$	$(C-1) + Cp + d\{p - (d+1)/2\} + d + 2$
$[a_cb_cQd]$	$(C-1) + Cp + d\{p - (d+1)/2\} + 2C + 1$
$[ab_cQd]$	$(C-1) + Cp + d\{p - (d+1)/2\} + C + 2$
$[a_cbQd]$	$(C-1) + Cp + d\{p - (d+1)/2\} + C + 2$
$[abQd]$	$(C-1) + Cp + d\{p - (d+1)/2\} + 3$

best of our knowledge), sequential variable selection methods, such as forward additions and backward eliminations, are unrealistic. In addition, given the current computation power, best subset selection methods are prohibitive.

As an alternative, it has been proposed to apply dimension reduction techniques, such as PCA, prior to clustering (Liu and Shao, 2003). However, the dimension reduction and subsequent clustering are separated in these methods, which may cause undesirable results. As pointed out by many articles (Chang, 1983; Raftery, 2003), the using of the first few principal components in clustering may destroy the clustering structure of the original data.

There is plenty of work in the Bayesian framework. One commonly used idea is to parameterize the mean as $\boldsymbol{\mu}_c = \boldsymbol{\mu} + \boldsymbol{\delta}_c$, where $\boldsymbol{\mu}$ is the global mean. If some elements of $\boldsymbol{\delta}_c$ are equal to 0 for all c, then the corresponding attributes are considered to be noninformative for clustering. Tadesse et al. (2005) focused on the clustering problem in terms of a multivariate normal mixture model with an unknown number of components. A reversible jump Markov chain Monte Carlo methodology is applied to define a sampler that could move between different dimensional spaces, and the variable selection was handled through the introduction of a binary inclusion/exclusion latent vector. Another type of method focuses on sequentially comparing two nested models, and determining the addition or elimination of an attribute from the current model based on a greedy search algorithm (Raftery and Dean, 2006). However, time issue is the main drawback of these types of methods. Generally speaking, by allowing more general covariance matrices, the Bayesian approaches are more flexible yet also time-consuming, due to the MCMC for stochastic searches.

Another branch of methods applies the penalized likelihood approach to select variables. Assuming the covariance matrices to be a common diagonal matrix, Pan and Shen (2007) employed the L_1 penalty to shrink mixture component means towards the global mean, automatically realizing variable selection and delivering a sparse solution. The objective function for this method is:

$$\ell_{gmm}(\boldsymbol{\theta}) - \lambda \sum_{c=1}^{C} \sum_{j=1}^{p} |\mu_{cj}|.$$

A modified BIC is proposed as a model selection criterion to choose C and λ. The above methods focus on regularizing the mean parameters which represent the centers of clusters. However, the dependencies among variables within each cluster are ignored, which may lead to incorrect orientations or shapes of the resulting clusters.

Zhou et al. (2009) studied a regularized Gaussian mixture model with general covariance matrices, and developed a method that shrinks the means and covariance matrices, simultaneously, to solve the aforementioned issue. The penalty function in this case is

$$\mathbf{p}(\boldsymbol{\theta}) = \lambda_1 \sum_{c=1}^{C} \sum_{j=1}^{p} |\mu_{cj}| + \lambda_2 \sum_{c=1}^{C} \sum_{j,l=1}^{p} |W_{cjl}|,$$

where $\mathbf{W}_c = (W_{jl})$ is the inverse of the covariance matrix $\boldsymbol{\Sigma}_c$. To estimate the model, the graphical lasso method, developed by Friedman et al. (2008) is utilized for covariance estimation and embedded into the EM algorithm.

There are several other methods in the literature that combine clustering and variable selection together. Friedman and Meulman (2004) proposed a hierarchical clustering procedure that uncovers cluster structure on separate subsets of variables, rather than on all of them simultaneously. The algorithm does not explicitly select variables but rather assigns them different weights, which can be used to extract informative variables.

8.7 High-dimensional mixture modeling through random projections

Another type of method uses random projections. Dasgupta (1999, 2000) has identified random projection as a promising dimension reduction technique for learning mixtures of Gaussians. The key idea of random mapping arises from the Johnson-Lindenstrauss lemma: if points in a vector space are projected onto a randomly selected subspace of suitably high dimension, then the distances between the points are approximately preserved. In random projection, the original p-dimensional data is projected to a d-dimensional subspace $(d << p)$ through a random $d \times p$ random matrix R whose columns are orthonormal to each other. If we use $\mathbf{X}_{n \times p}$ to denote the original set of n p-dimensional observations, then

$$\mathbf{X}_{n \times d}^{RP} = \mathbf{X}_{n \times p} \mathbf{R}_{p \times d}$$

is the projection of the data onto a lower d-dimensional subspace. Dasgupta (1999) showed that there are two main theoretical results about random projection. First, data from a mixture of C Gaussians can be projected into just $O(\log C)$ dimensions while still maintaining the approximate level of separation between the clusters. In the numerical studies, the authors showed that $d = 10 \log C$ works nicely, which is independent of the original dimension of the data p and the number of data points n. Secondly, even if the original clusters are highly eccentric, random projection will make them more spherical.

Hecht-Nielsen (1994) showed that in a high-dimensional space, there exists a much larger number of almost orthogonal than orthogonal directions. That is to say, vectors of random directions might be sufficiently close to orthogonal, or $\mathbf{R}^\top \mathbf{R}$ is approximate to an identity matrix. Therefore, Bingham and Mannila (2001) argued that the random matrix $\mathbf{R}$ does not need to be orthogonal, to save computation time.

Due to the randomness of the projection matrix, different random projections may lead to radically different clustering results. Fern and Brodley (2003) proposed a three-step process to combine the idea of random projection and clustering. First, generate multiple clustering results using random projection and EM algorithm; secondly, aggregate the clustering results of a GMM on different random projections of the data into a similarity matrix containing the probability "estimates" that any two data points belong to the same cluster; thirdly, perform an agglomerative clustering procedure on such a matrix to produce the final groups.

Anderlucci et al. (2022) on the other hand, proposed a BIC-type criteria, to choose between different random projections. Then, the cluster membership vector of the "best" few projections is aggregated via consensus.

8.8 Multivariate generalized hyperbolic mixtures

Browne and McNicholas (2015) proposed a multivariate generalized hyperbolic mixture model (GHMM) defined as

$$f(\mathbf{x}; \boldsymbol{\theta}) = \sum_{c=1}^{C} \pi_c g_c(\mathbf{x}; \lambda_c, \omega_c, \boldsymbol{\mu}_c, \boldsymbol{\Sigma}_c, \boldsymbol{\alpha}_c), \tag{8.8}$$

where the c-th component density is

$$g_c(\mathbf{x}; \lambda_c, \omega_c, \boldsymbol{\mu}_c, \boldsymbol{\Sigma}_c, \boldsymbol{\alpha}_c) = \left[\frac{\omega_c + \delta(\mathbf{x}; \boldsymbol{\mu}_c, \boldsymbol{\Sigma}_c)}{\omega_c + \boldsymbol{\alpha}_c^\top \boldsymbol{\Sigma}_c^{-1} \boldsymbol{\alpha}_c} \right]^{(\lambda_c - p/2)/2}$$

$$\times \frac{K_{\lambda_c - p/2} \left(\sqrt{(\omega_c + \boldsymbol{\alpha}_c^\top \boldsymbol{\Sigma}_c^{-1} \boldsymbol{\alpha}_c)[\omega_c + \delta(\mathbf{x}; \boldsymbol{\mu}_c, \boldsymbol{\Sigma}_c)]} \right)}{(2\pi)^{p/2} |\boldsymbol{\Sigma}_c|^{1/2} K_{\lambda_c}(\omega_c) \exp\{-(\mathbf{x} - \boldsymbol{\mu})^\top \boldsymbol{\Sigma}_c^{-1} \boldsymbol{\alpha}_c\}},$$

with the index parameter λ_c, concentration parameter ω_c, skewness parameter $\boldsymbol{\alpha}_c$, location parameter $\boldsymbol{\mu}_c$, and scale matrix $\boldsymbol{\Sigma}_c$. Note that $\delta(\mathbf{x}; \boldsymbol{\mu}_c, \boldsymbol{\Sigma}_c) = (\mathbf{x} - \boldsymbol{\mu}_c)^\top \boldsymbol{\Sigma}_c^{-1} (\mathbf{x} - \boldsymbol{\mu}_c)$ denotes the squared Mahalanobis distance between $\mathbf{x}$ and $\boldsymbol{\mu}_c$, and K_{λ_c} denotes the modified Bessel function of the third kind with index λ_c. Note that, in this case, the p-dimensional random vector $\mathbf{x}$ can be considered as generated by combining a generalized inverse Gaussian (GIG) random variable Y, where the density of $Y \sim \mathrm{GIG}(\omega, \eta, \lambda)$ is

$$f(y; \omega, \eta, \lambda) = \frac{(y/\eta)^{\lambda - 1}}{2\eta K_\lambda(\omega)} \exp\left\{ -\frac{\omega}{2} \left(\frac{y}{\eta} + \frac{\eta}{y} \right) \right\},$$

with a latent multivariate Gaussian random variable $U \sim N(\mathbf{0}, \boldsymbol{\Sigma})$, through the relationship $\mathbf{x} = \boldsymbol{\mu} + Y\boldsymbol{\alpha} + \sqrt{Y}U$ (by fixing $\eta = 1$).

The complete log-likelihood of this model is

$$\ell_c(\boldsymbol{\theta}) = \sum_{i=1}^{n} \sum_{c=1}^{C} z_{ic} \left[\log \pi_c + \log\{\phi(\mathbf{x}_i | \boldsymbol{\mu}_c + y_{ic}\boldsymbol{\alpha}_c, y_{ic}\boldsymbol{\Sigma}_c)\} + \log\{f(y_{ic}; \omega_c, \lambda_c)\} \right],$$

where $\mathbf{x}|y_{ic}, z_{ic} = 1 \sim N(\boldsymbol{\mu}_c + y_{ic}\boldsymbol{\alpha}_c, y_{ic}\boldsymbol{\Sigma}_c)$ and $Y_{ic} \sim \mathrm{GIG}(\omega_c, 1, \lambda_c)$. The details of the EM algorithm for generalized hyperbolic mixtures are given in Browne and McNicholas (2015).

Morris and McNicholas (2016) proposed a new mixture model-based approach based on fitting generalized hyperbolic mixtures on a reduced subspace within the paradigm of model-based clustering, classification, and discriminant analysis. Given a generalized hyperbolic mixture (8.8), the authors aim to find a subspace $\mathcal{S}(\boldsymbol{\beta})$ where the cluster means and covariances vary the most. In order to do so, a kernel matrix M_{GHMM} for a generalized hyperbolic mixture is defined as

$$M_{GHMM} = \sum_{c=1}^{C} \pi_c(\tilde{\boldsymbol{\mu}}_c - \boldsymbol{\mu})(\tilde{\boldsymbol{\mu}}_c - \boldsymbol{\mu})^\top \boldsymbol{\Sigma}^{-1} \sum_{c=1}^{C} \pi_c(\tilde{\boldsymbol{\mu}}_c - \boldsymbol{\mu})(\tilde{\boldsymbol{\mu}}_c - \boldsymbol{\mu})^\top$$
$$+ \sum_{c=1}^{C} \pi_c(\tilde{\boldsymbol{\Sigma}}_c - \tilde{\boldsymbol{\Sigma}})\boldsymbol{\Sigma}^{-1}(\tilde{\boldsymbol{\Sigma}}_c - \tilde{\boldsymbol{\Sigma}})^\top,$$

where $\tilde{\boldsymbol{\mu}}_c = \boldsymbol{\mu}_c + \boldsymbol{\alpha}_c$ and $\tilde{\boldsymbol{\Sigma}}_c = \boldsymbol{\Sigma}_c + \boldsymbol{\alpha}_c\boldsymbol{\alpha}_c^\top$ are the mean and covariance of $\mathbf{X}$ in (8.8), $\boldsymbol{\Sigma} = \sum_{i=1}^{n}(\mathbf{x}_i - \boldsymbol{\mu})(\mathbf{x}_i - \boldsymbol{\mu})^\top/n$ denotes the overall covariance matrix and $\tilde{\boldsymbol{\Sigma}} = \sum_{c=1}^{C} \pi_c\tilde{\boldsymbol{\Sigma}}_c$ is the pooled within-cluster covariance matrix. Then, an Eigen-decomposition of M_{GHMM} is

$$M_{GHMM}\nu_i = t_i\boldsymbol{\Sigma}\nu_i, \qquad (8.9)$$

where $t_1 \geq t_2 \geq \ldots \geq t_d > 0$ and $\nu_i^\top\boldsymbol{\Sigma}\nu_j = 1$ for $i = j$ and 0 otherwise. Then, the eigenvectors $[\nu_1, \ldots, \nu_d]$ where $d \leq p$ is the spanning set of the dimension reduced subspace $\mathcal{S}(\boldsymbol{\beta})$.

The method of Morris and McNicholas (2016) can be summarized as follows:

Step 1. Fit a GHMM (8.8) to the data through an EM algorithm;

Step 2. Find the GHMM dimension reduction directions, through (8.9);

Step 3. Project the data onto the estimated subspace $\mathcal{S}(\boldsymbol{\beta})$, and apply the greedy search algorithm (Scrucca, 2010) to discard those that do not provide clustering information;

Step 4. Fit a GHMM (8.8) to the selected variables and repeat to Step 2;

Step 5. Repeat Step 2-4 until none of the features can be discarded.

8.9 High-dimensional mixture regression models

In the regression settings, Khalili and Chen (2007) systematically studied the variable selection of finite mixtures of regression (FMR) models through the penalized likelihood approach. We say that $(\mathbf{x}, Y)$ follows an FMR model of complexity C if the conditional distribution of Y given $\mathbf{x}$ is

$$f(y; \mathbf{x}, \boldsymbol{\theta}) = \sum_{c=1}^{C} \pi_c g(y; \mathbf{x}^\top\boldsymbol{\beta}_c, \phi_c),$$

where $\boldsymbol{\theta} = (\boldsymbol{\beta}_1^\top, \ldots, \boldsymbol{\beta}_C^\top, \boldsymbol{\phi}^\top, \boldsymbol{\pi}^\top)^\top$ with $\boldsymbol{\beta}_c = (\beta_{c1}, \ldots, \beta_{cp})^\top, \boldsymbol{\phi} = (\phi_1, \ldots, \phi_C)^\top, \boldsymbol{\pi} = (\pi_1, \ldots, \pi_{C-1})^\top$ with $\pi_c > 0$ and $\sum_{c=1}^{C} \pi_c = 1$. The log-likelihood of $\boldsymbol{\theta}$ is given by

$$\ell_{fmr}(\boldsymbol{\theta}) = \sum_{i=1}^{n} \left\{ \sum_{c=1}^{C} \pi_c g(y_i; \mathbf{x}_i^\top \boldsymbol{\beta}_c, \phi_c) \right\}.$$

The penalized log-likelihood function of Khalili and Chen (2007) is defined as:

$$\ell_{fmr}(\boldsymbol{\theta}) - p(\boldsymbol{\theta}),$$

where $p(\boldsymbol{\theta})$ is the penalty function defined as

$$p(\boldsymbol{\theta}) = \sum_{c=1}^{C} \pi_c \left\{ \sum_{j=1}^{p} p_c(\beta_{cj}) \right\},$$

where $p_c(\beta_{cj})$ are nonnegative and nondecreasing functions of $|\beta_{cj}|$. In the article of Khalili and Chen (2007), L_1 penalty, HARD, and SCAD penalty (Fan et al., 2001) are all considered. The asymptotic properties of the estimators are also thoroughly studied in Khalili and Chen (2007).

Khalili et al. (2011) considered the variable selection problem under a finite mixture of sparse normal linear (FMSL) model, where the number of candidate covariates, p, is comparable to the sample size n or even greater than n. Direct use of the modified EM algorithm of Khalili and Chen (2007) to FMSL models is too computationally intensive. In fact, due to the huge number of predictors, the penalized log-likelihood surface is very likely to be too flat to find a global maximizer. A two-stage procedure is proposed to estimate the FMSL. In the first stage, a likelihood-based boosting technique is proposed to screen variables, where variables are being selected but the final model is not fitted. Let $s^{(k)}$ be the currently active set after the k-th iteration. For $j = 1, \ldots, p$, fit single-variable mixture regression models through the adjusted log-likelihood function:

$$\tilde{\ell}(\boldsymbol{\pi}, \boldsymbol{\alpha}, \boldsymbol{\beta}, \boldsymbol{\sigma}^2; j) = \sum_{i=1}^{n} \log \left\{ \sum_{c=1}^{C} \pi_c \phi(y_i; \hat{\mu}_{ci}^{(k)} + \alpha_c + \beta_c \mathbf{x}_i[j], \sigma_c^2) \right\} - \sum_{c=1}^{C} p_n(\sigma_c),$$

where $\hat{\mu}_{ci}^{(k)}$ refers to the i-th element of $\hat{\boldsymbol{\mu}}_c^{(k)}$ in the k-th iteration, $\mathbf{x}_i[j]$ is the jth value of $\mathbf{x}_i$, and

$$p_n(\sigma_c) = a \left\{ \frac{S_n^2}{\sigma_c^2} - \log \left(\frac{S_n^2}{\sigma_c^2} \right) \right\}$$

is a nonnegative penalty function to avoid the unboundedness of the likelihood with S_n^2 being the sample variance of y_i. Let $j_0 = \arg\max_{j} \tilde{\ell}(\boldsymbol{\pi}, \boldsymbol{\alpha}, \boldsymbol{\beta}, \boldsymbol{\sigma}^2; j)$, and $\ell_n(j_0) = \max \tilde{\ell}(\boldsymbol{\pi}, \boldsymbol{\alpha}, \boldsymbol{\beta}, \boldsymbol{\sigma}^2; j) = \tilde{\ell}(\boldsymbol{\pi}, \boldsymbol{\alpha}, \boldsymbol{\beta}, \boldsymbol{\sigma}^2; j_0)$, then the active set is updated by $s^{(k+1)} = s^{(k)} \cup \{j_0\}$, the fit is boosted by

$$\hat{\mu}_{ci}^{(k+1)} = \hat{\mu}_{ci}^{(k)} + \nu(\hat{\alpha}_c + \hat{\beta}_c \mathbf{x}_i[j_0])$$

for $c = 1, \ldots, C$, where $\hat{\alpha}_c$ and $\hat{\beta}_c$ are the AMLEs corresponding to $\ell_n(j_0)$, $0 < \nu \leq 1$ a prespecified step size parameter. Then, in the second stage, the penalized likelihood technique of Khalili and Chen (2007) is applied to select the variables and fit the model.

8.10   Fitting high-dimensional mixture models using R

There are several packages and functions available for mixture model analysis for high-dimensional data.

The mclust package (Fraley et al., 2012) is suitable to fit mixtures of parsimonious GMM. The users could specify the covariance structures by themselves, or call BIC for comparison. For example, we apply parsimonious GMM to the iris data.

```
library(mclust)
Class = iris$Species
X = subset(iris, select = - Species)

# Common EEE covariance structure (equivalent to LDA)
irisMclustDA = MclustDA(X, Class, modelType = "EDDA",
                        modelNames = "EEE")
# Default 10-fold CV
cv10fold = cv.MclustDA(irisMclustDA)
cv10fold[c("error", "se")]
# LOO-CV
cvloo = cv.MclustDA(irisMclustDA, nfold = length(Class))
cvloo[c("error", "se")]

# General covariance structure selected by BIC
irisMclustDA = MclustDA(X, Class)
# Default 10-fold CV
cv = cv.MclustDA(irisMclustDA)
cv[c("error", "se")]

library(HDclassif)
data(Crabs, package = "HDclassif")
CrabsClass = Crabs$class
CrabsX = subset(Crabs, select = - class)
```

```
HDDC4 = hddc(data = CrabsX, K = 4, algo = "EM",
        init = "mini-em")
HDDC4pred = predict(HDDC4, CrabsX, CrabsClass)
# Users can specify the models to be fitted
HDDC8 = hddc(data = CrabsX, K = 1:8, model = c(1, 2, 7, 9))
HDDC8pred = predict(HDDC8, CrabsX, CrabsClass)
```

The EMMIXmfa (McLachlan and Peel, 2000; Baek et al., 2010a) package is designed to fit finite mixture models of factor analyzers and mixtures of common factor analyzers, where the matrix of factor loadings is common to components before the component-specific rotation of the component factors makes them white noise. The error distributions could be either multivariate normal or t-distributions. For example, to fit the iris data,

```
library(EMMIXmfa)
Y = subset(iris, select = - Species)
mfa_model = mfa(Y, g = 3, q = 3)
mtfa_model = mtfa(Y, g = 3, q = 3)
mcfa_model = mcfa(Y, g = 3, q = 3)
mctfa_model = mctfa(Y, g = 3, q = 3)
```

The CLEMM (Wang et al., 2020) package is suitable for model-based clustering based on envelopes.

```
install.packages("devtools")
library(devtools)
install_github("kusakehan/CLEMM")
library(CLEMM)
# Selection of dimension
dim_res = env_dim_selection(1:10, dat, K, typ = "G")
# Fit through CLEMM
res_clemm = clemm_em(dat, K, u = dim_res$u, typ = "G")
```

The clustvarsel (Scrucca, 2018) and fmerPack packages perform variable selection in Gaussian mixture models. We apply both packages to the prostate data, which is designed to examine the correlation between the level of prostate-specific antigen and a number of clinical measures in men who were about to receive radical prostatectomy. The dataset contains 97 observations and 10 variables, namely *lcavol* for log cancer volume, *lweight* for log prostate weight, *age* in years, *lbph* for log of the amount of benign prostatic hyperplasia, *svi* for seminal vesicle invasion, *lcp* for log of capsular penetration, *gleason*, *pgg45* for percent of Gleason score 4 or 5, and *lpsa* for response. The methods considered are:

(i). fmerPack for FMR model with lasso penalty;

(ii). clustvarsel1 for sequential backward greedy search;

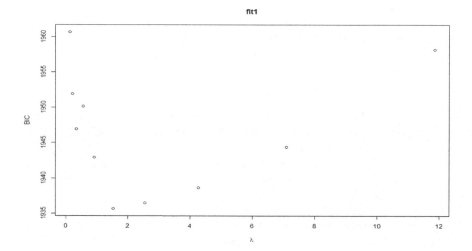

FIGURE 8.2
BIC plot for the tuning parameter selection on prostate data.

TABLE 8.4
Variable selection result for the `prostate` data

Variable	`fmerPack`	`clustvarsel1`	`clustvarsel2`	`clustvarsel3`
lcavol	-	-	-	✓
lweight	-	✓	✓	-
age	✓	✓	✓	-
lbph	✓	✓	✓	-
svi	-	✓	✓	-
lcp	✓	✓	✓	-
gleason	✓	✓	✓	-
pgg45	✓	✓	✓	✓

(iii). `clustvarsel2` for sequential backward greedy search with subsampling at hierarchical intialization stage;

(iv). `clustvarsel3` for headlong search.

The results are shown in Figure 8.2 and Table 8.4.

```
library(faraway)
data(prostate, package = "faraway")
X = as.matrix(subset(prostate, select = - lpsa))
y = as.vector(prostate$lpsa)
library(fmerPack)
m=3
```

```
fit1 <- path.fmrReg(y, X, m = m, modstr = list(nlambda = 10),
    control = list(n.ini = 1))
lam.sel = select.tuning(fit1, figure = TRUE, criteria =
        "BIC")
fmerPack = fmrReg(y, X, m = m, intercept = FALSE,lambda
        = lam.sel$info[1])
clustvarsel1 = clustvarsel(cbind(y, x), G = 1:5, direction
        = "backward")
clustvarsel2 = clustvarsel(cbind(y, x), G = 1:5, direction
        = "backward", samp = TRUE, sampsize = 50)
clustvarsel3 = clustvarsel(cbind(y, x), G = 1:5, search
        = "headlong")
```

9

Semiparametric mixture models

9.1 Why semiparametric mixture models?

Finite mixture models have been important tools for exploring complex data structures in many scientific areas. Parametric mixture models, being easy to interpret, quick to estimate, and having well-studied theoretical properties, have been popularly used ever since. However, similar to general parametric statistical inference tools, traditional mixture models are based on parametric model assumptions, such as linearity and normality, which are unrealistic or hard to satisfy in practice. Besides this, model misspecification could be disastrous in parametric mixture models and might lead to misleading results and inferences.

The need for semiparametric mixture models not only comes from the model point of view but also from practice. Bordes et al. (2006a) used a gene detection example to show the application of a two-component mixture model, in which one component is assumed to be known. To be more specific, in order to detect differentially expressed genes under two or more conditions in microarray data, a real-valued test statistic was calculated for each gene. Each test statistic is assumed to have a known distribution F_0 under the null hypothesis H_0 and an unknown distribution F under the alternative hypothesis H_1. The collected test statistics, each of which could be either from H_0 or H_1, should then form a two-component mixture model with F_0 and F as its component distributions. The researcher could then detect the differentially expressed genes from those nondifferentially expressed by estimating the mixing proportions and the unknown distribution. As another example, since the earnings included the Return on Equity (ROE) is comprised of real earnings and manipulated earnings, Huang et al. (2018c) proposed a two-component semiparametric mixture model with one known component, where the known component F_0 is assumed to be Pareto.

DOI: 10.1201/9781003038511-9

9.2 Semiparametric location shifted mixture models

Consider a C-component mixture model

$$f(x) = \sum_{c=1}^{C} \pi_c g(x - \mu_c), \qquad x \in \mathbb{R}, \tag{9.1}$$

where g is symmetric about the origin, $\boldsymbol{\pi} = (\pi_1, \ldots, \pi_C)^\top$ is a vector of unknown mixture proportions satisfying $\pi_c > 0$ for all c and $\sum_{c=1}^{C} \pi_c = 1$, and $\boldsymbol{\theta} = (\pi_1, \ldots, \pi_C, \mu_1, \ldots, \mu_C)^\top$ is the vector of unknown parameters. It has been shown by Bordes et al. (2006c) and Hunter et al. (2007), that model (9.1) is identifiable for $C \leq 3$ under some conditions. One of the most important conditions is the symmetry of g about the origin. For example, when $C = 2, \pi \notin \{0, 1/2, 1\}$ and $\mu_1 \neq \mu_2$, the model is identifiable. Several articles have focused on the estimation of the model (9.1), and their ideas are briefly discussed below.

Bordes et al. (2006c) proposed a cumulative distribution function (cdf) based M-estimation method to estimate $\boldsymbol{\theta}$ and g separately, showing that the estimator is $n^{-1/4+\alpha}$ a.s. consistent for all $\alpha > 0$. Assume $F(\cdot)$ and $G(\cdot)$ to be the cdf's of $f(\cdot)$ and $g(\cdot)$, respectively. Let $A_\theta = \pi \tau_{\mu_1} + (1 - \pi) \tau_{\mu_2}$ (von Neumann, 1931) with τ_μ being an invertible operator from L_1 to L_1. Then, the cdf version of (9.1) is equivalent to $F = A_\theta G$. Define $S_r\{G(\cdot)\} = 1 - G(-\cdot)$ to be a symmetry operator, then $F = A_\theta S_r A_\theta^{-1}$ is satisfied if and only if $\boldsymbol{\theta} = \boldsymbol{\theta}_0$ ($\boldsymbol{\theta}_0$ is the true value), which is consistent with the above identifiability result. Then, $\boldsymbol{\theta}$ is estimated by $\arg \min_{\theta \in \Theta} K(\boldsymbol{\theta}; \hat{F}_n)$, where $K(\boldsymbol{\theta}; F) = \int_{\mathbb{R}} \{F_\theta(X) - F(X)\}^2 dF(x)$, Θ is a compact parametric space and $\hat{F}_n$ is the empirical cdf of the sample drawn from F_{θ_0}. By symmetry, one can further estimate the unknown distribution G by $\hat{G}_n = \frac{1}{2}(I + S_r) A_{\hat{\theta}_n}^{-1} \hat{F}_n$, where I is the identity operator. However, as pointed out by Bordes et al. (2007), this direct estimator of g may not satisfy the conditions of a probability density function (pdf), making the numerical calculation a time-consuming process.

As a result, Bordes et al. (2007) proposed to estimate the unknown pdf g by

$$g_h(x) = \frac{1}{2n} \sum_{i=1}^{n} \sum_{c=1}^{C} p_{ic}\{K_h(x - x_i + \mu_c) + K_h(x + x_i - \mu_c)\}, \tag{9.2}$$

where p_{ic} is the probability that x_i comes from component c, $K_h(t) = h^{-1}K(t/h)$, $K(t)$ is a kernel function symmetric about 0, and h is a tuning parameter going to 0. Here, we suppress the dependence of h on n when there is no ambiguity. To achieve such an estimator, an algorithm generalized from the EM is proposed. However, the asymptotic behavior of the estimator (9.2) was not discussed.

Assume that $\mathcal{D}\{F_1, F_2\}$ denotes the distance of some kind between distributions F_1 and F_2, then, Hunter et al. (2007) proposed to estimate $\boldsymbol{\theta}$ by minimizing $d_n(\boldsymbol{\theta}, \hat{F}_n) = \mathcal{D}[\sum_{c=1}^{C} \pi_c \hat{F}_n(x + \mu_c), \sum_{c=1}^{C} \pi_c\{1 - \hat{F}_n(x - \mu_c)\}]$. The authors showed that under certain technical conditions, the estimator of $\boldsymbol{\theta}$ is asymptotically normally distributed with $\sqrt{n}$-rate. A decade later, Balabdaoui (2017) also studied this estimation methodology, formally proving the existence of such an estimator and showing its asymptotic distribution.

Let

$$\tilde{g}_h(x; Q) = \frac{1}{2} \int \{K_h(x - \sigma) + K_h(x + \sigma)\} dQ(\sigma), \qquad (9.3)$$

be a generalization of the kernel-based method, where Q is an unknown mixing distribution, Chee and Wang (2013) proposed to estimate $\boldsymbol{\theta}$ through a semiparametric MLE approach. However, since Q is unspecified, this estimation method is not easy to compute. In fact, even with a known and fixed $\boldsymbol{\theta}$ value, the estimation of Q involves an optimization problem over an infinite dimensional space, and therefore, is a challenging task. Fortunately, it has been shown by Lindsay (1983), that the nonparametric MLE (NPMLE) of Q is discrete with finite support points which are no more than the number of distinct observations. Denote the NPMLE of Q by

$$\hat{Q}_n = \sum_{j=1}^{m} w_j \delta_{\sigma_j}, \qquad (9.4)$$

which has mass at σ_j with probability w_j for $j = 1, \ldots, m$. Replacing Q by $\hat{Q}_n$, (9.3) then becomes

$$\tilde{g}_h(x; \mathbf{w}, \boldsymbol{\sigma}) = \frac{1}{2} \sum_{j=1}^{m} w_j \{K_h(x - \sigma_j) + K_h(x + \sigma_j)\}, \qquad (9.5)$$

where $\mathbf{w} = (w_1, \ldots, w_m)^\top$ and $\boldsymbol{\sigma} = (\sigma_1, \ldots, \sigma_m)^\top$. It could be easily seen that estimator (9.2) is indeed a special case of (9.5). With the above preparation work, model (9.1) then becomes

$$\tilde{f}_h(x; \boldsymbol{\theta}, \mathbf{w}, \boldsymbol{\sigma}) = \sum_{c=1}^{C} \pi_c \tilde{g}_h(x - \mu_c; \mathbf{w}, \boldsymbol{\sigma}), \qquad (9.6)$$

and the estimation of $\boldsymbol{\theta}$ and g can then be achieved through the maximization of log-likelihood of (9.6), which can be done by algorithms of Wang (2010).

A similar idea is also applied by Xiang et al. (2016). But instead of (9.3), the unknown density g is assumed to be

$$\breve{g}(x; Q) = \int_{\mathbb{R}^+} \frac{1}{\sigma} \phi\left(\frac{x}{\sigma}\right) dQ(\sigma), \qquad (9.7)$$

where $\phi(x)$ is the pdf of a standard normal distribution, and Q is also estimated by (9.4). Compared to (9.3), (9.7) includes a richer class of continuous distributions, and the resulting estimators are robust against outliers. The algorithm to estimate $\boldsymbol{\theta}$ and Q works by iterating between the estimation of $\boldsymbol{\theta}$ and Q. With a fixed $\boldsymbol{\theta}$, the parameters $\mathbf{w}, \boldsymbol{\sigma}$, and m are estimated through a gradient-based algorithm. At a given $\hat{Q}_n$, $\boldsymbol{\theta}$ could then be updated by a regular EM algorithm. Since no tuning parameter is to be selected, this method is simple to use.

Butucea and Vandekerkhove (2014) proposed to estimate model (9.1) from a completely different point of view. The Fourier Analysis is applied to invert the mixture operator, making the connection between the symmetry of g and that its Fourier transform has no imaginary part. Specifically, define $g^*(u) = \int_{\mathbb{R}} e^{ixu} g(x) dx$ as the Fourier transformation of $g(x)$, then model (9.1) implies

$$f^*(u) = \{\pi e^{iu\mu_1} + (1 - \pi)e^{iu\mu_2}\}g^*(u) \overset{def}{=} M(\boldsymbol{\theta}, u)g^*(u). \qquad (9.8)$$

Based on the symmetry of g, $Im\{f^*(u)/M(\boldsymbol{\theta}, u)\} = 0$ holds if and only if $\boldsymbol{\theta} = \boldsymbol{\theta}_0$. Finally, let $S(\boldsymbol{\theta}) = \int_{\mathbb{R}} \{f^*(u)/M(\boldsymbol{\theta}, u)\}^2 dW(u)$ be a contrast function based on (9.8), where W is a Lebesgue absolutely continuous probability measure supported by $\mathbb{R}$, and $S_n(\boldsymbol{\theta})$ is an estimator of $S(\boldsymbol{\theta})$, then $\boldsymbol{\theta}$ is estimated by $\arg\min_{\theta \in \Theta} S_n(\boldsymbol{\theta})$. Butucea and Vandekerkhove (2014) proved the asymptotic distribution of the estimators $\beta > 1/2$, under conditions simpler than Hunter et al. (2007). If identifiability assumptions could be satisfied, the estimators and convergence results could be further extended to the $C \geq 3$ cases.

Wu et al. (2017) proposed to estimate model (9.1) by minimizing a profile Hellinger distance. To be more specific, for $C = 2$ (the result could be easily extended to the cases with $C > 2$), model (9.1) can be written as

$$\mathcal{H} = \{h_{\boldsymbol{\theta}, g}(x) = \pi g(x - \mu_1) + (1 - \pi)g(x - \mu_2) : \boldsymbol{\theta} \in \Theta, g \in \mathcal{G}\},$$

where

$$\Theta = \{\boldsymbol{\theta} = (\pi, \mu_1, \mu_2) : \pi \in (0, 1/2) \cup (1/2, 1), \mu_1 < \mu_2\},$$

$$\mathcal{G} = \left\{g : g \geq 0, g(-x) = g(x), \int g(x) dx = 1\right\}.$$

Define a nonparametric kernel density estimator of (9.1), denoted by

$$\hat{h}_n(x) = \frac{1}{n} \sum_{i=1}^{n} K_h(X_i - x). \qquad (9.9)$$

Then, Wu et al. (2017) proposed an MPHD estimator of model (9.1) by

$$\hat{\boldsymbol{\theta}} = \arg\min_{\boldsymbol{\theta} \in \Theta} |h_{\boldsymbol{\theta}, g_{\boldsymbol{\theta}}}^{1/2} - \hat{h}_n^{1/2}|,$$

where $g_{\boldsymbol{\theta}}$ is defined as

$$g_{\boldsymbol{\theta}} = \arg\min_{g \in \mathcal{G}} |h_{\boldsymbol{\theta}, g}^{1/2} - \hat{h}_n^{1/2}|.$$

9.3   Two-component semiparametric mixture models with one known component

In addition to the above location-shifted mixture models, there is a two-component semiparametric mixture model that has been studied by dozens of articles, allowing one to see its importance.

The model is assumed to be

$$f(x) = (1 - \pi)g_0(x) + \pi g(x - \mu), \qquad x \in \mathbb{R}, \tag{9.10}$$

where g_0 is a known pdf, $g \in \mathcal{G}$ is unknown, where $\mathcal{G} = \{g : g \geq 0, \int g(x)dx = 1 \text{ and } g(-x) = g(x))\}$, and $\boldsymbol{\theta} = (\pi, \mu)^\top$ is the vector of unknown parameters. This model is sometimes referred to as a contamination model and is an extension of the classical two-component mixture model, in the sense that g is not restricted to any distribution families, but only symmetric. It has been shown by Bordes et al. (2006a) that model (9.10) is identifiable when g has a third-order moment and is zero-symmetric.

The inspiration of such a model comes from different application fields. For example, Bordes et al. (2006a) applied such a model in microarray data analysis to detect differentially expressed genes under two or more conditions. Song et al. (2010) used it for a sequential clustering algorithm, and Patra and Sen (2016), among others, introduced this model to astronomy and biology for contamination problems.

Bordes et al. (2006a) applied the cdf-based estimation method that is somewhat similar to the estimation of the model (9.1) by Bordes et al. (2006c). The inversion of cdf of model (9.10) leads to

$$G(x) = \frac{1}{\pi}\{F(x + \mu) - (1 - \pi)G_0(x + \mu)\}, \tag{9.11}$$

where G_0 is the cdf of g_0. Next, let

$$H_{11}(x; \mu, m, F) = \frac{\mu}{m}F(x + \mu) + \frac{m - \mu}{m}G_0(x + \mu),$$

$$H_{21}(x; \mu, m, F) = 1 - \frac{\mu}{m}F(\mu - x) + \frac{\mu - m}{m}G_0(\mu - x),$$

where m is the first-order moment of F. Then, the symmetry of g implies that μ should minimize the distance between H_{11} and H_{21}, i.e., the estimator $\hat{\mu}_n$ should minimize $d_1\{H_{11}(\cdot; \mu, \hat{m}_n, \hat{F}_n), H_{21}(\cdot; \mu, \hat{m}_n, \hat{F}_n)\}$, where d_1 is the L_q distance, and $\hat{F}_n$ and $\hat{m}_n$ are the empirical versions of F and m, respectively, derived from a sample of size n. The mixing proportion π could then be estimated by $\hat{\pi}_n = \hat{m}_n / \hat{\mu}_n$. However, simulations show that the estimator is not numerically stable. In addition, there are lack of theoretical properties.

Bordes and Vandekerkhove (2010) offered a modification to the above methodology. They also used the inversion equation (9.11), but instead of d_1, the following distance is considered:

$$d_2(\boldsymbol{\theta}) = \int_{\mathbb{R}} \{H_{12}(x;\theta,F) - H_{22}(x;\theta,F)\}^2 dF(x),$$

where

$$H_{12}(x;\theta,F) = \frac{1}{\pi}F(x+\mu) + \frac{1-\pi}{\pi}G_0(x+\mu),$$
$$H_{22}(x;\theta,F) = 1 - \frac{1}{\pi}F(\mu-x) + \frac{1-\pi}{\pi}G_0(\mu-x).$$

In numerical study, an empirical version of d_2 is defined as

$$d_{2n}(\boldsymbol{\theta}) = \frac{1}{n}\sum_{i=1}^{n}\{H_{12}(X_i;\theta,\tilde{F}_n) - H_{22}(X_i;\theta,\tilde{F}_n)\}^2,$$

where $\tilde{F}_n = \int_{-\infty}^{x}\hat{f}_n(t)dt$, a smoothed version of the empirical cdf $\hat{F}_n$, and $\hat{f}_n(x)$ is as defined in (9.9).

Maiboroda and Sugakova (2011) proposed a generalized estimating equations (GEE) method to estimate the model (9.10). Denote $(X_1, \ldots, X_n)$ as a sample generated from (9.10), and z, z_0 and δ be three random variables such that $z \sim g$, $z_0 \sim g_0$ and $\delta \sim \text{Bin}(1, \pi)$. Then, $X_i \sim \delta(z+\mu) + (1-\delta)z_0$. Define two odd functions, h_1, h_2, and $H_{j3}(\mu) = Eh_j(z_0-\mu)$ for any $\mu \in \mathbb{R}$ and $j = 1, 2$. Then, the expectation $Eh_j(X_i - \mu) = \pi Eh_j(z) + (1-\pi)H_{j3}(\mu)$. However, since h_js are odd and g is symmetric, $Eh_j(z) = 0$. Therefore, $Eh_j(X_i - \mu) = (1-\pi)H_{j3}(\mu)$. Based on the aforementioned results, the authors proposed to estimate the Euclidean parameter $\boldsymbol{\theta}$ of (9.10) through the following unbiased estimating equations $\hat{h}_j(\mu) - (1-\pi)H_{j3}(\mu) = 0$ for $j = 1, 2$, where $\hat{h}_j(\mu) = n^{-1}\sum_{i=1}^{n}h_j(X_i - \mu)$. Under mild conditions, the consistency and asymptotic normality of this estimator are then proved by the authors.

Among those who studied model (9.10), most assume the unknown density g to be symmetric and use this condition as a key element in their estimation processes. Patra and Sen (2016) proposed a new estimation procedure based on shape-restricted function estimation, without assuming the symmetry of g. They started with one naive estimator of G as

$$\hat{G}(x;\pi) = \frac{\hat{F}_n(x) - (1-\pi)G_0(x)}{\pi}.$$

This estimator is simple to calculate and easy to understand, but not guaranteed to be nondecreasing. In addition, this estimator does not necessarily lie between 0 and 1. That is, it may not be qualified to be a cdf. The authors then proposed a modified estimator of G, denoted by $\tilde{G}(x;\pi)$, which minimizes $n^{-1}\sum_{i=1}^{n}\{W(X_i) - \hat{G}(X_i;\pi)\}^2$ over all distribution functions W. Note that both $\hat{G}$ and $\tilde{G}$ depend on the unknown mixing proportion π. Therefore, they suggested to estimate π first by

$$\hat{\pi}_n = \inf\left\{\pi \in (0,1) : d_n\{\hat{G}(x;\pi), \tilde{G}(x;\pi)\} \le \frac{a_n}{\sqrt{n}}\right\},$$

where a_n is a sequence of constants and d_n denotes the L_2 distance, i.e., $d_n(g,h) = \int\{g(x) - h(x)\}^2 d\hat{F}_n(x)$. The choice of a_n is crucial for the performance of $\hat{\pi}_n$. It has been shown that if $a_n = o(\sqrt{n})$, then $\hat{\pi}_n$ is a consistent estimator of π. In application, cross-validation is recommended for a proper choice of a_n. Once π is estimated, one could then estimate G by $\tilde{G}(x;\hat{\pi}_n)$.

There have been some generalizations of the model (9.10) since it has been introduced. For example, Hohmann and Holzmann (2013) studied

$$f(x) = (1-\pi)g_0(x-\nu) + \pi g(x-\mu), \quad x \in \mathbb{R},$$

where ν is another nonnull location parameter. The model is identifiable under assumptions made on the tails of the characteristic function for the true underlying mixture. The estimation process is not too different from that of Bordes and Vandekerkhove (2010), and the estimator is shown to be asymptotically normally distributed.

Xiang et al. (2014) and Ma and Yao (2015) studied another extension of model (9.10), assuming g_0 to be known with an unknown parameter. That is,

$$f_{\theta,g}(x) = (1-\pi)g_0(x;\xi) + \pi g(x-\mu), \quad x \in \mathbb{R}, \tag{9.12}$$

where ξ is an unknown parameter and $\boldsymbol{\theta} = (\pi, \mu, \xi)^\top$. Model (9.12) is shown to be identifiable under some mild conditions by Ma and Yao (2015). In order to estimate (9.12), Xiang et al. (2014) proposed a minimum profile Hellinger distance (MPHD) estimation procedure. The Hellinger distance between two functions f_1 and f_2 is defined as

$$d_H(f_1, f_2) = \|f_1^{1/2} - f_2^{1/2}\|,$$

where, for a and b in $L_2(\nu)$, $< a, b >$ is their inner product in $L_2(\nu)$ and $|a|$ is the $L_2(\nu)$-norm. With the nonparametric kernel density estimator $\hat{f}_n$ (9.9) in hand, the authors proposed to estimate (9.12) by minimizing $d_H(f_{\theta,g}, \hat{f}_n)$ Since the optimization involves both the parametric and nonparametric parts, the profile idea is implemented to facilitate the calculation.

For any density function h and t, define functional $g(t, h)$ as

$$g(t, h) = \arg\min_{l \in \mathcal{G}}\|f_{t,l}^{1/2} - h^{1/2}\|,$$

and then define the profile Hellinger distance as

$$d_{PH}(t, h) = \|f_{t,g(t,h)}^{1/2} - h^{1/2}\|.$$

Now the MPHD functional $T(h)$ is defined as

$$T(h) = \arg\min_{t \in \Theta} d_{PH}(t, h) = \arg\min_{t \in \Theta} \| f_{t,g(t,h)}^{1/2} - h^{1/2} \|. \tag{9.13}$$

Then, the MPHD estimator of $\boldsymbol{\theta}$ is defined as $T(\hat{f}_n)$. Xiang et al. (2014) established asymptotic properties of the MPHD estimator for the model (9.10) for the reason of model identifiability, and further claimed that the results should also hold for model (9.12) when it is identifiable.

The next theorem gives results on the existence and uniqueness of the proposed estimator, and the continuity of the function defined in (9.13).

Theorem 9.1. With T defined by (9.13), if model (9.10) is identifiable, then we have

(i) For every $f_{\theta,g}$, there exists $T(f_{\theta,g}) \in \Theta$ satisfying (9.13);

(ii) $T(f_{\theta,g}) = \boldsymbol{\theta}$ uniquely for any $\boldsymbol{\theta} \in \Theta$;

(iii) $T(f_n) \to T(f_{\theta,g})$ for any sequences $\{f_n\}_{n \in \mathbb{N}}$ such that $\| f_n^{1/2} - g_{\theta,f}^{1/2} \| \to 0$ and

$$\sup_{t \in \Theta} \| f_{t,g(t,f_n)} - f_{t,g(t,f_{\theta,g})} \| \to 0$$

as $n \to \infty$.

Under further conditions on the kernel density estimator defined in (9.9), the consistency of the MPHD estimator is established in the next theorem.

Theorem 9.2. Suppose that

(i) The kernel function $K(\cdot)$ is absolutely continuous and bounded with compact support.

(ii) $\lim_{n \to \infty} h = 0$ and $\lim_{n \to \infty} n^{1/2} h = \infty$.

(iii) The model (9.10) is identifiable and $f_{\theta,g}$ is uniformly continuous.

Then $d_H(\hat{f}_n, f_{\theta,g}) \xrightarrow{p} 0$ as $n \to \infty$, and therefore $T(\hat{f}_n) \xrightarrow{p} T(f_{\theta,g})$ as $n \to \infty$.

Define a map $\boldsymbol{\theta} \mapsto s_{\theta,h}$ as $s_{\theta,h} = f_{\theta,g(\theta,h)}^{1/2}$, and suppose that for $\boldsymbol{\theta} \in \Theta$ there exists a 2×1 vector $\dot{s}_{\theta,h}$ with components in $L_2(\nu)$ and a 2×2 matrix $\ddot{s}_{\theta,h}$ with components in $L_2(\nu)$ such that

$$s_{\theta+\alpha e,h}(x) = s_{\theta,h}(x) + \alpha e^\top \dot{s}_{\theta,h}(x) + \alpha e^\top u_{\alpha,h}(x), \tag{9.14}$$

$$\dot{s}_{\theta+\alpha e,h}(x) = \dot{s}_{\theta,h}(x) + \alpha \ddot{s}_{\theta,h}(x) e + \alpha v_{\alpha,h}(x) e, \tag{9.15}$$

where e is a 2×1 real vector of unit Euclidean length and α is a scalar in a neighborhood of zero, $u_{\alpha,h}(x)$ is 2×1, $v_{\alpha,h}(x)$ is 2×2, and the components of $u_{\alpha,h}$ and $v_{\alpha,h}$ tend to zero in $L_2(\nu)$ as $\alpha \to 0$. The next theorem shows the asymptotic distribution of the MPHD estimator.

Theorem 9.3. Suppose that the following conditions hold.

(i) Model (9.10) is identifiable.

(ii) The conditions in Theorem 9.2 hold.

(iii) The map $\boldsymbol{\theta} \mapsto s_{\theta,h}$ satisfies (9.14) and (9.15) with continuous gradient vector $\dot{s}_{\theta,h}$ and continuous Hessian matrix $\ddot{s}_{\theta,h}$ in the sense that $\|\dot{s}_{\theta_n,h_n} - \dot{s}_{\theta,h}\| \to 0$ and $\|\ddot{s}_{\theta_n,h_n} - \ddot{s}_{\theta,h}\| \to 0$ whenever $\boldsymbol{\theta}_n \to \boldsymbol{\theta}$ and $\|h_n^{1/2} - h^{1/2}\| \to 0$ as $n \to \infty$.

(iv) $< \ddot{s}_{\theta,g_\theta,f}, f_{\theta,g}^{1/2} >$ is invertible.

Then, with T defined in (9.13) for model (9.10), the asymptotic distribution of $n^{1/2}(T(\hat{f}_n) - T(f_{\theta,g}))$ is $N(0, \Sigma)$ with variance matrix Σ defined by

$$\Sigma = \langle \ddot{s}_{\theta,f_{\theta,g}}, f_{\theta,g}^{1/2} \rangle^{-1} \langle \dot{s}_{\theta,f_{\theta,g}}, \dot{s}_{\theta,f_{\theta,g}}^{\top} \rangle \langle \ddot{s}_{\theta,f_{\theta,g}}, f_{\theta,g}^{1/2} \rangle^{-1}.$$

The proofs of Theorems 9.2 and 9.3 are provided in Section 9.9.

Huang et al. (2018b) studied another special case of model (9.10) where g_0 follows a Pareto distribution with an unknown parameter ξ. The identifiability of such a model is established and a smoothed likelihood as well as a profile-likelihood-based estimation technique is proposed. A μ-symmetric smoothing kernel is defined as $K_{h,\mu}(x,t) = (2h)^{-1}[K\{(x-t)/h\} + K\{(2\mu - x - t)/h\}]$, and correspondingly a nonlinear smoothing operator for $g(\cdot)$ is

$$\mathcal{N}_\mu g(x) = \exp\left\{ \int K_{h,\mu}(x,t) \log g(t) dt \right\}.$$

Then, the authors proposed to estimate the parameters through the following smoothed log-likelihood:

$$\ell(\mu, \pi, \xi, g) = \sum_{i=1}^{n} \log\{(1-\pi)g_0(X_i; \xi) + \pi \mathcal{N}_\mu g(X_i)\}.$$

In the estimating procedure, μ is estimated separately from π, ξ, and g. If μ is known or has been estimated, say μ_0, then π, ξ, and g are estimated through a regular EM algorithm by maximizing $\ell(\mu, \pi, \xi, g)$, denoted by $\hat{\pi}_\mu, \hat{\xi}_\mu$ and $\hat{g}_\mu$. Then, the estimator of μ can be updated by maximizing the profile likelihood $\ell(\mu, \hat{\pi}_\mu, \hat{\xi}_\mu, \hat{g}_\mu)$, using some numerical methods.

Nguyen and Mattias (2014) considered a special case of (9.10) by assuming $g_0(\cdot) = 1$. The authors showed that the quadratic risk of any estimator of π does not have a parametric convergence rate when g is not 0 on any nonempty interval, which mainly occurs because the Fisher information of the model is 0 when g is bounded away from 0 for all nonempty intervals. We conjecture that such results may also hold for model (9.1), which is an interesting topic for future work.

9.4 Semiparametric mixture models with shape constraints

Consider the following C-component mixture model:

$$f(x) = \sum_{c=1}^{C} \pi_c g_c(x), \quad x \in \mathbb{R}, \tag{9.16}$$

where g_c's are unknown component densities.

In semiparametric mixture models literature, nonparametric shape constraints are becoming increasingly popular. Log-concave distributions, including normal, Laplace, logistic, and gamma and beta with certain parameter constraints, have been most commonly used. Such distributions are not restricted to any parametric assumptions, as a result, the estimation results will not suffer from model misspecification. More importantly, the Log-concave Maximum Likelihood Estimator (LCMLE) is shown to be tuning-free, which makes the methodology even more user-friendly. Specifically, given i.i.d. data $X_1, \ldots, X_n$ from a log-concave density function g, the Log-concave Maximum Likelihood Estimator (LCMLE) $\widehat{g}$ exists uniquely and has support on the convex hull of the dataset (by Theorem 2 of Cule et al. (2010b)). In addition, $\log \widehat{g}$ is a piecewise linear function whose knots are a subset of $\{X_1, \ldots, X_n\}$. Walther (2002) and Rufibach (2007) provided algorithms for computing $\widehat{g}(X_i), i = 1, \ldots, n$. The entire log-density $\log \widehat{g}$ can then be computed by the linear interpolation between $\log \widehat{g}(X_{(i)})$ and $\log \widehat{g}(X_{(i+1)})$.

Chang and Walther (2007) assumed each component in (9.16) to be log-concave, that is, $\log g_c(x)$ is a concave function for each c. In the first stage of the estimation process, the MLE of a Gaussian mixture model is computed through a regular EM algorithm, denoted by $\hat{\pi}_c$ and $\hat{g}_c$. The second stage also involves an EM algorithm. In the expectation step, the log-concave MLE $\tilde{g}_c(\cdot)$ is used to calculate the classification probability for observation i belonging to component c:

$$\hat{p}_{ic} = \frac{\hat{\pi}_c \tilde{g}_c(X_i)}{\sum_{c'=1}^{C} \hat{\pi}_{c'} \tilde{g}_{c'}(X_i)}.$$

Then, in the M-step, the mixing proportion is calculated as $\tilde{\pi}_c = \sum_{i=1}^{n} p_{ic}/n$, and the estimator of unknown density function $\tilde{g}_c(\cdot)$ is computed through a method developed in Walther (2002) and Rufibach (2007), where $\{p_{ic}, c = 1, \ldots, C\}$ are used as weights for the i-th observation. Simulation shows that only five iterations are needed in the second part of the algorithm, assuring that the methodology is computationally efficient.

The authors also extended the model and the estimation technique to the *multivariate* scenario. Let $(N_1, \ldots, N_d)$ denote a multivariate normal distribution with mean $\mathbf{0}$ and covariance Σ, and $G_1, \ldots, G_d$ be cdf's of arbitrary univariate log-concave distributions. Then, within a component,

$(X_{i1}, \ldots, X_{id})^\top \in \mathbb{R}^d$ are assumed to have the density $(G_1^{-1}\Phi(N_1), \ldots,$ $G_d^{-1}\Phi(N_d))^\top$, where Φ stands for the cdf of the standard normal distribution. Then, the joint density of the c-th component is

$$g_c(x_1, \ldots, x_d) = \phi_{0,\Sigma}\{\Phi^{-1}G_1(x_1), \ldots, \Phi^{-1}G_d(x_d)\} \times \prod_{j=1}^{d} \frac{g_j(x_j)}{\phi_{0,I}\{\Phi^{-1}G_j(x_j)\}},$$

where $\phi_{\mu,\Sigma}$ is the multivariate normal density function with mean μ and covariance Σ. The estimation method is quite similar to the univariate case, hence is omitted.

Hu et al. (2016) proposed a log-concave MLE (LCMLE) to estimate the unknown densities $g_c(\cdot)$'s, and showed its theoretical properties. They assumed that the mixture distribution belongs to

$$\mathcal{F}_\alpha = \left\{ f : f(x) = \sum_{c=1}^{C} \pi_c \exp\{\psi_c(x)\} \right\},$$

where $\psi = (\psi_1, \ldots, \psi_C) \in \Psi_\alpha$ and $\Psi_\alpha = \{(\psi_1, \ldots, \psi_C) : \psi_c$ is a d-dimensional concave function, and $|S(\psi)| \geq \alpha > 0$ for some $\alpha \in (0, 1]\}$, where $S(\psi) = M_{(1)}(\psi)/M_{(C)}(\psi)$, $M_c(\psi) = \max_{x \in \mathbb{R}^d}\{\psi_c(x)\}, M_{(1)}(\psi) = \min_c\{M_c(\psi)\}$, and $M_{(C)}(\psi) = \max_c\{M_c(\psi)\}$. Then, the LCMLE is defined as $\arg\max_{f \in \mathcal{F}_\alpha} \int \log(f) d\hat{F}_n$. The existence of such an estimator is proved by Hu et al. (2016). Please also refer to Section 7.5 to learn more about using a similar idea to provide robust mixture regression modeling.

To ensure identifiability, Balabdaoui and Doss (2018) took symmetry into account, and studied the estimation and inference for mixtures of log-concave distributions assuming each component to be symmetric. They further proved, under some technical conditions, the nonparametric log-concave MLE converges to the true density at the usual $n^{-2/5}$-rate in the L_1 distance, assuming the estimators of Euclidean parameters are $\sqrt{n}$-consistent.

In addition to log-concave densities, Al Mohamad and Boumahdaf (2018) considered another type of shape constraint. A semiparametric two-component mixture model was proposed, where one component is parametric, and the other is from a distribution family with linear constraints. With prior linear information about the unknown distribution, they proposed to estimate the model based on ϕ-divergences. Simulation studies show that this method performs better than some existing methods when the signal of the parametric component is low.

9.5 Semiparametric multivariate mixtures

When modeling the multivariate covariates $\mathbf{x} \in \mathbb{R}^d$ $(d > 1)$ using a mixture model, one commonly used assumption is that the joint density $g_c(\cdot)$ is equal to the product of its marginal densities. That is to say, the coordinates of $\mathbf{x}$ are independent, conditional on the component from which $\mathbf{x}$ is drawn. With the aforementioned assumptions in hand, in this section, the following model is considered:

$$f(\mathbf{x}) = \sum_{c=1}^{C} \pi_c \prod_{j=1}^{d} g_{cj}(x_j). \tag{9.17}$$

When $C = 2$ and $d > 2$, Hall and Zhou (2003) showed that the identifiability of the model (9.17) can typically be achieved. Allman et al. (2009) further proved that when $d > 2$, if the density functions $g_{1j}, \ldots, g_{Cj}$ are linearly independent except for on a Lebesgue-measure-zero set, the parameters in (9.17) are identifiable.

Assuming the coordinates of $\mathbf{x}$ are conditionally independent and that there are blocks of coordinates with identical density, Benaglia et al. (2009) proposed a more general semiparametric multivariate mixtures model. If all the blocks are of size 1, then it becomes the model (9.17), whose coordinates in $\mathbf{x}$ are conditionally independent but with different distributions. If there exists one block, then the coordinates are not only conditionally independent, but also identically distributed. In more general cases, let b_j denote the block to which the j-th coordinate belongs, then $1 \leq b_j \leq B$, and B is the total number of such blocks. Denote the model as

$$f(\mathbf{x}) = \sum_{c=1}^{C} \pi_c \prod_{j=1}^{d} g_{cb_j}(x_j).$$

To estimate the model, an EM-like algorithm is proposed. In the E-step, a classification probability $p_{ic}^{(t)}$ is calculated conditional on the current estimators, where $t = 1, 2, \ldots$ denotes the number of iterations. Then, in the M-step, the mixing proportions are updated as $\pi_c^{(t+1)} = n^{-1} \sum_{i=1}^{n} p_{ic}^{(t)}$, and the densities are updated as

$$g_{cl}^{(t+1)}(u) = \frac{1}{nC_l \pi_c^{(t+1)}} \sum_{j=1}^{d} \sum_{i=1}^{n} p_{ic}^{(t)} I\{b_j = l\} K_h(u - x_{ij}),$$

for $c = 1, \ldots, C, l = 1, \ldots, B$, where $C_l = \sum_{j=1}^{d} I\{b_j = l\}$ is the number of coordinates in the l-th block.

Levine et al. (2011) improved the work of Benaglia et al. (2009) and introduced a smoothed log-likelihood function. Let $\mathcal{N}g_c(\mathbf{x}) = \exp \int K_h^d(\mathbf{x} - \mathbf{u}) \log g_c(\mathbf{u}) d\mathbf{u}$, where $K_h^d(\mathbf{u}) = h^{-d} \prod_{j=1}^{d} K(u_j/h)$, and $K^d(\mathbf{u}) = \prod_{j=1}^{d} K(u_j)$. Based on this modification, a new EM algorithm is proposed and proved to have the monotonicity property similar to the manner of a maximization-minimization (MM) algorithm.

Similar to Benaglia et al. (2009), Chauveau et al. (2015) studied a semiparametric multivariate mixture model assuming conditionally independent and identically distributed coordinates belong to the same block. An algorithm is proposed, which is quite similar to the one in Benaglia et al. (2009) but with different bandwidths for each component and block. The algorithm is then proved to attain the monotone ascent property. The methodology is further extended to the univariate model (9.1).

9.6   Semiparametric hidden Markov models

Hidden Markov models have been proposed to model data coming from heterogeneous populations when the observed phenomenon is driven by a latent Markov chain. Suppose that $\{S_t, Y_t\}_0^\infty$ is a Markov process and that S_t is unobserved latent homogeneous irreducible Markov chain with finite state space $\{1, \dots, S\}$. Conditional on S_t's, Y_t's are independently distributed, and for each t, the distribution of Y_t depends only on the current latent variable S_t.

Since parametric modeling may lead to poor results in particular applications, Gassiat and Rousseau (2016); Yau et al. (2011), among others, proposed several semiparametric and nonparametric finite HMMs. Define a location model

$$Y_t = m_{S_t} + \epsilon_t, \quad t \in \mathbb{N}, \tag{9.18}$$

where $\{\epsilon_t\}_{t \in \mathbb{N}}$ is a sequence of iid random variables with distribution F, and $m_j \in \mathbb{R}, j = 1, \dots, S$. Let $\mathbf{\Gamma}$ be the $S \times S$ transition matrix of S_t, with elements $\gamma_{jk} = P(S_{t+1} = k | S_t = j)$. Then, under model (9.18), the joint distribution of (Y_1, Y_2) is given by

$$P_{\theta,F}(A \times B) = \sum_{j=1}^S \sum_{k=1}^S \gamma_{jk} F(A - m_j) F(B - m_k), \quad \forall A, B \in \mathcal{B}_\mathbb{R},$$

where $\mathcal{B}_\mathbb{R}$ denotes the Borel σ-field of $\mathbb{R}$, $\boldsymbol{\theta} = (\mathbf{m}^\top, (\gamma_{jk})_{1 \leq j,k \leq S, (j,k) \neq (S,S)})^\top$, $\mathbf{m} = (m_1, \dots, m_S)^\top$, and $P_{\theta,F}$ to be the joint distribution of (Y_1, Y_2) under model (9.18). The following theorem (Gassiat and Rousseau, 2016) proves that as long as $\mathbf{\Gamma}$ is nonsingular, the knowledge of the distribution of (Y_1, Y_2) allows the identification of $S, \mathbf{m}, \mathbf{\Gamma}$ and F.

Theorem 9.4. Let F and $\tilde{F}$ be any probability distributions on $\mathbb{R}$, and S and $\tilde{S}$ be positive integers. If $\theta \in \Theta_S^0$ and $\tilde{\theta} \in \Theta_{\tilde{S}}^0$, where Θ_S denotes the set of parameters θ such that $m_1 = 0 \leq m_2 \leq \dots \leq m_S$, then,

$$P_{\theta,F} = P_{\tilde{\theta},\tilde{F}} \text{ if and only if } S = \tilde{S}, \boldsymbol{\theta} = \tilde{\boldsymbol{\theta}}, \text{ and } F = \tilde{F}.$$

9.6.1 Estimation methods

The parametric part of the model can be estimated through the identifiability result. Denote by ϕ_F and $\phi_{\tilde{F}}$ the characteristic function of F and $\tilde{F}$, respectively. Let $\phi_{\theta,1}$ and $\phi_{\theta,2}$ be the characteristic function of m_{S_1} and m_{S_2} under $P_{\theta,F}$, respectively, and $\Phi(\theta)$ the characteristic function of the distribution of (m_{S_1}, m_{S_2}) under $P_{\theta,F}$. If the distribution of Y_1 and Y_2 is the same under $P_{\theta,F}$ and $P_{\tilde{\theta},\tilde{F}}$, then for any $t \in \mathbb{R}$:

$$\phi_F(t)\phi_{\theta,1}(t) = \phi_{\tilde{F}}(t)\phi_{\tilde{\theta},1}(t),$$

$$\phi_F(t)\phi_{\theta,2}(t) = \phi_{\tilde{F}}(t)\phi_{\tilde{\theta},2}(t).$$

Similarly, if the distribution of (Y_1, Y_2) is the same under $P_{\theta,F}$ and $P_{\tilde{\theta},\tilde{F}}$, then for any $\mathbf{t} = (t_1, t_2) \in \mathbb{R}^2$

$$\phi_F(t_1)\phi_F(t_2)\Phi_\theta(\mathbf{t}) = \phi_{\tilde{F}}(t_1)\phi_{\tilde{F}}(t_2)\Phi_{\tilde{\theta}}(\mathbf{t}).$$

As a result, for any probability density w on $\mathbb{R}^2$, integer S and $\boldsymbol{\theta}$, define a contrast function

$$M(S, \boldsymbol{\theta}) = \int_{\mathbb{R}^2} \left| \Phi_{\theta^\star}(t_1, t_2)\, \phi_{\theta,1}(t_1)\, \phi_{\theta,2}(t_2) - \Phi_\theta(t_1, t_2)\, \phi_{\theta^\star}(t_1)\, \phi_{\theta^\star}(t_2) \right|^2$$

$$\times |\phi_{F^\star}(t_1)\, \phi_{F^\star}(t_2)|^2\, w(t_1, t_2)\, \mathrm{d}t_1\, \mathrm{d}t_2,$$

Then, by Theorem 9.4, $M(S, \boldsymbol{\theta}) = 0$ if and only if $S = S^\star$ and $\boldsymbol{\theta} = \boldsymbol{\theta}^\star$. One can then estimate $M(S, \boldsymbol{\theta})$ by the following:

$$M_n(S, \boldsymbol{\theta}) = \int_{\mathbb{R}^2} \left| \widehat{\Phi}_n(t_1, t_2)\, \phi_{\theta,1}(t_1)\, \phi_{\theta,2}(t_2) - \Phi_\theta(t_1, t_2)\, \widehat{\phi}_{n,1}(t_1)\, \widehat{\phi}_{n,2}(t_2) \right|^2$$

$$\times w(t_1, t_2)\, \mathrm{d}t_1\, \mathrm{d}t_2,$$

where

$$\widehat{\Phi}_n(t_1, t_2) = \frac{1}{n} \sum_{j=1}^{n-1} \exp i\, (t_1 Y_j + t_2 Y_{j+1}), \quad \widehat{\phi}_{n,1}(t) = \widehat{\Phi}_n(t, 0)$$

and $\widehat{\phi}_{n,2}(t) = \widehat{\Phi}_n(0, t)$.

Furthermore, define an increasing function $J(S)$, which goes to infinity as S goes to infinity, and a positive continuous function $I_S(\boldsymbol{\theta})$, which goes to infinity on the boundary of Θ_S or whenever $\|m\|$ tends to infinity. In addition, let $\{\lambda_n\}_{n \in \mathbb{N}}$ be a decreasing sequence of real numbers tending to 0 and

$$\lim_{n \to +\infty} \sqrt{n}\lambda_n = +\infty.$$

Then, the preliminary estimator $(S_n, \tilde{\theta}_n)$ is calculated as the minimizer of $C_n(S, \boldsymbol{\theta})$, where

$$C_n(S, \boldsymbol{\theta}) = M_n(S, \boldsymbol{\theta}) + \lambda_n \left[J(S) + I_S(\boldsymbol{\theta}) \right],$$

and the estimator $\hat{\boldsymbol{\theta}}_n$ is defined as the minimizer of $M_n(S_n, \cdot)$ over the following compact subset

$$\left\{ \boldsymbol{\theta} \in \Theta_{S_n} : I_{S_n}(\boldsymbol{\theta}) \leq 2I_{S_n}(\tilde{\boldsymbol{\theta}}_n) \right\}.$$

Gassiat and Rousseau (2016) proved that, under some mild conditions, the estimator $\hat{\boldsymbol{\theta}}_n$ is root-n consistent with an asymptotic Gaussian distribution.

Given a consistent estimator, $\hat{\boldsymbol{\theta}}_n$, of the parametric part $\boldsymbol{\theta}$, Gassiat and Rousseau (2016) further proposed to estimate the distribution F by assuming that it is absolutely continuous with respect to Lebesgue measure. To be more specific, assume that the unknown distribution $F^\star$ has density $f^\star$ with respect to the Lebesgue measure. Then, the density $h^\star$ of Y_1 can be written as

$$h^\star(y) = \sum_{j=1}^{S^\star} \mu^\star(j) f^\star\left(y - m_j^\star\right),$$

where $\mu^\star(j) = \sum_{i=1}^{S^\star} \gamma_{i,j}^\star, 1 \leq j \leq S^\star$. Gassiat and Rousseau (2016) proposed to estimate $h^\star$ and $f^\star$ nonparametrically based on a penalized composite maximum likelihood estimator. For any density function f, define

$$\ell_n(f) = \frac{1}{n} \sum_{i=1}^{n} \log \left[\sum_{j=1}^{S^\star} \hat{\mu}(j) f\left(Y_i - \hat{m}_j\right) \right],$$

where $\hat{\mu}(j) = \sum_{i=1}^{S^\star} \hat{Q}_{i,j}$ and $\hat{m}_j$ are estimators of $\mu^\star(j)$, and $m_j^\star$, respectively, and $Q_{i,j}$ is the probability of $S_1 = i$ and $S_2 = j$. Then, we can define an estimator $\hat{f}_p$, which maximize $\ell_n(f)$ over $\mathcal{F}_p$, where $\mathcal{F}_p$ is the family of p-component Gaussian mixture densities defined as

$$\mathcal{F}_p = \left\{ \sum_{j=1}^{p} \pi_j \phi_{\sigma_j}\left(x - \mu_j\right) \right\},$$

where $\phi_\sigma(x)$ denotes the Gaussian density with mean 0 and variance σ^2. The order p is chosen by minimizing

$$D_n(p) = -\ell_n\left(\hat{f}_p\right) + \text{pen}(p, n),$$

where $\text{pen}(p, n)$ is some penalty term. Let $\hat{p}$ be the minimizer of D_n. Then, the estimator of $f^\star$ is defined as $\hat{f} = \hat{f}_{\hat{p}}$.

For Gaussian regression, the penalty term could be chosen as the slope heuristics, and in cases of Gaussian mixtures, the computation of $\hat{f}_p$ could be simply calculated through an EM algorithm. The authors proved that the estimator is adaptive over regular classes of densities.

9.7 Bayesian nonparametric mixture models

Note that the mixture model can be written as

$$f(x; Q) = \int g(x; \lambda) dQ(\lambda). \tag{9.19}$$

The above model (9.19) is equivalent to the following hierarchical model by considering λ as a latent variable:

$$x_i \mid \lambda_i \sim g(x_i; \lambda_i), \lambda_i \sim Q.$$

The nonparametric mixture model treats Q completely unspecified as to whether it is discrete, continuous, or in any particular family of distributions. Bayesian nonparametric mixture models impose a prior probability model, say $p(Q)$, on infinite-dimensional random mixing measure $Q(\cdot)$. See, for example, Hjort et al. (2010); Phadia (2015); Ghosal and Van der Vaart (2017); Müller et al. (2015) for a good introduction to Bayesian nonparametric inference.

The commonly used models for $p(Q)$ include the Dirichlet process prior, normalized random measures with independent increments, and the determinantal point process. The Dirichlet process (DP) prior proposed by Ferguson (1973), denoted by $Q \sim \mathcal{DP}(\alpha, Q_0)$, contains two parameters $\alpha > 0$ and a probability distribution Q_0, and assumes that $Q = \sum_j w_j \delta_{\mu_j}$, where

$$w_j = v_j \prod_{l<j}(1 - v_l), \text{ with } v_l \sim Beta(1, \alpha), \text{ i.i.d. and } \mu_j \sim Q_0, \text{ i.i.d.} \tag{9.20}$$

The model (9.20) is also called the stick-breaking representation (Sethuraman, 1994) and the resulting mixture model (9.19) is known as a DP mixture, which is one of the most widely used Bayesian nonparametric mixture models. Bayesian nonparametric mixture models can also be used for nonparametric density estimation.

The Dirichlet process is closely related to the Dirichlet distribution. Let Θ be the parameter space for λ. If $Q \sim \mathcal{DP}(\alpha, Q_0)$, if and only if for any partition $\{A_1, \ldots, A_d\}$ of Θ,

$$(Q(A_1), \ldots, Q(A_d)) \sim \mathcal{D}(\alpha Q_0(A_1), \ldots, \alpha Q_0(A_d)).$$

One main advantage of using the DP prior for Q is that it leads to a simple posterior distribution. More specifically, if $\boldsymbol{\lambda} = (\lambda_1, \ldots, \lambda_n)$ is an i.i.d. sample with $\lambda_i \sim Q$ and $Q \sim \mathcal{DP}(\alpha, Q_0)$, then

$$Q \mid \boldsymbol{\lambda} \sim \mathcal{DP}(\alpha + n, Q_1),$$

where $Q_1 \propto \alpha Q_0 + \sum_{i=1}^n \delta_{\lambda_i}$, a weighted average between the prior measure H_0 and the empirical distribution of the λ_i.

Let $\{\lambda_1^*, \ldots, \lambda_C^*\}$ be distinct values among $\{\lambda_1, \ldots, \lambda_n\}$, $C_k = \{i : \lambda_i = \lambda_k^*\}$, and $\mathbf{x}_k^* = \{x_i : i \in C_k\}$. Let $z_i = k$ if $i \in C_k$ and $\mathbf{z} = (z_1, \ldots, z_n)$. Bush and MacEachern (1996) proposed Gibbs sampling based on conditional distributions of $p(\lambda_k^*|\mathbf{z}, \mathbf{x})$ and $p(z_k|\mathbf{z}_{-i}, \mathbf{x})$:

$$p(\lambda_k^*|\mathbf{z}, \mathbf{x}) \propto Q_0(\lambda_k^*) \prod_{i \in C_k} g(x_i; \lambda_k^*),$$

$$p(z_k|\mathbf{z}_{-i}, \mathbf{x}) \propto \begin{cases} n_{k,-i} g(x_i|z_i = k, \mathbf{x}_{k,-i}^*), & \text{for } k \in C_{-i}, \\ \alpha q_0(x_i), & \text{for } k \notin C_{-i}, \end{cases}$$

where $q_0(x) = \int g(x; \lambda) dQ_0(\lambda)$, and $\mathbf{z}_{-i}$ denotes $\mathbf{z}$ with the ith element removed. Similar notations for $n_{k,-i}$, $\mathbf{x}_{k,-i}^*$ and C_{-i}.

Some other priors alternative to the Dirichlet process for Bayesian nonparametric mixtures include the product partition models (Hartigan, 1990, PPM) and determinantal point process (Xu et al., 2016). Similar ideas of Bayesian nonparametric mixtures can also be extended to the regression settings if we let x contain both independent and dependent variables in the model (9.19). Please see, for example, Müller et al. (1996) and Park and Dunson (2010).

9.8 Fitting semiparametric mixture models using R

In Section 9.2, a semiparametric location-shifted mixture model (9.1) was considered, and Xiang et al. (2016) proposed a continuous-scale mixture approach to estimate the parameters. We apply the continuous scale approach to the elbow diameter data, which contains the elbow diameters of 507 physically active people, and due to the gender difference, it is highly likely that there are two clusters of observations. Cross-validation shows that a two-component mixture model is suitable for the data. Figure 9.1 shows the histogram of the elbow data and the CDFs of the unknown distribution $f(x)$ for different methods: SMCSM (blue dotted line, the method by Xiang et al. (2016)), MLE (red solid line, mixture model with normally distributed errors), SPEM (black dashed line, semiparametric EM algorithm by Benaglia et al. (2009)) and empirical CDF (black dash-dotted line). It can be seen that the CDFs of SPEM and SMCSM are closer to the empirical CDF than that of MLE.

```
library(MixSemiRob)
data(elbow, package = "MixSemiRob")
ini = mixnorm(elbow)
res = mixScale(elbow, ini)
```

In Section 9.3, for the two-component mixture model (with one known component) (9.10), the MPHD estimator is proposed by Xiang et al. (2014).

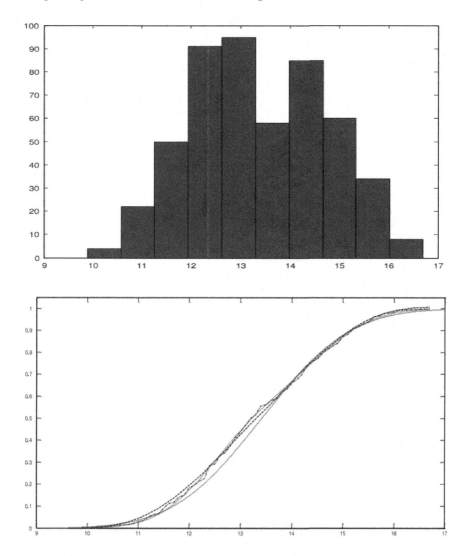

FIGURE 9.1
Histogram and fitted CDFs of the elbow data.

We apply this method for the `iris` data, which is perhaps one of the best-known data sets in pattern recognition literature. This data contains four attributes: sepal length (in cm), sepal width (in cm), petal length (in cm), and petal width (in cm), and there are three classes of 50 instances each, where each class refers to a type of iris plant. One class is linearly separable from the other two and the latter are not linearly separable from each other. Results show a mixing proportion of 0.3195, which is close to the true mixing proportion of 0.3333, given the fact that there are actually 3 instances with 50 each.

```
data = subset(iris, select = - Species)
pca = prcomp(data, center = TRUE, scale. = TRUE)$x[, 1]
data_adjusted = data.frame(data) - unlist(data[8, ])
pca_adjusted = pca - pca[8]

library(MixSemiRob)
resp = resmu = ressigma = numeric()
for(i in 1:4){
  temp = mixOnekn(x = data_adjusted[, i])
  mhdeest = mixMPHD(x = data_adjusted[, i], ini = temp)
  resp[i] = mhdeest$pi
  resmu[i] = mhdeest$mu
  ressigma[i] = mhdeest$sigma
}
temp = mixOnekn(x = pca_adjusted)
mhdeest = mixMPHD(x = pca_adjusted, ini = temp)
resp[5] = mhdeest$pi
resmu[5] = mhdeest$mu
ressigma[5] = mhdeest$sigma
```

Most of those who studied model (9.10) assume the unknown density g to be symmetric and use this condition as a key element in their estimation processes. Oppositely, Patra and Sen (2016) proposed a new estimation procedure based on shape-restricted function estimation, without assuming the symmetry of g. The admix is available to carry out the procedure proposed in Patra and Sen (2016). For example, we can apply the method to a simulated data:

```
library(admix)
lcomp = list(f1 = "norm", g1 = "norm", f2 = "norm",
        g2 = "norm")
lparam = list(f1 = list(mean = 0, sd = 1),
              g1 = list(mean = 2, sd = 0.7),
              f2 = list(mean = 0, sd = 1),
              g2 = list(mean = -3, sd = 1.1))
sim1 = rsimmix(n = 1000, unknownComp_weight = 0.8,
              comp.dist = list(lcomp$f1, lcomp$g1),
              comp.param = list(lparam$f1, lparam$g1))
sim2 = rsimmix(n = 900, unknownComp_weight = 0.85,
              comp.dist = list(lcomp$f2, lcomp$g2),
              comp.param = list(lparam$f2, lparam$g2))
list.comp = list(f1 = NULL, g1 = "norm", f2 = NULL,
            g2 = "norm")
```

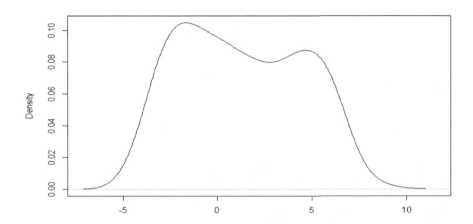

FIGURE 9.2
Density of the ethanol data.

```
list.param = list(f1 = NULL, g1 = list(mean = 2, sd = 0.7),
                  f2 = NULL, g2 = list(mean = -3, sd = 1.1))
est = admix_estim(samples = list(sim1$mixt.data, sim2$mixt.
    data),
                    sym.f = TRUE, est.method = "IBM",
                    comp.dist = lcomp, comp.param = lparam)
est
```

In addition, Bordes and Vandekerkhove (2010) offered another cdf-based estimation method for the two-component mixture model (9.10). We apply this methodology to a set of ethanol data as an example, and the density curve shown in Figure 9.2 clearly indicates a mixture of distributions. After the parameter was estimated, the `gaussianity_test` function performs a one-sample Gaussianity test in the mixture model using Bordes and Vandekerkhove (2010) estimation method. The result shows a p-value of almost 0, indicating that there is strong evidence to reject the unknown component density to be Gaussian.

```
library(MixSemiRob)
data(ethanol, package = "MixSemiRob")
plot(density(data),main="",xlab="")
library(admix)
lcomp = list(f = NULL, g = "norm")
lparam = list(f = NULL, g = list(mean = 0, sd = 1))
out = BVdk_estimParam(data, method = "L-BFGS-B", lcomp,
```

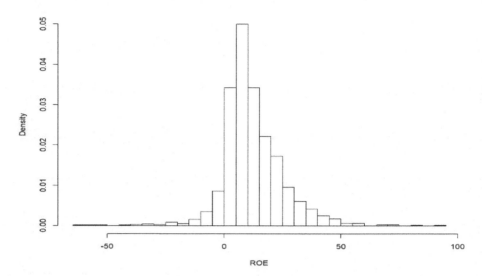

FIGURE 9.3
Histogram of the ROE data.

```
    lparam)
test = gaussianity_test(sample1 = data, comp.dist = lcomp,
                        comp.param = lparam, K = 3,
                        lambda = 0.1, support = "Real")
```

What's more, also in Section 9.3, Huang et al. (2018c) studied another special case of model (9.10) where g_0 follows a Pareto distribution with unknown parameter ξ. We apply this method to the ROE data. This data set, shown in Figure 9.3 contains a total of 2,110 Chinese listed companies on their Return on Equity (ROE), meaning the amount of net income returned as a percentage of shareholders' equity, which is an important index to measure a corporation's profitability. It is also a useful indicator for fundamental analysts to price the value of stocks.

```
library(MixSemiRob)
data(ROE)
numc=2;n=dim(data[1]);
  est=paretomix1(data$x,n,numc,est_alpha,est_beta,est_p,
  est_h,h);
```

The estimated parameters for nonparametric components are $\hat{\lambda} = 0.898$ and $\hat{\mu} = 0.117$, indicating that around 89.8% of the listed companies reported true data on their ROE, with a mean of 11.72%.

In Section 9.4, Chang and Walther (2007) extended EM algorithm to work with the flexible, nonparametric class of log-concave component distributions.

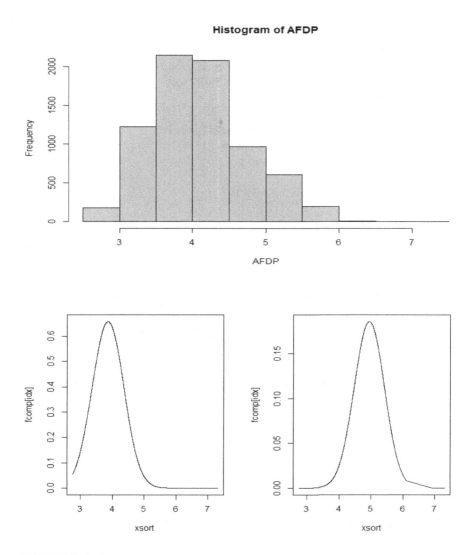

FIGURE 9.4
Histogram and density of the AFDP data.

We apply this method to the AFDP data, which contains the 11 sensor measures aggregated over one hour (by means of average or sum) from a gas turbine located in Turkey's northwestern region for the purpose of studying flue gas emissions. The estimation results are shown in Figure 9.4.

```
set.seed(666)
library(MixSemiRob)
data(AFDP, package = "MixSemiRob")
```

```
EMnorm = EMnormal(x = AFDP, C = 2)
EMlogc = mixLogconc(x = AFDP, C = 2, ini = EMnorm)
fcomp = rep(0, length(AFDP))
xsort = sort(AFDP, index.return = TRUE)$x
idx = sort(AFDP, index.return = TRUE)$ix
par(mfrow = c(1, 2))
for(cc in 1:2){
   fcomp = EMlogc$pi[cc] * EMlogc$f[, cc]
   plot(xsort, fcomp[idx], type = "l")
}
```

9.9   Proofs

In this section, the proofs of Theorems 9.1–9.3 are presented.

Proof of Theorem 9.1.

(i) Let $d(t) = \|f_{t,g(t,f_{\theta,g})}^{1/2} - f_{\theta,g}^{1/2}\|$. For any sequence $\{t_n : t_n \in \Theta, t_n \to t$ as $n \to \infty\}$,

$$
\begin{aligned}
|d^2(t_n) - d^2(t)| &= \left| \int (f_{t_n,g(t_n,f_{\theta,g})}^{1/2}(x) - f_{\theta,g}^{1/2}(x))^2 dx \right. \\
&\qquad \left. - \int (f_{t,g(t,f_{\theta,g})}^{1/2}(x) - f_{\theta,g}^{1/2}(x))^2 dx \right| \\
&= 2\left| \int (f_{t_n,g(t_n,f_{\theta,g})}^{1/2}(x) - f_{t,g(t,f_{\theta,g})}^{1/2}(x)) f_{\theta,g}^{1/2}(x) dx \right| \\
&\leq 2\|f_{t_n,g(t_n,f_{\theta,g})}^{1/2} - f_{t,g(t,f_{\theta,g})}^{1/2}\|.
\end{aligned}
$$

Since $\int f_{t_n,g(t_n,f_{\theta,g})}(x)dx = \int f_{t,g(t,f_{\theta,g})}(x)dx = 1$, we have

$$
\begin{aligned}
\|f_{t_n,g(t_n,f_{\theta,g})}^{1/2} - f_{t,g(t,f_{\theta,g})}^{1/2}\|^2 &= \int \left[f_{t_n,g(t_n,f_{\theta,g})}^{1/2}(x) - f_{t,g(t,f_{\theta,g})}^{1/2}(x) \right]^2 dx \\
&\leq \int \|f_{t,g(t,f_{\theta,g})}(x) - f_{t_n,g(t_n,f_{\theta,g})}(x)\| dx \\
&= 2 \int \left[f_{t,g(t,f_{\theta,g})}(x) - f_{t_n,g(t_n,f_{\theta,g})}(x) \right]^+ dx.
\end{aligned}
$$

Also, $[f_{t,g(t,f_{\theta,g})}(x) - f_{t_n,g(t_n,f_{\theta,g})}(x)]^+ \leq f_{t,g(t,f_{\theta,g})}(x)$, and $f_{t,g(t,f_{\theta,g})}(x)$ is continuous in t for every x. Thus, by the Dominated Convergence Theorem, $\|f_{t_n,g(t_n,f_{\theta,g})}^{1/2} - f_{t,g(t,f_{\theta,g})}^{1/2}\| \to 0$ as $n \to \infty$. So, $d(t_n) \to d(t)$ as $n \to \infty$, i.e., d is continuous on Θ and achieves a minimum for $t \in \Theta$.

(ii) By assumption, $f_{\theta,g}$ is identifiable. Immediately, we have $T(f_{\theta,g}) = \theta$ uniquely.

(iii) Let $d_n(t) = \|f^{1/2}_{t,g(t,f_n)} - f^{1/2}_n\|$ and $d(t) = \|f^{1/2}_{t,g(t,f_{\theta,g})} - f^{1/2}_{\theta,g}\|$. By Minkowski's inequality,

$$
|d_n(t) - d(t)|
$$

$$
= \left| \left[\int (f^{1/2}_{t,g(t,f_n)}(x) - f^{1/2}_n(x))^2 dx \right]^{1/2} - \left[\int (f^{1/2}_{t,g(t,f_{\theta,g})}(x) - f^{1/2}_{\theta,g}(x))^2 dx \right]^{1/2} \right|
$$

$$
\leq \left\{ \int \left[f^{1/2}_{t,g(t,f_n)}(x) - f^{1/2}_n(x) - f^{1/2}_{t,g(t,f_{\theta,g})}(x) + f^{1/2}_{\theta,g}(x) \right]^2 dx \right\}^{1/2}
$$

$$
\leq \left\{ 2 \int \left[f^{1/2}_{t,g(t,f_n)}(x) - f^{1/2}_{t,g(t,f_{\theta,g})}(x) \right]^2 dx + 2 \int \left[f^{1/2}_n(x) - f^{1/2}_{\theta,g}(x) \right]^2 dx \right\}^{1/2}
$$

Consequently,

$$
\sup_{t \in \Theta} |d_n(t) - d(t)|
$$

$$
\leq \left\{ 2 \sup_{t \in \Theta} \int \left[f^{1/2}_{t,g(t,f_n)}(x) - f^{1/2}_{t,g(t,f_{\theta,g})}(x) \right]^2 dx + 2 \int \left[f^{1/2}_n(x) - f^{1/2}_{\theta,g}(x) \right]^2 \right\}^{1/2},
$$

and the right-hand side of the above equation goes to zero as $n \to \infty$ by assumptions. Then with $\boldsymbol{\theta}_0 = T(f_{\theta,g})$ and $\boldsymbol{\theta}_n = T(f_n)$, we have $d_n(\boldsymbol{\theta}_0) \to d(\boldsymbol{\theta}_0)$ and $d_n(\boldsymbol{\theta}_n) - d(\boldsymbol{\theta}_n) \to 0$ as $n \to \infty$.

If $\boldsymbol{\theta}_n \not\to \boldsymbol{\theta}_0$, then there exists a sub-sequence $\{\boldsymbol{\theta}_m\} \subseteq \{\boldsymbol{\theta}_n\}$ such that $\boldsymbol{\theta}_m \to \boldsymbol{\theta}' \neq \boldsymbol{\theta}_0$, implying that $\boldsymbol{\theta}' \in \Theta$ and $d(\boldsymbol{\theta}_m) \to d(\boldsymbol{\theta}')$ by the continuity of d. From the above result, we have $d_m(\boldsymbol{\theta}_m) - d_m(\boldsymbol{\theta}_0) \to d(\boldsymbol{\theta}') - d(\boldsymbol{\theta}_0)$. By the definition of $\boldsymbol{\theta}_m$, $d_m(\boldsymbol{\theta}_m) - d_m(\boldsymbol{\theta}_0) \leq 0$, we can conclude that $d(\boldsymbol{\theta}') - d(\boldsymbol{\theta}_0) \leq 0$. However, by the definition of $\boldsymbol{\theta}_0$ and the uniqueness of it, $d(\boldsymbol{\theta}') > d(\boldsymbol{\theta}_0)$. This is a contradiction, and therefore $\boldsymbol{\theta}_n \to \boldsymbol{\theta}_0$.

Proof of Theorem 9.2.

Let F_n denote the empirical cdf of $X_1, X_2, ..., X_n$, which are assumed i.i.d. with density $f_{\theta,g}$ and cdf F. Let

$$
\tilde{f}_n(x) = (c_n s_n)^{-1} \int K((c_n s_n)^{-1}(x - y)) dF(y).
$$

Let $B_n(x) = n^{1/2}[F_n(x) - F(x)]$, then

$$
\sup_x |\hat{f}_n(x) - \tilde{f}_n(x)| = \sup_x n^{-1/2}(c_n s_n)^{-1} \left| \int K((c_n s_n)^{-1}(x - y)) dB_n(y) \right|
$$

$$
\leq n^{-1/2}(c_n s_n)^{-1} \sup_x |B_n(x)| \int |K'(x)| dx \xrightarrow{P} 0. \quad (9.21)
$$

Suppose $[a, b]$ is an interval that contains the support of K, then

$$\sup_x |\tilde{f}_n(x) - f_{\theta,g}(x)| = \sup_x \left| \int K(t) f_{\theta,g}(x - c_n s_n t) dt - f_{\theta,g}(x) \right|$$

$$= \sup_x \left| f_{\theta,g}(x - c_n s_n \xi) \int K(t) dt - f_{\theta,g}(x) \right|, \text{ with } \xi \in [a, b]$$

$$\leq \sup_x \sup_{t \in [a,b]} |f_{\theta,g}(x - c_n s_n t) - f_{\theta,g}(x)| \xrightarrow{P} 0 \qquad (9.22)$$

From (9.21) and (9.22), we have

$$\sup_x |\hat{f}_n(x) - f_{\theta,g}(x)| \xrightarrow{P} 0.$$

From an argument similar to the proof of Theorem 9.1, $\|\hat{f}_n^{1/2}(x) - f_{\theta,g}^{1/2}(x)\| \xrightarrow{P}$ 0 and $\sup_{t \in \Theta} \|f_{t,g(t,\hat{f}_n)} - f_{t,g(t,f_{\theta,g})}\| \to 0$ as $n \to \infty$. By Theorem 9.1, $T(\hat{f}_n) \xrightarrow{P}$ $T(f_{\theta,g})$ as $n \to \infty$.

Proof of Theorem 9.3.

Let

$$D(\theta, h) = \int \dot{s}_{\theta,h}(x) h^{1/2}(x) dx = \langle \dot{s}_{\theta,h}, h^{1/2} \rangle,$$

and it follows that $D(T(f_{\theta,g}), f_{\theta,g}) = 0$, $D(T(\hat{f}_n), \hat{f}_n) = 0$, and therefore

$$0 = D(T(\hat{f}_n), \hat{f}_n) - D(T(f_{\theta,g}), f_{\theta,g})$$
$$= [D(T(\hat{f}_n), \hat{f}_n) - D(T(f_{\theta,g}), \hat{f}_n)] + [D(T(f_{\theta,g}), \hat{f}_n) - D(T(f_{\theta,g}), f_{\theta,g})].$$

Since the map $\theta \mapsto s_{\theta,h}$ satisfies (9.14) and (9.15), $D(\theta, h)$ is differentiable in θ with derivative

$$\dot{D}(\theta, h) = \langle \ddot{s}_{\theta,h}, h^{1/2} \rangle$$

that is continuous in θ. Then,

$$D(T(\hat{f}_n), \hat{f}_n) - D(T(f_{\theta,g}), \hat{f}_n) = (T(\hat{f}_n) - T(f_{\theta,g})) \dot{D}(T(f_{\theta,g}), \hat{f}_n)$$
$$+ o_p(T(\hat{f}_n) - T(f_{\theta,g})).$$

With $\theta = T(f_{\theta,g})$,

$$D(T(f_{\theta,g}), \hat{f}_n) - D(T(f_{\theta,g}), f_{\theta,g}) = \langle \dot{s}_{\theta,\hat{f}_n}, \hat{f}_n^{1/2} \rangle - \langle \dot{s}_{\theta,f_{\theta,g}}, f_{\theta,g}^{1/2} \rangle$$

$$= 2 \langle \dot{s}_{\theta,f_{\theta,g}}, \hat{f}_n^{1/2} - f_{\theta,g}^{1/2} \rangle + \langle \dot{s}_{\theta,\hat{f}_n} - \dot{s}_{\theta,f_{\theta,g}}, \hat{f}_n^{1/2} - f_{\theta,g}^{1/2} \rangle$$

$$+ \langle \dot{s}_{\theta,\hat{f}_n}, f_{\theta,g}^{1/2} \rangle - \langle \hat{f}_n^{1/2}, \dot{s}_{\theta,f_{\theta,g}} \rangle$$

$$= 2 \langle \dot{s}_{\theta,f_{\theta,g}}, \hat{f}_n^{1/2} - f_{\theta,g}^{1/2} \rangle + [\langle \dot{s}_{\theta,\hat{f}_n}, f_{\theta,g}^{1/2} \rangle - \langle \hat{f}_n^{1/2}, \dot{s}_{\theta,f_{\theta,g}} \rangle]$$

$$+ O(\|\dot{s}_{\theta,\hat{f}_n} - \dot{s}_{\theta,f_{\theta,g}}\| \cdot \|\hat{f}_n^{1/2} - f_{\theta,g}^{1/2}\|)$$

$$= 2 \langle \dot{s}_{\theta,f_{\theta,g}}, \hat{f}_n^{1/2} - f_{\theta,g}^{1/2} \rangle + o_p(\|\hat{f}_n^{1/2} - f_{\theta,g}^{1/2}\|).$$

Applying the algebraic identity

$$b^{1/2} - a^{1/2} = (b-a)/(2a^{1/2}) - (b-a)^2/[2a^{1/2}(b^{1/2} + a^{1/2})^2],$$

we have that

$$n^{1/2}\langle \dot{s}_{\theta,f_{\theta,g}}, \hat{f}_n^{1/2} - f_{\theta,g}^{1/2}\rangle = n^{1/2}\int \dot{s}_{\theta,f_{\theta,g}}(x)\frac{\hat{f}_n(x) - f_{\theta,g}(x)}{2f_{\theta,g}^{1/2}(x)}dx + R_n$$

$$= n^{1/2}\int \dot{s}_{\theta,f_{\theta,g}}(x)\frac{\hat{f}_n(x)}{2f_{\theta,g}^{1/2}(x)}dx + R_n$$

$$= n^{1/2}\cdot\frac{1}{n}\sum_{i=1}^{n}\frac{\dot{s}_{\theta,f_{\theta,g}}(X_i)}{2f_{\theta,g}^{1/2}(X_i)} + o_p(1) + R_n$$

with $|R_n| \leq n^{1/2}\int\frac{|\dot{s}_{\theta,f_{\theta,g}}(x)|}{2f_{\theta,g}^{3/2}(x)}[\hat{f}_n(x) - f_{\theta,g}(x)]^2dx \xrightarrow{p} 0$. Since $\langle \ddot{s}_{\theta,f_{\theta,g}}, f_{\theta,g}^{1/2}\rangle$ is assumed to be invertible, then

$$T(\hat{f}_n) - T(f_{\theta,g}) = -[\langle \ddot{s}_{\theta,f_{\theta,g}}, f_{\theta,g}^{1/2}\rangle^{-1} + o_p(1)]\frac{1}{n}\sum_{i=1}^{n}\frac{\dot{s}_{\theta,f_{\theta,g}}(X_i)}{f_{\theta,g}^{1/2}(X_i)} + o_p(n^{-1/2})$$

and therefore, the asymptotic distribution of $n^{1/2}(T(\hat{f}_n) - T(f_{\theta,g}))$ is $N(0, \Sigma)$ with variance matrix Σ defined by

$$\Sigma = \langle \ddot{s}_{\theta,f_{\theta,g}}, f_{\theta,g}^{1/2}\rangle^{-1}\langle \dot{s}_{\theta,f_{\theta,g}}, \dot{s}_{\theta,f_{\theta,g}}^{\top}\rangle\langle \ddot{s}_{\theta,f_{\theta,g}}, f_{\theta,g}^{1/2}\rangle^{-1}.$$

10

Semiparametric mixture regression models

10.1 Why semiparametric regression models?

Since the parametric model assumption, such as linearity might not be satisfied, both theoretically and practically, many semiparametric finite mixture of regressions (FMR) models have been proposed and demonstrated to have superior performance during the last few years. By allowing the mixing proportions to depend on a covariate, Young and Hunter (2010) and Huang and Yao (2012) studied semiparametric mixtures of regression models with varying proportions. Huang et al. (2013) and Xiang and Yao (2018) relaxed the parametric assumptions on the mean functions and/or variances to accommodate for complicated data structures. However, due to the application of kernel regression in the estimation procedure, the models are not suitable for data with moderate to high-dimensional predictors. Xiang and Yao (2020) proposed using the idea of single-index to transfer the multivariate predictors to univariate. Hunter and Young (2012) and Ma et al. (2021) studied an FMR model where linearity was still assumed within each component, but the error terms were modeled fully nonparametrically.

10.2 Mixtures of nonparametric regression models

In the traditional FMR model, the mean functions are assumed to be linear. In this section, several models are introduced to relax this assumption.

The first model was proposed by Huang et al. (2013) as:

$$Y|_{X=x} \sim \sum_{c=1}^{C} \pi_c(x) N\{m_c(x), \sigma_c^2(x)\}, \quad x \in \mathbb{R}, \qquad (10.1)$$

where $\pi_c(\cdot), m_c(\cdot)$, and $\sigma_c^2(\cdot)$ are smooth nonparametric functions satisfying $\sum_{c=1}^{C} \pi_c(\cdot) = 1$. This model was motivated by a US house price index data example. Because the scatter plot of house price index (HPI) change and GDP growth demonstrates different nonlinear patterns in various macroeconomic

DOI: 10.1201/9781003038511-10

cycles, Huang et al. (2013) proposed the semiparametric mixture of regressions model (10.1) to fit this data set. Since normality is still assumed for the error distribution, the model is considered semiparametric. To estimate the unknown functions, kernel regression is incorporated into a modified EM algorithm. At the $(t+1)$-th iteration $(t = 1, 2, \ldots)$, a classification probability is calculated, like a regular EM algorithm, and denoted by $p_{ic}^{(t+1)}$, based on the estimates from the t-th iteration. In the M-step, the function estimators are updated by maximizing the following objective function:

$$\sum_{i=1}^{n} \sum_{c=1}^{C} p_{ic}^{(t+1)} [\log \pi_c + \log \phi\{Y_i | m_c, \sigma_c^2\}] K_h(X_i - x),$$

with respect to π_c, m_c and σ_c. Note that the kernel wights $K_h(X_i - x)$ force a parametric mixture model fit to local data.

Although model (10.1) is flexible enough, its efficiency is not very satisfactory. Taking both flexibility and estimation efficiency into account, Xiang and Yao (2018) proposed a model which assumes the mean functions to be nonparametric, while the mixing proportions and variances are assumed to be constant:

$$Y|_{X=x} \sim \sum_{c=1}^{C} \pi_c N\{m_c(x), \sigma_c^2\}, \qquad x \in \mathbb{R}. \tag{10.2}$$

Since both parametric and nonparametric parts coexist, model (10.2) is more difficult than model (10.1) to estimate. The authors proposed a one-step back-fitting estimation procedure.

First, $\boldsymbol{\pi}$, $\mathbf{m}$, and $\boldsymbol{\sigma}^2$ are estimated locally by maximizing the following local log-likelihood function:

$$\ell_1(\boldsymbol{\pi}(x), \mathbf{m}(x), \boldsymbol{\sigma}^2(x)) = \sum_{i=1}^{n} \log\{\sum_{c=1}^{C} \pi_c \phi(Y_i | m_c, \sigma_c^2)\} K_h(X_i - x). \tag{10.3}$$

Denote $\tilde{\boldsymbol{\pi}}(x)$, $\tilde{\mathbf{m}}(x)$, and $\tilde{\boldsymbol{\sigma}}^2(x)$ as the maximizer. These estimators are in fact the model estimates of (10.1) proposed by Huang et al. (2013). Note that, in (10.3), the global parameters $\boldsymbol{\pi}$ and $\boldsymbol{\sigma}^2$ are estimated locally due to the kernel weights. To improve the efficiency, the authors proposed to update the estimates of $\boldsymbol{\pi}$ and $\boldsymbol{\sigma}^2$ by maximizing the following log-likelihood function:

$$\ell_2(\boldsymbol{\pi}, \boldsymbol{\sigma}^2) = \sum_{i=1}^{n} \log\{\sum_{c=1}^{C} \pi_c \phi(Y_i | \tilde{m}_c(X_i), \sigma_c^2)\}. \tag{10.4}$$

Let $\hat{\boldsymbol{\pi}}$ and $\hat{\boldsymbol{\sigma}}^2$ be the estimators of this step. The estimate of $\mathbf{m}(\cdot)$ is then further improved by maximizing the following local log-likelihood function:

$$\ell_3(\mathbf{m}(x)) = \sum_{i=1}^{n} \log\{\sum_{c=1}^{C} \hat{\pi}_c \phi(Y_i | m_c, \hat{\sigma}_c^2)\} K_h(X_i - x), \tag{10.5}$$

which, compared to (10.3), replaces π_j and σ_j^2 by $\hat{\pi}_j$ and $\hat{\sigma}_j^2$, respectively. Let $\hat{\mathbf{m}}(x)$ be the solution of (10.5). Since $\hat{\beta}$ has a faster convergence rate than $\hat{\mathbf{m}}(x)$, $\hat{\mathbf{m}}(x)$ is shown, in Theorem 10.2, to have the same asymptotic properties as if β were known. Then, $\hat{\pi}$, $\hat{\mathbf{m}}(x)$, and $\hat{\sigma}^2$ are referred to as the one-step backfitting estimates.

In the following theorems (Xiang and Yao, 2018), the estimators of both the global and local parameters are proven to achieve the optimal convergence rate. Let $\beta = (\pi^\top, (\sigma^2)^\top)^\top$, $\theta = (\mathbf{m}\top, \pi^\top, (\sigma^2)^\top)^\top = (\mathbf{m}^\top, \beta^\top)^\top$. Define

$$\ell(\theta, y) = \log \sum_{c=1}^{C} \pi_c \phi(y|m_c, \sigma_c^2),$$

and let

$$\mathbf{I}_\theta(x) = -\mathrm{E}\left[\frac{\partial^2 \ell(\theta, y)}{\partial\theta\partial\theta^T}|X = x\right], \quad \mathbf{I}_\beta(x) = -\mathrm{E}\left[\frac{\partial^2 \ell(\theta, y)}{\partial\beta\partial\beta^T}|X = x\right],$$

$$\mathbf{I}_m(x) = -\mathrm{E}\left[\frac{\partial^2 \ell(\theta, y)}{\partial\mathbf{m}\partial\mathbf{m}^T}|X = x\right], \quad \mathbf{I}_{\beta m}(x) = -\mathrm{E}\left[\frac{\partial^2 \ell(\theta, y)}{\partial\beta\partial\mathbf{m}^T}|X = x\right],$$

$$\lambda(u|x) = \mathrm{E}\left[\frac{\partial \ell(\theta(x), y)}{\partial\mathbf{m}}|X = u\right].$$

Under further conditions described in Section 10.10, the consistency and asymptotic normality of $\hat{\pi}$ and $\hat{\sigma}^2$ are established in the next theorem.

Theorem 10.1. Suppose that conditions (C1) and (C3)–(C10) in Section 10.10 are satisfied, then

$$\sqrt{n}(\hat{\beta} - \beta) \xrightarrow{D} N(0, \mathbf{B}^{-1}\Sigma\mathbf{B}^{-1}),$$

where $\mathbf{B} = \mathrm{E}\{I_\beta(X)\}$, $\Sigma = \mathrm{Var}\{\partial\ell(\theta(X), Y)/\partial\beta - \varpi(X, Y)\}$, $\varpi(x, y) = \mathbf{I}_{\beta m}\varphi(x, y)$, and $\varphi(x, y)$ is a $k \times 1$ vector consisting of the first k elements of $\mathbf{I}_\theta^{-1}(x)\partial\ell(\theta(x), y)/\partial\theta$.

The above theorem shows that the estimator of the global parameters achieved the optimal root-n convergence rate.

Theorem 10.2. Suppose that conditions (C2)–(C10) in Section 10.10 are satisfied, then

$$\sqrt{nh}(\hat{\mathbf{m}}(x) - \mathbf{m}(x) - \delta_m(x) + o_p(h^2)) \xrightarrow{D} N(0, f^{-1}(x)\mathbf{I}_m^{-1}(x)\nu_0),$$

where $f(\cdot)$ is the density of X, $\delta_m(x)$ is a $k \times 1$ vector consisting of the first k elements of $\Delta(x)$ with

$$\Delta(x) = \mathbf{I}_m^{-1}(x)\{\frac{1}{2}\lambda''(x|x) + f^{-1}(x)f'(x)\lambda'(x|x)\}\kappa_2 h^2.$$

To compare between model (10.1) and model (10.2), a generalized likelihood ratio test was proposed to determine whether or not the mixing proportions and variances indeed depend on the covariates. The following hypotheses are being tested:

$$H_0 : \pi_c(x) \equiv \pi_c, c = 1, ..., C - 1;$$
$$\sigma_c^2(x) \equiv \sigma_c^2, c = 1, ..., C;$$
$$\pi_c \text{ and } \sigma_c^2 \text{ are unknown in } (0, 1) \text{ and } \mathbb{R}^+.$$
$$H_1 : \pi_c(x) \text{ or } \sigma_c^2(x) \text{ is not constant for some } c.$$

Let $\ell_n(H_0)$ and $\ell_n(H_1)$ be the log-likelihood functions computed under the null and alternative hypotheses, respectively. Then, we can construct a likelihood ratio test statistic:

$$T = \ell_n(H_1) - \ell_n(H_0). \tag{10.6}$$

Since the models under both H_0 and H_1 are semiparametric, this likelihood ratio statistic is different from the parametric ones. The following theorem establishes the Wilks types of results for (10.6), that is, the asymptotic null distribution is independent of the nuisance parameters π and σ, and the nuisance nonparametric mean functions $\mathbf{m}(x)$.

Theorem 10.3. Suppose that conditions (C9)–(C13) in the Section 10.10 hold and that $nh^4 \to 0$ and $nh^2 \log(1/h) \to \infty$, then

$$r_K T \overset{a}{\sim} \chi_\delta^2,$$

where $r_K = [K(0) - 0.5 \int K^2(t)dt]/ \int [K(t) - 0.5K * K(t)]^2 dt$, $\delta = r_K(2k - 1)|\mathcal{X}|[K(0) - 0.5 \int K^2(t)dt]/h$, $|\mathcal{X}|$ denotes the length of the support of X, and $K * K$ is the 2nd convolution of $K(\cdot)$.

Theorem 10.3 unveils a new Wilks type of phenomenon, providing a simple and useful method for semiparametric inferences.

However, model (10.1) and model (10.2) cannot handle predictors of high dimension, due to the application of kernel regression in the estimation process. As a result, Xiang and Yao (2020) proposed a series of semiparametric FMR models with single-index. The first model is defined as:

$$Y|_{\mathbf{x}} \sim \sum_{c=1}^{C} \pi_c(\boldsymbol{\alpha}^\top \mathbf{x}) N(m_c(\boldsymbol{\alpha}^\top \mathbf{x}), \sigma_c^2(\boldsymbol{\alpha}^\top \mathbf{x})), \tag{10.7}$$

where the single index $\boldsymbol{\alpha}^\top \mathbf{x}$ transfers the multivariate nonparametric problem into a univariate one. When $C = 1$, model (10.7) becomes a single index model (Ichimura, 1993; Härdle et al., 1993). If $\mathbf{x}$ is indeed a scalar, then model (10.7) reduces to model (10.1). Zeng (2012) also applied the single index idea to the component means and variances, assuming the component proportions do not depend on the predictor $\mathbf{x}$. However, the theoretical properties of their estimates are not discussed.

Below is the log-likelihood of collected data $(\mathbf{x}_i, Y_i), i = 1, \ldots, n$ drawn from model (10.7):

$$\sum_{i=1}^{n} \log \left\{ \sum_{c=1}^{C} \pi_c(\boldsymbol{\alpha}^\top \mathbf{x}_i) \phi(Y_i | m_c(\boldsymbol{\alpha}^\top \mathbf{x}_i), \sigma_c^2(\boldsymbol{\alpha}^\top \mathbf{x}_i)) \right\},$$

which is not ready for maximization yet. By existing sufficient dimension reduction methods, such as Sliced Inverse Regression (Li, 1991, SIR), one can find an initial estimate of $\boldsymbol{\alpha}$, denoted by $\tilde{\boldsymbol{\alpha}}$. Then, the authors proposed to estimate the nonparametric functions by maximizing the following local log-likelihood:

$$\sum_{i=1}^{n} \log \left\{ \sum_{c=1}^{C} \pi_c(\tilde{\boldsymbol{\alpha}}^\top \mathbf{x}_i) \phi(Y_i | m_c(\tilde{\boldsymbol{\alpha}}^\top \mathbf{x}_i), \sigma_c^2(\tilde{\boldsymbol{\alpha}}^\top \mathbf{x}_i)) \right\} K_h(\tilde{\boldsymbol{\alpha}}^\top \mathbf{x}_i - z).$$

Denote the estimates of this step as $\hat{\boldsymbol{\pi}}(\cdot)$, $\hat{\mathbf{m}}(\cdot)$, and $\hat{\boldsymbol{\sigma}}^2(\cdot)$, which are referred to as the one-step (OS) estimator. The authors proposed to further update the estimator of $\boldsymbol{\alpha}$ by maximizing

$$\sum_{i=1}^{n} \log \left\{ \sum_{c=1}^{C} \hat{\pi}_c(\boldsymbol{\alpha}^\top \mathbf{x}_i) \phi(Y_i | \hat{m}_c(\boldsymbol{\alpha}^\top \mathbf{x}_i), \hat{\sigma}_c^2(\boldsymbol{\alpha}^\top \mathbf{x}_i)) \right\}, \qquad (10.8)$$

with respect to $\boldsymbol{\alpha}$, through some optimization algorithm. The proposed fully iterative backfitting (FIB) estimator $\hat{\boldsymbol{\alpha}}$ iterates the above two steps until convergence.

The asymptotic properties of the proposed estimates are investigated below. Let $\boldsymbol{\theta}(z) = (\boldsymbol{\pi}^\top(z), \mathbf{m}^\top(z), (\boldsymbol{\sigma}^2)^\top(z))^\top$, and define

$$\ell(\boldsymbol{\theta}(z), y) = \log \sum_{c=1}^{C} \pi_c(z) \phi\{y | m_c(z), \sigma_c^2(z)\},$$

$$\mathbf{q}(z) = \frac{\partial \ell(\boldsymbol{\theta}(z), y)}{\partial \boldsymbol{\theta}}, \quad \mathbf{Q}(z) = \frac{\partial^2 \ell(\boldsymbol{\theta}(z), y)}{\partial \boldsymbol{\theta} \partial \boldsymbol{\theta}^\top},$$

$$\mathbf{I}_\theta^{(1)}(z) = -\mathrm{E}[\mathbf{Q}(Z) | Z = z], \quad \boldsymbol{\lambda}_1(u|z) = \mathrm{E}[\mathbf{q}(z) | Z = u].$$

Under further conditions given in Section 10.10, the asymptotic properties of the OS estimators $\hat{\boldsymbol{\pi}}(\cdot)$, $\hat{\mathbf{m}}(\cdot)$, and $\hat{\boldsymbol{\sigma}}^2(\cdot)$ are given in the following theorem.

Theorem 10.4. Assume that conditions (C14)–(C20) in Section 10.10 hold. Then, as $n \to \infty$, $h \to 0$ and $nh \to \infty$, we have

$$\sqrt{nh}\{\hat{\boldsymbol{\theta}}(z) - \boldsymbol{\theta}(z) - \mathbf{b}_1 + o_p(h^2)\} \xrightarrow{D} N\{0, \nu_0 f^{-1}(z) \mathbf{I}_\theta^{(1)}(z)\},$$

where

$$\mathbf{b}_1(z) = \mathbf{I}_\theta^{(1)-1} \left\{ \frac{f'(z) \boldsymbol{\lambda}_1'(z|z)}{f(z)} + \frac{1}{2} \boldsymbol{\lambda}_1''(z|z) \right\} \kappa_2 h^2,$$

with $f(\cdot)$ being the marginal density function of $\boldsymbol{\alpha}^\top \mathbf{x}$.

The asymptotic variance of $\hat{\boldsymbol{\theta}}(z)$ is the same as those given in Huang et al. (2013), indicating that the nonparametric functions can be estimated with the same accuracy as if the single index $\boldsymbol{\alpha}^\top \mathbf{x}$ were known, which is the result of $\boldsymbol{\alpha}$ being estimated at a faster convergence rate than $\hat{\boldsymbol{\theta}}(z)$.

The next theorem gives the asymptotic results of the FIB estimator $\hat{\boldsymbol{\alpha}}$.

Theorem 10.5. Assume that conditions (C14)–(C21) in Section 10.10 hold, then, as $n \to \infty$, $nh^4 \to 0$, and $nh^2/\log(1/h) \to \infty$,

$$\sqrt{n}(\hat{\boldsymbol{\alpha}} - \boldsymbol{\alpha}) \xrightarrow{D} N(0, \mathbf{Q}_1^{-1}),$$

where

$$\mathbf{Q}_1 = \mathrm{E}\left[\{\mathbf{x}\boldsymbol{\theta}'(Z)\}\mathbf{Q}(Z)\{\mathbf{x}\boldsymbol{\theta}'(Z)\}^\top \right.$$
$$\left. -\mathbf{x}\boldsymbol{\theta}'(Z)\mathbf{Q}(Z)\mathbf{I}_\theta^{(1)-1}(Z)\mathrm{E}\{\mathbf{Q}(Z)[\mathbf{x}\boldsymbol{\theta}'(Z)]^\top|Z\}\right].$$

Based on Theorem 10.5, we know that the single index parameter $\boldsymbol{\alpha}$ can be estimated at $\sqrt{n}$ convergence rate.

10.3 Mixtures of regression models with varying proportions

In a parametric FMR model, the mixing proportions are assumed to be constant. However, the model could be mistakenly specified providing misleading results if the covariates $\mathbf{x}$ contain some information about the relative weights. In this section, several FMR models with varying proportions will be discussed, where the error density is assumed to be known.

First, consider the model:

$$Y|\mathbf{x} \sim \sum_{c=1}^{C} \pi_c(\mathbf{x}) N(\mathbf{x}^\top \boldsymbol{\beta}_c, \sigma_c^2), \tag{10.9}$$

where $\pi_c(\cdot)$s are unknown smooth functions, $\mathbf{x}$ is a p-dimensional vector, and $\boldsymbol{\beta}_c$ and σ_c^2 are cth component parameters. Under some mild conditions, model (10.9) has been proved by Huang and Yao (2012) to be identifiable. If $\pi_c(\mathbf{x})$ is assumed to be logistic, then model (10.9) becomes the hierarchical mixtures of experts (Jacobs et al., 1997, HME) in the neural network. Young and Hunter (2010), however, modeled $\pi_c(\mathbf{x})$ from a different point of view. They assumed that

$$\pi_c(\mathbf{x}_i) = \mathrm{E}[z_{ic}|\mathbf{x}_i], \tag{10.10}$$

where z_{ic} is an indicator variable indicating whether or not the i-th observation is from the c-th component. If z_{ic} is treated as a response, then (10.10) indicates nothing but a mean structure in regression analysis. As a result, Young and Hunter (2010) estimated the mixing proportions by:

$$\arg\min_{\boldsymbol{\alpha}} \sum_{i=1}^{n} K_{\mathbf{h}}(\mathbf{x}_i - \mathbf{x}_l) \left\{ z_{ic} - \left(\alpha_0 + \sum_{t=1}^{p} \alpha_t(x_{i,t} - x_{l,t}) \right) \right\}^2, \qquad (10.11)$$

where $K_{\mathbf{h}}(\cdot)$ is a scaled multivariate kernel density function. In reality, however, z_{ic}'s are not known beforehand, hence the authors proposed to estimate them through the following algorithm. First, run the EM algorithm for a regular FMR model for a few repetitions, then, the converged posterior probability, denoted by p_{ic}^{∞}, is used to replace z_{ic} in (10.11). If $\boldsymbol{\pi}_c(\mathbf{x})$'s are known, then the $\boldsymbol{\beta}_c$'s and σ_c^2's can be readily estimated by a regular EM algorithm. Note that due to the "curse of dimensionality," the multivariate kernel density function does not work very well for predictors with high dimensions, and therefore, the authors themselves were extremely cautious about cases with high-dimensional predictors.

Huang and Yao (2012) also studied model (10.9), but estimated $\boldsymbol{\pi}_c(\mathbf{x})$ fully nonparametrically. A one-step backfitting algorithm was proposed to achieve the optimal convergence rate for both the global and local parameters. The asymptotic properties of the estimators were thoroughly studied.

Wang et al. (2014) extended model (10.9) to a mixture of GLMs with varying proportions model:

$$Y|\mathbf{x} \sim \sum_{c=1}^{C} \pi_c(\mathbf{x}) g_c(y|\mathbf{x}, \boldsymbol{\theta}_c), \qquad (10.12)$$

where g_c is a function of the exponential family whose mean is $\mu_c(\mathbf{x}) = f_c^{-1}(\mathbf{x}^\top \boldsymbol{\beta}_c)$ and $f_c(\cdot)$ is a component-specific link function. Wang et al. (2014) established the identifiability of the model (10.12) under very general conditions. As a special case, Cao and Yao (2012) studied a model where the response variable is binomial. In this model, both the component proportions and the success probabilities depend on the predictor nonparametrically. That is,

$$Y|_{X=x} \sim \pi_1(x)\mathrm{Bin}(y; N, 0) + \pi_2(x)\mathrm{Bin}(y; N, p(x)), \qquad (10.13)$$

where $\pi_1(x) + \pi_2(x) = 1$, and $\mathrm{Bin}(Y; N, p)$ denotes the probability mass function of a binomially distributed random variable Y with the number of trials N and a success probability p. Since the first component is indeed a degenerate distribution with mass 1 on 0, the model (10.13) can be widely used in data with an extra number of zeros. In the article, the authors applied model (10.13) to a rain dataset from a global climate model and a historical rain data set from Edmonton, Canada.

To better model the nonparametric effects of multivariate predictors, Xiang and Yao (2020) proposed another model, which introduces single-index into the component proportions of the traditional FMR model. The model is defined as:

$$Y|_{\mathbf{x}} \sim \sum_{c=1}^{C} \pi_c(\boldsymbol{\alpha}^{\top}\mathbf{x})N(\mathbf{x}^{\top}\boldsymbol{\beta}_c, \sigma_c^2).$$

Given the parametric estimates of $(\boldsymbol{\alpha}, \boldsymbol{\beta}, \boldsymbol{\sigma}^2)$, say $(\hat{\boldsymbol{\alpha}}, \hat{\boldsymbol{\beta}}, \hat{\boldsymbol{\sigma}}^2)$, then $\boldsymbol{\pi}(\cdot)$ can be estimated locally by maximizing the following local log-likelihood function:

$$\ell_1(\boldsymbol{\pi}) = \sum_{i=1}^{n} \log\{\sum_{c=1}^{C} \pi_c(\hat{\boldsymbol{\alpha}}^{\top}\mathbf{x}_i)\phi(Y_i|\mathbf{x}_i^{\top}\hat{\boldsymbol{\beta}}_c, \hat{\sigma}_c^2)\}K_h(\hat{\boldsymbol{\alpha}}^{\top}\mathbf{x}_i - z). \qquad (10.14)$$

Let $\hat{\boldsymbol{\pi}}(\cdot)$ be the estimate that maximizes (10.14). One can then further update the estimate of $(\boldsymbol{\alpha}, \boldsymbol{\beta}, \boldsymbol{\sigma}^2)$ by maximizing

$$\ell_2(\boldsymbol{\alpha}, \boldsymbol{\beta}, \boldsymbol{\sigma}^2) = \sum_{i=1}^{n} \log\{\sum_{c=1}^{C} \hat{\pi}_c(\boldsymbol{\alpha}^{\top}\mathbf{x}_i)\phi(Y_i|\mathbf{x}_i^{\top}\boldsymbol{\beta}_c, \sigma_c^2)\}. \qquad (10.15)$$

Therefore, the authors propose a backfitting algorithm to iterate between estimating $(\boldsymbol{\alpha}, \boldsymbol{\beta}, \boldsymbol{\sigma}^2)$ and estimating $\boldsymbol{\pi}(\cdot)$.

Let $\boldsymbol{\eta} = (\boldsymbol{\beta}^{\top}, (\boldsymbol{\sigma}^2)^{\top})^{\top}$ and $\boldsymbol{\omega} = (\boldsymbol{\alpha}^{\top}, \boldsymbol{\eta}^{\top})^{\top}$. Define

$$\ell(\boldsymbol{\pi}(z), \boldsymbol{\omega}, \mathbf{x}, y) = \log \sum_{c=1}^{C} \pi_c(z)\phi\{y|\mathbf{x}^{\top}\boldsymbol{\beta}_c, \sigma_c^2\},$$

$$\mathbf{q}_{\pi}(z) = \frac{\partial\ell(\boldsymbol{\pi}(z), \boldsymbol{\omega}, x, y)}{\partial\boldsymbol{\pi}}, \quad \mathbf{Q}_{\pi\pi}(z) = \frac{\partial^2\ell(\boldsymbol{\pi}(z), \boldsymbol{\omega}, x, y)}{\partial\boldsymbol{\pi}\partial\boldsymbol{\pi}^{\top}}.$$

Similarly, define $\mathbf{q}_{\omega}$, $\mathbf{Q}_{\omega\omega}$, and $\mathbf{Q}_{\pi\eta}$. Denote $\mathbf{I}_{\pi}(z) = -\mathrm{E}[\mathbf{Q}_{\pi\pi}(Z)|Z = z]$ and $\boldsymbol{\lambda}_2(u|z) = \mathrm{E}[\mathbf{q}_{\pi}(z)|Z = u]$.

Under some regularity conditions, the asymptotic properties of $\hat{\boldsymbol{\pi}}(z)$ are given in the following theorem and its proof is given in Section 10.10.

Theorem 10.6. Assume that conditions (C14)-(C17) and (C22)-(C24) in Section 10.10 hold. Then, as $n \to \infty$, $h \to 0$ and $nh \to \infty$, we have

$$\sqrt{nh}\{\hat{\boldsymbol{\pi}}(z) - \boldsymbol{\pi}(z) - \mathbf{B}_2(z) + o_p(h^2)\} \xrightarrow{D} N\{0, \nu_0 f^{-1}(z)\mathbf{I}_{\pi}^{(2)}(z)\},$$

where

$$\mathbf{B}_2(z) = \mathbf{I}_{\pi}^{(2)-1}\left\{\frac{f'(z)\boldsymbol{\lambda}_2'(z|z)}{f(z)} + \frac{1}{2}\boldsymbol{\lambda}_2''(z|z)\right\}\kappa_2 h^2.$$

The asymptotic property of the parametric estimate $\hat{\boldsymbol{\omega}}$ is given in the following theorem.

Theorem 10.7. Assume that conditions (C14)-(C17) and (C22)-(C25) in Section 10.10 hold. Then, as $n \to \infty$, $nh^4 \to 0$, and $nh^2/\log(1/h) \to \infty$,

$$\sqrt{n}(\hat{\boldsymbol{\omega}} - \boldsymbol{\omega}) \xrightarrow{D} N(0, \mathbf{Q}_2^{-1}),$$

where,

$$\mathbf{Q}_2 = \mathrm{E}\left[\mathbf{Q}_{\pi\pi}(Z)\begin{pmatrix}\mathbf{x}\boldsymbol{\pi}'(Z) \\ \mathbf{I}\end{pmatrix}\left\{\begin{pmatrix}\mathbf{x}\boldsymbol{\pi}'(Z) \\ \mathbf{I}\end{pmatrix} - \begin{pmatrix}\mathbf{I}_{\pi}^{(2)-1}(Z)\mathrm{E}\{\mathbf{Q}_{\pi\pi}(Z)(\mathbf{x}\boldsymbol{\pi}'(Z))^{\top}|Z\} \\ \mathbf{I}_{\pi}^{(2)-1}(Z)\mathrm{E}\{\mathbf{Q}_{\pi\eta}(Z)|Z\}\end{pmatrix}\right\}^{\top}\right].$$

10.4   Machine learning embedded semiparametric mixtures of regressions

Xue and Yao (2021) proposed a new class of semiparametric estimation methods by combining the ideas of machine learning, such as Neural Networks, for nonparametric functions and maximum likelihood estimation (MLE) for parametric parts. The proposed hybrid semiparametric estimation method could better handle multivariate covariates than the traditional kernel regression-based methods, which suffer from the well-known "curse of dimensionality." In addition, compared to the traditional machine learning methods, the new hybrid estimation method can retain a nice interpretation of parametric statistical models since the new method allows keeping some parametric parts such as the linear model assumption for each component regression. Moreover, the hybrid idea of this new method can be easily extended to other semiparametric statistical models as well as other machine learning methods.

Neural networks developed by McCulloch and Pitts (1943) are one of the most popular approaches in machine learning. The idea of neural networks originates from biological brains. The fundamental element of a neural network is a "neuron," which is capable of receiving input signals, as well as processing and transmitting output signals. In addition, each neuron is connected to at least one other neuron through weights based on the degrees of importance of given connections in the neural network. The feed-forward neural network and the recurrent neural network are two main categories of neural network architectures based on the type of connections between neurons. The common part of the two neural networks is that the connections between neurons are from distinct layers so that each neuron in one layer is connected to every other neuron in the next layer. In addition, the signal flow will be transmitted across the network. However, in the feed-forward neural network, signals flow only in one way from input neurons to output neurons. On the other hand, if there exists a "feedback" signal, the network is called a "recurrent neural network." In this section, we mainly employ the feed-forward neural network and train the network using a back-propagation algorithm (Rumelhart et al., 1986).

Figure 10.1 shows an example of a feed-forward neural network containing input, hidden, and output layers. The first layer is the input layer, which receives the input of raw data and passes it to the hidden layer. The feed-forward neural network can have zero or multiple hidden layers. The hidden layers perform computations and transfer information from the input neurons to the output neurons. The output layer is the last layer in a neural network, which receives inputs from the hidden layer and performs similar computations as shown in the hidden layers. Each layer is composed of multiple neurons. The neurons in the network receive inputs from the previous layer and produce outputs by applying certain transformations, called activation functions,

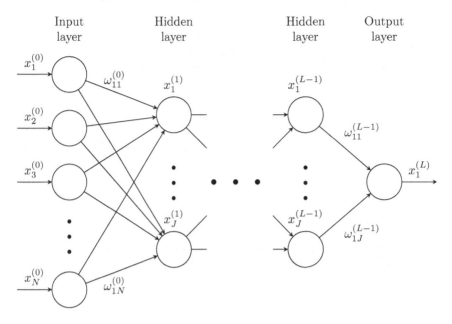

FIGURE 10.1
Feed-forward neural network with L layers.

on the inputs provided. The connection between the jth neuron in the lth layer and ith neuron in the $(l + 1)$th layer is characterized by the synaptic weight coefficient $\omega_{ij}^{(l)}$ and the bias threshold $b_i^{(l)}$. Each neuron in the network receives "weighted" information from the connected neurons and produces an output by transmitting the weighted sum of those input signals through an "activation function." The output in the ith neuron of the $(l + 1)$th layer is $x_i^{(l+1)} = f(b_i^{(l)} + \sum_j \omega_{ij}^{(l)} x_j^{(l)})$, where f is an activation function. Table 10.1 shows some examples of commonly used nonlinear activation functions including the traditional function Sigmoid, Tanh and Rectified Linear Unit function (ReLu) (Nair and Hinton, 2010). The feed-forward neural network is trained by the back-propagation algorithm (Rumelhart et al., 1986), which is a stochastic gradient descent algorithm and updates the synaptic weights by calculating the derivative of a cost function with respect to the synaptic weights $\mathbf{w}$ and the bias $\mathbf{b}$, where $\mathbf{w}$ collects all the weight coefficients and $\mathbf{b}$ collects all the bias thresholds.

To illustrate the idea of Xue and Yao (2021), we mainly consider the semiparametric mixture regression model (10.9) studied by Huang and Yao (2012), where the mixture proportions are unknown smooth functions while the component regression functions are linear functions of covariates. But the hybrid idea can be easily extended to estimate the mixtures of nonparametric regression models (10.1) proposed by Huang et al. (2013). Given the

TABLE 10.1
Some commonly used nonlinear activation functions

Name	Function	Derivative	Figure
Sigmoid	$f(x) = \frac{1}{1+e^{-x}}$	$f'(x) = f(x)(1 - f(x))^2$	
Tanh	$f(x) = \frac{e^x - e^{-x}}{e^x + e^{-x}}$	$f'(x) = 1 - f(x)^2$	
ReLU	$f(x) = \begin{cases} 0 & \text{if } x < 0 \\ x & \text{if } x \geq 0. \end{cases}$	$f'(x) = \begin{cases} 0 & \text{if } x < 0 \\ 1 & \text{if } x \geq 0. \end{cases}$	

observations $\{(\mathbf{x}_1, y_1), \ldots, (\mathbf{x}_n, y_n)\}$ from the mixture regression model (10.9), the log-likelihood function is

$$\ell(\boldsymbol{\theta}|\mathbf{X}, \mathbf{y}) = \sum_{i=1}^{n} \log \left\{ \sum_{c=1}^{C} \pi_c(\boldsymbol{x_i}) \phi \left(y_i; \mathbf{x}_i^\top \boldsymbol{\beta}_c, \sigma_c^2 \right) \right\},$$

where $\mathbf{X} = (\mathbf{x}_1, \ldots, \mathbf{x}_n), \mathbf{y} = (y_1, \ldots, y_n)$, and $\boldsymbol{\theta}$ collects all unknown parameters including unknown mixing proportion functions. Xue and Yao (2021) proposed a modified EM algorithm of Algorithm 10.4.1 to estimate $\boldsymbol{\theta}$.

In the M-step, one could also replace the soft weights $\{p_{ij}^{(t)}, i = 1, \ldots, n, c = 1, \ldots, C\}$ by hard clustering labels as the input for the response. The neural network along with the hyperparameter tuning in the M-step can be done using the R packages "*nnet*" and "*caret*."

10.5   Mixture of regression models with nonparametric errors

In a traditional FMR model, the errors are assumed to be coming from a normal distribution. However, if this assumption is unreasonable, it could lead to biased or even misleading results. Many models and/or methodologies were proposed to solve this problem.

Hunter and Young (2012) studied the following model:

$$Y|\mathbf{x} \sim \sum_{c=1}^{C} \pi_c g(y - \mathbf{x}^\top \boldsymbol{\beta}_c), \tag{10.16}$$

Algorithm 10.4.1 Semiparatric EM type algorithm (Xue and Yao, 2021)

Given the initial values $\hat{\boldsymbol{\theta}}^{(0)}$, iterate the following E-step and M-step until convergence.

E-Step: Given the current parameter estimates $\boldsymbol{\theta}^{(t)}$, compute

$$
p_{ic}^{(t)} = \frac{\pi_c^{(t)}(\boldsymbol{x}_i)\,\phi\left(y_i; \mathbf{x}_i^T \boldsymbol{\beta}_c^{(t)}, \sigma_c^{2(t)}\right)}{\sum_{c'=1}^{C} \pi_{c'}^{(t)}(\boldsymbol{x}_i)\,\phi\left(y_i; \mathbf{x}_i^\top \boldsymbol{\beta}_{c'}^{(t)}, \sigma_{c'}^{2(t)}\right)},
$$

where $i = 1, \ldots, n, c = 1, \ldots, C$.

M-Step: Update $\boldsymbol{\theta}$ given $\{p_{ic}^{(t)}, i = 1, \ldots, n, c = 1, \ldots, C\}$.

a. Update β_c and σ_c by

$$
\hat{\boldsymbol{\beta}}_c^{(t+1)} = \arg\min_{\boldsymbol{\beta}_c} \sum_{i=1}^{n} p_{ic}^{(t)}(y_i - \boldsymbol{x}_i^\top \boldsymbol{\beta}_c)^2
$$

$$
= \left(\sum_{i=1}^{n} p_{ic}^{(t)} \mathbf{x}_i \mathbf{x}_i^\top\right)^{-1} \left(\sum_{i=1}^{n} p_{ic}^{(t)} \mathbf{x}_i y_i\right),
$$

$$
\hat{\sigma}_c^{2(t+1)} = \frac{\sum_{i=1}^{n} p_{ic}^{(t)}\left(y_i - \mathbf{x}_i^\top \hat{\boldsymbol{\beta}}_c^{(t+1)}\right)^2}{\sum_{i=1}^{n} p_{ic}^{(t)}}, c = 1, \ldots, C.
$$

b. Update $\hat{\pi}_c^{(t+1)}(\mathbf{x}_i) = f_{\mathbf{w},\mathbf{b}}(\mathbf{x}_i)$ based on the neural network using $\{p_{ic}^{(t)}, i = 1, \ldots, n, c = 1, \ldots, C\}$ as the input for the response and $\mathbf{x}_i$ as the input for the covariates, where $f_{\mathbf{w},\mathbf{b}}(\cdot)$ is the function of the neural network mapping the input to the output.

---

where linearity was still assumed within each component, but the error terms are only assumed to be from $g(\cdot)$, which could be any density function with median 0. The authors proved that as long as the regression planes are not parallel, model (10.16) is identifiable. Hunter and Young (2012) proposed estimating the model (10.16) by maximizing the following smoothed log-likelihood :

$$
\ell_s(\boldsymbol{\pi}, \boldsymbol{\beta}, g) = \sum_{i=1}^{n} \log \left\{ \sum_{c=1}^{C} \pi_c \mathcal{N}_h g(y_i - \mathbf{x}_i^\top \boldsymbol{\beta}_c) \right\},
$$

where $\mathcal{N}_h g = \exp \int h^{-1} K\left\{(x - u)/h\right\} \log g(u) du$ is a nonlinear smoother. The effectiveness of the new methods was demonstrated through numerical studies.

Ma et al. (2021) extended model (10.16) by allowing the error density for each component to be different and unknown:

$$Y|\mathbf{x} \sim \sum_{c=1}^{C} \pi_c g_c(y - \mathbf{x}^\top \boldsymbol{\beta}_c). \tag{10.17}$$

The model (10.17), but not the model (10.16), also includes the popular normal mixture models with unequal component variances as special cases. Ma et al. (2021) proved the identifiability of this more general model of (10.17) without requiring the symmetry of $g_c(\cdot)$.

Ma et al. (2021) proposed an EM-type algorithm to estimate the model (10.17). More specifically, at the $(t+1)$th iteration, E-step computes classification probabilities:

$$p_{ic}^{(t+1)} = \frac{\pi_c^{(t)} g_c^{(t)}(r_{ic}^{(t)})}{\sum_{c=1}^{C} \pi_c^{(t)} g_c^{(t)}(r_{ic}^{(t)})}, \quad i = 1, \ldots, n, \ c = 1, \ldots, C,$$

where $r_{ic}^{(t)} = y_i - \mathbf{x}_i^\top \boldsymbol{\beta}_c^{(t)}$. In the M-step, we update the component parameters π_c and $\boldsymbol{\beta}_c$ by

$$\pi_c^{(t+1)} = \frac{1}{n} \sum_{i=1}^{n} p_{ic}^{(t+1)},$$

$$\boldsymbol{\beta}_c^{(t+1)} = \arg\max_{\boldsymbol{\beta}_c} \sum_{i=1}^{n} p_{ic}^{(t+1)} \log[g_c^{(t)}(y_i - \mathbf{x}_i^\top \boldsymbol{\beta}_c)],$$

and update the nonparametric density function g_c by a modified kernel density estimator:

$$g_c^{(k+1)}(t) = \frac{1}{\sum_{i=1}^{n} p_{ic}^{(k+1)}} \sum_{i=1}^{n} p_{ic}^{(k+1)} K_h(r_{ic}^{(k+1)} - t),$$

where $c = 1, \ldots, C$, $K_h(t) = h^{-1} K(t/h)$, and $K(t)$ is a kernel function, such as the Epanechnikov kernel. Ma et al. (2021) established the consistency and asymptotic normality results of the proposed estimators.

On the other hand, Hu et al. (2017) studied a model where the error densities are assumed to be $g_c(\cdot) = \exp\{\phi_c(\cdot)\}$, where $\phi_c(\cdot)$'s are some unknown concave functions. That is, the errors are assumed to be coming from log-concave distributions. Hu et al. (2017) proposed estimating $g_c(\cdot)$ in M-step by

$$g_c^{(t+1)} = \arg\max_{g_c} \sum_{i=1}^{n} p_{ic}^{(t+1)} \log g_c(y_i - \mathbf{x}_i^\top \hat{\boldsymbol{\beta}}_c^{(t+1)})$$

$$= \arg\max_{\phi_c} \sum_{i=1}^{n} p_{ic}^{(t+1)} \phi_c(y_i - \mathbf{x}_i^\top \hat{\boldsymbol{\beta}}_c^{(t+1)}).$$

The error density g_c is updated through the function `mlelcd` in the R package `LogConcDEAD` through n fitted residuals $y_i - \mathbf{x}_i^\top \boldsymbol{\beta}_c^{(t+1)}$ with weights $p_{ic}^{(t+1)}$, $i = 1, \ldots, n$.

Wu and Yao (2016) proposed a semiparametric mixture of quantile regressions model, where the conditional quantiles (such as median) of the response variable were regressed on the covariates. That is, given $Z = c$,

$$Y = \mathbf{x}^\top \boldsymbol{\beta}_c(\tau) + \varepsilon_c(\tau), \tag{10.18}$$

where $\boldsymbol{\beta}_c(\tau) = (\beta_{0c}(\tau), \ldots, \beta_{pc}(\tau))^\top$ is the τ-th quantile regression coefficient for the c-th component, and the τ-th quantile of the error density was assumed to be zero. Note that there is no parametric assumption about the error density. Due to the robustness of quantiles, model (10.18) is more robust than the traditional FMR models, and may reveal more details of the data structure. The parameters and the error densities were estimated through an EM-type algorithm incorporated with kernel regression.

10.6   Semiparametric regression models for longitudinal/functional data

In this section, we introduce some semiparametric mixture of regression models designed for longitudinal and functional data.

Yao et al. (2011) is one of the earliest articles looking into this problem. A classical functional linear model (FLM) is defined as

$$E(Y|X) = \int \beta(t)X(t)dt,$$

where $\beta(\cdot)$ is a regression function assumed to be smooth and square-integrable. However, since the subjects may belong to different mutually exclusive groups that possess different mechanisms to produce the response, a functional mixture regression model (FMR) was then introduced as

$$E(Y|X) = \int \beta_c(t)X(t)dt, \tag{10.19}$$

if the subject belongs to the cth group. By the well-known Karhunen-Loéve expansion, a process X can be written as

$$X(t) = \mu(t) + \sum_{m=1}^{\infty} \xi_m \phi_m(t), \tag{10.20}$$

where $\{\phi_m\}$ is a sequence of orthonormal eigenfunctions of X, $\xi_m = \int \{X(t) - \mu(t)\} \phi_m(t)dt$ are the functional principal component (FPC) scores of X. Apply

(10.20) into model (10.19), the model is then refined to be

$$E(Y_i|X_i, M, i \in K_c) = b_{c0} + \sum_{m=1}^{\infty} b_{cm}\xi_{im}, \qquad (10.21)$$

where $K_c = \{i: \text{the } i\text{th subject belongs to the } c\text{th group}\}$, $c = 1, \ldots, C$, $b_{c0} = \int \beta_c(t)\mu(t)dt$, $b_{cm} = \int \beta_c(t)\phi_m(t)dt$. Of course, regularization is needed for the estimation to truncate the infinite sum to a finite sum of M terms, and model (10.21) becomes

$$E(Y_i|X_i, M, i \in K_c) = b_{c0} + \sum_{m=1}^{M} b_{cm}\xi_{im}, \qquad (10.22)$$

where $M = \min\{l : \sum_{m=1}^{l} \lambda_m / \sum_{m=1}^{\infty} \lambda_m \geq \tau\}$, τ is a pre-decided threshold value, λ_ms are the eigenvalues corresponding to ϕ_m. Since the model (10.22) is quite similar to the classical mixture of linear regression models, EM algorithms can be applied to the estimation process.

Intensive longitudinal data (ILD), which is rich in information, has become increasingly popular in behavioral science. However, ILD is quite difficult to analyze, due to heterogeneity and nonlinearity in its data structure. As a result, Dziak et al. (2015) proposed a mixture of time-varying effect models (MixTVEM), which incorporate time-varying effect model (TVEM) into a mixture model framework. The model is summarized as follows. Conditional on time-invariant subject-level covariates $s_1, \ldots, s_Q$, the probability that i-th individual comes from c-th class is

$$\pi_{ic} = P(Z_i = c) = \frac{\exp\left(\gamma_{0,c} + \sum_{q=1}^{Q} \gamma_{qc} s_q\right)}{\sum_{t=1}^{C} \exp\left(\gamma_{0,t} + \sum_{q=1}^{Q} \gamma_{qt} s_q\right)},$$

and within each component, the means are assumed to be the same as the TVEM model in Tan et al. (2012):

$$\mu_{ij} = E(y_{ij}|Z_i = c) = \beta_{0c}(t_{ij}) + \beta_{1c}(t_{ij})x_{ij1} + \ldots + \beta_{pc}(t_{ij})x_{ijp},$$

where $x_1, \ldots, x_p$ are the observation-level covariates. In addition, the covariance structure of Y_{ij} is assumed to be of the form

$$\text{cov}(y_{ij}, y_{ij'}) = \sigma_a^2 \rho^{|t_{ij} - t_{ij'}|} + \sigma_e^2,$$

where σ_a^2 and σ_e^2 denote the variances of subject-level and observation-level errors, respectively. Since the error terms are still assumed to be normally distributed, although nonparametric in means, the MixTVEM is still treated as semiparametric models. Individuals are assumed to be clustered into one and only one of those latent classes to guarantee the identifiability of the model. EM algorithm is applied for estimation, and a penalized B-spline is used to approximate $\beta(\cdot)$'s, where the penalization is applied to ensure a smooth and parsimonious curve. Huang et al. (2018c) also investigated the identifiability and statistical inference for mixtures of time-varying coefficient models, in which each mixture component follows a varying coefficient model, and the mixing proportions and dispersion parameters are unknown smooth functions.

In functional data analysis, data is in-homogeneous, collected at irregular, possibly subject-depending time points. Huang et al. (2014) proposed a new estimation procedure for the mixture of Gaussian processes to accommodate this type of data. Conditional on $Z_i = c$, the model assumes

$$y_{ij} = \mu_c(t_{ij}) + \sum_{q=1}^{\infty} \xi_{iqc} \nu_{qc}(t_{ij}) + \varepsilon_{ij}, \qquad i = 1, \ldots, n; j = 1, \ldots, N_i,$$

where N_i is the number of observations from the ith subject, ε_{ij}s are independent and $N(0, \sigma^2)$ distributed, $\mu_c(t)$ is the cth component mean of a Gaussian process with covariance function $G_c(s, t)$, and ξ_{iqc} and $\nu_{qc}(t)$ are the functional principal component (FPC) score and eigenfunctions of $G_c(s, t)$, respectively(Roger and Pol, 1991, Karhunen-Loève theorem).

To analyze heterogeneous functional data with functional covariates, given $Z = c$, Wang et al. (2016) proposed to model $\{y(t), t \in T\}$ in a functional-linear way:

$$y(t) = \mathbf{X}(t)^{\top} \boldsymbol{\beta}_c(t) + \varepsilon_c(t), \tag{10.23}$$

where $\mathbf{X}(t)$ is a random covariate process of dimension p, $\boldsymbol{\beta}_c(t)$ is a smooth regression coefficient function of the c-th component, and $\varepsilon_c(t)$ is a Gaussian process with mean zero, independent of $\mathbf{X}(t)$, and is assumed to be of the form

$$\varepsilon_c(t) = \zeta_c(t) + e(t),$$

where $\zeta_c(t)$ denotes a trajectory process with covariance $\Gamma_c(s, t) = \text{cov}\{\xi_c(s), \xi_c(t)\}$, and $e(t)$ is the measurement error with constant variance σ^2. For ease of notation, define $y_{ij} = y_i(t_{ij})$, $j = 1, \ldots, N_i$, and similarly define $\varepsilon_{cij}, e_{ij}$, etc. Similar to Huang et al. (2014), by the Karhunen-Loève theorem, model (10.23) can be represented as

$$y_{ij} = \mathbf{X}_i(t_{ij})^{\top} \boldsymbol{\beta}_c(t_{ij}) + \sum_{q=1}^{\infty} \xi_{iqc} \nu_{qc}(t_{ij}) + e_{ij},$$

where $\nu_{qc}(\cdot)$'s are the eigenfunctions of $\Gamma_c(s, t)$, λ_{qc}'s are the corresponding eigenvalues, and ξ_{iqc}'s are the uncorrelated FPC of $\zeta_c(t)$ satisfying $E(\xi_{iqc}) = 0$ and $\text{var}(\xi_{iqc}) = \lambda_{qc}$. Ignoring the correlation structure, y_{ij} can be thought to be coming from the following mixture of Gaussian process:

$$y(t) \sim \sum_{c=1}^{C} \pi_c N\{\mathbf{X}(t)^{\top} \boldsymbol{\beta}_c(t), \sigma_c^{*2}(t)\},$$

where $\sigma_c^{*2}(t) = \Gamma_c(t,t) + \sigma^2$. Then, the parameters $\pi_c, \boldsymbol{\beta}_c(\cdot)$, and $\sigma_c^{*2}(\cdot)$ can be estimated by an EM-type algorithm, which is very close to the one discussed in Huang et al. (2014).

10.7   Semiparmetric hidden Markov models with covariates

The hidden Markov model regression (HMMR) has been popularly used in many fields such as gene expression and activity recognition. However, the traditional HMMR requires a strong linearity assumption for the emission model. Huang et al. (2018a) proposed a semiparametric hidden Markov model with nonparametric regression (HMM-NR) in which the mean and variance of emission model are unknown smooth functions. The new semiparametric model reduces the modeling bias, thus enhancing the applicability of the traditional hidden Markov model regression.

Suppose that the stochastic triplet sequence $\{(Z_t, X_t, Y_t), t = 1, \ldots, T\}$ is a finite realization of a Markov process, and Z_t is unobserved latent homogeneous irreducible Markov chain with finite state space $\{1, 2, \cdots, C\}$. Let Γ be the $C \times C$ transition matrix of Z_t, with elements $\gamma_{jk} = P(Z_{t+1} = k | Z_t = j)$, and $\boldsymbol{\pi} = (\pi_1, \ldots, \pi_C)$ be the initial probabilities of the states, where $\sum_{k=1}^{C} \pi_k = 1$ and $\sum_{k=1}^{C} \gamma_{jk} = 1$ for all j. Huang et al. (2018a) proposed the following HMM-NR model:

$$p(Z_{t+1} | Z_t, X_t, Y_t) = p(Z_{t+1} | Z_t),$$
$$Y_t | X_t, Z_t = k \sim N\{\mu_k(X_t), \sigma_k^2(X_t)\}, \qquad (10.24)$$

where $\mu_k(\cdot)$ is an unknown smooth mean function, $\sigma_k^2(\cdot)$ is a positive unknown smooth variance function, and $N(\mu, \sigma^2)$ is the normal distribution with mean μ and variance σ^2. Note that the traditional HMMR is a special case of the proposed HMM-NR if $\mu_k(x)$ is assumed to be a linear function of x and $\sigma_k^2(x)$ is constant.

Denote by $p_k(x, y) = P(Y_t = y | X_t = x, Z_t = k) = \phi\{y; \mu_k(x), \sigma_k^2(x)\}$ for $k = 1, \ldots, C$, where $\phi(\cdot; \mu, \sigma^2)$ is a normal density function with mean μ and variance σ^2. Let $P(x, y)$ be a $C \times C$ diagonal matrix with diagonal elements $p_1(x, y), \ldots, p_C(x, y)$, i.e.,

$$P(x, y) = \mathrm{diag}(p_1(x, y), \ldots, p_C(x, y)).$$

The likelihood function for the observed data is $\ell(\boldsymbol{\pi}, \Gamma, \boldsymbol{\theta}(\cdot))$,

$$\ell(\boldsymbol{\pi}, \Gamma, \boldsymbol{\theta}(\cdot)) = \boldsymbol{\pi} P(x_1, y_1) \Gamma P(x_2, y_2) \Gamma P(x_3, y_3) \cdots \Gamma P(x_T, y_T) \mathbf{1}', \qquad (10.25)$$

where $\boldsymbol{\theta}(x) = \{\mu_k(x), \sigma_k^2(x), k = 1, \cdots, C\}$.

Borrowing the idea from the parametric HMM, Huang et al. (2018a) proposed using an EM algorithm to simplify the computation. Define the $1 \times C$ vector of forward probabilities $\boldsymbol{\alpha}_t = (\alpha_{t1}, \dots, \alpha_{tC})$ as

$$\boldsymbol{\alpha}_t = \boldsymbol{\pi} P(x_1, y_1) \Gamma P(x_2, y_2) \cdots \Gamma P(x_t, y_t), \qquad t = 1, 2, \dots, T,$$

and define the $C \times 1$ vector of backward probabilities $\boldsymbol{\beta}_t = (\beta_{t1}, \dots, \beta_{tC})'$ as

$$\boldsymbol{\beta}_t = \Gamma P(x_{t+1}, y_{t+1}) \Gamma P(x_{t+2}, y_{t+2}) \cdots \Gamma P(x_T, y_T) \mathbf{1}', \qquad t = 1, 2, \dots, T.$$

Then, in the E step, we compute

$$r_{tk} = \alpha_{tk} \beta_{tk} / \ell(\boldsymbol{\pi}, \Gamma, \boldsymbol{\theta}(\cdot)),$$

and

$$h_{tjk} = \alpha_{t-1,j} \gamma_{jk} p_k(x_t, y_t) \beta_{tk} / \ell(\boldsymbol{\pi}, \Gamma, \boldsymbol{\theta}(\cdot)),$$

where $\ell(\boldsymbol{\pi}, \Gamma, \boldsymbol{\theta}(\cdot))$ is defined in (10.25).

In the M step, we update the parametric parts by

$$\pi_k = r_{1k}$$

$$\gamma_{jk} = \frac{\sum_{t=2}^{T} h_{tjk}}{\sum_{t=2}^{T} \sum_{k=1}^{S} h_{tjk}} = \frac{\sum_{t=2}^{T} h_{tjk}}{\sum_{t=2}^{T} r_{t-1,j}},$$

and update the nonparametric function $\mu_k(x)$ and $\sigma_k(x)$ by kernel regression:

$$\mu_k(x) = \frac{\sum_{t=1}^{T} \omega_{tk}(x) \cdot y_t}{\sum_{t=1}^{T} \omega_{tk}(x)},$$

$$\sigma_k^2(x) = \frac{\sum_{t=1}^{T} \omega_{tk}(x) \cdot (y_t - \mu_k(x))^2}{\sum_{t=1}^{T} \omega_{tk}(x)},$$

where $\omega_{tk}(x) = r_{tk} K_h(x_t - x), t = 1, \dots, T, k = 1, \dots, C$. The computation is performed at a set of grid points, and then the interpolation is used to obtain a functional estimate of $\boldsymbol{\theta}(x)$.

Note that HMM-NR model (10.24) assumes that the transition matrix Γ is constant and does not depend on covariates. To make use of the covariates to better estimate Γ, Huang et al. (2021) proposed a nonparametric hidden Markov model (HMM), in which both the emission model and transition matrix are nonparametric functions of covariates. More specifically, conditioning on $X_t = x$, the transition matrix is denoted by a $C \times C$ matrix $\Gamma(x)$ with the elements $\gamma_{jk}(x) = P(S_{t+1} = k | S_t = j, X_t = x)$, for $j, k = 1, 2, \dots, C$. The transition functions are nonnegative and satisfy the constraints $\sum_{k=1}^{C} \gamma_{jk}(x) = 1$ for $j = 1, 2, \dots, C$. The emission model has the same assumption as (10.24).

Huang et al. (2021) proposed applying the composite likelihood (CL) function (Lindsay, 1988; Chen et al., 2016a) for the proposed nonparametric HMM to simplify the computation. Similar ideas can also be applied to traditional fully parametric HMM and the semiparametric MM-NR model (10.24). The CL approach, which can also be considered as a type of partial likelihood, could provide significant computational and modeling advantages in some complex situations, while still being consistent and asymptotically normal.

Since X_{t+1} is independent of (S_t, S_{t+1}), let

$$
\begin{aligned}
\pi_{jk}(x_t) &= P(S_t = j, S_{t+1} = k | X_t = x_t, X_{t+1} = x_{t+1}) \\
&= P(S_t = j, S_{t+1} = k | X_t = x_t) \\
&= P(S_{t+1} = k | S_t = j, X_t = x_t) P(S_t = j | X_t = x_t) \\
&= \gamma_{jk}(x_t) \sum_m \pi_{jm}(x_t).
\end{aligned}
$$

Hence, $\gamma_{jk}(x) = \pi_{jk}(x) / \sum_{m=1}^{C} \pi_{jm}(x)$. Note that the joint density function of (y_t, y_{t+1}) given covariate (x_t, x_{t+1}) is

$$
f(y_t, y_{t+1} | x_t, x_{t+1}) = \sum_{j=1}^{C} \sum_{k=1}^{C} \pi_{jk}(x_t) f(y_t, y_{t+1} | x_t, x_{t+1}, S_t = j, S_{t+1} = k)
$$

$$
= \sum_{j=1}^{C} \sum_{k=1}^{C} \pi_{jk}(x_t) \phi\{y_t | \mu_j(x_t), \sigma_j^2(x_t)\} \phi\{y_{t+1} | \mu_k(x_{t+1}), \sigma_k^2(x_{t+1})\}.
$$

In the composite likelihood approach, the T pairs of consecutive observations $\{(y_t, x_t; y_{t+1}, x_{t+1}), t = 1, \cdots, T\}$ are treated as independent bivariate random variables, and the composite likelihood function can be written as

$$
L(\boldsymbol{\pi}(\cdot), \boldsymbol{\mu}(\cdot), \boldsymbol{\sigma}^2(\cdot)) = \prod_{t=1}^{T} f(y_t, y_{t+1} | x_t, x_{t+1})
$$

$$
= \prod_{t=1}^{T} \left(\sum_{j=1}^{C} \sum_{k=1}^{C} [\pi_{jk}(x_t) \phi\{y_t | \mu_j(x_t), \sigma_j^2(x_t)\} \phi\{y_{t+1} | \mu_k(x_{t+1}), \sigma_k^2(x_{t+1})\}] \right),
$$

(10.26)

where $\boldsymbol{\pi}(\cdot) = \{\pi_{jk}(\cdot), j, k = 1, 2, \ldots, C\}$. Note that since the pairs of observations are not really independent, the composite likelihood constructed in (10.26) is a pseudo-likelihood, thus the corresponding estimator will suffer some loss of efficiency. However, the advantage of the composite likelihood is to transform the complex nonparametric HMM to a much simpler nonparametric bivariate mixture regression model with mixing proportions $\pi_{jk}(x_t)$, for which the estimation and inference can be studied easily.

Note that (10.26) is not ready to be maximized due to the nonparametric smoothing functions. Huang et al. (2021) proposed estimating the model

(10.26) by maximizing the following local composite likelihood function:

$$\ell_T(\boldsymbol{\pi}, \boldsymbol{\mu}, \boldsymbol{\sigma}^2; x) = \sum_{t=1}^{T} \log \left\{ \sum_{j=1}^{C} \sum_{k=1}^{C} \pi_{jk} \phi\{y_t | \mu_j, \sigma_j^2\} \phi\{y_{t+1} | \mu_k, \sigma_k^2\} \right\}$$

$$\times K_h(x_t - x) K_h(x_{t+1} - x), \qquad (10.27)$$

where $\boldsymbol{\pi}$ is a $C(C-1) \times 1$ vector

$$(\pi_{11}, \ldots, \pi_{1C}, \pi_{21}, \ldots, \pi_{2C}, \ldots, \pi_{C-1,1}, \ldots, \pi_{C-1,C})^\top,$$

$\boldsymbol{\mu} = (\mu_1, \ldots, \mu_C)^\top$, $\boldsymbol{\sigma}^2 = (\sigma_1^2, \ldots, \sigma_C^2)^\top$, and $K_h = h^{-1} K(\cdot/h)$ is a rescaled kernel of a kernel function $K(\cdot)$ with a bandwidth h. Note that $\{\pi_{jk}, j, k = 1, 2, \ldots, C\}$ satisfy the constraints $\sum_{j=1}^{C} \sum_{k=1}^{C} \pi_{jk} = 1$, and $\pi_{jk} \geq 0$, for $j = 1, 2, \ldots, C$, $k = 1, 2, \ldots, C$.

The maximization of (10.27) can be done by an EM algorithm. In the E-step, we compute

$$r_{tjk} = \frac{\pi_{jk}(x_t) \phi\{y_t | \mu_j(x_t), \sigma_j^2(x_t)\} \phi\{y_{t+1} | \mu_k(x_{t+1}), \sigma_k^2(x_{t+1})\}}{\sum_{j=1}^{C} \sum_{k=1}^{C} \pi_{jk}(x_t) \phi\{y_t | \mu_j(x_t), \sigma_j^2(x_t)\} \phi\{y_{t+1} | \mu_k(x_{t+1}), \sigma_k^2(x_{t+1})\}}.$$

In the M-step, we update $\boldsymbol{\pi}$ by,

$$\hat{\pi}_{jk}(x) = \frac{\sum_{t=1}^{T} r_{tjk} K_h(x_t - x) K_h(x_{t+1} - x)}{\sum_{t=1}^{T} K_h(x_t - x) K_h(x_{t+1} - x)},$$

and update $\boldsymbol{\mu}$ and $\boldsymbol{\sigma}$ by

$$\hat{\mu}_k(x) = \frac{\sum_{t=1}^{T} \sum_{j=1}^{C} (y_t r_{tjk} + y_{t+1} r_{tkj}) K_h(x_t - x) K_h(x_{t+1} - x)}{\sum_{t=1}^{T} \sum_{j=1}^{N} (r_{tjk} + r_{tkj}) K_h(x_t - x) K_h(x_{t+1} - x)},$$

$$\hat{\sigma}_k^2(x) = \frac{\sum_{t=1}^{T} \sum_{j=1}^{C} \left\{ r_{tjk}(y_t - \mu_k)^2 + r_{tkj}(y_{t+1} - \mu_k)^2 \right\} K_h(x_t - x) K_h(x_{t+1} - x)}{\sum_{t=1}^{T} \sum_{j=1}^{C} (r_{tjk} + r_{tkj}) K_h(x_t - x) K_h(x_{t+1} - x)}.$$

Huang et al. (2021) proved the consistency of the above estimators and provided their asymptotic normality results. In addition, they also proposed a generalized likelihood ratio test statistic to formally test whether the transition probabilities depend on covariates, i.e.,

$$H_0 : \boldsymbol{\pi}(x) = \boldsymbol{\pi} \qquad versus \qquad H_1 : \boldsymbol{\pi}(x) \neq \boldsymbol{\pi}$$

where $\boldsymbol{\pi}$ is an unknown constant vector that does not depend on covariates.

10.8 Some other semiparametric mixture regression models

There are many more semiparametric mixture regression models that have been proposed. For example, Vandekerkhove (2013) studied a two-component mixture of regression models where the mixing proportion, slope, intercept and error distribution of one component is unknown while the other is known. The method proposed by Vandekerkhove (2013) performs well for datasets of reasonable sizes. However, its computation is challenging when the sample size is large since this method is based on the optimization of a contrast function of size $O(n^2)$. Bordes et al. (2013) also studied the same model as Vandekerkhove (2013), and proposed a new method-of-moments estimator whose computation order is of $O(n)$. Young (2014) extended the mixture of linear regression models to incorporate changepoints by assuming one or more of the components as piecewise linear. Such a model is a combination of the traditional mixture of linear regression models and the standard changepoint regression model. Faicel (2016) proposed a new fully unsupervised algorithm to learn regression mixture models with an unknown number of components. Unlike the standard EM for the mixture of regressions, this method does not require accurate initialization. Montuelle and le Pennec (2014) studied a mixture of Gaussian regressions model with logistic weights, and proposed to estimate the number of components and other parameters through a penalized maximum likelihood approach. Butucea et al. (2017) considered a nonlinear mixture of regression models with one known component. A local estimation procedure based on the symmetry of local noise is proposed to estimate the proportion and location functions.

10.9 Fitting semiparametric mixture regression models using R

In Section 10.2, Huang et al. (2013) proposed a mixture of nonparametric regression models (10.1), in which the regression functions are linear functions of the predictors, but the mixing proportions are smoothing functions of a covariate. As an example, we apply this method to the CO2-GNP data. The dataset contains the gross national product (GNP) and estimated carbon dioxide (CO2) emission per capita of 28 countries in the same year. It is believed that GNP and CO2 are highly correlated, but the way CO2 affects GNP might be different among many countries.

```
library(mixtools)
data(CO2data, package = "mixtools")
x = CO2data$GNP
y = CO2data$CO2
n = nrow(CO2data)

library(MixSemiRob)
u = seq(from = min(x), to = max(x), length = 10)
```

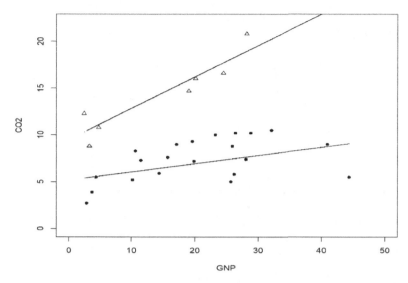

FIGURE 10.2
Estimation and hard-clustering of CO2-GNP data.

```
est = mixregPvary(x = x, y = y, C = 2, z = x, u = u, h = 1.5)
p = est$pi_z
beta = est$beta
var = est$var
```

The estimation and hard-clustering results are shown in Figure 10.2. The lower component includes countries like the US, the UK, Canada, and Australia, which are mainly developed countries trying to increase their GNP at a lower price of CO_2 emission. On the other hand, the countries in the upper component include Kuwait, Saudi Arabia, and Qatar, which, at their current status, would rather pay a higher price to increase their GNP.

In addition, Xiang and Yao (2018) proposed a new class of semiparametric mixture of regression models (10.2), where the mixing proportions and variances are constants, but the component regression functions are smooth functions of a covariate. We apply this model to the US house price index data, which contains the monthly change of S&P/Case-Schiller House Price Index (HPI) and the monthly growth rate United States Gross Domestic Product (GDP) from January 1990 to December 2002. It is well known that GDP measures the size of a nation's economy, and HPI is a measure of a nation's average housing price. As a result, it is natural to believe that HPI and GDP are related to each other.

```
library(MixSemiRob)
data(UShouse, package = "MixSemiRob")
x = (data(2:end,1) - data(1:end-1,1))./data(1:end-1,1);
```

```
y = (data(2:end,2) - data(1:end-1,2))./data(1:end-1,2);
n = length(y);


estLEM = semimrLocal(x = x, y = y)  # local EM algorithm
estGEM = semimrGlobal(x = x, y = y) # global EM algorithm
phat = estLEM$pi
varhat = estLEM$var
muhat = estLEM$mu
class = cbind(phat[1] * dnorm(y, muhat[, 1],
                  sqrt(varhat[, 1])),
              phat[2] * dnorm(y, muhat[, 2],
                  sqrt(varhat[, 2])))
group = class[, 1] > class[, 2]
xsort = sort(x, index.return = TRUE)$x
idx = sort(x, index.return = TRUE)$ix
mat = matrix(1:2, 1, 2)
layout(mat)
layout.show(2)
plot(x, y, pch = 19, cex = 0.8, xlab = "GDP", ylab = "HPI")
plot(x[group], y[group], pch = 19, cex = 0.8, xlab = "GDP",
     ylab = "HPI", ylim = c(-0.3, 1))
points(x[group == FALSE], y[group == FALSE], pch = 17,
       cex = 0.9)
lines(xsort, muhat[idx, 1], lty = 1, lwd = 2)
lines(xsort, muhat[idx, 2], lty = 1, lwd = 2)
```

The estimation results are summarized in Figure 10.3. It turns out that the circle points (lower cluster) are mainly from January 1990 to September 1997, whereas the triangle points (upper cluster) are mainly from October 1997 to December 2002, during which the economy experienced an Internet boom and bust.

In case there are high-dimensional covariates, Xiang and Yao (2020) proposed a class of semiparametric mixture regression models with single-index for model-based clustering, the single index transfers the high-dimensional nonparametric problem to a univariate nonparametric problem. We apply this method to the NBA data, which contains some descriptive statistics of all 105 guards from the 1992-1993 season. We are interested in how points per game (PPM) is affected by the height of a player (Height), minutes per game (MPG) and free throw percentage (FTP).

```
library(MixSemiRob)
data(NBA, package = "MixSemiRob")
x = NBA[, c("Height", "MPG", "FTP")]
y = NBA[, "PPM"]
x = t(t(x)/apply(x, 2, sd))
```

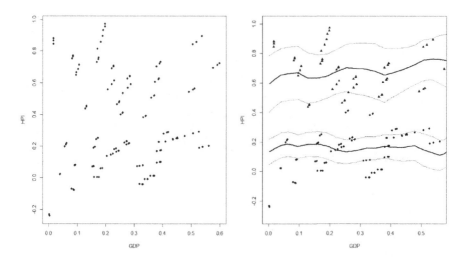

FIGURE 10.3
Estimation and classification of housing data.

```
y = y/sd(y)
est_b = sinvreg(x = x, y = y)$direction[, 1]
est = semimrFull(x = x, y = y, h = 0.3442, coef = est_b)
est_p = est$pi
est_var = est$var
est_mu = est$mu
est_b = est$coef
class = cbind(est_p[, 1] * dnorm(y, est_mu[, 1],
                sqrt(est_var[, 1])),
              est_p[, 2] * dnorm(y, est_mu[, 2],
                sqrt(est_var[, 2])))
group = class[, 1] > class[, 2]
mat = matrix(1:2, 1, 2)
layout(mat)
layout.show(2)
z = x %*% est_b
plot(z, y, pch = 19, cex = 0.8, xlab = "index", ylab = "PPM",
     ylim = c(1.5, 6.5), xlim = c(14, 19.5))
zsort = sort(z, index.return = TRUE)$x
idx = sort(z, index.return = TRUE)$ix
plot(z[group], y[group], pch = 19, cex = 0.8, xlab = "index",
     ylab = "PPM", ylim = c(1.5, 6.5), xlim = c(14, 19.5))
points(z[group == FALSE], y[group == FALSE], pch = 2,
   cex = 0.9)
```

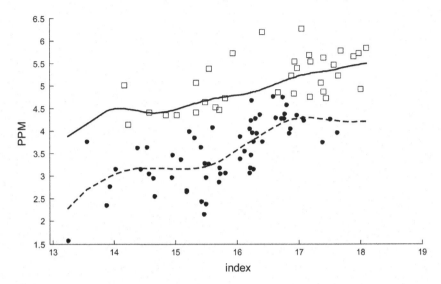

FIGURE 10.4
Estimation and hard-clustering of NBA data.

```
lines(zsort, est_mu[idx, 1], lty = 2, lwd = 2)
lines(zsort, est_mu[idx, 2], lty = 1, lwd = 2)
```

The estimation and hard clustering results are shown in Figure 10.4. The 95% confidence interval for $\hat{\alpha}$ are $(0.134, 0.541)$, $(0.715, 0.949)$ and $(0.202, 0.679)$, indicating that, MPG is the most influential factor on PPM. This could possibly be explained by coaches that tend to let good players with higher PPM play longer minutes per game (i.e., higher MPG). The two groups of guards our new models found might be explained by the difference between shooting guards and passing guards.

In Section 10.3, Huang and Yao (2012) proposed a mixture of regressions model with varying proportions (10.9). In order to apply the model, one can apply the following function:

```
res = mixregPvary(x, y)
```
where x is a matrix of independent variables and y is the response.

As a special case, Cao and Yao (2012) studied the model (10.13), which is a nonparametric mixture of binomial distribution with one degenerate component. To apply this model, one can run the following function:

```
res = semimrBin(t, x, N)
```
where t is a vector time variable, x is a vector of the observed number of successes, and N is a scalar, specifying the number of trials for the Binomial distribution.

TABLE 10.2
Mixture of regressions with nonparametric errors on tone data

Parameters	Mixture2	Mixture3
β_{11}	1.919	1.901
β_{12}	0.041	0.049
β_{21}	0.025	0.010
β_{22}	0.977	0.966
λ_1	0.659	-

In Section 10.5, Hunter and Young (2012) studied a mixture of regression model with nonparametric errors (10.16), where the error terms are only assumed to be from an unknown but same density $g(\cdot)$. In addition, Ma et al. (2021) extended (10.16) by allowing the error density for each component to be different, resulting in the model (10.17). In order to apply this method, run the following code:

```
# same component error density (Hunter and Young, 2012)
res = kdeem.h(x, y, C)
# different component error density (Ma et al., 2021)
res = kdeem(x, y, C)
```
where C indicates the number of components (with default value 2).

Let's apply the two methods to the **tonedata**, which compares perceived tone and actual tone for a trained musician. A pure fundamental tone was played to a trained musician. Electronically generated overtones were added, determined by a stretching ratio of 2.0, which corresponds to the harmonic pattern usually heard in traditional definite-pitched instruments. The musician was asked to tune an adjustable tone to the octave above the fundamental tone. The variable "tuned" gives the ratio of the adjusted tone to the fundamental, i.e. tuned=2.0 would be the correct tuning for all stretch ratio values. The data analyzed here belongs to 150 trials with the same musician. The estimation results are summarized in Table 10.2 and Figure 10.5.

```
library(MixSemiRob)
data(tone, packages = "MixSemiRob")
y = tone$tuned
x = tone$stretchratio
Result.kdeem.lse = kdeem.lse(x = x, y = y, C = 2)
Result.kdeem = kdeem(x = x, y = y, C = 2,
    ini = Result.kdeem.lse)
Result.kdeem.h = kdeem.h(x = x, y = y, C = 2,
                    ini = Result.kdeem,
```

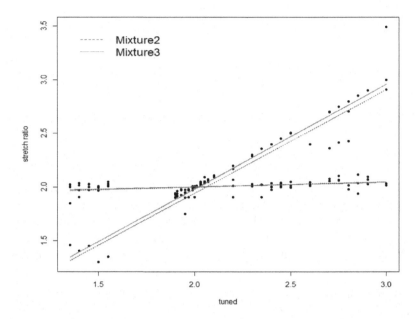

FIGURE 10.5
Mixture of regressions with nonparametric errors on tone data.

```
                              maxiter = 3)
plot(x, y, xlab = "tuned", ylab = "stretch ratio", pch = 19,
     cex = 0.7, main = "")
BET11 = Result.kdeem$beta
l11 = BET11[2, 2] * x + BET11[1, 2]
l12 = BET11[2, 1] * x + BET11[1, 1]
lines(x, l11, lty = 2, col = "red")
lines(x, l12, lty = 2, col = "red")
BET12 = Result.kdeem.h$beta
l21 = BET12[2, 2] * x + BET12[1, 2]
l22 = BET12[2, 1] * x + BET12[1, 1]
lines(x, l21, lty = 3, col = "blue")
lines(x, l22, lty = 3, col = "blue")
legend(min(x), max(y), legend = c("Mixture2", "Mixture3"),
       col = c("red", "blue"), lty = 2:3, cex = 0.8,
       bty = "n")
```

10.10 Proofs

In this section, the conditions required by the theorems are listed below. They are not the weakest sufficient conditions, but could easily facilitate the proofs. Brief proofs of the theorems are also presented.

Technical Conditions:

(C1) $nh^4 \to 0$ and $nh^2 \log(1/h) \to \infty$ as $n \to \infty$ and $h \to 0$.

(C2) $nh \to \infty$ as $n \to \infty$ and $h \to 0$.

(C3) The sample $\{(X_i, Y_i), i = 1, \ldots, n\}$ are independently and identically distributed from $f(x, y)$ with finite sixth moments. The support for x, denoted by $\mathcal{X} \in \mathbb{R}$, is bounded and closed.

(C4) $f(x, y) > 0$ in its support and has a continuous first derivative.

(C5) $|\partial^3 \ell(\boldsymbol{\theta}, X, Y)/\partial \theta_i \partial \theta_j \partial \theta_k| \leq M_{ijk}(X, Y)$, where $\mathrm{E}(M_{ijk}(X, Y))$ is bounded for all i, j, k and all X, Y.

(C6) The unknown functions $m_c(x)$, $c = 1, \ldots, C$, have continuous second derivative.

(C7) $\sigma_c^2 > 0$ and $\pi_c > 0$ for $c = 1, \ldots, C$ and $\sum_{c=1}^{C} \pi_c = 1$.

(C8) $\mathrm{E}(X^{2r}) < \infty$ for some $\epsilon < 1 - r^{-1}$, $n^{2\epsilon-1}h \to \infty$.

(C9) $I_\theta(x)$ and $I_m(x)$ are positive definite.

(C10) The kernel function $K(\cdot)$ is symmetric, continuous with compact support.

(C11) The marginal density $f(x)$ of X is Lipschitz continuous and bounded away from 0. X has a bounded support $\mathcal{X}$.

(C12) $t^3 K(t)$ and $t^3 K'(t)$ are bounded and $\int t^4 K(t)dt < \infty$.

(C13) $\mathrm{E}|q_\theta|^4 < \infty$, $\mathrm{E}|q_m|^4 < \infty$, where $\frac{\partial \ell(\theta(X),Y)}{\partial \theta} = q_\theta$, and $\frac{\partial \ell(\theta(X),Y)}{\partial m} = q_m$.

(C14) The sample $\{(\mathbf{x}_i, Y_i), i = 1, \ldots, n\}$ is independent and identically distributed from its population $(\mathbf{x}, Y)$. The support for $\mathbf{x}$, denoted by $\mathcal{X}$, is a compact subset of $\mathbb{R}^3$.

(C15) The marginal density of $\boldsymbol{\alpha}^\top \mathbf{x}$, denoted by $f(\cdot)$, is twice continuously differentiable and positive at the point z.

(C16) The kernel function $K(\cdot)$ has a bounded support, and satisfies that

$$\int K(t)dt = 1, \qquad \int tK(t)dt = 0, \qquad \int t^2 K(t)dt < \infty,$$

$$\int K^2(t)dt < \infty, \qquad \int |K^3(t)|dt < \infty.$$

(C17) $h \to 0$, $nh \to 0$, and $nh^5 = O(1)$ as $n \to \infty$.

(C18) The third derivative $|\partial^3 \ell(\boldsymbol{\theta}, y)/\partial\theta_i\partial\theta_j\partial\theta_k| \le M(y)$ for all y and all $\boldsymbol{\theta}$ in a neighborhood of $\boldsymbol{\theta}(z)$, and $E[M(y)] < \infty$.

(C19) The unknown functions $\boldsymbol{\theta}(z)$ have a continuous second derivative. For $j = 1,\ldots,k$, $\sigma_j^2(z) > 0$, and $\pi_j(z) > 0$ for all $\mathbf{x} \in \mathcal{X}$.

(C20) For all i and j, the following conditions hold:

$$E\left[\left|\frac{\partial\ell(\boldsymbol{\theta}(z), Y)}{\partial\theta_i}\right|^3\right] < \infty \qquad E\left[\left(\frac{\partial^2\ell(\boldsymbol{\theta}(z), Y)}{\partial\theta_i\partial\theta_j}\right)^2\right] < \infty$$

(C21) $\boldsymbol{\theta}_0^{''}(\cdot)$ is continuous at the point z.

(C22) The third derivative $|\partial^3\ell(\boldsymbol{\pi}, y)/\partial\pi_i\partial\pi_j\partial\pi_k| \le M(y)$ for all y and all $\boldsymbol{\pi}$ in a neighborhood of $\boldsymbol{\pi}(z)$, and $E[M(y)] < \infty$.

(C23) The unknown functions $\boldsymbol{\pi}(z)$ have continuous second derivative. For $j = 1,...,k$, $\pi_j(z) > 0$ for all $\mathbf{x} \in \mathcal{X}$.

(C24) For all i and j, the following conditions hold:

$$E\left[\left|\frac{\partial\ell(\boldsymbol{\pi}(z), Y)}{\partial\pi_i}\right|^3\right] < \infty \qquad E\left[\left(\frac{\partial^2\ell(\boldsymbol{\pi}(z), Y)}{\partial\pi_i\partial\pi_j}\right)^2\right] < \infty$$

(C25) $\boldsymbol{\pi}''(\cdot)$ is continuous at the point z.

Proof of Theorem 10.1.

Define $\hat{\boldsymbol{\beta}}^* = \sqrt{n}(\hat{\boldsymbol{\beta}} - \boldsymbol{\beta})$, where $\hat{\boldsymbol{\beta}}$ maximizes $\ell_2(\boldsymbol{\beta})$ in (10.4). Let

$$\ell(\tilde{\mathbf{m}}(X_i), \boldsymbol{\beta}, Y_i) = \log\{\sum_{c=1}^{C}\pi_c\phi(Y_i|\tilde{m}_c(X_i), \sigma_c^2\},$$

$$\ell(\tilde{\mathbf{m}}(X_i), \boldsymbol{\beta} + \boldsymbol{\beta}^*/\sqrt{n}, Y_i) = \log\{\sum_{c=1}^{C}(\pi_c + \pi_c^*/\sqrt{n})\phi(Y_i|\tilde{m}_c(X_i), \sigma_c^2 + \sigma_c^{2*}/\sqrt{n}\}.$$

Since $\hat{\boldsymbol{\beta}}$ maximizes ℓ_2, it is easy to see that $\hat{\boldsymbol{\beta}}^*$ maximizes

$$\ell_n(\boldsymbol{\beta}^*) = \sum_{i=1}^{n}\{\ell(\tilde{\mathbf{m}}(X_i),\boldsymbol{\beta}+\boldsymbol{\beta}^*/\sqrt{n},Y_i) - \ell(\tilde{\mathbf{m}}(X_i),\boldsymbol{\beta},Y_i\}$$

$$=\mathbf{a}_n\boldsymbol{\beta}^* + \frac{1}{2}\boldsymbol{\beta}^{*\top}\mathbf{B}_n\boldsymbol{\beta}^* + o_p(\|\boldsymbol{\beta}^*\|^2),$$

where $\mathbf{a}_n = \sqrt{\frac{1}{n}}\sum_{i=1}^{n}\frac{\partial\ell(\tilde{\mathbf{m}}(X_i),\boldsymbol{\beta},Y_i)}{\partial\boldsymbol{\beta}}$ and $\mathbf{B}_n = \frac{1}{n}\sum_{i=1}^{n}\frac{\partial^2\ell(\tilde{\mathbf{m}}(X_i),\boldsymbol{\beta},Y_i)}{\partial\boldsymbol{\beta}\partial\boldsymbol{\beta}^{\top}}$. It can be easily seen that $\mathbf{B}_n = -\mathbf{B} + o_p(1)$ with $\mathbf{B} = \mathrm{E}\{\mathbf{I}_\beta(X)\}$, therefore, by quadratic approximation lemma,

$$\hat{\boldsymbol{\beta}}^* = \mathbf{B}^{-1}\mathbf{A}_n + o_p(1). \tag{10.28}$$

Define $\mathbf{r}_{1n} = \sqrt{\frac{1}{n}}\sum_{i=1}^{n}\frac{\partial^2\ell(\mathbf{m}(X_i),\boldsymbol{\beta},Y_i)}{\partial\boldsymbol{\beta}\partial\mathbf{m}^T}(\tilde{\mathbf{m}}(X_i)-\mathbf{m}(X_i))$, then $\mathbf{a}_n = \sqrt{\frac{1}{n}}\sum_{i=1}^{n}\frac{\partial\ell(\mathbf{m}(X_i),\boldsymbol{\beta},Y_i)}{\partial\boldsymbol{\beta}}+\mathbf{r}_{1n}+O_p(\sqrt{\frac{1}{n}}\|\tilde{\mathbf{m}}-\mathbf{m}\|_\infty^2)$. Let $\boldsymbol{\varphi}(X_t,Y_t)$ be a $k\times1$ vector whose elements are the first k entries of $\mathbf{I}_\theta^{-1}(X_t)\frac{\partial\ell(\boldsymbol{\theta}(X_t),Y_t)}{\partial\boldsymbol{\theta}}$. From assumption (C1), we know that $O_p\{n^{1/2}[\gamma_n h^2 + \gamma_n^2\log^{1/2}(1/h)]\} = o_p(1)$, where $\gamma_n = (nh)^{-1/2}$. By similar argument as the proof of Theorem 2 in Huang et al. (2013), it can be shown that $\tilde{\boldsymbol{\theta}}(X_i) - \boldsymbol{\theta}(X_i) = \frac{1}{n}f^{-1}(X_i)\mathbf{I}_\theta^{-1}(X_i)\sum_{t=1}^{n}\frac{\partial\ell(\boldsymbol{\theta}(X_i),Y_t)}{\partial\boldsymbol{\theta}}K_h(X_t - X_i) + O_p\{\gamma_n h^2 + \gamma_n^2\log^{1/2}(1/h)\}$. Since $\mathbf{m}(X_i) - \mathbf{m}(X_t) = O(X_i - X_t)$,

$$\mathbf{r}_{1n} = n^{-3/2}\sum_{t=1}^{n}\sum_{i=1}^{n}\frac{\partial^2\ell(\mathbf{m}(X_i),\boldsymbol{\beta},Y_i)}{\partial\boldsymbol{\beta}\partial\mathbf{m}^T}\frac{\boldsymbol{\varphi}(X_t,Y_t)}{f(X_i)}K_h(X_i - X_t) + O_p(n^{1/2}h^2)$$

$$= \mathbf{r}_{2n} + O_p(n^{1/2}h^2).$$

It can also be shown that $\mathrm{E}[\frac{1}{n}\sum_{i=1}^{n}\frac{\partial^2\ell(\mathbf{m}(X_i),\boldsymbol{\beta},Y_i)}{\partial\boldsymbol{\beta}\partial\mathbf{m}^T}f^{-1}(X_i)K_h(X_i - X_t)] = \mathbf{I}_{\beta m}(X_t)$. Let $\boldsymbol{\varpi}(X_t,Y_t) = \mathbf{I}_{\beta m}(X_t)\boldsymbol{\varphi}(X_t,Y_t)$, and $\mathbf{r}_{n3} = -n^{-1/2}\sum_{t=1}^{n}\boldsymbol{\varpi}(X_t,Y_t)$, then $\mathbf{r}_{n2} - \mathbf{r}_{n3} \xrightarrow{P} 0$, and therefore

$$\mathbf{a}_n = \sqrt{\frac{1}{n}}\sum_{i=1}^{n}\{\frac{\partial\ell(\mathbf{m}(X_i),\boldsymbol{\beta},Y_i)}{\partial\boldsymbol{\beta}} - \boldsymbol{\varpi}(X_i,Y_i)\} + o_p(1),$$

given $nh^4 \to 0$. Let $\boldsymbol{\Sigma} = \mathrm{Var}\{\frac{\partial\ell(\boldsymbol{\theta}(X),Y)}{\partial\boldsymbol{\beta}} - \boldsymbol{\varpi}(X,Y)\}$, then $\mathrm{Var}(\mathbf{a}_n) = \boldsymbol{\Sigma}$. It can be easily seen that $\mathrm{E}(b a_n) = 0$, therefore by (10.28),

$$\sqrt{n}(\hat{\boldsymbol{\beta}} - \boldsymbol{\beta}) \xrightarrow{D} N(0, \mathbf{B}^{-1}\boldsymbol{\Sigma}\mathbf{B}^{-1}).$$

Proof of Theorem 10.2.

Define $\hat{\mathbf{m}}^* = \sqrt{nh}(\hat{\mathbf{m}}(x) - \mathbf{m}(x))$, where $\hat{\mathbf{m}}(x)$ maximizes (10.5). It can be shown that

$$\hat{\mathbf{m}}^*(x) = f(x)^{-1}\mathbf{I}_m(x)^{-1}\hat{\mathbf{s}}_n + o_p(1), \tag{10.29}$$

where

$$\hat{\mathbf{s}}_n = \sqrt{\frac{h}{n}} \sum_{i=1}^{n} \frac{\partial \ell(\mathbf{m}(x), \hat{\boldsymbol{\beta}}, Y_i)}{\partial \mathbf{m}} K_h(X_i - x).$$

Notice that

$$\hat{\mathbf{s}}_n = \sqrt{\frac{h}{n}} \sum_{i=1}^{n} \frac{\partial \ell(\mathbf{m}(x), \boldsymbol{\beta}, Y_i)}{\partial \mathbf{m}} K_h(X_i - x)$$

$$+ \sqrt{\frac{h}{n}}(\hat{\boldsymbol{\beta}} - \boldsymbol{\beta}) \sum_{i=1}^{n} \frac{\partial^2 \ell(\mathbf{m}(x), \boldsymbol{\beta}, Y_i)}{\partial \mathbf{m} \partial \boldsymbol{\beta}^{\top}} K_h(X_i - x) + o_p(1)$$

$$\equiv \mathbf{s}_n + \mathbf{d}_n + o_p(1).$$

where $\mathbf{s}_n = \sqrt{\frac{h}{n}} \sum_{i=1}^{n} \frac{\partial \ell(\boldsymbol{\theta}(x), Y_i)}{\partial \boldsymbol{\theta}} K_h(X_i - x)$. Since $\sqrt{n}(\hat{\boldsymbol{\beta}} - \boldsymbol{\beta}) = O_p(1)$ and $\frac{1}{n} \sum_{i=1}^{n} \frac{\partial^2 \ell(\mathbf{m}(x), \boldsymbol{\beta}, Y_i)}{\partial \mathbf{m} \partial \boldsymbol{\beta}^{\top}} K_h(X_i - x) = -f(x) \mathbf{I}_{\beta m}^{\top}(x) + o_p(1)$, then $\mathbf{d}_n = \sqrt{nh}(\hat{\boldsymbol{\beta}} - \boldsymbol{\beta}) \frac{1}{n} \sum_{i=1}^{n} \frac{\partial^2 \ell(\mathbf{m}(x), \boldsymbol{\beta}, Y_i)}{\partial \mathbf{m} \partial \boldsymbol{\beta}^{\top}} K_h(X_i - x) = -\sqrt{h} f(x) \mathbf{I}_{\beta m}^{\top}(x) + o_p(1)$. Thus, from (10.29), $\hat{\mathbf{m}}^*(x) = f(x)^{-1} \mathbf{I}_m(x)^{-1} \mathbf{s}_n + o_p(1)$. Let $\boldsymbol{\lambda}(u|x) = \mathrm{E}[\frac{\partial \ell(\mathbf{m}(x), \boldsymbol{\beta}, Y)}{\partial \mathbf{m}} | X = u]$, it can be shown that

$$\mathrm{E}(\mathbf{s}_n) = \sqrt{nh}[\frac{1}{2} f(x) \boldsymbol{\lambda}''(x|x) + f'(x) \boldsymbol{\lambda}'(x|x)] \kappa_2 h^2, \mathrm{Var}(\mathbf{s}_n) = f(x) \mathbf{I}_m(x) \nu_0,$$

where $\nu_0 = \int K^2(t) dt$. To complete the proof, let $\boldsymbol{\Delta}(x) = \mathbf{I}_m^{-1}(x)[\frac{1}{2} \boldsymbol{\lambda}''(x|x) + f^{-1}(x) f'(x) \boldsymbol{\lambda}'(x|x)] \kappa_2 h^2$, and $\boldsymbol{\delta}_m(x)$ be a $k \times 1$ vector whose elements are the first k entries of $\boldsymbol{\Delta}(x)$, then

$$\sqrt{nh}(\hat{\mathbf{m}}(x) - \mathbf{m}(x) - \boldsymbol{\delta}_m(x) + o_p(h^2)) \xrightarrow{D} N(0, f^{-1}(x) \mathbf{I}_m^{-1}(x) \nu_0).$$

Proof of Theorem 10.3.

Since $\hat{\boldsymbol{\beta}}$ has a faster convergence rate than $\hat{\mathbf{m}}(\cdot)$, $\hat{\mathbf{m}}(\cdot)$ has the same asymptotic properties as if $\boldsymbol{\beta}$ were known. Therefore, in the following proof, we study the property of $\hat{\mathbf{m}}(\cdot)$ assuming $\boldsymbol{\beta}$ to be known.

Define $\frac{\partial \ell(\boldsymbol{\theta}(X_i), Y_i)}{\partial \boldsymbol{\theta}} = \mathbf{q}_{\theta i}$, $\frac{\partial^2 \ell(\boldsymbol{\theta}(X_i), Y_i)}{\partial \boldsymbol{\theta} \partial \boldsymbol{\theta}^{\top}} = \mathbf{Q}_{\theta i}$ and similarly, define $\mathbf{q}_{mi}$, $\mathbf{Q}_{mi}$ and so on. Let $\tilde{\boldsymbol{\theta}}$ be the estimator under H_1 (model 10.1), and $\hat{\mathbf{m}}$ be the estimator under H_0 (model 10.2). From previous proof, we have

$$\tilde{\boldsymbol{\theta}}(X_i) - \boldsymbol{\theta}(X_i) = \frac{1}{n} f^{-1}(X_i) \mathbf{I}_{\theta}^{-1}(X_i) \sum_{t=1}^{n} \mathbf{q}_{\theta t} K_h(X_t - X_i)(1 + o_p(1)),$$

$$(10.30)$$

$$\hat{\mathbf{m}}(X_i) - \mathbf{m}(X_i) = \frac{1}{n} f^{-1}(X_i) \mathbf{I}_m^{-1}(X_i) \sum_{t=1}^{n} \mathbf{q}_{mt} K_h(X_t - X_i)(1 + o_p(1)).$$

$$(10.31)$$

By (10.30) and (10.31), we can obtain that

$$
\sum_{i=1}^{n} \ell(\tilde{\boldsymbol{\theta}}(X_i), Y_i) - \sum_{i=1}^{n} \ell(\boldsymbol{\theta}(X_i), Y_i)
$$

$$
= \left\{ \frac{1}{n} \sum_{i,l} \mathbf{q}_{\theta i}^{\top} f^{-1}(X_l) \mathbf{I}_{\theta}^{-1}(X_l) \mathbf{q}_{\theta l} K_h(X_i - X_l) \right.
$$

$$
\left. + \frac{1}{2n^2} \sum_{i,j,l} \mathbf{q}_{\theta i}^{\top} f^{-2}(X_l) \mathbf{I}_{\theta}^{-1}(X_l) \mathbf{Q}_{\theta l} \mathbf{I}_{\theta}^{-1}(X_l) \mathbf{q}_{\theta j} K_h(X_i - X_l) K_h(X_j - X_l) \right\}
$$

$$
\times (1 + o_p(1)),
$$

$$
\sum_{i=1}^{n} \ell(\hat{\mathbf{m}}(X_i), Y_i) - \sum_{i=1}^{n} \ell(\mathbf{m}(X_i), Y_i)
$$

$$
= \left\{ \frac{1}{n} \sum_{i,l} \mathbf{q}_{mi}^{\top} f^{-1}(X_l) \mathbf{I}_{m}^{-1}(X_l) \mathbf{q}_{ml} K_h(X_i - X_l) \right.
$$

$$
\left. + \frac{1}{2n^2} \sum_{i,j,l} \mathbf{q}_{mi}^{\top} f^{-2}(X_l) \mathbf{I}_{m}^{-1}(X_l) \mathbf{Q}_{mml} \mathbf{I}_{m}^{-1}(X_l) \mathbf{q}_{mj} K_h(X_i - X_l) K_h(X_j - X_l) \right\}
$$

$$
\times (1 + o_p(1)),
$$

and so,

$$
T = \frac{1}{n} \sum_{i,l} [\mathbf{q}_{\theta i}^{\top} \mathbf{I}_{\theta}^{-1}(X_l) \mathbf{q}_{\theta l} - \mathbf{q}_{mi}^{\top} \mathbf{I}_{m}^{-1}(X_l) \mathbf{q}_{ml}] f^{-1}(X_l) K_h(X_i - X_l)
$$

$$
+ \frac{1}{2n^2} \sum_{i,j,l} [\mathbf{q}_{\theta i}^{\top} \mathbf{I}_{\theta}^{-1}(X_l) \mathbf{Q}_{\theta l} \mathbf{I}_{\theta}^{-1}(X_l) \mathbf{q}_{\theta j} - \mathbf{q}_{mi}^{\top} \mathbf{I}_{m}^{-1}(X_l) \mathbf{Q}_{ml} \mathbf{I}_{m}^{-1}(X_l) \mathbf{q}_{mj}]
$$

$$
\times f^{-2}(X_l) K_h(X_i - X_l) K_h(X_j - X_l)
$$

$$
\equiv \Lambda_n + \frac{1}{2} \Gamma_n.
$$

By similar argument as Fan et al. (2001), it can be shown that under conditions (C9)-(C12), as $h \to 0$, $nh^{3/2} \to \infty$,

$$
\Lambda_n = \frac{2k-1}{h} K(0) \mathrm{E} f(X)^{-1}
$$

$$
+ \frac{1}{n} \sum_{l \neq i} [\mathbf{q}_{\theta i}^{\top} \mathbf{I}_{\theta}^{-1}(X_l) \mathbf{q}_{\theta l} - \mathbf{q}_{mi}^{\top} \mathbf{I}_{m}^{-1}(X_l) \mathbf{q}_{ml}] \frac{K_h(X_i - X_l)}{f(X_l)} + o_p(h^{-1/2}),
$$

$$
\Gamma_n = - \frac{(2k-1)}{h} \mathrm{E} f(X)^{-1} \int K^2(t) dt
$$

$$-\frac{2}{n}\sum_{i<j}[\mathbf{q}_{\theta i}^{\top}\mathbf{I}_{\theta}^{-1}(X_i)\mathbf{q}_{\theta j} - \mathbf{q}_{mi}^{\top}\mathbf{I}\mathbf{q}_m^{-1}(X_i)\mathbf{q}_{mj}]\frac{K_h * K_h(X_i - X_j)}{f(X_i)}$$
$$+ o_p(h^{-1/2}).$$

Therefore, $T = \mu_n + W_n/2\sqrt{h} + o_p(h^{-1/2})$, where $\mu_n = \frac{(2k-1)|\mathcal{X}|}{h}[K(0) - 0.5\int K^2(t)dt]$,

$$W_n = \frac{\sqrt{h}}{n}\sum_{i\neq j}\{\mathbf{q}_{\theta i}^{\top}\mathbf{I}_{\theta}^{-1}(X_j)[2K_h(X_i - X_j) - K_h * K_h(X_i - X_j)]f^{-1}(X_j)\mathbf{q}_{\theta j}$$
$$- \mathbf{q}_{mi}^{\top}\mathbf{I}_m^{-1}(X_j)[2K_h(X_i - X_j) - K_h * K_h(X_i - X_j)]f^{-1}(X_j)\mathbf{q}_{mj}\}.$$

It can be shown that $\mathrm{Var}(W_n) \to \zeta$, where $\zeta = 2(2k-1)\mathrm{E}f^{-1}(X)\int[2K(t) - K * K(t)]^2dt$. Apply Proposition 3.2 in de Jong (1987), we obtain that

$$W_n \xrightarrow{D} N(0, \zeta),$$

and completes the proof.

Proof of Theorem 10.4.

Let

$$\hat{\pi}_c^* = \sqrt{nh}\{\hat{\pi}_c - \pi_c(z)\}, \quad c = 1, \ldots, C - 1.$$
$$\hat{m}_c^* = \sqrt{nh}\{\hat{m}_c - m_c(z)\}, \quad c = 1, \ldots, C,$$
$$\hat{\sigma}_c^{2*} = \sqrt{nh}\{\hat{\sigma}_c^2 - \sigma_C^2(z)\}, \quad c = 1, \ldots, C.$$

Define $\hat{\boldsymbol{\pi}}^* = (\hat{\pi}_1^*, \ldots, \hat{\pi}_{C-1}^*)^{\top}$, $\hat{\mathbf{m}}^* = (\hat{m}_1^*, \ldots, \hat{m}_C^*)^{\top}$, $\hat{\boldsymbol{\sigma}}^* = (\hat{\sigma}_1^*, \ldots, \hat{\sigma}_C^*)^{\top}$ and denote $\hat{\boldsymbol{\theta}}^* = (\hat{\boldsymbol{\pi}}^{*\top}, \hat{\mathbf{m}}^{*\top}, (\hat{\boldsymbol{\sigma}}^{*2})^{\top})^{\top}$. Let $a_n = (nh)^{-1/2}$ and

$$\ell(\boldsymbol{\theta}(z), \tilde{\boldsymbol{\alpha}}, \mathbf{x}_i, Y_i) = \log\left\{\sum_{c=1}^{C}\pi_c(\tilde{\boldsymbol{\alpha}}^{\top}\mathbf{x}_i)\phi(Y_i|m_c(\tilde{\boldsymbol{\alpha}}^{\top}\mathbf{x}_i), \sigma_c^2(\tilde{\boldsymbol{\alpha}}^{\top}\mathbf{x}_i))\right\} \quad (10.32)$$
$$\times K_h(\tilde{\boldsymbol{\alpha}}^{\top}\mathbf{x}_i - z).$$

If $(\hat{\boldsymbol{\pi}}, \hat{\mathbf{m}}, \hat{\boldsymbol{\sigma}}^2)^{\top}$ maximizes (10.2), then $\hat{\boldsymbol{\theta}}^*$ maximizes

$$\ell_n^*(\boldsymbol{\theta}^*) = h\sum_{i=1}^{n}[\ell(\boldsymbol{\theta}(z) + a_n\boldsymbol{\theta}^*, \tilde{\boldsymbol{\alpha}}, \mathbf{x}_i, Y_i) - \ell(\boldsymbol{\theta}(z), \tilde{\boldsymbol{\alpha}}, \mathbf{x}_i, Y_i)]K_h(\hat{Z}_i - z)$$

with respect to $\boldsymbol{\theta}^*$. By a Taylor expansion,

$$\ell_n^*(\boldsymbol{\theta}^*) = \mathbf{W}_{1n}^{\top}\boldsymbol{\theta}^* + \frac{1}{2}\boldsymbol{\theta}^{*\top}\mathbf{A}_{1n}\boldsymbol{\theta}^* + o_p(1),$$

where

$$\mathbf{W}_{1n} = \sqrt{\frac{h}{n}} \sum_{i=1}^{n} \frac{\partial \ell(\boldsymbol{\theta}(z), \tilde{\boldsymbol{\alpha}}, \mathbf{x}_i, Y_i)}{\partial \boldsymbol{\theta}} K_h(\hat{Z}_i - z),$$

and

$$\mathbf{A}_{1n} = \frac{1}{n} \sum_{i=1}^{n} \frac{\partial^2 \ell(\boldsymbol{\theta}(z), \tilde{\boldsymbol{\alpha}}, \mathbf{x}_i, Y_i)}{\partial \boldsymbol{\theta} \partial \boldsymbol{\theta}^\top} K_h(\hat{Z}_i - z).$$

By WLLN, it can be shown that $\mathbf{A}_{1n} = -f(z)\mathcal{I}_\theta^{(1)}(z) + o_p(1)$. Therefore,

$$\ell_n^*(\boldsymbol{\theta}^*) = \mathbf{W}_{1n}^\top \boldsymbol{\theta}^* - \frac{1}{2} f(z) \boldsymbol{\theta}^{*\top} \mathbf{I}_\theta^{(1)}(z) \boldsymbol{\theta}^* + o_p(1).$$

Using the quadratic approximation lemma (see, for example, Fan and Gijbels (1996)), we have that

$$\hat{\boldsymbol{\theta}}^* = f(z)^{-1} \mathcal{I}_\theta^{(1)}(z)^{-1} \mathbf{W}_{1n} + o_p(1).$$

Note that

$$\mathbf{W}_{1n} = \sqrt{\frac{h}{n}} \sum_{i=1}^{n} \frac{\partial \ell(\boldsymbol{\theta}(z), \boldsymbol{\alpha}, \mathbf{x}_i, Y_i)}{\partial \boldsymbol{\theta}} K_h(Z_i - z) + \mathbf{D}_{1n} + O_p(\sqrt{\frac{h}{n}} \|\tilde{\boldsymbol{\alpha}} - \boldsymbol{\alpha}\|^2)$$

where

$$\mathbf{D}_{1n} = \sqrt{\frac{h}{n}} \sum_{i=1}^{n} \left\{ \frac{\partial^2 \ell(\boldsymbol{\theta}(z), \boldsymbol{\alpha}, \mathbf{x}_i, Y_i)}{\partial \boldsymbol{\theta} \partial \boldsymbol{\theta}^\top} [\mathbf{x}_i \boldsymbol{\theta}'(Z_i)]^\top K_h(Z_i - z) \right\} (\tilde{\boldsymbol{\alpha}} - \boldsymbol{\alpha}).$$

Since $\sqrt{n}(\tilde{\boldsymbol{\alpha}} - \boldsymbol{\alpha}) = O_p(1)$, it can be shown that

$$\mathbf{D}_{1n} = -\sqrt{h} f(z) \mathrm{E}[\frac{\partial^2 \ell(\boldsymbol{\theta}(z), \boldsymbol{\alpha}, \mathbf{x}, Y)}{\partial \boldsymbol{\theta} \partial \boldsymbol{\theta}^\top} [\mathbf{x} \boldsymbol{\theta}'(Z)]^\top] = o_p(1),$$

and

$$O_p(\sqrt{\frac{h}{n}} \|\tilde{\boldsymbol{\alpha}} - \boldsymbol{\alpha}\|^2) = o_p(1).$$

Therefore,

$$\mathbf{W}_{1n} = \sqrt{\frac{h}{n}} \sum_{i=1}^{n} \frac{\partial \ell(\boldsymbol{\theta}, \boldsymbol{\alpha}, \mathbf{x}_i, Y_i)}{\partial \boldsymbol{\theta}} K_h(Z_i - z) + o_p(1).$$

To complete the proof, we now calculate the mean and variance of $\mathbf{W}_n$. Note that

$$\mathrm{E}(\mathbf{W}_{1n}) = \sqrt{nh} E \left[\mathrm{E}[\frac{\partial \ell(\boldsymbol{\theta}, \boldsymbol{\alpha}, \mathbf{x}_i, Y_i)}{\partial \boldsymbol{\theta}} K_h(Z_i - z)|Z = z_0] \right]$$

$$= \sqrt{nh}[\frac{1}{2} f(z) \boldsymbol{\lambda}_1''(z|z) + f'(z) \boldsymbol{\lambda}_1'(z|z)] \kappa_2 h^2.$$

Similarly, we can show that

$$\text{Cov}(\mathbf{W}_{1n}) = f(z)\mathbf{I}_\theta^{(1)}(z)\nu_0 + o_p(1),$$

The rest of the proof follows a standard argument.

Proof of Theorem 10.5.

Denote $Z = \boldsymbol{\alpha}^\top \mathbf{x}$ and $\hat{Z} = \hat{\boldsymbol{\alpha}}^\top \mathbf{x}$. Let

$$\ell(\boldsymbol{\theta}(z), X, Y) = \log \sum_{c=1}^{C} \pi_c(z)\phi(Y|m_c(z), \sigma_c^2(z)).$$

If $\hat{\boldsymbol{\theta}}(z_0; \hat{\boldsymbol{\alpha}})$ maximizes (10.2), then it solves

$$\mathbf{0} = n^{-1} \sum_{i=1}^{n} \frac{\partial\ell(\hat{\boldsymbol{\theta}}(z_0; \hat{\boldsymbol{\alpha}}), X_i, Y_i)}{\partial\boldsymbol{\theta}} K_h(\hat{Z}_i - z_0).$$

Apply a Taylor expansion and use the conditions on h, we obtain

$$\begin{aligned}
\mathbf{0} = & n^{-1} \sum_{i=1}^{n} \mathbf{q}_i(Z_i)K_h(Z_i - z_0) + n^{-1}\sum_{i=1}^{n}[\mathbf{Q}_i(Z_i)K_h(Z_i - z_0)]\,(\,\hat{\boldsymbol{\theta}}(z_0; \hat{\boldsymbol{\alpha}}) - \boldsymbol{\theta}(z_0)) \\
& + n^{-1}\sum_{i=1}^{n}\mathbf{Q}_i(Z_i)[\mathbf{x}_i\boldsymbol{\theta}'(Z_i)]^\top K_h(Z_i - z_0)(\hat{\boldsymbol{\alpha}} - \boldsymbol{\alpha}) + o_p(n^{-1/2}) + O_p(h^2).
\end{aligned}$$

By a similar argument as in the previous proof,

$$\begin{aligned}
& \hat{\boldsymbol{\theta}}(z_0; \hat{\boldsymbol{\alpha}}) - \boldsymbol{\theta}(z_0) \\
= & n^{-1}f^{-1}(z_0)\mathbf{I}_\theta^{(1)-1}(z_0)\sum_{i=1}^{n}\mathbf{q}_i(Z_i)K_h(Z_i - z_0) \\
& - \mathbf{I}_\theta^{(1)-1}(z_0)\mathrm{E}\{\mathbf{Q}(Z)[\mathbf{x}\boldsymbol{\theta}'(Z)]^\top|Z = z_0\}(\hat{\boldsymbol{\alpha}} - \boldsymbol{\alpha}) + o_p(n^{-1/2}).
\end{aligned}$$
$$(10.33)$$

Note that

$$\begin{aligned}
\hat{\boldsymbol{\theta}}(\hat{\boldsymbol{\alpha}}^\top\mathbf{x}_i; \hat{\boldsymbol{\alpha}}) - \boldsymbol{\theta}(\boldsymbol{\alpha}^\top\mathbf{x}_i) & = \hat{\boldsymbol{\theta}}(\hat{\boldsymbol{\alpha}}^\top\mathbf{x}_i; \hat{\boldsymbol{\alpha}}) - \hat{\boldsymbol{\theta}}(\boldsymbol{\alpha}^\top\mathbf{x}_i; \hat{\boldsymbol{\alpha}}) + \hat{\boldsymbol{\theta}}(\boldsymbol{\alpha}^\top\mathbf{x}_i; \hat{\boldsymbol{\alpha}}) - \boldsymbol{\theta}(\boldsymbol{\alpha}^\top\mathbf{x}_i) \\
& = (\hat{\boldsymbol{\theta}}'(\boldsymbol{\alpha}^\top\mathbf{x}_i; \hat{\boldsymbol{\alpha}}))^\top(\hat{\boldsymbol{\alpha}}^\top - \boldsymbol{\alpha}^\top)\mathbf{x}_i + \hat{\boldsymbol{\theta}}(\boldsymbol{\alpha}^\top\mathbf{x}_i; \hat{\boldsymbol{\alpha}}) - \boldsymbol{\theta}(\boldsymbol{\alpha}_0^\top\mathbf{x}_i) + o_p(n^{-1/2}) \\
& = (\boldsymbol{\theta}'(\boldsymbol{\alpha}^\top\mathbf{x}_i))^\top(\hat{\boldsymbol{\alpha}}^\top - \boldsymbol{\alpha}^\top)\mathbf{x}_i + \hat{\boldsymbol{\theta}}(\boldsymbol{\alpha}^\top\mathbf{x}_i; \hat{\boldsymbol{\alpha}}) - \boldsymbol{\theta}(\boldsymbol{\alpha}^\top\mathbf{x}_i) + o_p(n^{-1/2}),
\end{aligned}$$
$$(10.34)$$

where the second part is handled by (10.33).

Since $\hat{\boldsymbol{\alpha}}$ maximizes (10.8), it is the solution to

$$\mathbf{0} = \lambda\hat{\boldsymbol{\alpha}} + n^{-1/2}\sum_{i=1}^{n}\mathbf{x}_i\hat{\boldsymbol{\theta}}'(\hat{\boldsymbol{\alpha}}^\top\mathbf{x}_i; \hat{\boldsymbol{\alpha}})\frac{\partial\ell(\hat{\boldsymbol{\theta}}(\hat{\boldsymbol{\alpha}}^\top\mathbf{x}_i; \hat{\boldsymbol{\alpha}}), X_i, Y_i)}{\partial\boldsymbol{\theta}},$$

where λ is the Lagrange multiplier. By the Taylor expansion and using (10.34), we have that

$$
\begin{aligned}
\mathbf{0} =& \lambda\hat{\boldsymbol{\alpha}} + n^{-1/2}\sum_{i=1}^{n}\mathbf{x}_i\boldsymbol{\theta}'(Z_i)\mathbf{q}_i(Z_i)\\
&+ n^{-1/2}\sum_{i=1}^{n}\mathbf{x}_i\boldsymbol{\theta}'(Z_i)\mathbf{Q}_i(Z_i)[\hat{\boldsymbol{\theta}}(\hat{\boldsymbol{\alpha}}^{\top}\mathbf{x}_i) - \boldsymbol{\theta}(\boldsymbol{\alpha}^{\top}\mathbf{x}_i)] + o_p(1)\\
=& \lambda\hat{\boldsymbol{\alpha}} + n^{-1/2}\sum_{i=1}^{n}\mathbf{x}_i\boldsymbol{\theta}'(Z_i)\mathbf{q}_i(Z_i)\\
&+ n^{-1/2}\sum_{i=1}^{n}\mathbf{x}_i\boldsymbol{\theta}'(Z_i)\mathbf{Q}_i(Z_i)(\mathbf{x}_i\boldsymbol{\theta}'(Z_i))^{\top}(\hat{\boldsymbol{\alpha}} - \boldsymbol{\alpha})\\
&+ n^{-1/2}\sum_{i=1}^{n}\mathbf{x}_i\boldsymbol{\theta}'(Z_i)\mathbf{Q}_i(Z_i)[\hat{\boldsymbol{\theta}}(Z_i) - \boldsymbol{\theta}(Z_i)]) + o_p(1).
\end{aligned}
$$

Define

$$
\mathbf{A}_{\alpha} = \mathrm{E}\{[\mathbf{x}\boldsymbol{\theta}'(Z)]\mathbf{Q}(Z)[\mathbf{x}\boldsymbol{\theta}'(Z)]^{\top}\},
$$

and apply (10.33),

$$
\begin{aligned}
\mathbf{0} =& \lambda\hat{\boldsymbol{\alpha}} + n^{-1/2}\sum_{i=1}^{n}\mathbf{x}_i\boldsymbol{\theta}'(Z_i)\mathbf{q}_i(Z_i) + n^{1/2}A_{\alpha}(\hat{\boldsymbol{\alpha}} - \boldsymbol{\alpha})\\
&- n^{-1/2}\sum_{i=1}^{n}\mathbf{x}_i\boldsymbol{\theta}'(Z_i)\mathbf{Q}_i(Z_i)\mathbf{I}_{\theta}^{-1}(Z_i)\mathrm{E}\{\mathbf{Q}(Z)[\mathbf{x}\boldsymbol{\theta}'(Z)]^{\top}|Z = Z_i\}(\hat{\boldsymbol{\alpha}} - \boldsymbol{\alpha})\\
&+ n^{-3/2}\sum_{i=1}^{n}\mathbf{x}_i\boldsymbol{\theta}'(Z_i)\mathbf{Q}_i(Z_i)f^{-1}(Z_i)\mathbf{I}_{\theta}^{-1}(Z_i)\sum_{t=1}^{n}\mathbf{q}_t(Z_t)K_h(Z_t - Z_i) + o_p(1)\\
=& \lambda\hat{\boldsymbol{\alpha}} + n^{-1/2}\sum_{i=1}^{n}\mathbf{x}_i\boldsymbol{\theta}'(Z_i)q_{1i}(Z_i) + \mathbf{Q}_1 n^{1/2}(\hat{\boldsymbol{\alpha}} - \boldsymbol{\alpha})\\
&+ n^{-3/2}\sum_{i=1}^{n}\frac{\mathbf{x}_i\boldsymbol{\theta}'(Z_i)\mathbf{Q}_i(Z_i)}{f(Z_i)}\mathbf{I}_{\theta}^{(1)-1}(Z_i)\sum_{t=1}^{n}\mathbf{q}_t(Z_t)K_h(Z_t - Z_i) + o_p(1).
\end{aligned}
$$

(10.35)

Interchanging the summations in the last term, we get

$$
\begin{aligned}
&n^{-1/2}\sum_{i=1}^{n}\left[n^{-1}\sum_{t=1}^{n}\mathbf{x}_t\boldsymbol{\theta}'(Z_t)\mathbf{Q}_t(Z_t)K_h(Z_t - Z_i)f^{-1}(Z_t)\mathbf{I}_{\theta}^{-1}(Z_t)\mathbf{q}_i(Z_i)\right]\\
=& n^{-1/2}\sum_{i=1}^{n}\mathrm{E}[\mathbf{x}\boldsymbol{\theta}'(Z)\mathbf{Q}(Z)|Z_i]\mathbf{I}_{\theta}^{(1)-1}(Z_i)\mathbf{q}_i(Z_i) + o_p(1).
\end{aligned}
$$

(10.36)

Let $\Gamma_\alpha = \mathbf{I} - \alpha\alpha^\top + o_p(1)$. Combining (10.35) and (10.36), and multiply by Γ_α, we have

$$\Gamma_\alpha \mathbf{Q}_1 n^{1/2}(\hat{\alpha} - \alpha)$$

$$= n^{-1/2} \sum_{i=1}^{n} \Gamma_\alpha \{ \mathbf{x}_i \boldsymbol{\theta}'(Z_i) + \mathrm{E}[\mathbf{x}\boldsymbol{\theta}'(Z)\mathbf{Q}(Z)|Z_i]\mathbf{I}_\theta^{(1)-1}(Z_i)\}\mathbf{q}_i(Z_i) + o_p(1).$$

$$(10.37)$$

It can be shown that the right-hand side of (10.37) has the covariance matrix $\Gamma_\alpha \mathbf{Q}_1 \Gamma_\alpha$, therefore completing the proof.

Proof of Theorem 10.6.

This proof is similar to the proof of Theorem 10.4.

Let $\hat{\pi}_c^* = \sqrt{nh}\{\hat{\pi}_c - \pi_c(z)\}$, $c = 1, \ldots, C - 1$, and $\hat{\boldsymbol{\pi}}^* = (\hat{\pi}_1^*, \ldots, \hat{\pi}_{C-1}^*)^\top$. It can be shown that

$$\hat{\boldsymbol{\pi}}^* = f(z)^{-1}\mathbf{I}_\pi^{(2)-1}(z)\mathbf{W}_{2n} + o_p(1),$$

where

$$\mathbf{W}_{2n} = \sqrt{\frac{h}{n}} \sum_{i=1}^{n} \frac{\partial \ell(\boldsymbol{\pi}(z), \hat{\boldsymbol{\omega}}, \mathbf{x}_i, Y_i)}{\partial \boldsymbol{\pi}} K_h(\hat{Z}_i - z).$$

To complete the proof, notice that

$$\mathrm{E}(\mathbf{W}_{2n}) = \sqrt{nh}\,\mathrm{E}\left\{ E[\frac{\partial \ell(\boldsymbol{\pi}, \boldsymbol{\lambda}, \mathbf{x}_i, Y_i)}{\partial \boldsymbol{\pi}} K_h(Z_i - z)|Z = z_0] \right\}$$

$$= \sqrt{nh}[\frac{1}{2}f(z)\boldsymbol{\lambda}_2''(z|z) + f'(z)\boldsymbol{\lambda}_2'(z|z)]\kappa_2 h^2,$$

and $\mathrm{Var}(\mathbf{W}_{2n}) = f(z)\mathbf{I}_\pi^{(2)}(z)\nu_0 + o_p(1)$. The rest of the proof follows a standard argument.

Proof of Theorem 10.7.

The proof is similar to the proof of Theorem 10.5. It can be shown that

$$\hat{\boldsymbol{\pi}}(z_0; \hat{\boldsymbol{\omega}}) - \boldsymbol{\pi}(z_0) = n^{-1}f^{-1}(z_0)\mathbf{I}_\pi^{(2)-1}(z_0) \sum_{i=1}^{n} \mathbf{q}_{\pi i}(Z_i)K_h(Z_i - z_0)$$

$$- \mathbf{I}_\pi^{(2)-1}(z_0)\mathrm{E}\{\mathbf{Q}_{\pi\pi}(Z)[\mathbf{x}\boldsymbol{\pi}'(Z)]^\top|Z = z_0\}(\hat{\alpha} - \alpha)$$

$$- \mathbf{I}_\pi^{(2)-1}(z_0)\mathrm{E}\{\mathbf{q}_{\pi\eta}(Z)|Z = z_0\}(\hat{\boldsymbol{\eta}} - \boldsymbol{\eta}) + o_p(n^{-1/2}),$$

and therefore,

$$\hat{\boldsymbol{\pi}}(\hat{Z}_i; \hat{\boldsymbol{\omega}}) - \boldsymbol{\pi}(Z_i) = \{\mathbf{x}_i \boldsymbol{\pi}'(Z_i)\}^\top(\hat{\alpha} - \alpha) + \hat{\boldsymbol{\pi}}(Z_i; \hat{\boldsymbol{\omega}}) - \boldsymbol{\pi}(Z_i) + o_p(n^{-\frac{1}{2}}). \quad (10.38)$$

Since $\hat{\omega}$ maximizes (10.15), it is the solution to

$$0 = \gamma \begin{pmatrix} \hat{\alpha} \\ 0 \end{pmatrix} + n^{-\frac{1}{2}} \sum_{i=1}^{n} \begin{pmatrix} \mathbf{x}_i \hat{\pi}'(\hat{Z}_i; \hat{\omega}) \\ \mathbf{I} \end{pmatrix} \mathbf{q}_{\pi}(\hat{\pi}(\hat{Z}_i; \hat{\omega}), \hat{\omega}),$$

where γ is the Lagrange multiplier. By Taylor series and (10.38)

$$0 = \gamma \begin{pmatrix} \hat{\alpha} \\ 0 \end{pmatrix} + n^{-\frac{1}{2}} \sum_{i=1}^{n} \boldsymbol{\lambda}_{1i} \mathbf{q}_{\pi i}(Z_i) + n^{\frac{1}{2}} \mathbf{Q}_2 \begin{pmatrix} \hat{\alpha} - \alpha \\ \hat{\eta} - \eta \end{pmatrix}$$

$$+ n^{-\frac{3}{2}} \sum_{i=1}^{n} \boldsymbol{\lambda}_{1i} \mathbf{Q}_{\pi\pi i}(Z_i) f^{-1}(Z_i) \mathbf{I}_{\pi}^{(2)-1}(Z_i) \sum_{j=1}^{n} \mathbf{q}_{\pi j}(Z_j) K_h(Z_j - Z_i) + o_p(1)$$

$$= \gamma \begin{pmatrix} \hat{\alpha} \\ 0 \end{pmatrix} + n^{-\frac{1}{2}} \sum_{i=1}^{n} \boldsymbol{\lambda}_{1i} \mathbf{q}_{\pi i}(Z_i) + n^{\frac{1}{2}} \mathbf{Q}_2 \begin{pmatrix} \hat{\alpha} - \alpha \\ \hat{\eta} - \eta \end{pmatrix}$$

$$+ n^{-\frac{1}{2}} \sum_{i=1}^{n} \mathrm{E}[\boldsymbol{\lambda}_{1i} \mathbf{Q}_{\pi\pi}(Z_i)] \mathbf{I}_{\pi}^{(2)-1}(Z_i) \mathbf{q}_{\pi i}(Z_i) + o_p(1), \quad (10.39)$$

where $\boldsymbol{\lambda}_{1i} = \begin{pmatrix} \mathbf{x}_i \boldsymbol{\pi}'(Z_i) \\ \mathbf{I} \end{pmatrix}$, and the last equation is the result of interchanging

the summations. Let $\boldsymbol{\Gamma}_{\alpha} = \begin{pmatrix} \mathbf{I} - \alpha\alpha^{\top} & 0 \\ 0 & \mathbf{I} \end{pmatrix} + o_p(1)$. By (10.39), and multiply

by $\boldsymbol{\Gamma}_{\alpha}$, we have

$$n^{\frac{1}{2}} \boldsymbol{\Gamma}_{\alpha} \mathbf{Q}_2 \begin{pmatrix} \hat{\alpha} - \alpha \\ \hat{\eta} - \eta \end{pmatrix}$$

$$= n^{-\frac{1}{2}} \sum_{i=1}^{n} \boldsymbol{\Gamma}_{\alpha} \left\{ \boldsymbol{\lambda}_{1i} - \mathcal{I}_{\pi}^{(2)-1}(Z_i) \mathrm{E}[\boldsymbol{\lambda}_{1i}(Z_i) \mathbf{Q}_{\pi\pi}(Z_i) | Z_i] \right\} \mathbf{q}_{\pi i}(Z_i) + o_p(1).$$

$$(10.40)$$

It can be shown that the right-hand side of (10.40) has the covariance matrix $\boldsymbol{\Gamma}_{\alpha} \mathbf{Q}_2 \boldsymbol{\Gamma}_{\alpha}$, thus completing the proof.

Bibliography

Aggarwal, C., Wolf, J., Yu, P., Procopiuc, C., and Park, J. (1999). Fast algorithms for projected clustering. In *The 1999 ACM SIGMOD international conference on Management of data*, pages 61–72. ACM Press. Philadelphia, Pennsylvaniaa, USA.

Aggarwal, C. and Yu, P. (2000). Finding generalized projected clusters in high dimensional spaces. In *The 2000 ACM SIGMOD international conference on Management of data*, pages 70–81. ACM Press. Dallas, Texas, USA.

Agrawal, R., Gehrke, J., Gunopulos, D., and Raghavan, P. (1998). Automatic subspace clustering of high dimensional data for data mining applications. In *The 1998 ACM SIGMOD international conference on Management of data*, pages 94–105. ACM Press. Seattle, Washington, USA.

Agresti, A. (2003). *Categorical data analysis*. John Wiley & Sons.

Akaike, H. (1974). A new look at the statistical identification model. *IEEE Trans Autom Control*, 19:716–723.

Al Mohamad, D. and Boumahdaf, A. (2018). Semiparametric two-component mixture models when one component is defined through linear constraints. *IEEE Trans. Inform. Theory*, 64:795–830.

Alexandrovich, G., Holzmann, H., and Ray, S. (2013). On the number of modes of finite mixtures of elliptical distributions. In *Algorithms from and for Nature and Life*, pages 49–57. Springer.

Allman, E., Matias, C., and Rhodes, J. (2009). Identifiability of parameters in latent structure models with many observed variables. *The Annals of Statistics*, 37:3099–3132.

Améndola, C., Drton, M., and Sturmfels, B. (2016). Maximum likelihood estimates for gaussian mixtures are transcendental. In I. S. Kotsireas, S. M. Rump, & C. K. Yap (Eds.), Mathematical Aspects of Computer and Information Sciences - 6th International Conference, MACIS 2015, Revised Selected Papers (pp. 579–590).

Anderlucci, L., Fortunato, F., and Montanari, A. (2022). High-dimensional clustering via random projections. *Journal of Classification*, 39:191–216.

Anderson, T. W. and Darling, D. A. (1952). Asymptotic theory of certain "goodness of fit" criteria based on stochastic processes. *The Annals of Mathematical Statistics*, 23:193–212.

Anderson, T. W. and Darling, D. A. (1954). A test of goodness of fit. *Journal of the American Statistical Association*, 49(268):765–769.

Andrews, D. F. and Mallows, C. L. (1974). Scale mixtures of normal distributions. *Journal of the Royal Statistical Society: Series B (Methodological)*, 36(1):99–102.

Andrews, J. and McNicholas, P. (2011). Extending mixtures of multivariate t-factor analyzers. *Statistical Computation*, 21:361–373.

Antoniadis, A. (1997). Wavelets in statistics: a review. *Journal of the Italian Statistical Society*, 6(2):97.

Arellano-Valle, R. and Azzalini, A. (2006). On the unification of families of skew-normal distributions. *Scandinavian Journal of Statistics*, 33:561–574.

Arellano-Valle, R. and Genton, M. (2005). On fundamental skew distributions. *Journal of Multivariate Analysis*, 96:93–116.

Asparouhov, T., Hamaker, E. L., and Muthén, B. (2017). Dynamic latent class analysis. *Structural Equation Modeling: A Multidisciplinary Journal*, 24(2):257–269.

Asparouhov, T. and Muthén, B. (2010). Bayesian analysis using mplus: Technical implementation.

Asparouhov, T. and Muthén, B. (2011). Using Bayesian priors for more flexible latent class analysis. In *proceedings of the 2011 joint statistical meeting*, Miami Beach, FL. American Statistical Association Alexandria, VA.

Asparouhov, T. and Muthén, B. (2008). Multilevel mixture models. *In G.R. Hancock & K.M. Samuelsen (Eds.), Advances in latent variable mixture models*, pages 27–51. Information Age Publishing.

Atwood, C. L. et al. (1976). Convergent design sequences, for sufficiently regular optimality criteria. *The Annals of Statistics*, 4(6):1124–1138.

Azaïs, J. M., Gassiat, E., and Mercadier, C. (2009). The likelihood ratio test for general mixture models with or without structural parameter. *ESAIM: Probabiility and Statistics*, 13:301–327.

Azzalini, A. (1985). A class of distributions which includes the normal ones. *Scandinavian Journal of Statistics*, 12:171–178.

Azzalini, A. and Capitanio, A. (2003). Distribution generated by perturbation of symmetry with emphasis on a multivariate skew *t*-distribution. *Journal of the Royal Statistical Society: Series B*, 65:367–389.

Azzalini, A. and Dalla Valle, A. (1996). The multivariate skew-normal distribution. *Biometrika*, 83:715–726.

Babu, G. J. and Rao, C. R. (2004). Goodness-of-fit tests when parameters are estimated. *Sankhyā: The Indian Journal of Statistics*, 66:63–74.

Baek, J., McLachlan, G. J., and Flack, L. K. (2010a). Mixtures of factor analyzers with common factor loadings: applications to the clustering and visualization of high-dimensional data. *IEEE Transactions on Pattern Analysis and Machine Intelligence*, 32:1298–1309.

Baek, J., McLachlan, G. J., and Flack, L. K. (2010b). Mixtures of factor analyzers with common factor loadings: applications to the clustering and visualization of high-dimensional data. Technical report.

Bai, X., Chen, K., and Yao, W. (2016). Mixture of linear mixed models using multivariate t distribution. *Journal of Statistical Computation and Simulation*, 86:771–787.

Bai, X., Yao, W., and Boyer, J. E. (2012). Robust fitting of mixture regression models. *Computational Statistics & Data Analysis*, 56(7):2347–2359.

Balabdaoui, F. (2017). Revisiting the Hodges Lehmann estimator in a location mixture model: Is asymptotic normality good enough? *Electronic Journal of Statistics*, 11:4563–4595.

Balabdaoui, F. and Doss, C. (2018). Inference for a two-component mixture of symmetric distributions under log-concavity. *Bernoulli*, 24:1053–1071.

Bandeen-Roche, K., Miglioretti, D. L., Zeger, S. L., and Rathouz, P. J. (1997). Latent variable regression for multiple discrete outcomes. *Journal of the American Statistical Association*, 92(440):1375–1386.

Banfield, J. D. and Raftery, A. E. (1993). Model-based gaussian and non-gaussian clustering. *Biometrics*, 49:803–821.

Barndorff, N. (1965). Identifiability of mixtures of exponential families. *Journal of Mathematical Analysis and Applications*, 12:115–121.

Bartolucci, F., Farcomeni, A., and Pennoni, F. (2012). *Latent Markov models for longitudinal data*. CRC Press.

Basso, R. M., Lachos, V. H., Cabral, C. R. B., and Ghosh, P. (2010). Robust mixture modeling based on scale mixtures of skew-normal distributions. *Computational Statistics & Data Analysis*, 54(12):2926–2941.

Basu, S. (1996). Existence of a normal scale mixture with a given variance and a percentile. *Statistics & Probability Letters*, 28(2):115–120.

Baum, L. E., Petrie, T., Soules, G., and Weiss, N. (1970). A maximization technique occurring in the statistical analysis of probabilistic functions of markov chains. *The Annals of Mathematical Statistics*, 41(1):164–171.

Beath, K. J. (2017). randomlca: An r package for latent class with random effects analysis. *Journal of Statistical Software*, 81:1–25.

Beath, K. J. and Heller, G. Z. (2009). Latent trajectory modelling of multivariate binary data. *Statistical Modelling*, 9(3):199–213.

Behboodian, J. (1970). On the modes of a mixture of two normal distributions. *Technometrics*, 12(1):131–139.

Benaglia, T., Chauveau, D., and Hunter, D. (2009). An em-like algorithm for semi- and nonparametric estimation in multivariate mixtures. *Journal of Computational and Graphical Statistics*, 18:505–526.

Bensmail, H., Celeux, G., Raftery, A. E., and Robert, C. P. (1997). Inference in model-based cluster analysis. *Statistics and Computing*, 7(1):1–10.

Berger, J. O. (2013). *Statistical decision theory and Bayesian analysis.* Springer Science & Business Media.

Bezdek, J. C., Hathaway, R. M., and Huggins, V. J. (1985). Parametric estimation for normal mixtures. *Pattern Recognition*, 3:79–84.

Bickel, P. J. and Chernoff, H. (1993). Asymptotic distribution of the likelihood ratio statistic in a prototypical non regular problem. In: Ghosh JK, Mitra SK, Parthasarthy KR, Prakasa Rao BLS (eds) Statistics and probability: a Raghu Raj Bahadur festschrift. Wiley Eastern Limited, New Delhi, pp 83–96

Biernacki, C., Celeux, G., and Govaert, G. (2000). Assessing a mixture model for clustering with the integrated completed likelihood. *IEEE Trans Pattern Anal Mach Intell*, 22:719–725.

Biernacki, C., Celeux, G., and Govaert, G. (2003). Choosing starting values for the em algorithm for getting the highest likelihood in multivariate gaussian mixture models. *Computational Statistics & Data Analysis*, 41(3-4): 561–575.

Bingham, E. and Mannila, H. (2001). Random projection in dimensionality reduction: applications to image and text data. *In KDD '01: Proceedings of the seventh ACM SIGKDD international conference on Knowledge discovery and data mining*, pages 245–250. San Francisco, California.

Bohning, D. (1982). Convergence of simar's algorithm for finding the maximum likelihood estimate of a compound poisson process. *The Annals of Statistics*, 10(3):1006–1008.

Böhning, D. (1986). A vertex-exchange-method in d-optimal design theory. *Metrika*, 33(1):337–347.

Böhning, D. (1999). *Computer-Assisted Analysis of Mixtures and Applications.* Chapman and Hall/CRC, Boca Raton, FL.

Böhning, D., Dietz, E., Schaub, R., Schlattmann, P., and Lindsay, B. G. (1994). The distribution of the likelihood ratio for mixtures of densities from the one-parameter exponential family. *Annals of the Institute of Statistical Mathematics*, 46(2):373–388.

Bolck, A., Croon, M., and Hagenaars, J. (2004). Estimating latent structure models with categorical variables: One-step versus three-step estimators. *Political Analysis*, 12(1):3–27.

Bordes, L., Chauveau, D., and Vandekerkhove, P. (2007). A stochastic EM algorithm for a semiparametric mixture model. *Computational Statistics and Data Analysis*, 51:5429–5443.

Bordes, L., Delmas, C., and Vandekerkhove, P. (2006a). Semiparametric estimation of a two-component mixture model where one component is known. *Scandinavian Journal of Statistics*, 33:733–752.

Bordes, L., Kojadinovic, I., and Vandekerkhove, P. (2013). Semiparametric estimation of a two-component mixture of linear regressions in which one component is known. *Electronic Journal of Statistics*, 7:2603–2644.

Bordes, L., Mottelet, S., and Vandekerkhove, P. (2006b). Semiparametric estimation of a two-component mixture model. *The Annals of Statistics*, 34:1204–1232.

Bordes, L., Mottelet, S., and Vandekerkhove, P. (2006c). Semiparametric estimation of a two-component mixture model. *The Annals of Statistics*, 34:1204–1232.

Bordes, L. and Vandekerkhove, P. (2010). Semiparametric two-component mixture model with a known component: An asymptotically normal estimator. *Mathematical Methods of Statistics*, 19:22–41.

Bouveyron, C., Girard, S., and Schmid, C. (2007). High-dimensional data clustering. *Computational Statistics and Data Analysis*, 52:502–519.

Bozdogan, H. (1993). Choosing the number of component clusters in the mixture-model using a new informational complexity criterion of the inverse-fisher information matrix. *In: Opitz O, Lausen B, Klar R (eds) Information and classification: concepts, methods and applications.* Springer, Berlin, pages 44–54.

Bozdogan, H. and Sclove, S. (1984). Multi-sample cluster analysis using akaike's information criterion. *Annals of the Institute of Statistical Mathematics*, 36:163–180.

Branco, M. and Key, D. (2001). A general class of multivariate skew-elliptical distributions. *Journal of Multivariate Analysis*, 79:99–113.

Brijs, T., Karlis, D., Swinnen, G., Vanhoof, K., Wets, G., and Manchanda, P. (2004). A multivariate poisson mixture model for marketing applications. *Statistica Neerlandica*, 58:322–348.

Brooks, S. P. and Roberts, G. O. (1998). Convergence assessment techniques for markov chain monte carlo. *Statistics and Computing*, 8(4):319–335.

Browne, R. P. and McNicholas, P. D. (2012). Model-based clustering, classification, and discriminant analysis of data with mixed type. *Journal of Statistical Planning and Inference*, 142:2976–2984.

Browne, R. P. and McNicholas, P. D. (2015). A mixture of generalized hyperbolic distributions. *Canadian Journal of Statistics*, 43(2):176–198.

Bryant, P. and Williamson, J. A. (1978). Asymptotic behaviour of classification maximum likelihood estimates. *Biometrika*, 65(2):273–281.

Bryant, P. G. (1991). Large-sample results for optimization-based clustering methods. *Journal of Classification*, 8(1):31–44.

Bühlmann, P. and Van De Geer, S. (2011). *Statistics for high-dimensional data: methods, theory and applications*. Springer Science & Business Media.

Burnham, K. P. and Anderson, D. R. (2002). *Model selection and multimodel inference: a practical information theoretic approach*, 2nd ed. Springer, Berlin.

Bush, C. A. and MacEachern, S. N. (1996). A semiparametric bayesian model for randomised block designs. *Biometrika*, 83(2):275–285.

Busse, L. M., Orbanz, P., and Buhmann, J. M. (2007). Cluster analysis of heterogeneous rank data. *In Proceedings of the 24th International Conference on Machine Learning*. Corvallis, Oregon, USA.

Butucea, C., Ngueyep Tzoumpe, R., and Vandekerkhove, P. (2017). Semiparametric topographical mixture models with symmetric errors. *Bernoulli*, 23:825–862.

Butucea, C. and Vandekerkhove, P. (2014). Semiparametric mixtures of symmetric distributions. *Scandinavian Journal of Statistics*, 41:227–239.

Cadavez, V. and Henningse, A. (2012). The use of seemingly unrelated regression (sur) to predict the carcass composition of lambs. *Meat Sciences*, 92:548–553.

Cai, J., Song, X., Lam, K., and Ip, E. (2011). A mixture of generalized latent variable models for mixed mode and heterogeneous data. *Computational Statistics & Data Analysis*, 55:2889–2907.

Cao, J. and Yao, W. (2012). Semiparametric mixture of binomial regression with a degenerate component. *Statistica Sinica*, 22:27–46.

Carroll, R. J. and Welsh, A. H. (1988). A note on asymmetry and robustness in linear regression. *The American Statistician*, 42(4):285–287.

Celeux, G. (1998). Bayesian inference for mixtures: The label switching problem. In Payne, R. and Green, P., editors, *In Compstat 98-Proc. in Computational Statistics*, pages 227–232, Physica, Heidelberg.

Celeux, G. and Govaert, G. (1995). Gaussian parsimonious clustering models. *Pattern Recognition*, 28(5):781–793.

Celeux, G., Hurn, M., and Robert, C. (2000). Computational and inferential difficulties with mixture posterior distributions. *Journal of American Statistical Association*, 95:957–970.

Celeux, G., Martin, O., and Lavergne, C. (2005). Mixture of linear mixed models for clustering gene expression profiles from repeated microarray experiments. *Statistical Modelling*, 5(3):243–267.

Chakravarty, I. M., Roy, J., and Laha, R. G. (1967). *Handbook of methods of applied statistics*. McGraw-Hill.

Chandra, S. (1977). On the mixtures of probability distributions. *Scandinavian Journal of Statistics*, 4:105–112.

Chang, G. and Walther, G. (2007). Clustering with mixtures of log-concave distributions. *Computational Statistics and Data Analysis*, 51:6242–6251.

Chang, J. and Jin, D. (2002). A new cell-based clustering method for large, high-dimensional data in data mining applications. In *The 2002 ACM symposium on Applied computing*, pages 503–507. ACM Press. Madrid, Spain.

Chang, W. (1983). On using principal components before separating a mixture of two multivariate normal distributions. *Applied Statistics*, 32:267–275.

Chauveau, D., D.R., Hunter, and Levinez, M. (2015). Estimation for conditional independence multivariate finite mixture models. *Statistical Surveys*, 9:1–31.

Chee, C. and Wang, Y. (2013). Estimation of finite mixtures with symmetric components. *Statistical Computation*, 23:233–249.

Chen, H. and Chen, J. (2001). The likelihood ratio test for homogeneity in finite mixture models. *Canadian Journal of Statistics*, 29(2):201–215.

Chen, H. and Chen, J. (2003). Tests for homogeneity in normal mixtures with presence of a structural parameter. *Statistica Sinica*, 13:351–365.

Chen, H., Chen, J., and Kalbfleisch, J. (2001). A modified likelihood ratio test for homogeneity in finite mixture models. *Journal of the Royal Statistical Society, B.*, 63:19–29.

Chen, J. and Cheng, P. (2000). The limiting distribution of the restricted likelihood ratio statistic for finite mixture models. *Chinese Journal of Applied Probability and Statistics*, 2:159–167.

Chen, J., Huang, Y., and Wang, P. (2016a). Composite likelihood under hidden markov model. *Statistica Sinica*, 26:1569–1586.

Chen, J. and Kalbeisch, J. (1996). Penalized minimum-distance estimates in finite mixture models. *Canadian Journal of Statistics*, 24:167–175.

Chen, J. and Khalili, A. (2008). Order selection in finite mixture models with a nonsmooth penalty. *Journal of the American Statistical Association*, 103:1674–1683.

Chen, J. and Li, P. (2009). Hypothesis test for normal mixture models: The EM approach. *The Annals of Statistics*, 37:2523–2542.

Chen, J., Li, P., and Fu, Y. (2012). Inference on the order of a normal mixture. *Journal of the American Statistical Association*, 107(499):1096–1105.

Chen, J., Li, S., and Tan, X. (2016b). Consistency of the penalized mle for two-parameter gamma mixture models. *Science China Mathematics*, 59(12):2301–2318.

Chen, J. and Tan, X. (2009). Inference for multivariate normal mixtures. *Journal of Multivariate Analysis*, 100(7):1367–1383.

Chen, J., Tan, X., and Zhang, R. (2008). Inference for normal mixtures in mean and variance. *Statistica Sinica*, 18:443–465.

Chen, Y. and Samworth, R. J. (2013). Smoothed log-concave maximum likelihood estimation with applications. *Statistica Sinica*, 23:1373–1398.

Cheng, C. H., Fu, A., and Zhang, Y. (1999). Entropy-based subspace clustering for mining numerical data. In *The fifth ACM SIGKDD international conference on Knowledge discovery and data mining*, pages 84–93. ACM Press. San Diego, California, USA.

Cheng, R. and Liu, W. (2001). The consistency of estimators in finite mixture models. *Scandinavian Journal of Statistics*, 28(4):603–616.

Chib, S. and Winkelmann, R. (2001). Markov chain monte carlo analysis of correlated count data. *Journal of Business & Economic Statistics*, 19:428–435.

Chu, C., Glad, I., Godtliebsen, F., and Marron, J. (1998). Edge-preserving smoothers for image processing. *Journal of the American Statistical Association*, 93(442):526–541.

Chung, H., Loken, E., and Schafer, J. (2004). Difficulties in drawing inferences with finite-mixture models: a simple example with a simple solution. *Journal of the American Statistical Association*, 58:152–158.

Claeskens, G. and Hjort, N. J. (2008). *Model selection and model averaging*. Cambridge University Press, Cambridge.

Coakley, C. W. and Hettmansperger, T. P. (1993). A bounded influence, high breakdown, efficient regression estimator. *Journal of the American Statistical Association*, 88(423):872–880.

Collins, L. M. and Lanza, S. T. (2009). *Latent class and latent transition analysis: With applications in the social, behavioral, and health sciences*, volume 718. John Wiley & Sons.

Collins, L. M. and Wugalter, S. E. (1992). Latent class models for stage-sequential dynamic latent variables. *Multivariate Behavioral Research*, 27(1):131–157.

Cowles, M. K. and Carlin, B. P. (1996). Markov chain monte carlo convergence diagnostics: a comparative review. *Journal of the American Statistical Association*, 91(434):883–904.

Cramér, H. (1928). On the composition of elementary errors: First paper: Mathematical deductions. *Scandinavian Actuarial Journal*, 1928(1):13–74.

Cron, A. J. and West, M. (2011). Efficient classification-based relabeling in mixture models. *Journal of the American Statistical Association*, 65:16–20.

Croux, C., Rousseeuw, P. J., and Hössjer, O. (1994). Generalized s-estimators. *Journal of the American Statistical Association*, 89(428):1271–1281.

Cule, M., Gramacy, R., Samworth, R., et al. (2009). Logconcdead: An r package for maximum likelihood estimation of a multivariate log-concave density. *Journal of Statistical Software*, 29(2):1–20.

Cule, M., Samworth, R., et al. (2010a). Theoretical properties of the log-concave maximum likelihood estimator of a multidimensional density. *Electronic Journal of Statistics*, 4:254–270.

Cule, M., Samworth, R., and Stewart, M. (2010b). Maximum likelihood estimation of a multi-dimensional log-concave density. *Journal of the Royal Statistical Society: Series B (Statistical Methodology)*, 72(5):545–607.

Dacunha-Castelle, D. and Gassiat, E. (1997). The estimation of the order of a mixture model. *Bernoulli*, 3:279–299.

Dacunha-Castelle, D. and Gassiat, E. (1999). Testing the order of a model using locally conic parametrization: Population mixtures and stationary ARMA processes. *The Annals of Statistics*, 27:1178–1209.

Dasgupta, S. (1999). Learning mixtures of gaussians. *In 40th Annual IEEE symp. on Foundations of Computer Sciences*, pages 634–644.

Dasgupta, S. (2000). Experiments with random projections. *In Proc. Uncertainty in Artificial Intelligence*. San Francisco, California, USA.

Dayton, C. M. and Macready, G. B. (1988). Concomitant-variable latent-class models. *Journal of the American Statistical Association*, 83(401):173–178.

de Jong, P. (1987). A central limit theorem for generalized quadratic forms. *Probability Theory and Related Fields*, 75:261–277.

Dellaportas, P., Stephens, D. A., Smith, A. F. M., and Guttman, I. (1996). A comparative study of perinatal mortality using a two-component mixture model. In Berry, D. and Stangl, D., editors, *Bayesian Biostatistics*, pages 601–616. CRC Press.

Dempster, A. P., Laird, N. M., and Rubin, D. B. (1977a). Maximum likelihood from incomplete data via the em algorithm. *Journal of the Royal Statistical Society: Series B (Methodological)*, 39(1):1–22.

Dempster, A. P., Laird, N. M., and Rubin, D. B. (1977b). Maximum likelihood from incomplete data via the em algorithm (with discussion). *Journal of Royal Statistical Association: Series B*, 39:1–38.

Diebolt, J. and Robert, C. P. (1994). Estimation of finite mixture distributions through bayesian sampling. *Journal of Royal Statistical Association: Series B*, 56:363–375.

Doğru, F. Z. and Arslan, O. (2018). Robust mixture regression modeling using the least trimmed squares (lts)-estimation method. *Communications in Statistics-Simulation and Computation*, 47(7):2184–2196.

Donoho, D. L. (1982). Breakdown properties of multivariate location estimators. Technical report, Technical report, Harvard University, Boston.

Donoho, D. L. and Johnstone, J. M. (1994). Ideal spatial adaptation by wavelet shrinkage. *Biometrika*, 81(3):425–455.

Dümbgen, L., Rufibach, K., et al. (2009). Maximum likelihood estimation of a log-concave density and its distribution function: Basic properties and uniform consistency. *Bernoulli*, 15(1):40–68.

Dümbgen, L., Samworth, R., Schuhmacher, D., et al. (2011). Approximation by log-concave distributions, with applications to regression. *The Annals of Statistics*, 39(2):702–730.

Dziak, J., Li, R., Tan, X. Shiffman, S., and Shiyko, M. (2015). Modeling intensive longitudinal data with mixtures of nonparametric trajectories and time-varying effects. *Psychological Methods*, 20:444–469.

D'Elia, A. and Piccolo, D. (2005). A mixture model for preferences data analysis. *Computational Statistics & Data Analysis*, 49:917–934.

Eddy, W. F. (1980). Optimum kernel estimators of the mode. *The Annals of Statistics*, 8:870–882.

Efron, B. and Olshen, R. A. (1978). How broad is the class of normal scale mixtures? *The Annals of Statistics*, 6:1159–1164.

Eisenberger, I. (1964). Genesis of bimodal distributions. *Technometrics*, 6(4): 357–363.

Elliott, M. R., Gallo, J. J., Ten Have, T. R., Bogner, H. R., and Katz, I. R. (2005). Using a bayesian latent growth curve model to identify trajectories of positive affect and negative events following myocardial infarction. *Biostatistics*, 6(1):119–143.

Escobar, M. D. and West, M. (1995). Bayesian density estimation and inference using mixtures. *Journal of the American Statistical Association*, 90(430):577–588.

Everitt, B. (1984). *An Introduction to Latent Variable Models*. London: Chapman & Hall.

Faicel, C. (2016). Unsupervised learning of regression mixture models with unknown number of components. *Journal of Statistical Computation and Simulation*, 86:2308–2334.

Fan, J. and Gijbels, I. (1996). Local Polynomial Modelling and Its Applications. Monographs on Statistics and Applied Probability, 66. CRC Press, London.

Fan, J., Zhang, C., and Zhang, J. (2001). Generalized likelihood ratio statistics and Wilks phenomenon. *The Annals of Statistics*, 29:153–193.

Feng, Z. D. and McCulloch, C. E. (1996). Using bootstrap likelihood ratios in finite mixture models. *Journal of the Royal Statistical Society: Series B (Methodological)*, 58(3):609–617.

Ferguson, T. S. (1973). A Bayesian analysis of some nonparametric problems. *The Annals of Statistics*, 1:209–230.

Fern, X. and Brodley, C. (2003). Random projection for high dimensional data clustering: a cluster ensemble approach. *Proc. of ICML*, pages 186–193.

Finch, S. J., Mendell, N. R., and Thode Jr, H. C. (1989). Probabilistic measures of adequacy of a numerical search for a global maximum. *Journal of the American Statistical Association*, 84(408):1020–1023.

Finch, W. H. and French, B. F. (2014). Multilevel latent class analysis: Parametric and nonparametric models. *The Journal of Experimental Education*, 82(3):307–333.

Fisher, R. (1936). The use of multiple measurements in taxonomic problems. *The Annals of Eugenics*, 7:179–188.

Fokoué, E. (2005). Mixtures of factor analyzers: An extension with covariates. *Journal of Multivariate Analysis*, 95:370–384.

Formann, A. K. (1988). Latent class models for nonmonotone dichotomous items. *Psychometrika*, 53:45–62.

Formann, A. K. (1992). Linear logistic latent class analysis for polytomous data. *Journal of the American Statistical Association*, 87(418):476–486.

Fraley, C. and Raftery, A. E. (2002). Model-based clustering, discriminant analysis, and density estimation. *Journal of the American Statistical Association*, 97(458):611–631.

Fraley, C., Raftery, A. E., Murphy, T. B., and Scrucca, L. (2012). mclust version 4 for r: Normal mixture modeling for model-based clustering, classification, and density estimation. Technical report, Department of Statistics, University of Washington.

Fridman, M. (1993). *Hidden markov model regression*. PhD thesis, University of Pennsylvania.

Friedman, J. (1989). Regularized discriminant analysis. *The Journal of the American Statistical Association*, 84:165–175.

Friedman, J.H. and Meulman, J. (2004). Clustering objects on subsets of attributes with discussion. *Journal of Royal Statistical Society, Series. B*, 66(4):815–849.

Friedman, J. H., Hastie, T. J., and Tibshirani, R. (2008). Sparse inverse covariance estimation with the graphical lasso. *Biostatistics*, 9:432–41.

Friedman, J. H. and Meulman, J. J. (2004). Clutering objects on subsets of attributes. *Journal of the Royal Statistical Society: Series B*, 66:815–849.

Frühwirth-Schnatter, S. (2001). Markov chain monte carlo estimation of classical and dynamic switching and mixture models. *Journal of American Statistical Association*, 96:194–209.

Frühwirth-Schnatter, S. (2006). *Finite Mixture and Markov Switching Models*. Springer.

Furman, W. D. and Lindsay, B. G. (1994). Testing for the number of components in a mixture of normal distributions using moment estimators. *Computational Statistics & Data Analysis*, 17(5):473–492.

Galimberti, G. and Soffritti, G. (2020). Seemingly unrelated clusterwise linear regression. *Advances in Data Analysis and Classification*, 14:235–260.

Galindo Garre, F. and Vermunt, J. K. (2006). Avoiding boundary estimates in latent class analysis by bayesian posterior mode estimation. *Behaviormetrika*, 33(1):43–59.

Gallaugher, M. and McNicholas, P. (2018). Finite mixtures of skewed matrix variate distributions. *Pattern Recognition*, 80:83–93.

García-Escudero, L. A., Gordaliza, A., Greselin, F., Ingrassia, S., and Mayo-Íscar, A. (2017). Robust estimation of mixtures of regressions with random covariates, via trimming and constraints. *Statistics and Computing*, 27(2):377–402.

García-Escudero, L. A., Gordaliza, A., Mayo-Íscar, A., and San Martín, R. (2010). Robust clusterwise linear regression through trimming. *Computational Statistics & Data Analysis*, 54(12):3057–3069.

Gassiat, E. and Rousseau, J. (2016). Nonparametric finite translation hidden Markov models and extensions. *Bernoulli*, 22:193–212.

Gelfand, A. E. and Smith, A. F. (1990). Sampling-based approaches to calculating marginal densities. *Journal of the American Statistical Association*, 85(410):398–409.

Gervini, D., Yohai, V. J., et al. (2002). A class of robust and fully efficient regression estimators. *The Annals of Statistics*, 30(2):583–616.

Geweke, J. (2007). Interpretation and inference in mixture models: Simple mcmc works. *Computational Statistics and Data Analysis*, 51:3529–3550.

Ghahramani, Z. and Beal, M. (2000). Variational inference for bayesian mixture of factor analysers. In *S.A. Solla, T.K. Leen, K.R. Muller (Eds.), Advances in Neural Information Processing Systems*, vol. 12, MIT Press, Cambridge, MA.

Ghahramani, Z. and Hinton, G. E. (1997). The em algorithm for mixture of factor analyzers. Technical report, Toronto, ON: University of Toronto.

Ghosal, S. and Van der Vaart, A. (2017). *Fundamentals of nonparametric Bayesian inference*, volume 44. Cambridge University Press.

Ghosh, J. K. and Sen, P. K. (1985). On the asymptotic performance of the log likelihood ratio statistic for the mixture model and related results. *In Proceedings of the Berkeley Conference in Honor of Jerzy Neyman and Jack Kiefer II* (Berkeley, Calif., 1983), pages 789–806.

Godwin, R. and Böhning, D. (2017). Estimation of the population size by using the one-flated positive poisson model. *Applied Statistics*, 66:425–448.

Goil, S., Nagesh, H., and Choudhary, A. (1999). Mafia: Efficient and scalable subspace clustering for very large data sets. Technical report, Northwestern University, Evanston, IL 60208.

Goldfeld, S. M. and Quandt, R. E. (1973). A markov model for switching regressions. *Journal of Econometrics*, 1(1):3–15.

Goodman, L. A. (1974). Exploratory latent structure analysis using both identifiable and unidentifiable models. *Biometrika*, 61(2):215–231.

Gordon, A.D., Vichi, M. (1998). Partitions of partitions. *Journal of Classification*, 15:265–285.

Gormley, I. C. and Murphy, T. B. (2008). A mixture of experts model for rank data with applications in election studies. *The Annals of Applied Statistics*, 2:1452–1477.

Grazian, C. and Robert, C. P. (2018). Jeffreys priors for mixture estimation: Properties and alternatives. *Computational Statistics & Data Analysis*, 121:149–163.

Gruet, M. A., Philippe, A., and & Robert, C. P. (1999). MCMC control spreadsheets for exponential mixture estimation. *Journal of Computational and Graphical Statistics*, 8:298–317.

Grün, B. and Hornik, K. (2012). Modelling human immunodeficiency virus ribonucleic acid levels with finite mixtures for censored longitudinal data. *Journal of the Royal Statistical Society: Series C (Applied Statistics)*, 61(2):201–218.

Grün, B. and Leisch, F. (2008). *Finite Mixtures of Generalized Linear Regression Models*, pages 205–230. Physica-Verlag HD, Heidelberg.

Grün, B. and Leisch, F. (2009). Dealing with label switching in mixture models under genuine multimodality. *Journal of Multivariate Analysis*, 100:851–861.

Guo, Y., Hastie, T. J., and Tibshirani, R. (2007). Regularized linear discriminant analysis and its application in microarrays. *Biostatistics*, 8:86–100.

Hagenaars, J. A. (1990). *Categorical longitudinal data: Log-linear panel, trend, and cohort analysis*. SAGE Publications, Incorporated.

Hagenaars, J. A. and McCutcheon, A. L. (2002). *Applied latent class analysis*. Cambridge University Press.

Hampel, F.R., Ronchetti, E., Rouseeuw, P.J., and Stahel, W.A. (1986). Robust statistics: The approach based on influence functions. New York: Wiley.

Hall, P. and Zhou, X. (2003). Nonparametric estimation of component distributions in a multivariate mixture. *The Annals of Statistics*, 31:201–224.

Handschin, E., Schweppe, F. C., Kohlas, J., and Fiechter, A. (1975). Bad data analysis for power system state estimation. *IEEE Transactions on Power Apparatus and Systems*, 94(2):329–337.

Harper, D. (1972). Local dependence latent structure models. *Psychometrika*, 37(1):53–59.

Harris, K. M. and Udry, J. R. (2018). National longitudinal study of adolescent to adult health (add health), 1994-2008 [public use]. Ann Arbor, MI: Carolina Population Center, University of North Carolina-Chapel Hill [distributor], Inter-university Consortium for Political and Social Research [distributor], pages 08–06.

Hartigan, J. A. (1985). A failure of likelihood asymptotics for normal mixtures. In *Proceedings of the Berkeley Conference in Honor of J. Neyman and J. Kiefer*, pages 807–810.

Hartigan, J. A. (1990). Partition models. *Communications in Statistics-Theory and Methods*, 19(8):2745–2756.

Hasnat, M., Velcin, J., Bonnevay, S., and Jacques, J. (2015). Simultaneous clustering and model selection for multinomial distribution: a comparative study. *In: Advances in Intelligent Data Analysis XIV. Springer.*

Hastie, T., Buja, A., and Tibshirani, R. (1995). Penalized discriminant analysis. *The Annals of Statistics*, 23:73–102.

Hastie, T. and Tibshirani, R. (1996). Discriminant analysis by gaussian mixtures. *Journal of the Royal Statistical Society Series B*, 58:155–176.

Hathaway, R. J. (1983). Constrained maximum likelihood estimation for a mixture of m univariate normal distribbutions. Technical report, University of South Carolina, Columbia, South Carolina.

Hathaway, R. J. (1985). A constrained formulation of maximum-likelihood estimation for nomral mixture distributions. *The Annals of Statistics*, 13:795–800.

Hathaway, R. J. (1986). A constrained em algorithm for univariate mixutres. *Journal of Statistical Computation and Simulation*, 23:211–230.

Härdle, W., Hall, P., and Ichimura, H. (1993). Optimal smoothing in single-index models. *Annals of Statistics*, 21:157–178.

He, X., Cai, D., Shao, Y., Bao, H., and Han, J. (2011). Laplacian regularized gaussian mixture model for data clustering. *IEEE Transactions on Knowledge and Data Engineering*, 23:1406–1418.

Hecht-Nielsen, R. (1994). Context vectors: general purpose approximate meaning representations self-organized from raw data. *In J.M. Zurada, R.J. Marks II, and C.U. Robinson, editors, Computational Intelligence, Imitating Life*, pages 43–56. IEEE Press.

Hennig, C. (2000). Identifiablity of models for clusterwise linear regression. *Journal of Classification*, 17(2).

Henry, K. L. and Muthén, B. (2010). Multilevel latent class analysis: An application of adolescent smoking typologies with individual and contextual predictors. *Structural Equation Modeling*, 17(2):193–215.

Hinton, G. E., Dayan, P., and Revow, M. (1997). Modeling the manifolds of images of handwritten digits. *IEEE Transactions on Neural Networks*, 8:65–73.

Hjort, N. L., Holmes, C., Müller, P., and Walker, S. G. (2010). *Bayesian nonparametrics*, volume 28. Cambridge University Press.

Hjorth, J. (1994). *Computer Intensive Statistical Methods: Validation, Model Selection and Bootstrap*. Chapman and Hall, UK.

Hoben, T., Arnold, L., and Brainerd, C. (1993). Modeling growth and individual differences in spatial tasks. *Monographs of the Society for Research in Child Development*, 58:1–191.

Hohmann, E. and Holzmann, H. (2013). Semiparametric location mixtures with distinct components. *Statistics*, 47:348–362.

Holzmann, H., Munk, A., and Gneiting, T. (2006). Identifiability of finite mixtures of elliptical distributions. *Scandinavian Journal of Statistics*, 33(4):753–763.

Holzmann, H. and Vollmer, S. (2008). A likelihood ratio test for bimodality in two-component mixtures with application to regional income distribution in the eu. *AStA Advances in Statistical Analysis*, 92(1):57–69.

Hu, H., Wu, Y., and Yao, W. (2016). Maximum likelihood estimation of the mixture of log-concave densities. *Computational Statistics and Data Analysis*, 101:137–147.

Hu, H., Yao, W., and Wu, Y. (2017). The robust EM-type algorithms for log-concave mixtures of regression models. *Computational Statistics and Data Analysis*, 111:14–26.

Huang, J., Ma, S., and Zhang, C.-H. (2008). Adaptive lasso for sparse high-dimensional regression models. *Statistica Sinica*, 18:1603–1618.

Huang, M., Huang, Y., and Yao, W. (2021). Statistical inference for nonparametric and semiparametric hidden markov model via composite likelihood approach. *Science China Mathematics*, 66:601–626.

Huang, M., Ji, Q., and Yao, W. (2018a). Semiparametric hidden markov model with non-parametric regression. *Communications in Statistics-Theory and Methods*, 47:5196–5204.

Huang, M., Li, R., Wang, H., and Yao, W. (2014). Estimating mixture of Gaussian processes by kernel smoothing. *Journal of Business and Economic Statistics*, 32:259–270.

Huang, M., Li, R., and Wang, S. (2013). Nonparametric mixture of regression models. *Journal of the American Statistical Association*, 108:929–941.

Huang, M., Wang, S., Wang, H., and Jin, T. (2018b). Maximum smoothed likelihood estimation for a class of semiparametric Pareto mixture densities. *Statistical Interface*, 11:31–40.

Huang, M., Wang, S., Yao, W., and Chen, Y. (2018c). Statistical inference and applications of mixture of varying coefficient models. *Scandinavian Journal of Statistics*, 45:618–643.

Huang, M. and Yao, W. (2012). Mixture of regression models with varying mixing proportions: A semiparametric approach. *Journal of the American Statistical Association*, 107:711–724.

Huang, T., Peng, H., and Zhang, K. (2017). Model selection for Gaussian mixture models. *Statistica Sinica*, 27:147–169.

Huber, P. J. (1981). *Robust Statistics*. New York: John Wiley and Sons.

Huber, P. J. et al. (1964). Robust estimation of a location parameter. *The Annals of Mathematical Statistics*, 35(1):73–101.

Humphreys, K. and Janson, H. (2000). Latent transition analysis with covariates, nonresponse, summary statistics and diagnostics: Modeling children's drawing development. *Multivariate Behavioral Research*, 35:89–119.

Hunt, L. and Jorgensen, M. (1999). Mixture model clustering using the multimix program. *Australian and New Zealand Journal of Statistics*, 41:154–171.

Hunt, L. and Jorgensen, M. (2003). Mixture model clustering for mixed data with missing information. *Computational Statistics & Data Analysis*, 41:429–440.

Hunter, D., Wang, S., and Hettmansperger, T. (2007). Inference for mixtures of symmetric distributions. *The Annals of Statistics*, 35:224–251.

Hunter, D. and Young, D. (2012). Semiparametric mixtures of regressions. *Journal of Nonparametric Statistics*, 24:19–38.

Hurn, M., Justel, A., and Robert, C. P. (2003). Estimating mixtures of regressions. *Journal of Computational and Graphical Statistics*, 12:55–79.

Ichimura, H. (1993). Semiparametric least squares (SLS) and weighted SLS estimation of single-index models. *Journal of Econometrics*, 58:71–120.

Ingrassia, S. (2004). A likelihood-based constrained algorithm for multivariate normal mixture models. *Statistical Methods & Applications*, 13(2):151–166.

Ishiguro, M., Sakamoto, Y., and Kitagawa, G. (1997). Bootstrapping log likelihood and EIC, an extension of AIC. *Annals of the institute of Statistical Mathematics*, 49:411–434.

Jacobs, R., Peng, F., and Tanner, M. (1997). A Bayesian approach to model selection in hierarchical mixtures-of-experts architectures. *Neural Networks*, 10:231–241.

Jacobs, R. A., Jordan, M. I., Nowlan, S. J., and Hinton, G. E. (1991). Adaptive mixtures of local experts. *Journal of Neural Computation*, 3:79–87.

Jacques, J. and Biernacki, C. (2014). Model-based clustering for multivariate partial ranking data. *Journal of Statistical Planning and Inference*, 149:201–217.

Jaeckel, L. A. (1972). Estimating regression coefficients by minimizing the dispersion of the residuals. *The Annals of Mathematical Statistics*, 43:1449–1458.

James, L. F., Priebe, C. E., and Marchette, D. J. (2001). Consistent estimation of mixture complexity. *The Annals of Statistics*, 29:1281–1296.

Janssen, P., Marron, J. S., Veraverbeke, N., and Sarle, W. (1995). Scale measures for bandwidth selection. *Journal of Nonparametric Statistics*, 5(4):359–380.

Jasra, A., Holmes, C. C., and Stephens, D. A. (2005). Markov chain monte carlo methods and the label switching problem in Bayesian mixture modeling. *Statistical Science*, 20:50–67.

Jeffreys, H. (1998). *The theory of probability*. OUP Oxford.

Jewell, N. P. (1982). Mixtures of exponential distributions. *The Annals of Statistics*, 10:479–484.

Jiang, W. and Tanner, M. A. (1999). Hierarchical mixtures-of-experts for exponential family regression models: approximation and maximum likelihood estimation. *The Annals of Statistics*, 27(3):987–1011.

Jin, C., Zhang, Y., Balakrishnan, S., Wainwright, M. J., and Jordan, M. I. (2016). Local maxima in the likelihood of gaussian mixture models: Structural results and algorithmic consequences. In *Advances in neural information processing systems*, pages 4116–4124.

Jorgensen, M. (2004). Using multinomial mixture models to cluster internet traffic. *Australian & New Zealand Journal of Statistics*, 46:205–218.

Jorgensen, M. (2013). Forming clusters from census areas with similar tabular statistics. *Communications in Statistics – Theory and Methods*, 42:2136–2151.

Jung, T. and Wickrama, K. A. (2008). An introduction to latent class growth analysis and growth mixture modeling. *Social and Personality Psychology Compass*, 2(1):302–317.

Kamakura, W. A. and Russell, G. (1989). A probabilistic choice model for market segmentation and elasticity structure. *Journal of Marketing Research*, 26:379–390.

Karlis, D. and Meligkotsidou, L. (2007). Finite mixtures of multivariate poisson distributions with applications. *Journal of Statistical Planning and Inference*, 137:1942–1960.

Karlis, D. and Santourian, A. (2009). Model-based clustering with non-ellipticaly contoured distributions. *Statistics and Computing*, 19:73–83.

Karlis, D. and Xekalaki, E. (2003). Choosing initial values for the em algorithm for finite mixtures. *Computational Statistics & Data Analysis*, 41(3-4):577–590.

Kasahara, H. and Shimotsu, K. (2015). Testing the number of components in normal mixture regression models. *Journal of the American Statistical Association*, 110(512):1632–1645.

Kaya, H. and Salah, A. A. (2015). Adaptive mixtures of factor analyzers. *ArXiv*, 1507.02801.

Kelava, A. and Brandt, H. (2019). A nonlinear dynamic latent class structural equation model. *Structural Equation Modeling: A Multidisciplinary Journal*, 26(4):509–528.

Kelker, D. (1971). Infinite divisibility and variance mixtures of the normal distribution. *The Annals of Mathematical Statistics*, 42(2):802–808.

Keribin, C. (2000). Consistent estimation of the order of mixture models. *Sankhyā: The Indian Journal of Statistics, Series A*, pages 49–66.

Keshavarzi, S., Ayatollahi, S., Zare, N., and Pakfetra, M. (2012). Application of seemingly unreated regression in medical data with intermittently observed time-dependent covariates. *Computational and Mathematical Methods in Medicine*, 2012:821–843.

Khalili, A. and Chen, J. (2007). Variable selection in finite mixture of regression models. *Journal of the American Statistical Association*, 102:1025–1038.

Khalili, A., Chen, J., and Lin, S. (2011). Feature selection in finite mixture of sparse normal linear models in high-dimensional feature space. *Biostatistics*, 12:156–172.

Kiefer, J. and Wolfowitz, J. (1956). Consistency of the maximum likelihood estimator in the presence of infinitely many incidental parameters. *The Annals of Mathematical Statistics*, 27:886–906.

Kiefer, N. M. (1978). Discrete parameter variation: efficient estimation of a switching regression model. *Econometrica: Journal of the Econometric Society*, 46:427–434.

Kim, D. and Seo, B. (2014). Assessment of the number of components in gaussian mixture models in the presence of multiple local maximizers. *Journal of Multivariate Analysis*, 125:100–120.

Kostantinos, N. (2000). Gaussian mixtures and their applications to signal processing. *Advanced Signal Processing Handbook: Theory and Implementation for Radar, Sonar, and Medical Imaging Real Time Systems*. CRC Press, Boca Raton, Florida.

Krasker, W. S. and Welsch, R. E. (1982). Efficient bounded-influence regression estimation. *Journal of the American statistical Association*, 77(379):595–604.

Kuiper, N. H. (1960). Tests concerning random points on a circle. *Nederl. Akad. Wetensch. Proc. Ser. A*, 63:38–47.

Lachos, V., Ghosh, P., and Arellano-Valle, R. (2010). Likelihood based inference for skew normal independent linear mixed model. *Statistica Sinica*, 20:303–322.

Laird, N. (1978). Nonparametric maximum likelihood estimation of a mixing distribution. *Journal of the American Statistical Association*, 73(364):805–811.

Lambert, D. (1992). Zero-inflated poisson regression, with an application to defects in manufacturing. *Technometrics*, 34:1–14.

Lanza, S. T., Bray, B. C., and Collins, L. M. (2013). An introduction to latent class and latent transition analysis. *Handbook of Psychology*, 2:691–716.

Lanza, S. T., Collins, L. M., Schafer, J. L., and Flaherty, B. P. (2005). Using data augmentation to obtain standard errors and conduct hypothesis tests in latent class and latent transition analysis. *Psychological Methods*, 10:84–100.

Lanza, S. T. and Cooper, B. R. (2016). Latent class analysis for developmental research. *Child Development Perspectives*, 10(1):59–64.

Lawrence, C. J. and Krzanowski, W. J. (1996). Mixture separation for mixed-mode data. *Statistics and Computing*, 6:85–92.

Lazarsfeld, P. F. (1950). The logical and mathematical foundation of latent structure analysis. *Studies in Social Psychology in World War II Vol. IV: Measurement and Prediction*, pages 362–412.

Lee, P. H. and Philip, L. (2012). Mixtures of weighted distance-based models for ranking data with applications in political studies. *Computational Statistics & Data Analysis*, 56:2486–2500.

Lee, S. and McLachlan, G. (2011). On the fitting of mixtures of multivariate skew t-distributions via the em algorithm. *arXiv:1109 4706*.

Lee, S. and McLachlan, G. (2013a). Model-based clustering and classification with non-normal mixture distributions. *Statistical Methods & Applications*, 22:427–454.

Lee, S. and McLachlan, G. (2013b). On mixtures of skew normal and skew t-distributions. *Advances in Data Analysis & Classification*, 7:241–266.

Lee, Y., MacEachern, S. N., Jung, Y., et al. (2012). Regularization of case-specific parameters for robustness and efficiency. *Statistical Science*, 27(3):350–372.

Lemdani, M. and Pons, O. (1999). Likelihood ratio tests in contamination models. *Bernoulli*, 5:705–719.

Leroux, B. and Puterman, M. (1992). Maximum-penalized-likelihood estimation for independent and markov-dependent mixture models. *Biometrics*, 48:545–558.

Leroux, B. G. (1992). Consistent estimation of a mixing distribution. *The Annals of Statistics*, 20:1350–1360.

Lesperance, M. L. and Kalbfleisch, J. D. (1992). An algorithm for computing the nonparametric mle of a mixing distribution. *Journal of the American Statistical Association*, 87(417):120–126.

Levine, M., Hunter, D., and Chauveau, D. (2011). Maximum smoothed likelihood for multivariate mixtures. *Biometrika*, 98:403–416.

Li, J., Ray, S., and Lindsay, B. G. (2007). A nonparametric statistical approach to clustering via mode identification. *Journal of Machine Learning Research*, 8:1687–1723.

Li, K.-C. (1991). Sliced inverse regression for dimension reduction. *Journal of the American Statistical Association*, 86(414):316–327.

Li, P., Chen, J., and Marriott, P. (2009). Non-finite fisher information and homogeneity: An em approach. *Biometrika*, pages 411–426.

Li, Y., Lord-Bessen, J., Shiyko, M., and Loeb, R. (2018). Bayesian latent class analysis tutorial. *Multivariate behavioral research*, 53(3):430–451.

Liang, F. (2008). Clustering gene expression profiles using mixture model ensemble averaging approach. *JP Journal of Biostatistics*, 2:57–80.

Lin, T. (2009). Maximum likelihood estimation for multivariate skew normal mixture models. *Journal of Multivariate Analysis*, 100:257–265.

Lindsay, B. (1983). The geometry of mixture likelihoods, part I: A general theory. *Annals of Statistics*, 11:86–94.

Lindsay, B. G. (1988). Composite likelihood methods. *Contemporary Mathematics*, 80(1):221–239.

Lindsay, B. G. (1995). Mixture models: Theory, geometry, and applications. In *NSF-CBMS Regional Conference Series in Probability and Statistics v 5*, Hayward, CA. Institure of Mathematical Statistics.

Lindsay, B. G. and Basak, P. (1993). Multivariate normal mixtures: A fast consistent method of moments. *Journal of the American Statistical Association*, 88(422):468–476.

Lindsay, B. G. et al. (1983). The geometry of mixture likelihoods, part ii: the exponential family. *The Annals of Statistics*, 11(3):783–792.

Lindsay, B. G. and Roeder, K. (1993). Uniqueness of estimation and identifiability in mixture models. *The Canadian Journal of Statistics*, 21:139–147.

Lindstrom, M. J. and Bates, D. M. (1988). Newton–Raphson and em algorithms for linear mixed-effects models for repeated-measures data. *Journal of the American Statistical Association*, 83(404):1014–1022.

Linton, O. and Xiao, Z. (2007). A nonparametric regression estimator that adapts to error distribution of unknown form. *Econometric Theory*, 23:371–413.

Linzer, D. A. and Lewis, J. B. (2011). polca: An r package for polytomous variable latent class analysis. *Journal of Statistical Software*, 42:1–29.

Liu, B., Xia, Y., and Yu, P. (2000). Clustering through decision tree construction. In *The ninth international conference on Information and knowledge management*, pages 20–29. ACM Press. McLean, Virginia, USA.

Liu, J., Cai, D., and He, X. (2010). Gaussian mixture model with local consistency. In *Proceedings of the Twenty-Fourth AAAI Conference on Artificial Intelligence, AAAI*. Atlanta, Georgia, USA.

Liu, S., Wu, H., and Meeker, W. Q. (2015). Understanding and addressing the unbounded likelihood problem. *Journal of the American Statistical Association*, 69(3):191–200.

Liu, X. and Shao, Y. (2003). Asymptotics for likelihood ratio tests under loss of identifiability. *The Annals of Statistics*, 31:807–832.

Ma, Y., Wang, S., Xu, L., and Yao, W. (2021). Semiparametric mixture regression with unspecified error distributions. *Test*, 30:429–444.

Ma, Y. and Yao, W. (2015). Flexible estimation of a semiparametric two-component mixture model with one parametric component. *Electronic Journal of Statistics*, 9:444–474.

MacQueen, J. et al. (1967). Some methods for classification and analysis of multivariate observations. In *Proceedings of the fifth Berkeley symposium on mathematical statistics and probability*, volume 1, pages 281–297. Oakland, CA, USA.

Magidson, J., Vermunt, J. K., and Madura, J. P. (2020). *Latent class analysis*. SAGE Publications Limited Thousand Oaks, CA, USA.

Maiboroda, R. and Sugakova, O. (2011). Generalized estimating equations for symmetric distributions observed with admixture. *Communications in Statistics: Theory and Methods*, 40:96–116.

Mallows, C. (1975). On some topics in robustness. unpublished memorandum, bell tel. Laboratories, Murray Hill.

Marin, J. M., Mengersen, K. L., and Robert, C. P. (2005). Bayesian modeling and inference on mixtures of distributions. In Dey, D. and Rao, C., editors, *Handbook of Statistics: Volume 25*, pages 459–507. North Holland, Amsterdam.

Maronna, R. A., Martin, R. D., Yohai, V. J., and Salibián-Barrera, M. (2019). *Robust statistics: theory and methods (with R)*. John Wiley & Sons.

Maronna, R. A. and Yohai, V. J. (1981). Asymptotic behavior of general m-estimates for regression and scale with random carriers. *Zeitschrift für Wahrscheinlichkeitstheorie und verwandte Gebiete*, 58(1):7–20.

Marron, J. S. and Wand, M. P. (1992). Exact mean integrated squared error. *The Annals of Statistics*, 20:712–736.

McCullagh, P. and Nelder, J. A. (2019). *Generalized linear models*. Routledge.

McCulloch, W. S. and Pitts, W. (1943). A logical calculus of the ideas immanent in nervous activity. *The Bulletin of Mathematical Biophysics*, 5(4):115–133.

McLachlan, G., Bean, R., and Peel, D. (2002). A mixture model-based approach to the clustering of microarray expression data. *Bioinformatics*, 18:413–422.

McLachlan, G. J. (1987). On bootstrapping the likelihood ratio test statistic for the number of components in a normal mixture. *Journal of the Royal Statistical Society, Series C (Applied Statistics)*, 36:318–324.

McLachlan, G. J. and Baek, J. (2010). Clustering of high-dimensional data via finite mixture models. In *Studies in Classification Data Analysis and Knowledge Organization*, pages 33–44.

McLachlan, G. J. and Basford, K. E. (1988). *Mixture models: Inference and applications to clustering*, volume 38. M. Dekker New York.

McLachlan, G. J., Bean, R. W., and Jones, L. B. (2007). Extension of the mixture of factor analyzers model to incorporate the multivariate t distribution. *Computational Statistics & Data Analysis*, 51:5327–5338.

McLachlan, G. J. and Krishnan, T. (2007). *The EM algorithm and extensions*, volume 382. John Wiley & Sons.

McLachlan, G. J. and Peel, D. (2000). *Finite Mixture Models*. Wiley, New York.

McLachlan, G. J., Peel, D., and Bean, R. W. (2003). Modelling high-dimensional data by mixtures of factor analyzers. *Computational Statistics and Data Analysis*, 41:379–388.

McNicholas, P. and Murphy, T. (2008). Parsimonious Gaussian mixture models. *Statistical Computations*, 18:295–296.

McNicholas, P. D. and Murphy, T. B. (2010). Model-based clustering of microarray expression data via latent gaussian mixture models. *Bioinformatics*, 26:2705–2712.

McNicholas, S., McNicholas, P., and Browne, R. (2017). A mixture of variance-gamma factor analyzers. *in S.E. Ahmed, ed, 'Big and Complex Data Analysis: Methdologies and Applications'*, Springer International Publishing, Cham, pages 369–385.

McParland, D. and Gormley, I. C. (2016). Model based clustering for mixed data: clustmd. *Advances in Data Analysis and Classification*, 10:155–169.

McParland, D. and Gormley, I. C. (2017). clustmd: Model based clustering for mixed data. r package version 1.2.1.

McParland, D., Gormley, I. C., McCormick, T. H., Clark, S. J., Kabudula, C. W., and Collinson, M. A. (2014). Clustering south african households based on their asset status using latent variable models. *Annals of Applied Statistics*, 8:747–776.

McParland, D., Phillips, C. M., Brennan, L., Roche, H. M., and Gormley, I. C. (2017). Clustering high dimensional mixed data to uncover sub-phenotypes: Joint analysis of phenotypic and genotypic data. *Statistics in Medicine*, 36:4548–4569.

Meilă, M. and Heckerman, D. (2001). An experimental comparison of model-based clustering methods. *Mach. Learn.*, 42:9–29.

Melnykov, V. and Zhu, X. (2018). On model-based clustering of skewed matrix data. *Journal of Multivariate Analysis*, 167:181–194.

Meng, X. and Dyk, D. (1997). The em algorithm - an old fold song sung to a fast new tune (with discussion. *Journal of the Royal Statistical Society B*, 59:511–567.

Meng, X. and Rubin, D. B. (1993). Maximum likelihood estimation via the ecm algorithm: A general framework. *Biometrika*, 80(2):267–278.

Mengersen , K. L. and Robert, C.P. (1996). Testing for Mixtures: A Bayesian Entropic Approach, in J M Bernardo, and others (eds), Bayesian Statistics 5: Proceedings of the Fifth Valencia International Meeting, Oxford.

Michael, S. and Melnykov, V. (2016). An effective strategy for initializing the em algorithm in finite mixture models. *Advances in Data Analysis and Classification*, 10(4):563–583.

Mises, R. v. (2013). *Wahrscheinlichkeit statistik und wahrheit*, volume 7. Springer-Verlag.

Mkhadri, A., Celeux, G., and Nasroallah, A. (1997). Regularization in discriminant analysis: an overview. *Computational Statistics and Data Analysis*, 23:403–423.

Mollica, C. and Tardella, L. (2014). Epitope profiling via mixture modeling of ranked data. *Statistics in Medicine*, 33:3738–3758.

Mollica, C. and Tardella, L. (2017). Bayesian plackett–luce mixture models for partially ranked data. *Psychometrika*, 82:442–458.

Montanari, A. and Viroli, C. (2011). Maximum likelihood estimation of mixtures of factor analyzers. *Computational Statistics and Data Analysis*, 55:2712–2723.

Montuelle, L. and le Pennec, E. (2014). Mixture of Gaussian regressions model with logistic weights, a penalized maximum likelihood approach. *Electronic Journal of Statistics*, 8:1661–1695.

Morlini, I. (2012). A latent variable approach for clustering mixed binary and continuous variables within a gaussian mixture model. *Advances in Data Analysis & Classification*, 6:5–28.

Morris, K. and McNicholas, P. (2003). Dimension reduction for model-based clustering via mixtures of shifted asymmetric laplace distributions. 83:2088–2093.

Morris, K. and McNicholas, P. D. (2016). Clustering, classification, discriminant analysis, and dimension reduction via generalized hyperbolic mixtures. *Computational Statistics & Data Analysis*, 97:133–150.

Müller, P., Erkanli, A., and West, M. (1996). Bayesian curve fitting using multivariate normal mixtures. *Biometrika*, 83(1):67–79.

Müller, P., Quintana, F. A., Jara, A., and Hanson, T. (2015). *Bayesian nonparametric data analysis*. Springer.

Murphy, K. and Murphy, T. B. (2017). Gaussian parsimonious clustering models with covariates. *arXiv preprint arXiv:1711.05632*.

Murphy, T. B. and Martin, D. (2003). Mixtures of distance-based models for ranking data. *Computational Statistics & Data Analysis*, 41:645–655.

Muthén, B. and Asparouhov, T. (2022). Latent transition analysis with random intercepts (ri-lta). *Psychological Methods*, 27(1):1.

Nair, V. and Hinton, G. E. (2010). Rectified linear units improve restricted boltzmann machines. In *Icml*. in Proceedings of the 27th International Conference on International Conference on Machine Learning, 801–814, Haifa, Israel.

Naranjo, J. D. and Hettmansperger, T. (1994). Bounded influence rank regression. *Journal of the Royal Statistical Society: Series B (Methodological)*, 56(1):209–220.

Neykov, N., Filzmoser, P., Dimova, R., and Neytchev, P. (2007). Robust fitting of mixtures using the trimmed likelihood estimator. *Computational Statistics & Data Analysis*, 52(1):299–308.

Nguyen, V. and Mattias, C. (2014). On efficient estimators of the proportion of true null hypotheses in a multiple testing setup. *Scandinavian Journal of Statistics*, 41:1167–1194.

Nylund-Gibson, K. and Choi, A. Y. (2018). Ten frequently asked questions about latent class analysis. *Translational Issues in Psychological Science*, 4(4):440.

O'Hagan, A., Murphy, T., Gormley, I., McNicholas, P., and Karlis, D. (2016). Clustering with the multivariate normal inverse gaussian distribution. *Computational Statistics and Data Analysis*, 93:18–30.

Pan, W. and Shen, X. (2007). Penalized model-based clustering with application to variable selection. *Journal of Machine Learning Research*, 8:1145–1164.

Papastamoulis, P. and Iliopoulos, G. (2010). An artificial allocations based solution to the label switching problem in Bayesian analysis of mixtures of distributions. *Journal of Computational and Graphical Statistics*, 19:313–331.

Park, J.-H. and Dunson, D. B. (2010). Bayesian generalized product partition model. *Statistica Sinica*, 20:1203–1226.

Parsons, L., Haque, E., and Liu, H. (2004). Subspace clustering for high dimensional data: a review. *SIGKDD Explorations*, 6:90–105.

Parzen, E. (1962). On estimation of a probability density function and mode. *The Annals of Mathematical Statistics*, 33: 1065–1076.

Patra, R. and Sen, B. (2016). Estimation of a two-component mixture model with applications to multiple testing. *Journal of the Royal Statistical Society: Series B*, 78:869–893.

Pearson, K. (1894). Contribution to the mathematical theory of evolution. *Philosophical Transactions of the Royal Society of London. A*, 185:71–110.

Pearson, K. (1915). On certain types of compound frequency distributions in which the components can e individually described by binomial series. *Biometrika*, 11:139–144.

Peters, Jr, B. C. and Walker, H. F. (1978). An iterative procedure for obtaining maximum-likelihood estimates of the parameters for a mixture of normal distributions. *SIAM Journal on Applied Mathematics*, 35(2):362–378.

Petersen, K. J., Qualter, P., and Humphrey, N. (2019). The application of latent class analysis for investigating population child mental health: A systematic review. *Frontiers in Psychology*, 10:1214.

Phadia, E. G. (2015). *Prior processes and their applications*. Springer.

Phillips, D. B. and Smith, A. F. (1996). Bayesian model comparison via jump diffusions. *Markov Chain Monte Carlo in Practice*, 215:239.

Pilla, R. S. and Lindsay, B. G. (2001). Alternative em methods for nonparametric finite mixture models. *Biometrika*, 88(2):535–550.

Preston, E. J. (1953). A graphical method for the analysis of statistical distributions into two normal components. *Biometrika*, 40(3/4):460–464.

Procopiuc, C., Jones, M., Agarwal, P., and Murali, T. (2002). A monte carlo algorithm for fast projective clustering. In *The 2002 ACM SIGMOD international conference on Management of data*, pages 418–427. ACM Press. Madison, Wisconsin.

Proust-Lima, C., Philipps, V., and Liquet, B. (2015). Estimation of extended mixed models using latent classes and latent processes: The r package lcmm. *arXiv preprint arXiv:1503.00890*.

Punzo, A. and McNicholas, P. D. (2017). Robust clustering in regression analysis via the contaminated Gaussian cluster-weighted model. *Journal of Classification*, 34(2):249–293.

Pyne, S., Hu, X., Wang, K., Rossin, E., Lin, T., Maier, L., Baecher-Allan, C., McLachlan, G., Tamayo, P., Hafler, D., De Jager, P., and Mesirow, J. (2009). Automated high-dimensional flow cytometric data analysis. *Proc Natl Acad Sci USA*, 106:8519–8542.

Qu, Y., Tan, M., and Kutner, M. H. (1996). Random effects models in latent class analysis for evaluating accuracy of diagnostic tests. *Biometrics*, 52:797–810.

Rabiner, L. R. (1989). A tutorial on hidden markov models and selected applications in speech recognition. *Proceedings of the IEEE*, 77(2):257–286.

Raftery, A. (2003). Bayesian clustering with variable and transformation selection. *Bayesian Statistics*, 7:266–271.

Raftery, A. E. and Dean, N. (2006). Variable selection for model-based clustering. *Journal of the American Statistical Association*, 101:168–178.

Rao, C.R. (1973). Linear Statistical Inference and its Applications: Second Edition. John Wiley & Sons, Inc.

Ray, S. and Lindsay, B. G. (2005). The topography of multivariate normal mixtures. *The Annals of Statistics*, 33(5):2042–2065.

Redner, R. A. and Walker, H. F. (1984). Mixture densities, maximum likelihood and the em algorithm. *SIAM Rev.*, 26:195–239.

Richardson, S. and Green, P. J. (1997). On bayesian analysis of mixtures with an unknown number of components (with discussion). *Journal of Royal Statistical Association: Series B*, 59:731–792.

Robert, C. P. (1996). *Mixtures of distributions: inference and estimation.* pages 441–464. Chapman and Hall London, London.

Robert, C. P., Ryden, T., and Titterington, D. M. (2000). Bayesian inference in hidden markov models through the reversible jump markov chain monte carlo method. *Journal of the Royal Statistical Society: Series B (Statistical Methodology)*, 62(1):57–75.

Rocci, R., Gattone, S. A., and Di Mari, R. (2017). A data driven equivariant approach to constrained gaussian mixture modeling. *Advances in Data Analysis and Classification*. 12:235–260. Change year to 2018.

Rodríguez, C. E. and Walker, S. G. (2012). Label switching in bayesian mixture models: Deterministic relabeling strategies. *Journal of Computational and Graphical Statistics*, 23(1):25–45.

Roeder, K. (1990). Density estimation with confidence sets exemplified by superclusters and voids in the galaxies. *Journal of the American Statistical Association*, 85(411):617–624.

Roeder, K. and Wasserman, L. (1997). Practical bayesian density estimation using mixtures of normals. *Journal of the American Statistical Association*, 92(439):894–902.

Roger, G. and Pol, S. (1991). *Stochastic Finite Elements: A Special Approach.* Springer.

Rousseeuw, P. and Yohai, V. (1984). Robust regression by means of s-estimators. In *Robust and nonlinear time series analysis*, pages 256–272. Springer.

Rousseeuw, P. J. (1984). Least median of squares regression. *Journal of the American statistical association*, 79(388):871–880.

Rousseeuw, P. J. and Leroy, A. M. (2005). *Robust regression and outlier detection*, volume 589. John wiley & sons.

Ruan, L. and Yuan, M. (2011). Regularized parameter estimation in high-dimensional gaussian mixture models. *Neural Computation*, 23:1605–1622.

Rubio, F. J. and Steel, M. F. (2014). Inference in two-piece location-scale models with jeffreys priors. *Bayesian Analysis*, 9(1):1–22.

Rufibach, K. (2007). Computing maximum likelihood estimators of a log-concave density function. *Journal of Statistical Computation and Simulation*, 77:561–574.

Rumelhart, D. E., Hinton, G. E., and Williams, R. J. (1986). Learning representations by back-propagating errors. *Nature*, 323(6088):533–536.

Sahu, S., Dey, D., and Branco, M. (2003). A new class of multivariate skew distributions with applications to Bayesian regression models. *Canadian Journal of Statistics*, 31:129–150.

Salah, A.A. and Alpaydin, E. (2004). Incremental mixtures of factor analysers, in Proceedings of the 17th International Conference on Pattern Recognition, Cambridge, UK, 276–279.

Sarkar, S., Zhu, X., Melnykov, V., and Ingrassia, S. (2020). On parsimonious models formodeling matix data. *Computational Statistics and Data Analysis*, 142:106–132.

Schilling, M. F., Watkins, A. E., and Watkins, W. (2002). Is human height bimodal? *The American Statistician*, 56(3):223–229.

Schwarz, G. (1978). Estimating the dimension of a model. *The Annals of Statistics*, 6:461–464.

Sclove, S. (1987). Application of some model-selection criteria to some problems in multivariate analysis. *Psychometrika*, 52:333–343.

Scott, D. W. (1992). *Multivariate Density Estimation*. John Wiley & Sons, New York, Chichester.

Scrucca, L. & Raftery, A. E. (2018). clustvarsel: A package implementing variable selection for gaussian model-based clustering in r. *Journal of Statistical Software*, 84(1):1–28.

Scrucca, L. (2010). Dimension reduction for model-based clustering. *Stat Comput*, 20:471–484.

Seidel, W., Mosler, K., and Alker, M. (2000). A cautionary note on likelihood ratio tests in mixture models. *Annals of the Institute of Statistical Mathematics*, 52(3):481–487.

Seo, B. and Kim, D. (2012). Root selection in normal mixture models. *Computational Statistics & Data Analysis*, 56(8):2454–2470.

Seo, B. and Lee, T. (2015). A new algorithm for maximum likelihood estimation in normal scale-mixture generalized autoregressive conditional heteroskedastic models. *Journal of Statistical Computation and Simulation*, 85(1):202–215.

Seo, B., Noh, J., Lee, T., and Yoon, Y. J. (2017). Adaptive robust regression with continuous gaussian scale mixture errors. *Journal of the Korean Statistical Society*, 46(1):113–125.

Sethuraman, J. (1994). A constructive definition of dirichlet priors. *Statistica Sinica*, 4:639–650.

She, Y. et al. (2009). Thresholding-based iterative selection procedures for model selection and shrinkage. *Electronic Journal of statistics*, 3:384–415.

She, Y. and Owen, A. B. (2011). Outlier detection using nonconvex penalized regression. *Journal of the American Statistical Association*, 106(494):626–639.

Shen, J. and He, X. (2015). Inference for subgroup analysis with a structured logistic-normal mixture model. *Journal of the American Statistical Association*, 110(509):303–312.

Siegel, A. F. (1982). Robust regression using repeated medians. *Biometrika*, 69(1):242–244.

Silverman, B. (1986). *Density Estimation for Statistics and Data Analysis*. Chapman and Hall.

Simpson, D. G., Yohai, V. J., et al. (1998). Functional stability of one-step gm-estimators in approximately linear regression. *The Annals of Statistics*, 26(3):1147–1169.

Skrondal, A. and Rabe-Hesketh, S. (2004). *Generalized latent variable modeling: multilevel, longitudinal, and structural equation models.* Chapman & Hall/CRC, USA.

Smyth, P. (2000). Model selection for probabilistic clustering using cross-validated likelihood. *Statistical Computations*, 10:63–72.

Song, S., Nicolae, D., and Song, J. (2010). Estimating the mixing proportion in a semiparametric mixture model. *Computational Statistics and Data Analysis*, 54:2276–2283.

Song, W., Yao, W., and Xing, Y. (2014). Robust mixture regression model fitting by laplace distribution. *Computational Statistics & Data Analysis*, 71:128–137.

Sperrin, M., Jaki, T., and Wit, E. (2010). Probabilistic relabelling strategies for the label switching problem in Bayesian mixture models. *Statistics and Computing*, 20(3):357–366.

Spezia, L., Coksley, S., Brewer, ., Donnelly, D., and Tree, A. (2014). Modelling species abundance in a river by negative binomial hidden markov models. *Computational Statistics and Data Analysis*, 71:599–614.

Stephens, M. (1972). Edf statistics for goodness-of-fit: Part 1. Technical report, Department of Statistics, Stanford University, California.

Stephens, M. (1997). *Bayesian methods for mixtures of normal distributions.* PhD thesis, Department of Statistics, University of Oxford.

Stephens, M. (2000a). Bayesian analysis of mixture models with an unknown number of components-an alternative to reversible jump methods. *Annals of Statistics*, 28:40–74.

Stephens, M. (2000b). Dealing with label switching in mixture models. *Journal of the Royal Statistical Society: Series B (Statistical Methodology)*, 62(4):795–809.

Stephens, M. A. (1974). Edf statistics for goodness of fit and some comparisons. *Journal of the American Statistical Association*, 69(347):730–737.

Stephens, M. A. (1976). Asymptotic results for goodness-of-fit statistics with unknown parameters. *The Annals of Statistics*, 4:357–369.

Stromberg, A. J., Hössjer, O., and Hawkins, D. M. (2000). The least trimmed differences regression estimator and alternatives. *Journal of the American Statistical Association*, 95(451):853–864.

Tadesse, M. G., Sha, N., and Vannucci, M. (2005). Bayesian variable selection in clustering high-dimensional data. *Journal of the American Statistical Association*, 100:602–617.

Tan, X., Shiyko, M., Li, R., Li, Y., and Dierker, L. (2012). A time-varying effect model for intensive longitudinal data. *Psychological Methods*, 17:61–77.

Tanaka, K. (2009). Strong consistency of the maximum likelihood estimator for finite mixtures of location–scale distributions when penalty is imposed on the ratios of the scale parameters. *Scandinavian Journal of Statistics*, 36(1):171–184.

Tanner, M. A. and Wong, W. H. (1987). The calculation of posterior distributions by data augmentation. *Journal of the American statistical Association*, 82(398):528–540.

Teicher, H. (1960). On the mixture of distributions. *The Annals of Mathematical Statistics*, 31:55–73.

Teicher, H. (1963). Identifiability of finite mixtures. *The Annals of Mathematical Statistics*, 34:1265–1269.

Tibshirani, R. (1996). Regression shrinkage and selection via the lasso. *Journal of the Royal Statistical Society: Series B (Methodological)*, 58(1):267–288.

Tibshirani, R., Hastie, T. J., Narasimhan, B., and Chu, G. (2003). Class prediction by nearest shrunken centroids, with applications to dna microarrays. *Statistical Science*, 18:104–117.

Tipping, M. and Bishop, C. (1999). Mixtures of probabilistic principal component analysers. *Neural Computation*, 11:443–482.

Titterington, D. M., Smith, A. F. M., and Makov, U. (1985). *Statistical Analysis of Finite Mixture Distributions*. Wiley.

Ueda, N., Nakano, R., Ghahramani, Z., and Hinton, E. (2000). Smem algorithm for mixture models. *Neural Comput.*, 12:2019–2128.

van den Bergh, M., Schmittmann, V. D., and Vermunt, J. K. (2017). Building latent class trees, with an application to a study of social capital. *Methodology: European Journal of Research Methods for the Behavioral and Social Sciences*, 13(S1):13.

Van Den Bergh, M. and Vermunt, J. K. (2018). Building latent class growth trees. *Structural Equation Modeling: A Multidisciplinary Journal*, 25(3):331–342.

Van Den Bergh, M. and Vermunt, J. K. (2019). Latent class trees with the three-step approach. *Structural Equation Modeling: A Multidisciplinary Journal*, 26(3):481–492.

Vandekerkhove, P. (2013). Estimation of a semiparametric mixture of regressions model. *Journal of Nonparametric Statistics*, 25:181–208.

Velicer, W. F., Martin, R. A., and Collins, L. M. (1996). Latent transition analysis for longitudinal data. *Addiction*, 91(12s1):197–210.

Venables, W. and Ripley, B. (2002). *Modern applied statistics with S*. Springer.

Vermunt, J. (2007). A hierarchical mixture model for clustering three-way data sets. *Computational Statistics and Data Analysis*, 51:5368–5376.

Vermunt, J. K. (2003). Multilevel latent class models. *Sociological methodology*, 33(1):213–239.

Vermunt, J. K. (2010). Latent class modeling with covariates: Two improved three-step approaches. *Political analysis*, 18(4):450–469.

Vermunt, J. K. and Magidson, J. (2004). Latent class analysis. *The Sage Encyclopedia of Social Sciences Research Methods*, 2:549–553.

Vichi, M. (1999). One mode classification of a three-way data set. *Journal of Classification*, 16:27–44.

Vichi, M., Rocci, R., and Kiers, A. (2007). Simultaneous component and clusering models for three-way data: within and between approaches. *Journal of Classification*, 24:71–98.

Viroli, C. (2010). Dimensionally reduced model-based clustering through mixtures of factor mixture analyzers. *Journal of Classification*, 27:363–388.

Viroli, C. (2011). Finite mixtures of matrix normal distributions for classifying three-way data. *Stat. Comput.*, 21:511–522.

von Neumann, J. (1931). Die Eindeutigkeit der Schrödingerschen Operatoren. *Math. Ann.*, 104:570–578.

Walker, A. M. (1969). On the asymptotic behaviour of posterior distributions. *Journal of the Royal Statistical Society: Series B (Methodological)*, 31(1):80–88.

Walther, G. (2002). Detecting the presence of mixing with multiscale maximum likelihood. *Journal of the American Statistical Association*, 97:508–513.

Wang, S., Huang, M., Wu, X., and Yao, W. (2016). Mixture of functional linear models and its application to CO2-GDP functional data. *Computational Statistics and Data Analysis*, 97:1–15.

Wang, S., Yao, W., and Huang, M. (2014). A note on the identifiability of nonparametric and semiparametric mixtures of GLMs. *Statistical and Probability Letters*, 93:41–45.

Wang, S. and Zhu, J. (2008). Variable selection for model-based high-dimensional clustering and its application to microarray data. *Biometrics*, 64:440–8.

Wang, W., Zhang, X., and Mai, Q. (2020). Model-based clustering with envelopes. *Electronic Journal of Statistics*, 14(1):82–109.

Wang, Y. (2007). On fast computation of the non-parametric maximum likelihood estimate of a mixing distribution. *Journal of the Royal Statistical Society: Series B*, 69(2):185–198.

Wang, Y. (2010). Maximum likelihood computation for fitting semiparametric mixture models. *Statics and Computing*, 20:75–86.

Wasserman, L. (2000). Asymptotic inference for mixture models by using data-dependent priors. *Journal of the Royal Statistical Society: Series B (Statistical Methodology)*, 62(1):159–180.

Watson, G. S. (1961). Goodness-of-fit tests on a circle. *Biometrika*, 48(1/2): 109–114.

Wedel, M. and Kamakura, W. A. (2000). *Market Segmentation: Conceptual and Methodological Foundations*. Springer Science and Business Media, USA.

Wei, G. C. and Tanner, M. A. (1990). A monte carlo implementation of the em algorithm and the poor man's data augmentation algorithms. *Journal of the American Statistical Association*, 85(411):699–704.

White, A. and Murphy, T. B. (2014). Bayeslca: An r package for bayesian latent class analysis. *Journal of Statiscal Software*, 61(13):1–28.

Wichitchan, S., Yao, W., and Yang, G. (2019a). Hypothesis testing for finite mixture models. *Computational Statistics & Data Analysis*, 132:180–189.

Wichitchan, S., Yao, W., and Yang, G. (2019b). A simple root selection method for univariate finite normal mixture models. *Communications in Statistics-Theory and Methods*, 48(15):3778–3794.

Woo, K. and Lee, J. (2002). *FINDIT: A Fast and Intelligent Subspace Clustering Algorithm using Dimension Voting*. PhD thesis, Korea Advanced Institute of Science and Technology.

Woo, M. and Sriram, T. N. (2006). Robust estimation of mixture complexity. *Journal of the American Statistical Association*, 101:1475–1485.

Wu, C.-F. (1978). Some algorithmic aspects of the theory of optimal designs. *The Annals of Statistics*, 6:1286–1301.

Wu, C. J. (1983). On the convergence properties of the em algorithm. *The Annals of Statistics*, pages 95–103.

Wu, J., Yao, W., and Xiang, S. (2017). Computation of an efficient and robust estimator in a semiparametric mixture model. *Journal of Statistical Computation and Simulation*, 87(11):2128–2137.

Wu, Q. and Yao, W. (2016). Mixtures of quantile regressions. *Computational Statistics and Data Analysis*, 93:162–176.

Wu, Q. and Yao, W. (2017). Relabel mixture models via modal clustering. *Communications in Statistics-Simulation and Computation*, 46(5):3406–3418.

Wynn, H. P. (1970). The sequential generation of d-optimum experimental designs. *The Annals of Mathematical Statistics*, 41:1655–1664.

Wynn, H. P. (1972). Results in the theory and construction of d-optimum experimental designs. *Journal of the Royal Statistical Society: Series B (Methodological)*, 34(2):133–147.

Xiang, S. and Yao, W. (2018). Semiparametric mixtures of nonparametric regressions. *Annals of the Institute of Statistical Mathematics*, 70:131–154.

Xiang, S. and Yao, W. (2020). Semiparametric mixtures of regressions with single-index for model based clustering. *Advances in Data Analysis and Classification*. 14:261–292.

Xiang, S., Yao, W., and Seo, B. (2016). Semiparametric mixture: Continuous scale mixture approach. *Computational Statistics and Data Analysis*, 103:413–425.

Xiang, S., Yao, W., and Wu, J. (2014). Minimum profile Hellinger distance estimation for a semiparametric mixture model. *Canadian Journal of Statistics*, 42:246–267.

Xie, B., Pan, W., and Shen, X. (2010). Penalized mixtures of factor analyzers with application to clustering high-dimensioanl microarray data. *Bioinformatics*, 26:501–508.

Xu, Y., Müller, P., and Telesca, D. (2016). Bayesian inference for latent biologic structure with determinantal point processes (dpp). *Biometrics*, 72(3):955–964.

Xue, J. and Yao, W. (2021). Machine learning embedded semiparametric mixtures of regressions with covariate-varying mixing proportions. *Econometrics and Statistics*, 22:159–171.

Yakowitz, S. J. and Spragins, J. D. (1968). On the identifiability of finite mixtures. *The Annals of Mathematical Statistics*, 39:209–214.

Yang, J., Wang, W., Wang, H., and Yu, P. (2002). δ-clusters: capturing subspace correlation in a large data set. In *In Data Engineering, 2002. Proceedings. 18th International Conference*, pages 517–528. San Jose, CA, USA.

Yang, L., Xiang, S., and Yao, W. (2017). Robust fitting of mixtures of factor analyzers using the trimmed likelihood estimator. 46:1280–1291. Communications in Statistics-Simulation and Computation.

Yao, F., Fu, Y., and Lee, T. (2011). Functional mixture regression. *Biostatistics*, 12:341–353.

Yao, W. (2010). A profile likelihood method for normal mixture with unequal variance. *Journal of Statistical Planning and Inference*, 140:2089–2098.

Yao, W. (2012a). Bayesian mixture labeling and clustering. *Communications in Statistics–Theory and Methods*, 41:403–421.

Yao, W. (2012b). Model based labeling for mixture models. *Statistics and Computing*, 22:337–347.

Yao, W. (2013a). A note on em algorithm for mixture models. *Statistics & Probability Letters*, 89:519–526.

Yao, W. (2013b). A simple solution to bayesian mixture labeling. *Communications in Statistics-Simulation and Computation*, 42(4):800–813.

Yao, W. (2015). Label switching and its solutions for frequentist mixture models. *Journal of Statistical Computation and Simulation*, 85(5):1000–1012.

Yao, W. and Li, L. (2014). An online bayesian mixture labelling method by minimizing deviance of classification probabilities to reference labels. *Journal of Statistical Computation and Simulation*, 84(2):310–323.

Yao, W. and Lindsay, B. G. (2009). Bayesian mixture labeling by highest posterior density. *Journal of American Statistical Association*, 104: 758–767.

Yao, W., Wei, Y., and Yu, C. (2014). Robust mixture regression using the t-distribution. *Computational Statistics & Data Analysis*, 71:116–127.

Yao, W. and Zhao, Z. (2013). Kernel density based linear regression estimate. *Communications in Statistics-Theory and Methods*, 42:4499–4512.

Yau, C., Papaspiliopoulos, O., Roberts, G., and Holmes, C. (2011). Bayesian nonparametric hidden markov models with pplications in genomics. *Journal of the Royal Statistical Society: Series B*, 73:37–57.

Yau, K. K., Lee, A. H., and Ng, A. S. (2003). Finite mixture regression model with random effects: application to neonatal hospital length of stay. *Computational statistics & data analysis*, 41(3-4):359–366.

Yeo, I. and Johnson, R. (2000). A new family of power transformations in improve normality or symmetry. *Biometrika*, 87:954–959.

Yohai, V. J. (1987). High breakdown-point and high efficiency robust estimates for regression. *The Annals of Statistics*, 15:642–656.

Yoshida, R., Higuchi, T., and Imoto, S. (2004). A mixed factors model for dimension reduction and extraction of a group structure in gene expression data. *Proceedings. 2004 IEEE Computational Systems Bioinformatics Conference, 2004. CSB 2004.*, pages 161–172.

Yoshida, R., Higuchi, T., Imoto, S., and Miyano, S. (2006). Arraycluster: an analytic tool for clustering, data visualization and module finder on gene expression profiles. *Bioinformatics*, 22 12:1538–9.

Young, D. (2014). Mixtures of regressions with changepoints. *Statistics and Computing*, 24:265–281.

Young, D. and Hunter, D. (2010). Mixtures of regressions with predictor-dependent mixing proportions. *Computational Statistics and Data Analysis*, 54:2253–2266.

Yu, C. and Yao, W. (2017). Robust linear regression: A review and comparison. *Communications in Statistics-Simulation and Computation*, 46(8):6261–6282.

Yu, C., Yao, W., and Chen, K. (2017). A new method for robust mixture regression. *Canadian Journal of Statistics*, 45(1):77–94.

Yu, C., Yao, W., and Yang, G. (2020). A selective overview and comparison of robust mixture regression estimators. *International Statistical Review*, 88(1):176–202.

Zeller, C. B., Cabral, C. R., and Lachos, V. H. (2016). Robust mixture regression modeling based on scale mixtures of skew-normal distributions. *Test*, 25(2):375–396.

Zeng, P. (2012). Finite mixture of heteroscedastic single-index models. *Open Journal of Statistics*, 2:12–20.

Zhang, C.-H. et al. (2010). Nearly unbiased variable selection under minimax concave penalty. *The Annals of Statistics*, 38(2):894–942.

Zhao, J. and Yu, P. (2008). Fast ml estimation for the mixture of factor analyzers via an ecm algorithm. *IEEE Transactions on Neural Network*, 19:1956–1961.

Zhou, H., Pan, W., and Shen, X. (2009). Penalized model-based clustering with unconstrained covariance matrices. *Electronic Journal of Statistics*, 3:1473–1496.

Zhou, M., Thayer, W. M., and Bridges, J. F. (2018). Using latent class analysis to model preference heterogeneity in health: a systematic review. *Pharmacoeconomics*, 36(2):175–187.

Zhu, H. and Zhang, H. (2004). Hypothesis testing in mixture regression models. *Journal of the Royal Statistical Society: Series B*, 66:3–16.

Zou, H. (2006). The adaptive lasso and its oracle properties. *Journal of the American Statistical Association*, 101(476):1418–1429.

Index

Printed in the United States
by Baker & Taylor Publisher Services